Climate Science: An Ecological Approach

Climate Science: An Ecological Approach

Edited by Mary D'souza

SYRAWOOD
PUBLISHING HOUSE

New York

Published by Syrawood Publishing House,
750 Third Avenue, 9th Floor,
New York, NY 10017, USA
www.syrawoodpublishinghouse.com

Climate Science: An Ecological Approach
Edited by Mary D'souza

© 2018 Syrawood Publishing House

International Standard Book Number: 978-1-68286-526-2 (Hardback)

Cataloging-in-Publication Data

Climate science : an ecological approach / edited by Mary D'Souza.
    p. cm.
Includes bibliographical references and index.
ISBN 978-1-68286-526-2
1. Climatology. 2. Ecology. 3. Climatic changes. I. D'Souza, Mary.
QC861.3 .C55 2018
551.6--dc23

TABLE OF CONTENTS

PREFACE

Every book is initially just a concept; it takes months of research and hard work to give it the final shape in which the readers receive it. In its early stages, this book also went through rigorous reviewing. The notable contributions made by experts from across the globe were first molded into patterned chapters and then arranged in a sensibly sequential manner to bring out the best results.

Climatology studies weather patterns over a defined period of time. This book on climate science discusses topics related to the mapping of weather as well as technologies that can predict possible weather patterns of a region. Weather can directly affect the supply of resources that are vital for sustained life and livelihood. A number of latest researches have been included in this book to keep the readers up-to-date with the global concepts in this area of study. Coherent flow of topics, student-friendly language and extensive use of examples make this book an invaluable source of knowledge. It will serve as a reference to climatologists, meteorologists, scientists, researchers and students associated with this field of study.

It has been my immense pleasure to be a part of this project and to contribute my years of learning in such a meaningful form. I would like to take this opportunity to thank all the people who have been associated with the completion of this book at any step.

Editor

A pervasive role for biomass burning in tropical high ozone/low water structures

Daniel C. Anderson[1], Julie M. Nicely[2], Ross J. Salawitch[1,2,3], Timothy P. Canty[1], Russell R. Dickerson[1], Thomas F. Hanisco[4], Glenn M. Wolfe[4,5], Eric C. Apel[6], Elliot Atlas[7], Thomas Bannan[8], Stephane Bauguitte[9], Nicola J. Blake[10], James F. Bresch[11], Teresa L. Campos[6], Lucy J. Carpenter[12], Mark D. Cohen[13], Mathew Evans[12,14], Rafael P. Fernandez[15,16], Brian H. Kahn[17], Douglas E. Kinnison[6], Samuel R. Hall[6], Neil R.P. Harris[18], Rebecca S. Hornbrook[6], Jean-Francois Lamarque[6,19], Michael Le Breton[8], James D. Lee[14], Carl Percival[8], Leonhard Pfister[20], R. Bradley Pierce[21], Daniel D. Riemer[7], Alfonso Saiz-Lopez[15], Barbara J.B. Stunder[13], Anne M. Thompson[4], Kirk Ullmann[6], Adam Vaughan[14] & Andrew J. Weinheimer[6]

Air parcels with mixing ratios of high O_3 and low H_2O (HOLW) are common features in the tropical western Pacific (TWP) mid-troposphere (300–700 hPa). Here, using data collected during aircraft sampling of the TWP in winter 2014, we find strong, positive correlations of O_3 with multiple biomass burning tracers in these HOLW structures. Ozone levels in these structures are about a factor of three larger than background. Models, satellite data and aircraft observations are used to show fires in tropical Africa and Southeast Asia are the dominant source of high O_3 and that low H_2O results from large-scale descent within the tropical troposphere. Previous explanations that attribute HOLW structures to transport from the stratosphere or mid-latitude troposphere are inconsistent with our observations. This study suggest a larger role for biomass burning in the radiative forcing of climate in the remote TWP than is commonly appreciated.

[1] Department of Atmospheric and Oceanic Science, University of Maryland, College Park, Maryland 20742, USA. [2] Department of Chemistry and Biochemistry, University of Maryland, College Park, Maryland 20742, USA. [3] Earth System Science Interdisciplinary Center, University of Maryland, College Park, Maryland 20742, USA. [4] NASA Goddard Space Flight Center, Greenbelt, Maryland 20771, USA. [5] Joint Center for Earth Systems Technology, University of Maryland Baltimore County, Baltimore, Maryland 21250, USA. [6] Atmospheric Chemistry Observation and Modeling Laboratory, National Center for Atmospheric Research, Boulder, Colorado 80305, USA. [7] Department of Atmospheric Sciences, Rosenstiel School of Marine and Atmospheric Science, University of Miami, Miami, Florida 33149, USA. [8] Centre for Atmospheric Science, School of Earth, Atmospheric, and Environmental Science, The University of Manchester, Manchester M13 9PL, UK. [9] Facility for Airborne Atmospheric Measurements, Cranfield MK43 0JR, UK. [10] Deparment of Chemistry, University of California, Irvine, California 92697, USA. [11] Mesoscale and Microscale Meteorology Laboratory, National Center for Atmospheric Research, Boulder, Colorado 80305, USA. [12] Wolfson Atmospheric Chemistry Laboratories, Department of Chemistry, University of York, York YO10 5DD, UK. [13] NOAA Air Resources Laboratory, College Park, Maryland 20740, USA. [14] National Centre for Atmospheric Science, Department of Chemistry, University of York, York YO10 5DD, UK. [15] Department of Atmospheric Chemistry and Climate, Institute of Physical Chemistry Rocasolano, CSIC, Madrid 28006, Spain. [16] Department of Natural Science, National Research Council (CONICET), FCEN-UNCuyo, Mendoza 5501, Argentina. [17] Jet Propulsion Laboratory, California Institute of Technology, Pasadena, California 91109, USA. [18] Department of Chemistry, Cambridge University, Cambridge CB2 1EW, UK. [19] Climate and Global Dynamics Laboratory, National Center for Atmospheric Research, Boulder, Colorado 80305, USA. [20] Earth Sciences Division, NASA Ames Research Center, Moffett Field, California 94035, USA. [21] NOAA/NESDIS Center for Satellite Applications and Research, Madison, Wisconsin 53706, USA. Correspondence and requests for materials should be addressed to D.C.A. (email: danderson@atmos.umd.edu).

Tropospheric O_3 is an important greenhouse gas. Ozone has exerted an increase in the global radiative forcing of climate of $\sim 0.4 \, Wm^{-2}$ between 1750 and 2011, almost equal to that of CH_4 over the same time period[1]. The largest contribution to the climatic influence of O_3 is due to enhancements over background in the tropical troposphere[2,3]. Elevated surface O_3 adversely affects human health and agriculture[4,5]. Legislation enacted to protect public health has significantly reduced emissions of O_3 precursors from automobiles, factories and power plants throughout the industrialized world, particularly in the Northern Hemisphere mid-latitudes[6,7]. Surface O_3 levels in the industrialized extra-tropics have plateaued or fallen dramatically in response to these actions[8,9]. It is unclear whether the measures taken to reduce surface O_3 in the extra-tropics will reduce the climatic impact of O_3, since its largest radiative influence is in the tropics.

In the marine boundary layer (MBL) of the tropical western Pacific (TWP), O_3 is removed by photochemical reactions involving halogen radicals of marine biogenic origin, resulting in O_3 abundances of $\sim 20 \, p.p.b.v.$ or lower[10,11]. Local convection can transmit this low O_3 air throughout the tropospheric column[12], resulting at times in O_3 profiles that have mixing ratios of $\sim 20 \, p.p.b.v.$ over an extended altitude range[13,14]. Air masses with elevated O_3 are frequently accompanied by water vapour mixing ratios depressed with respect to the local background, particularly in the mid-troposphere[15]. These high ozone/low water (HOLW) structures inhibit mixing and convection[16], alter the local radiative heating profile[17], and affect the atmosphere's oxidative capacity[13].

Large increases in O_3 relative to these low background values have frequently been observed in the TWP mid-troposphere[12,14,18–25]. Many of these studies note that water vapour tends to be depressed, with respect to the local background, for these high O_3 air parcels[12,14,18–22,24], often attributing the HOLW structures to sources outside the tropical troposphere. Stoller et al.[19] conclude that stratospheric air is the dominant source of HOLW air parcels observed over the TWP during multiple aircraft campaigns conducted in September and October 1991, February and March 1994 and September and October 1996. This conclusion was mainly based on the co-location of high potential vorticity (PV) with a few HOLW structures. These early campaigns lacked the instrumentation to measure the suite of chemical compounds sampled during modern campaigns, particularly the biomass burning tracer HCN. Newell et al.[18] reach a similar conclusion, analysing the same data as Stoller et al. in addition to observations from the MOZAIC campaign from between 1994 and 1997. They note elevated CO and CH_4 in some of their observed HOLW structures, citing entrainment of biomass burning emissions into stratospheric air. Using back trajectory analysis in addition to sonde measurements of O_3 and H_2O from 6 years of Southern Hemisphere ADditional OZonesonde Experiment (SHADOZ) data at three Pacific sites, Hayashi et al.[20] argue that transport from the mid-latitude upper troposphere (mlUT) is the dominant source of high mid-tropospheric O_3 in the TWP, with biomass burning only a minor contributor in some months. Kley et al.[12] likewise hypothesize a mlUT origin for HOLW features near 700 hPa measured by ozonesondes in the equatorial western Pacific during the CEPEX and TOGA-COARE campaigns conducted during fall 1992. Finally, Ridder et al.[25] attribute enhancements in O_3 and CO observed remotely in the Northern Hemisphere TWP during October and November 2009 to fossil fuel combustion in Asia, Europe and North America (that is, the mid-latitudes) and in the Southern Hemisphere (SH) to both fossil fuel combustion and biomass burning.

Numerous other studies claim that low H_2O in the TWP mid-troposphere, defined here as 300–700 hPa, is dominated by the transport of dry air from the mid-latitudes. Yoneyama and Parsons[26] conclude horizontal advection from the mid-latitudes, likely through Rossby wave breaking, creates the observed distribution of dry air during the TOGA-COARE experiment. Combining satellite data and trajectory simulations, Waugh[27] finds that high PV intrusions in the subtropics lead to low relative humidity (RH) in the intrusion itself and high RH ahead of the intrusion, consistent with transport from the lower stratosphere and the deep tropical troposphere, respectively. Cau et al.[28] conclude that a dominant dry air source in the TWP is dehydration of air parcels via poleward movement of tropical, upper tropospheric air into the jet followed by slow descent driven by radiative cooling. This study relied on 24-day back trajectories and did not account for the presence of convective precipitation, which is known to greatly alter air mass composition. Galewsky et al.[29] conclude, using a tagged tracer method and reanalysis data, that mid-latitude eddies and isentropic transport are the dominant source of low H_2O in the subtropics in December–February 2001/2002. This analysis focused on the zonal mean distribution of H_2O. The subtropical location of their water minimum is coincident with earth's major deserts, which means this region is decoupled from convective precipitation that rehydrates other regions of the tropics. Conversely, Dessler and Minschwaner[30] concluded that deep convective outflow associated with the Hadley circulation is the largest source of low H_2O in the tropical eastern Pacific. Our analysis, which focuses on the tropics (20° N–20° S) of the western Pacific, similarly suggests low water is controlled by outflow of the Hadley circulation.

Other studies cite the importance of biomass burning in controlling tropical O_3, particularly over the Atlantic Ocean and in the SH, but fail to account for the origin of low H_2O in HOLW air parcels. Jacob et al.[31] find that tropical processes alone, including in situ photochemical production from biomass burning emissions and lightning NO_x ($NO_x = NO + NO_2$), can reproduce observed tropical O_3 distributions using measurements over the South Atlantic during the TRACE-A campaign in September and October 1992. Other studies have attributed high O_3 observed in the TWP to photochemical production in biomass burning plumes. Oltmans et al.[14] investigated the O_3 climatology over three locations (Fiji, Samoa and Tahiti) in the TWP from the SHADOZ network. Elevated, mid-tropospheric O_3 was found at these sites, particularly in September and October. Back trajectory analysis connected some of these O_3 enhancements in Fiji and Samoa to biomass burning in Indonesia, and it was noted that air parcels descended $\sim 3 \, km$ in transit. In the PEM-Tropics A mission conducted in the Eastern and Central Pacific of the SH in August and September 1996, both Blake et al.[21] and Singh et al.[22] attribute elevated O_3 in HOLW parcels to photochemical production from biomass burning in Africa and South America based on elevated mixing ratios of biomass burning tracers (C_2H_2, NO_x, CO, CH_3Cl, PAN and C_2H_6) in the structures. Similar to Oltmans et al., Blake et al. and Singh et al. posit large-scale subsidence as a possible cause of the low H_2O but offer no quantitative support for this supposition. Folkins et al.[24] observed HOLW air over Fiji (18° S) at $\sim 120 \, hPa$ during the ASHOE/MAESA campaign in October 1994. These air parcels were enhanced in reactive nitrogen compounds and had high N_2O mixing ratios, indicative of a tropospheric origin. They attributed the low H_2O to transport from the mlUT using 10-day isentropic trajectories but did not provide any quantitative analysis of the effect of thermodynamics or descent on observed H_2O. Finally, Kondo et al.[23] attribute increases in O_3 of 26 p.p.b.v. over a 31 p.p.b.v. background in the western

Pacific to biomass burning during the TRACE-P campaign (February–April 2001). These studies either ignore the low H_2O (ref. 23), offer limited explanation[14,21,22], attribute low H_2O to extra-tropical transport[24], or do not describe H_2O (ref. 31). The source of high O_3 as well as the dynamical processes controlling the low H_2O in HOLW structures prevalent in the TWP is clearly unresolved.

Here we analyse aircraft and ozonesonde observations, model output, and satellite data to understand the origin of the HOLW air parcels observed in the TWP during winter 2014. The Coordinated Airborne Studies in the Tropics (CAST) and CONvective TRansport of Active Species in the Tropics (CONTRAST) aircraft campaigns based in Guam (13.5° N, 144.8° E) provide a comprehensive suite of chemical measurements. We examine the 17 CAST and 11 CONTRAST flights that provided unprecedented sampling of the TWP mid-troposphere during January and February 2014 (Fig. 1). These *in situ* observations are supplemented by satellite measurements of water vapour and outgoing longwave radiation (OLR) from the daily Atmospheric Infrared Sounder (AIRS) product as well as fire counts from the Moderate Resolution Imaging Spectroradiometer (MODIS) instrument. Ten day, kinematic back trajectories were initialized along the CAST and CONTRAST flight tracks using the Hybrid Single Particle Lagrangian Integrated Trajectory (HYSPLIT) model to connect observed air parcels to their source regions. The CAM-Chem chemical transport model was run with tagged biomass burning tracers to further elucidate ozone sources and was evaluated with ozonesonde observations from the SHADOZ network. Models, satellite data and aircraft observations are used to show that anthropogenic fires in tropical Africa and Southeast Asia are the dominant source of the high O_3 and that the low H_2O results from large-scale descent within the tropical troposphere, after detrainment of biomass burning plumes.

Results

Source of high ozone in the TWP. Representative O_3, CO and H_2O profiles from the CAST and CONTRAST campaigns are shown in Fig. 2. Unless otherwise specified, all references to H_2O in this paper refer to water vapour observations from an open-path laser hygrometer with complete discrimination against condensed water[32]. Mixing ratios of O_3 and H_2O in these profiles exhibit strong anti-correlation; that is, elevated O_3 is closely associated with low RH[15]. These HOLW structures were a pervasive feature seen throughout both campaigns. Here we define a HOLW structure as an air parcel satisfying the simultaneous criteria of O_3 >40 p.p.b.v. and RH <20%, where RH in this study is with respect to water for T > 273 K and to ice for T < 273 K. Relying solely on composition to define HOLW structures allows for the inclusion of both sharply defined, thin

features (for example, Fig. 2a) as well as those that occupy a large fraction of the tropospheric column (for example, Fig. 2d). HOLW features were primarily located between 300 and 700 hPa and observed during 45 of 104 independent profiles (that is, air parcels were not intentionally resampled). For air parcels with RH >20%, median O_3 was nearly constant with altitude at ~20 p.p.b.v. (Fig. 3a). For air with RH <20%, median O_3 peaked at 60 p.p.b.v. near 450 hPa, three times greater than background, and decreased to ~40 p.p.b.v. at 750 and 300 hPa.

Kinematic trajectories connect nearly all of these HOLW structures to tropical regions with active biomass burning. Figure 4a,b shows 10-day back trajectories found using the HYSPLIT model[33]. Trajectories are coloured by observed aircraft O_3 (invariant along each trajectory) and RH along the trajectory as output by HYSPLIT (see Methods section). Each trajectory stops at our estimate of the point of last precipitating convection, derived from a combination of cloud top height and precipitation

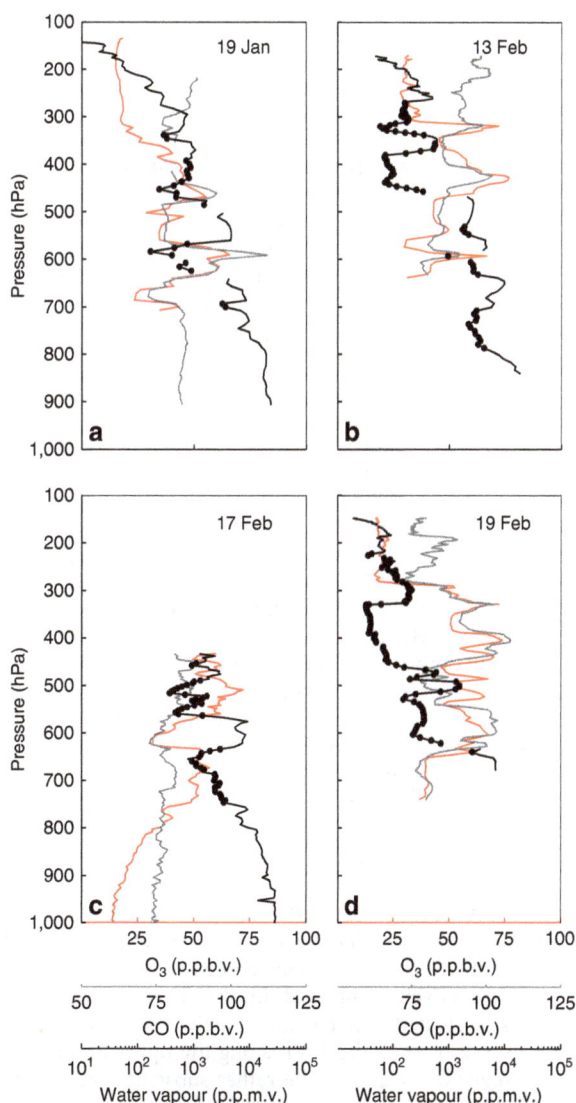

Figure 2 | Ozone, water vapour and CO profiles in the TWP. Sample profiles of 10 s averaged O_3 (red), CO (grey) and H_2O (blue) from four flights during CONTRAST (**a,b,d**) and CAST (**c**). Blue circles indicate measurements of H_2O mixing ratios for which RH <20%. RH is with respect to water and with respect to ice for temperatures above and below 273 K, respectively. Vertical profiles have a characteristic horizontal length of ~300 km.

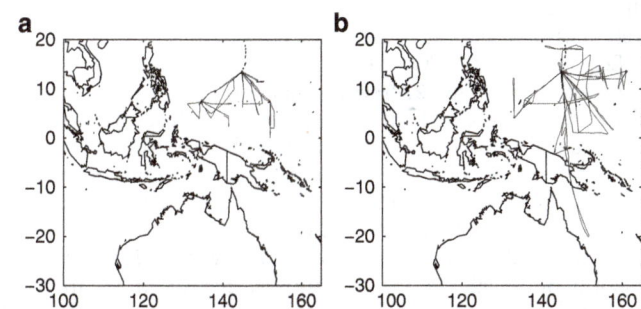

Figure 1 | Flight tracks for the CAST and CONTRAST campaigns. Tracks for (**a**) CAST and (**b**) CONTRAST flights analysed in this study.

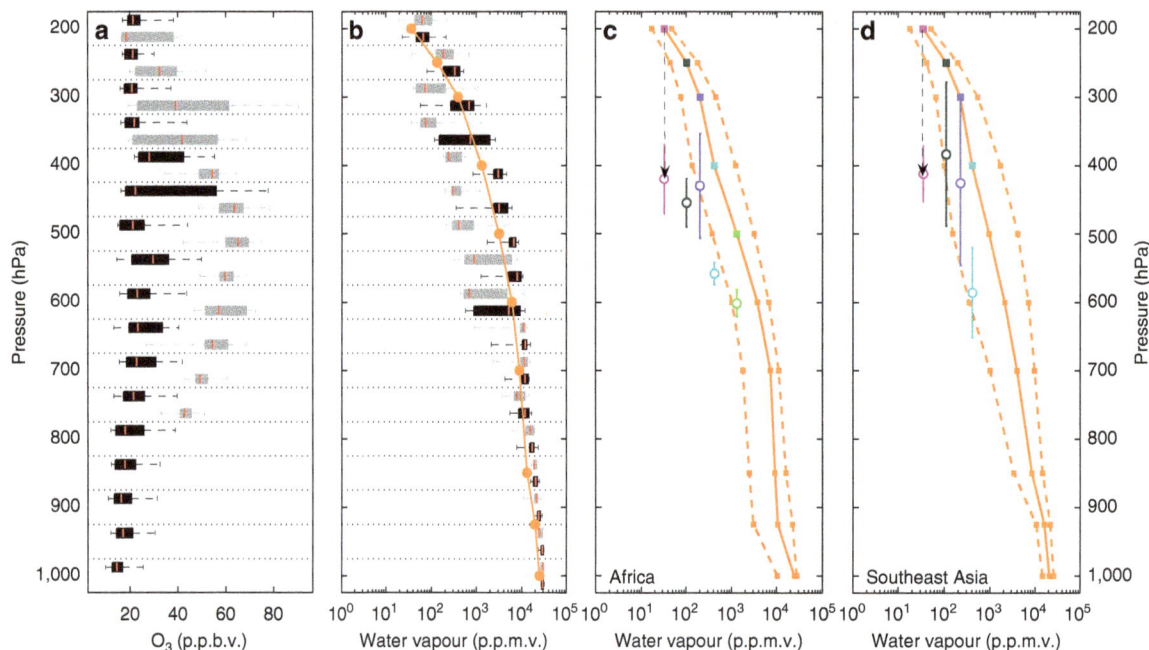

Figure 3 | Structure of ozone and water vapour in the TWP. (a) Profiles of *in situ* O_3 during CAST and CONTRAST for two modes of RH (blue, RH >20%; grey, RH <20%). Box and whisker plots show 5th, 25th, 50th, 75th and 95th percentiles for 50 hPa pressure bins. Each bin is delimited by a dotted line, and the two modes (grey and blue boxes) for a given pressure bin are offset for clarity. Between 400 and 700 hPa, the average (minimum) number of observations per blue and grey box is 650 (414) and 330 (43), respectively. Observations between 300 and 400 hPa were more limited with a minimum of 20 and a maximum of 56 observations. **(b)** Distribution of H_2O for two modes of HCN (grey, HCN >150 p.p.t.v.; blue, HCN <150 p.p.t.v.). Median AIRS H_2O over the ascending branch of the Hadley Cell (orange). Between 300 and 700 hPa, the average (minimum) number of observations per blue and grey box is 129 (28) and 111 (19), respectively. **(c)** Median (solid), 5th and 95th percentiles (dashed) of AIRS H_2O over the African biomass burning region (orange). Open circles represent the mean end point of descent ±1σ of trajectories starting over the African biomass burning region and arriving over the TWP, for various initial pressures (closed squares). **d** Same as **c** but for trajectories starting over Southeast Asia.

data from geostationary satellites and the Tropical Rainfall Measuring Mission satellite. We stop the trajectories at the point of last precipitating convection because convection lofts MBL air into the upper troposphere, where it detrains throughout the column, creating a nearly uniform O_3 profile and a water vapour profile controlled by the local saturation vapour pressure[15]. This convection effectively resets air parcel composition, removing any influence from outside the TWP. Further, global transport models are unable to reproduce the small-scale wind fields associated with deep convection, creating large uncertainty for back trajectories beyond this point. Air parcels originating from the clean SH and eastern Pacific had unimodal O_3 and RH distributions, with medians of 22 p.p.b.v. and 63%, respectively. Trajectories originating west of the study region, primarily over Africa and Southeast Asia, exhibit a bimodal distribution[15]. Assuming a mixture of two Gaussian distributions, the two modes are described by one with peak O_3 and RH of 21 p.p.b.v. and 60% and another with peaks of 55 p.p.b.v. and 4.5%. The low O_3 trajectories are indicative of air parcels under local influence (that is, convectively controlled in the TWP). The high O_3 trajectories originated over regions of active biomass burning in the tropics and reached the TWP along the upper branch of the Walker circulation (Fig. 4a). The rather substantial contribution of the HOLW structures to the overall atmospheric state of the TWP troposphere is quantified by Pan *et al.*[15].

Simultaneous elevation of CO and O_3 in the HOLW structures, in addition to regression analysis, suggests a tropospheric origin. Local maxima in O_3 and CO profiles closely track one another, with r^2 values between these two species ranging from 0.46 to 0.72 (Fig. 5) for the profiles shown in Fig. 2. A campaign-wide regression of CO and O_3 for air parcels that trace back to

continental regions yields $r^2 = 0.61$. Photochemical enhancement ratios of O_3 with respect to CO ($\Delta O_3/\Delta CO$) for the profiles shown in Fig. 2 range between 0.97 and 2.85 mol mol^{-1}. Mauzerall *et al.*[34] found fresh biomass burning plumes had enhancement ratios of ~0.15 mol mol^{-1} while plumes older than 6 days had values on the order of 0.75 mol mol^{-1}. These results are consistent with Parrington *et al.*[35] who found mean enhancement ratios of 0.81 mol mol^{-1} with a maximum ratio of 2.55 for boreal biomass burning plumes older than 5 days. Both comparisons suggest the air parcels analysed here are significantly aged and the enhancement ratios lie at the upper end of previous observations. This agrees with the back trajectories and photochemical aging analysis, discussed below, both of which yield air parcel ages of ~10 days. NO is also elevated in these air parcels (Supplementary Fig. 1), though the correlation with O_3 is not as prominent as for other species. This is likely due to the conversion of NO to other nitrogen-containing species, such as alkyl nitrates, peroxyacetyl nitrate and HNO_3, none of which were measured during CAST and CONTRAST.

Regression of CO against O_3 for all observations (Supplementary Fig. 2) reveals four distinct air mass types observed during CAST and CONTRAST: stratospheric, marine and two distinct polluted regimes, one with elevated CO, NO and O_3 and the other with only elevated CO. Known stratospheric air, characterized by high NO and a strong anti-correlation between CO and O_3, was encountered on two flights into the subtropics. Data collected during these two flights are excluded from the majority of the HOLW analysis, since our focus is on the tropical troposphere, but are used below to argue against a stratospheric origin of these structures. Parcels with marine characteristics (low CO, NO and O_3) all originated from the SH or eastern

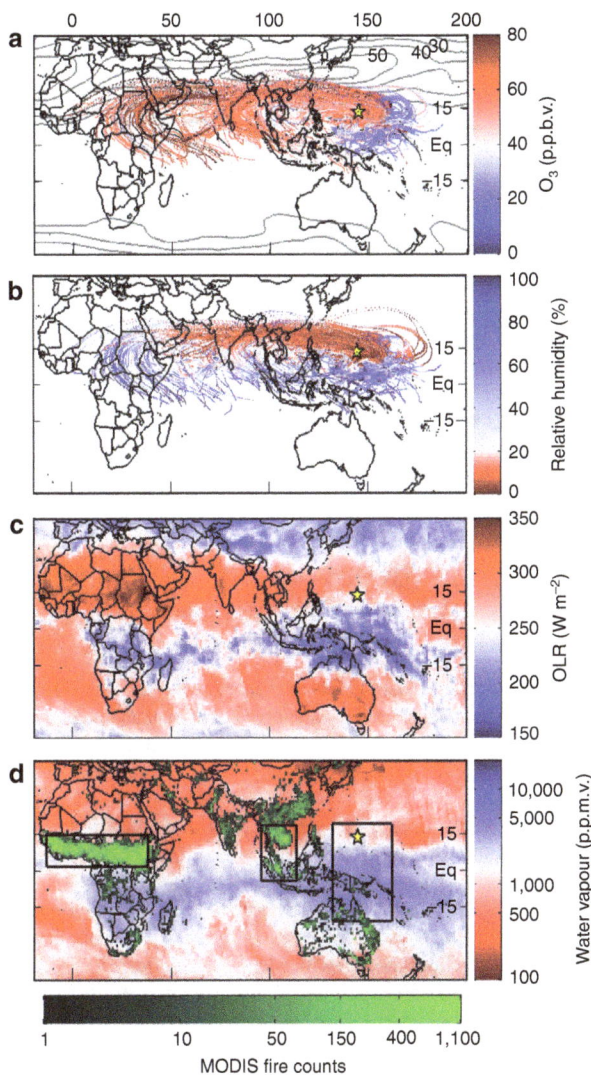

Figure 4 | Origins of air in the TWP. (**a**) 10-day, HYSPLIT back trajectories for all CAST flights analysed here and CONTRAST RF03–05 and RF07–14 for observed pressures between 300 and 700 hPa. Trajectories are stopped when encountering convective precipitation and coloured by observed O_3. For clarity, only every third is shown. Contours are zonal winds at 200 hPa averaged over January and February 2014 in 10 m s^{-1} intervals. The yellow star shows Guam. **b** Same as **a** but coloured by HYSPLIT RH along the trajectory (see methods). (**c**) AIRS daytime OLR averaged over CAST and CONTRAST flight days. **d** Same as **c** but for AIRS H_2O at 500 hPa. MODIS fire counts are the total for January and February 2014. Black rectangles represent the African and Southeast Asian tropical biomass burning regions (determined subjectively by the high fire counts) and the CAST/CONTRAST study region.

Pacific (Fig. 6). Back trajectory analysis of the two polluted regimes connects the high CO, NO and O_3 parcels to biomass burning regions in central Africa or Southeast Asia. The high CO/low NO/low O_3 parcels have various geographic origins in Southeast Asia, which could be reflective of emissions of high CO and low NO_x from two-stroke engines, dominant in this region of the world.

Analysis of other chemical tracers suggests that the observed O_3 in the HOLW structures is likely produced photochemically in biomass burning plumes. Both HCN and CH_3CN are emitted almost exclusively by biomass burning[36]. Elevated mixing ratios of either species in an air parcel therefore suggests a biomass

burning origin. Figure 7 shows the profiles of these two species as well as O_3 and the industrial tracer tetrachloroethylene (C_2Cl_4) for the same flights as Fig. 2. Panels b–d show very tight correlations of O_3 with HCN and CH_3CN, while, in panel a, HCN and CH_3CN are elevated between 400 and 700 hPa. Over the entire campaign, air parcels with back trajectories that trace back to Africa and Southeast Asia have a high correlation between O_3 and HCN ($r^2 = 0.80$), demonstrating significant biomass burning influence. The enhancement ratio of CH_3CN to CO (4.02 p.p.t.v./p.p.b.v.) for the HOLW structures is consistent with CH_3CN emissions from tropical forest burning (Supplementary Fig. 3)[37] and is significantly higher than the enhancement ratio for CH_3CN emissions from fossil fuel combustion (< 0.1 p.p.t.v./p.p.b.v.)[38]. The biomass burning tracers benzene (C_6H_6) and ethyne (C_2H_2) also show strong correlation with CO ($r^2 > 0.6$) as seen in Fig. 8, further confirming the origin of these air parcels.

Photochemical aging is consistent with the back trajectory analysis. Both C_6H_6 and C_2H_2 have lifetimes much shorter than that of CO (ref. 34). On the basis of modelled OH from CONTRAST, C_6H_6, C_2H_2 and CO in the TWP have lifetimes of ∼6, 12 and 35 days, respectively. All species are primarily removed by reaction with OH, so the change in CO with respect to either C_6H_6 or C_2H_2 relative to the initial emissions ratio allows determination of a photochemical age[39]. The photochemical age for C_6H_6 with respect to CO as determined by the Total Organic Gas Analyzer (TOGA) and Advanced Whole Air Sampler (AWAS) measurements were 13 and 8 days, respectively (see Methods section). The photochemical age for C_2H_2 was 11 days. These photochemical ages are broadly consistent with the elapsed time between detrainment of biomass burning plumes over Africa and Southeast Asia and transit to the TWP indicated by the back trajectory analysis (5 ± 4 and 7 ± 2 for Southeast Asia and Africa, respectively). The difference between the age indicated by back trajectory analysis and that of photochemical aging is likely due to dilution of biomass burning plumes with ambient air. Dilution would tend to artificially inflate the photochemical age since only CO is present in appreciable amounts in the background TWP. These relatively short values for photochemical age show the composition of air in the HOLW structures is of recent origin.

Tagged tracers for biomass burning CO in the CAM-Chem model[40], run using assimilated meteorology, support our interpretation. A strong African biomass burning influence is frequently seen in the upper troposphere over much of the TWP and extending as far east as Hawaii, at times accounting for 17% of total CO (Supplementary Fig. 4). Deep convection can loft emissions from fires into the upper troposphere, where strong westerlies transport pollutants long distances[41]. Southeast Asian emissions are prominent throughout the tropospheric column (Supplementary Fig. 4).

Photochemical ozone production. Analysis of ozonesonde observations, CAM-Chem O_3, and photochemical box model output quantitatively show the high O_3 likely originates within the tropical troposphere. Regions of tropical biomass burning have elevated O_3 as compared with the rest of the tropics, with an O_3 maximum over Africa and the Atlantic basin and a minimum over the TWP[42]. Median O_3 in central Africa and Southeast Asia from CAM-Chem was ∼50 p.p.b.v. (Fig. 9a) and ∼40 p.p.b.v. (Supplementary Fig. 5a), respectively, a factor of 2 greater than background O_3 in the TWP. We compare the CAM-Chem output to ozonesonde[43] measurements over Nairobi (Fig. 9b) and Hanoi (Supplementary Fig. 5b), both strongly influenced by biomass burning, to evaluate CAM-Chem model performance. Mean O_3 from the ozonesondes generally lie within 1σ of the mean

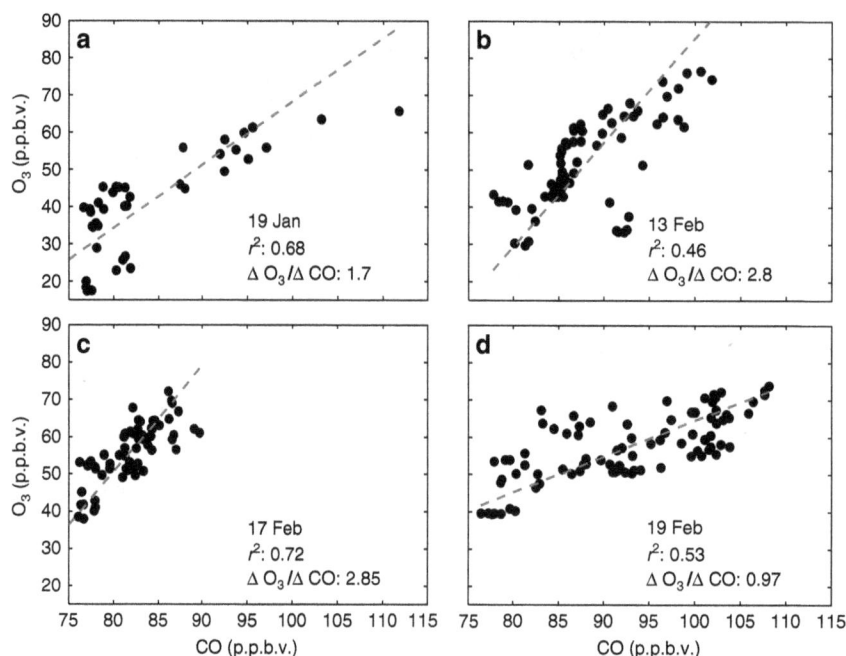

Figure 5 | Enhancement ratios of CO and ozone from CAST and CONTRAST. Regression of CO against O_3 for the data shown in Fig. 2 for pressures between 300 and 700 hPa (**a–d**). The $\Delta O_3/\Delta CO$ ratio for all profiles suggests significantly aged air, consistent with the back trajectory and photochemical aging analyses. The dashed red line is the best fit via orthogonal linear regression. Flight dates, slope and r^2 values are shown for all panels.

CAM-Chem value, suggesting the model accurately captures the O_3 profile in these locations. Means are used because only four or five ozonesonde profiles are available for the study period from each site. Median ozone profiles from CAM-Chem and SHADOZ show comparable agreement in the middle and upper troposphere and substantial differences near the surface (Supplementary Fig. 6). Transport of O_3 from these biomass burning regions cannot explain all of the observed O_3 in the HOLW structures, however, as values frequently peaked at ~75 p.p.b.v., implying there must be photochemical production as the air parcels travel from the biomass burning region.

To estimate the net O_3 production in the HOLW structures, a box model constrained by CONTRAST observations (see Methods section) was run for the profiles shown in Fig. 2. Net O_3 production in the HOLW structures was on the order of ~2 p.p.b.v. per day (Fig. 9c and Supplementary Fig. 7). The value calculated here is a lower bound to net O_3 production in the plume. It is likely that photochemical production along the flight track is significantly lower than in fresh biomass burning plumes, which would be enriched in both NO_x and volatile organic compounds (VOCs) as compared with the more aged air observed in CONTRAST. Lightning, a significant contributor to upper-tropospheric NO_x (ref. 44), would likely further enrich fresh plumes that were lofted into the upper troposphere through deep convection. These fresh plumes can have O_3 production rates of 7 p.p.b.v. per day or higher[42]. As the air ages, however, the abundance of HNO_3 in the plume increases[34], making NO_x unavailable for O_3 production. OH will also oxidize VOCs to less reactive species, and dilution with background air will tend to counteract photochemical O_3 production. Nevertheless, a net O_3 production rate of ~2 p.p.b.v. per day over 5–10 days, combined with initial O_3 of 40–50 p.p.b.v. in the outflow of biomass burning, is consistent with the observed O_3 in the HOLW structures.

Low water vapour origin. We now turn to the origin of low H_2O. As with O_3, H_2O observed during CAST and CONTRAST has two distinct modes, suggesting that H_2O in the TWP is controlled both locally and by large-scale processes from outside the study region. For air parcels with HCN <150 p.p.t.v., the H_2O profile follows the saturation vapour pressure (Fig. 3b, blue boxes), suggesting that the H_2O mixing ratio is controlled by local thermodynamics associated with deep convection. CONTRAST flights designed to measure fresh convective outflow were the only tropical flights where HOLW structures were not observed[15]. Parcels with HCN >150 p.p.t.v. had H_2O mixing ratios an order of magnitude lower than the local saturation vapour pressure. This is consistent with transport of dry air from outside of the TWP. Potential mechanisms to produce the dry air include horizontal advection from the mid-latitudes as well as large-scale descent in the tropics. The association of low H_2O with high HCN indicates these air parcels originate from biomass burning regions.

AIRS measurements[45] and trajectory analysis strongly support our supposition that the observed departures from background H_2O result from large-scale descent in the tropics. The ascending and descending branches of the Hadley Cell, as determined by AIRS OLR, are shown in Fig. 4c as regions with OLR <250 W m^{-2} (blue) and OLR >250 W m^{-2} (red), respectively[46]. AIRS H_2O averaged over the ascending branch agrees well with *in situ* H_2O in the low HCN air parcels (Fig. 3b). Regression of *in situ* H_2O observations from CONTRAST to co-located AIRS retrievals of H_2O (Supplementary Fig. 8) shows these data sets are in excellent agreement ($r^2 = 0.98$), allowing for direct comparison of the satellite-retrieved H_2O to the *in situ* observations. These profiles are characteristic of deep convection, leading to RH >70%. Back trajectories show the RH <20% air parcels frequently originate in the ascending branch of the Hadley cell, flow anti-cyclonically towards the descending branch, and then reach the TWP along the prevailing westerlies (Fig. 4). Air parcels originating over Africa descend 202 ± 64 hPa, on average, leading to significant decline in RH during transit to the TWP (Fig. 4b). Figure 3c shows the AIRS H_2O profile over Africa (orange lines and squares). The final H_2O profile over the TWP

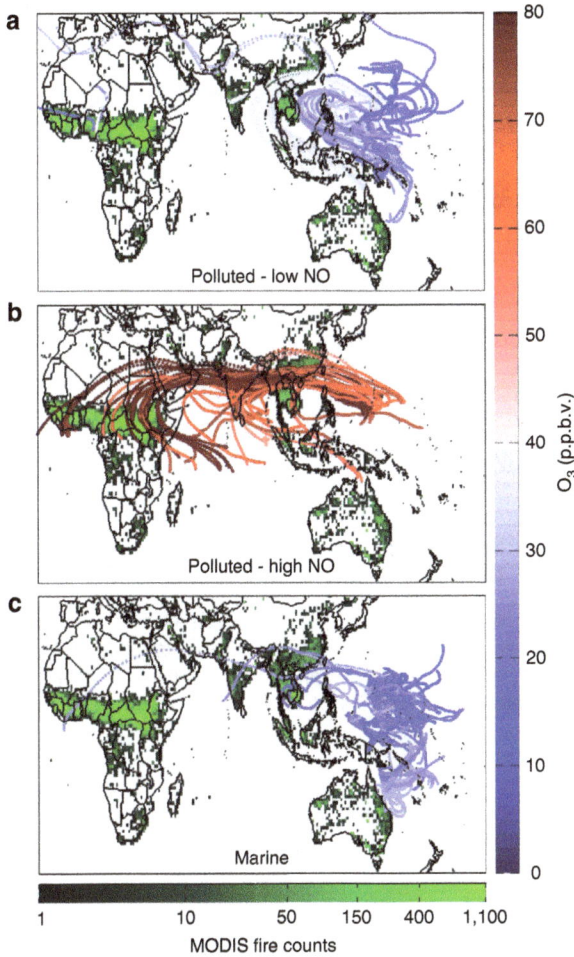

Figure 6 | Back trajectories by chemical regime. 10-day, HYSPLIT back trajectories for all CAST flights analysed here and CONTRAST RF03–05 and RF07–14 for observed pressures between 300 and 700 hPa. Trajectories are stopped when encountering convective precipitation and coloured by observed O_3. Trajectories are separated by the CO regimes illustrated in Supplementary Fig. 2: (**a**) Polluted—low NO (CO >95 p.p.b.v. and NO <40 p.p.t.v.), (**b**) Polluted—high NO (CO >95 p.p.b.v. and NO >50 p.p.t.v.), and (**c**) Marine (CO <80 p.p.b.v. and NO <30 p.p.t.v.). The sum of January and February 2014 MODIS fire counts is shown in green.

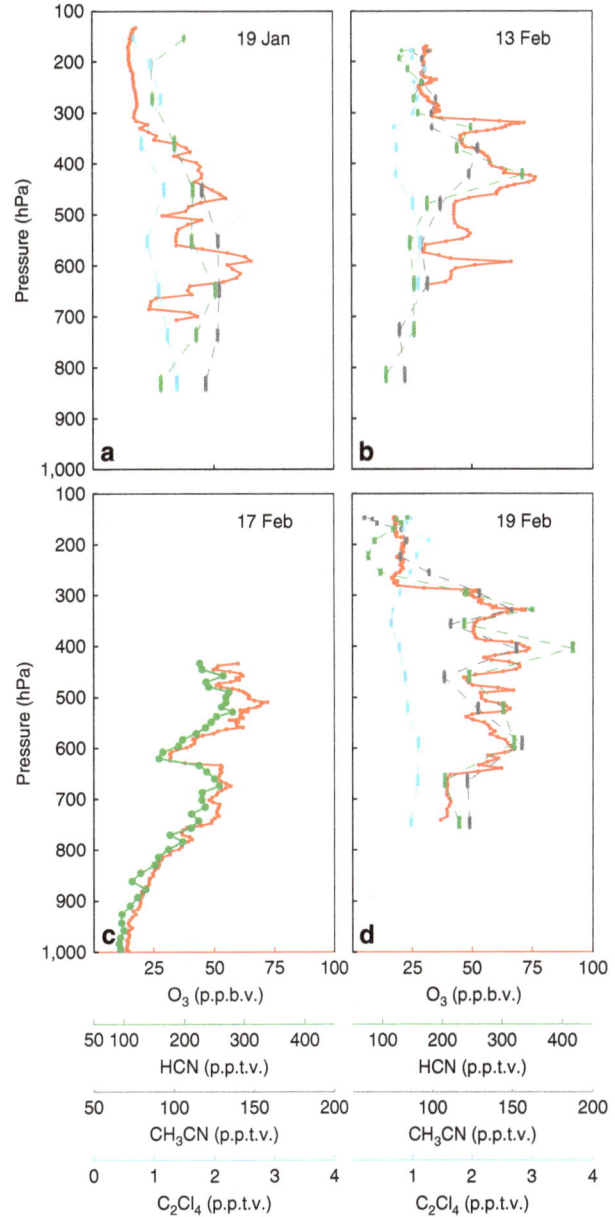

Figure 7 | Anthropogenic tracers in the HOLW structures. Sample profiles from CONTRAST (**a,b,d**) and CAST (**c**) from the four flights shown in Fig. 2. O_3 (red), CH_3CN (grey), HCN (green) and C_2Cl_4 (light blue) are shown. O_3 data are 10 s averages. CH_3CN and C_2Cl_4 were not measured during CAST. CONTRAST HCN, CH_3CN and C_2Cl_4 were sampled for 35 s at 2 min intervals. CAST HCN was sampled for 30 s. Vertical bars show the pressure range traversed during sampling.

after subsidence (circles), assuming conservation of the H_2O mixing ratio in transit, quantitatively agrees with *in situ* H_2O for the enhanced HCN mode (grey symbols, Fig. 3b). Similar agreement is found for AIRS H_2O profiles over Southeast Asia (Fig. 3d). This analysis demonstrates large-scale descent of tropical biomass burning plumes, on their transit to the TWP, can produce the dry element of the HOLW structures.

Drying of the air parcels by mixing with mid-latitude air or transport to higher latitudes is not supported by the back trajectory analysis. To determine whether air parcels were moistened or dried during transit, the difference between the trajectory starting point (that is, the trajectory initialization point along the flight track) and the H_2O mixing ratios at the trajectory end point (that is, 10 days before observation or the point of last precipitating convection) was calculated for all HOLW structures. Water vapour mixing ratios were calculated from the RH and temperature output by the HYSPLIT model along each trajectory. Supplementary Fig. 9 shows a histogram of these changes in H_2O. For the majority of air parcels (>80%), the H_2O mixing ratio either increased or did not change in transit to the TWP. This moistening indicates that the

majority of the HOLW structures encountered by the CAST and CONTRAST aircraft did not require additional condensation after detrainment to account for the low RH. In fact, these air parcels moistened (experienced a modest increase in H_2O mixing ratio) in transit, likely due to mixing with background air. The low RH is due to large-scale descent in the tropics and does not require condensation in the mlUT.

Potential origins outside the tropical troposphere. Mixing with mlUT air is inconsistent with the air mass history and observed composition of the HOLW structures. Parcel trajectories with

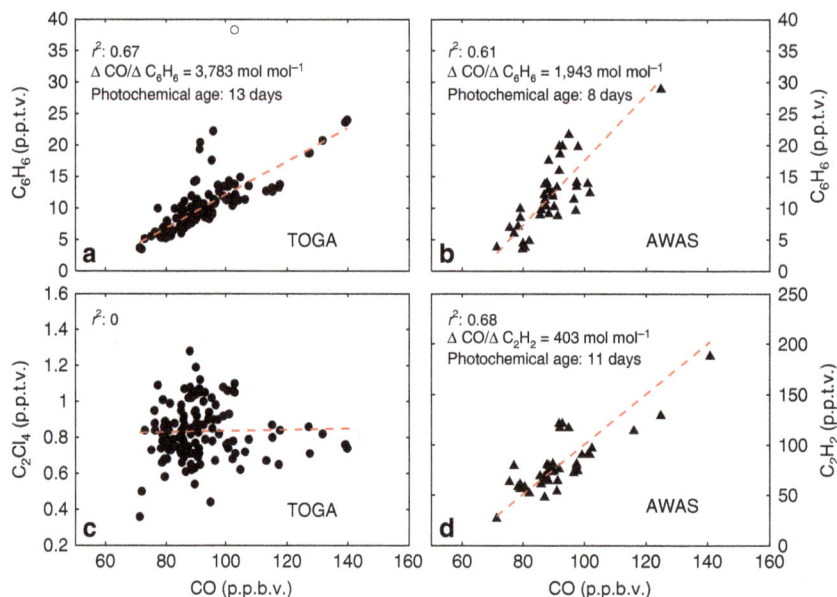

Figure 8 | Photochemical age of observed air parcels. Regression of CO against: TOGA C_6H_6 (**a**); AWAS C_6H_6 (**b**); TOGA C_2Cl_4 (**c**); and AWAS C_2H_2 (**d**). Data are only for the HOLW structures (O_3 >40 p.p.b.v. and RH <20%). The dashed red line is the best fit via orthogonal linear regression. A single data outlier (open circle, **a**) has been excluded from the analysis. Values of r^2 are shown for all panels; enhancement ratios and photochemical ages are shown for panels **a**, **b** and **d**. Excluding the data point for which C_6H_6 >25 p.p.t.v. for panel **b** results in an r^2 of 0.47, a CO to C_6H_6 enhancement ratio of 1,823 mol mol^{-1}, and a photochemical age of 7 days.

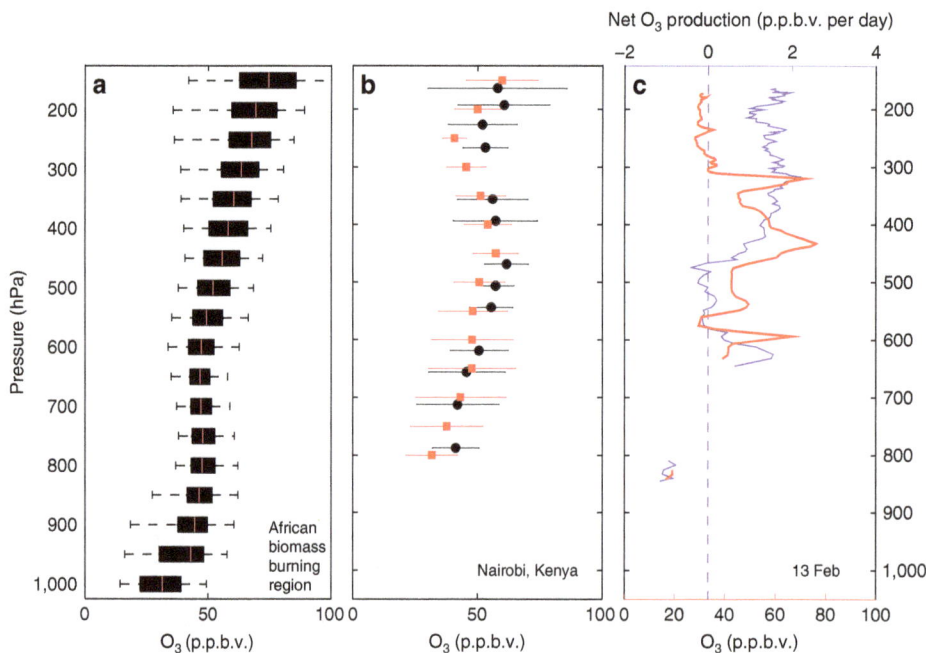

Figure 9 | Ozone over Africa and *in situ* potochemical ozone production in the TWP. (**a**) Vertical distribution of CAM-Chem O_3 in the African biomass burning region (that is, black box Fig. 4); 5th, 25th, median, 75th and 95th percentiles are shown. (**b**) Mean $\pm 1\sigma$ of SHADOZ ozonesonde observations over Nairobi, Kenya for January and February 2014 (red). Mean $\pm 1\sigma$ CAM-Chem O_3 modelled over Nairobi sampled on the same days as the ozonesondes (blue). (**c**) O_3 profile from Fig. 2b (red) and net O_3 production in the profile (purple, top axis), calculated using the DSMACC photochemical box model (see Methods section).

high O_3 began in the tropics and remained south of the jet core (Fig. 4a), indicating minimal contact with mid-latitude air. None of the back trajectories connect the HOLW air parcels to the mid-latitudes. This is consistent with the chemical composition of the filaments. Tetrachloroethylene (C_2Cl_4), a tracer of industrial pollution, has an atmospheric lifetime on the order of 4 months

and is primarily emitted in the Northern Hemisphere mid-latitudes, creating a strong latitudinal gradient. Air masses in the tropics influenced by mid-latitude emissions from China frequently have C_2Cl_4 mixing ratios >2 p.p.t.v. versus a tropical background of 1 p.p.t.v. or less[47]. Tetrachloroethylene mixing ratios in the HOLW structures are a factor of 1.7 times smaller

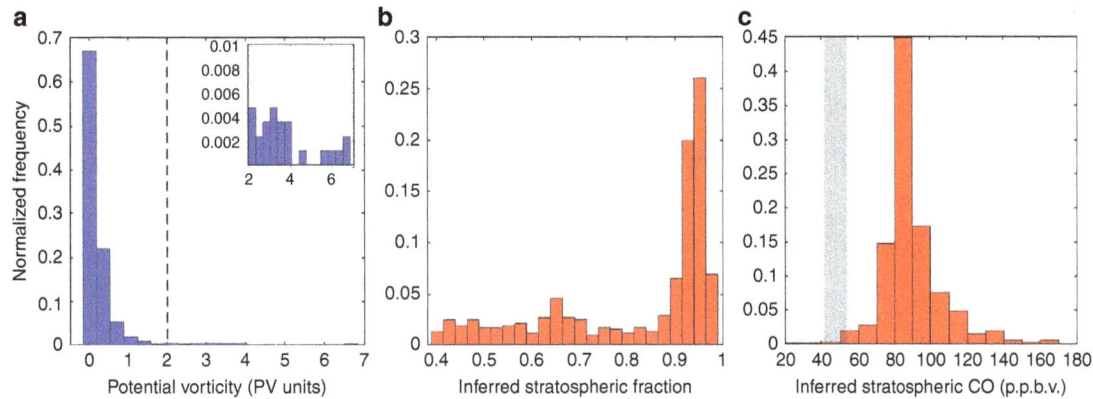

Figure 10 | Negligible stratospheric influence on the composition of the TWP troposphere. (**a**) Distribution of the maximum absolute value of PV along the back trajectories for all air parcels observed between 300 and 700 hPa that have travelled over the southeast Asian and/or African biomass burning regions. (**b**) Distribution of the inferred stratospheric fraction of air (f_{STRAT}) needed to explain the low H_2O in the HOLW filaments, if the depression in H_2O were solely due to stratospheric intrusion. (**c**) Distribution of inferred stratospheric mixing ratio of CO ($CO_{STRAT\ INFERRED}$) assuming the fraction of stratospheric air shown in **b**. Grey area represents the mean ± 2σ of CO observed in the stratosphere during CONTRAST RF15 ($O_3 > 200$ p.p.b.v.). This mixing line analysis suggests negligible stratospheric influence on the composition of the TWP mid-troposphere during CAST and CONTRAST.

than that observed in the mid-latitude free troposphere during CONTRAST (0.89 and 1.52 p.p.t.v., respectively), suggesting little influence from the mid-latitudes. This is further confirmed by the lack of correlation between C_2Cl_4 and either high O_3 (Fig. 7) or CO (Fig. 8c).

Stratospheric origin is also inconsistent with the observed composition of the HOLW structures. Elevated O_3 in the remote TWP can have plausible sources from the polluted troposphere or the O_3 rich stratosphere, resulting in enhancements of a similar magnitude. Mixing line analysis using measurements of at least two other tracers with distinctly different abundances in the troposphere and stratosphere can be used to assess the relative contribution of each potential source[48]. Here we use CO and H_2O as tracers of the polluted troposphere and the stratosphere, respectively (see Methods section). The mixing line analysis shows that to reproduce the observed H_2O profile, mixing of stratospheric and background air would require >90% stratospheric air for the majority of observed parcels (Fig. 10b) and, in turn, a stratospheric CO mixing ratio of >75 p.p.b.v. for the vast majority of the air parcels (red bars, Fig. 10c). Observed stratospheric CO was 47.6 ± 6 p.p.b.v. (2σ; grey bar, Fig. 10c). Only ~1% of parcels had an inferred stratospheric CO within the observed range of stratospheric CO, demonstrating a negligible role for stratospheric influence on the HOLW structures using mixing line analysis. Stratospheric influence was also estimated by interpolating PV to the calculated back trajectories. Only 4% of observed air parcels encountered stratospheric air, defined as intersecting a PV surface >2 PVU (ref. 49) along the trajectory (Fig. 10a). Relaxing the tropopause definition to 1.5 PVU does not significantly alter this percentage. The combination of low PV air along the trajectories and the mixing line analysis indicates negligible stratospheric influence on the composition of the TWP mid-troposphere during CAST and CONTRAST.

Frequent deep convection in the eastern Pacific likely prevents stratospheric air from reaching the study region. The westerly duct near 180–200° E, a region of preferential transport from the mlUT to the tropics, could potentially supply stratospheric HOLW air to the study area[50]. However, trajectories originating from the SH and eastern Pacific encountered precipitating convection ~2.2 days before observation, that is, these trajectories do not extend back to the westerly duct. Convection promotes mixing with MBL air, removing any extra-tropical

signature, as evidenced by the low O_3 mixing ratios for air originating from the eastern Pacific as discussed above.

Discussion

We have shown that the high O_3 in the HOLW structures sampled in the TWP during winter 2014 is quantitatively consistent with a tropical, biomass burning source and that the low H_2O mixing ratio is consistent with large-scale descent in the tropics. In a sense, low RH acts as a tropospheric age of air indicator in the tropics. Photochemical O_3 production driven by emissions from biomass burning regions, in combination with large-scale descent of tropical, tropospheric air parcels that do not experience active precipitation, leads to a strong anti-correlation of H_2O and O_3. Prior analyses of HOLW structures, which suggested an extra-tropical tropospheric origin[12,18–20,25], lacked the chemical sophistication of CAST and CONTRAST, relying primarily on ozonesondes or a limited number of chemical tracers in their analyses. Dynamical features suggested as possible mechanisms to bring dry air into the tropics—intrusions of high PV[27], Rossby wave breaking[26], and drying through mixing with the subtropical jet[28,29]—are inconsistent with the results presented here. These prior studies tend to focus on the subtropics rather than the tropics, use trajectories calculated without consideration of convective precipitation and at times in isentropic coordinates, or provide an interpretation for H_2O that is qualitative rather than quantitative. Our aircraft and satellite data indicate, in agreement with Dessler and Minschwaner[30], that large-scale descent in the tropics associated with the Hadley circulation exerts primary control on the H_2O composition of the TWP troposphere for air parcels that have not experienced recent convection.

The attribution of the high O_3 in these HOLW structures suggests a potentially larger role for biomass burning in the radiative forcing of climate in the remote TWP than is commonly appreciated. Tropical tropospheric O_3 is a greenhouse gas, exerting a strong radiative forcing on global climate[2,3]. However, present efforts to limit emissions of O_3 precursors are primarily focused on industrial activities and fossil fuel combustion that occur outside the tropics[51]. If the high O_3 in these structures is primarily of tropical origin, then present legislation to limit the emission of O_3 precursors in the extra-tropics may have little, if any, positive impact for the radiative forcing of climate due to tropospheric O_3. While it is beyond the scope of this paper to estimate the radiative effects of these HOLW structures, biomass burning[52] and HOLW[18] structures are common features of the tropics throughout the year, implying that these structures could have a substantial impact on both global and regional radiative forcing of climate.

Methods

Field campaigns. The CONTRAST campaign consisted of 13 research and 4 transit flights conducted using the National Center for Atmospheric Research (NCAR) Gulfstream V aircraft during January and February 2014. Objectives of the campaign included determining the budget and speciation of very short-lived halogen compounds in the TWP and investigating the transport pathways of these and other chemicals from the MBL to the tropical tropopause layer in a strongly convective environment. Research flights (RF) were either based out of Guam (13.5° N, 144.8° E) or conducted in transit between Broomfield, CO (39.95° N, 105.1° W), the home-base of the Gulfstream V, and Guam. Flights from Guam spanned latitudes from the northern coast of Australia to Japan and altitudes from ~0.5 to 15.5 km. Transit flights (RF01, RF02, RF16 and RF17) as well as flights that sampled primarily mid-latitude air (RF06 and RF15) have been excluded from our analysis, unless otherwise indicated. Tracks of the 11 CONTRAST flights considered here are shown in Fig. 1; these flights include all that sampled exclusively in the tropics. In all, 63 vertical profiles were conducted during these flights, offering unprecedented sampling in the middle and upper troposphere of the TWP. RF15 sampled the lower, mid-latitude stratosphere near Japan, as evidenced by O_3 mixing ratios that approached 1 p.p.m.v. The mixing ratio of stratospheric species quoted here are taken from this flight segment, where stratospheric air is defined as having $O_3 > 200$ p.p.b.v..

The NCAR Gulfstream V aircraft was outfitted to measure various trace gases, meteorological parameters and radiative flux. O_3, NO and NO_2 were measured by chemiluminescence at 1 Hz (ref. 53). The 1σ precisions of O_3, NO and NO_2 at 1 Hz sampling frequency, in the troposphere, are below 0.5 p.p.b.v., 10 p.p.t.v. and 20 p.p.t.v., respectively. Total uncertainties are 5% for NO and O_3 and is 20% for NO_2. O_3 has been corrected for quenching owing to ambient H_2O (ref. 54). CO was also measured at 1 Hz, with an Aero-Laser 5002 vacuum ultraviolet fluorescence instrument[55] with a 2σ uncertainty of 3 p.p.b.v. ± 3%. Water vapour was measured by an open path, laser hygrometer at two wavelengths (1,853.37 and 1,854.03 nm), allowing for the sampling of H_2O mixing ratios spanning 5 orders of magnitude[32]. Data were reported at 1 Hz with a 2σ precision of <3%. RH was calculated from observed H_2O and temperature. Reported RH is with respect to water for temperatures above 0 °C and with respect to ice for temperatures below 0 °C. Formaldehyde (HCHO), necessary for modelling OH and HO_2, was measured at high frequency using laser induced by the *in situ* airborne formaldehyde (ISAF) instrument[56]. HCHO was reported by ISAF at 1 Hz with a 2σ uncertainty of ± 20 p.p.t.v..

TOGA measured a suite of trace gases via gas chromatography/quadrupole mass spectrometry with a sampling time of 35 s and 2 min between sampling periods[57]. Measured species relevant to this study are hydrogen cyanide (HCN), acetonitrile (CH_3CN), tetrachloroethylene (C_2Cl_4), acetone (CH_3COCH_3), acetaldehyde (CH_3CHO), propane (C_3H_8) and HCHO. AWAS also measured a suite of trace gases, including ethyne (C_2H_2) and benzene (C_6H_6). AWAS acquires up to 60 samples of ambient air per flight in electropolished stainless-steel canisters. Sampling time is pressure dependent. Canisters were analysed post-flight using gas chromatography mass spectrometry. All CONTRAST data used in this study have been averaged over the TOGA observation time, unless otherwise indicated. The sampling resolution for vertical flight segments for data averaged over the TOGA sampling period is ~210 m.

Photolysis frequencies were calculated from up and downwelling, spectrally resolved actinic flux density by the HIAPER Airborne Radiation Package (HARP). The system uses independent, 2π steradian optical collectors connected via ultraviolet enhanced fiber optics to charged-coupled device detectors. Spectra were collected every 6 s at ~0.8 nm resolution between 280 and 600 nm with a full-width at half maximum of 1.7 and 2.4 nm in the ultraviolet and visible, respectively. Total photolysis frequencies were calculated from the actinic flux as well as laboratory determinations of molecular cross sections and quantum yields[58].

The CAST campaign was conducted simultaneously with CONTRAST. Whereas observations during CONTRAST were concentrated in the mid- to upper troposphere, the goal of CAST was to observe O_3, CO, NO, very short-lived organic halogen species and radicals in the MBL and lower troposphere. The two campaigns provide coverage of the TWP from just above the ocean surface to the base of the tropical tropopause layer. CAST flights were based out of Guam, Chuuk (7.4° N, 151.8° E), and Palau (7.5° N, 134.5° E), and covered altitudes from near the surface (30 m above mean sea level) to 8 km. The research portion of this campaign consisted of 23 flights during January and February 2014; 17 of these flights, as shown in Figure 1, provided observations between pressures of 300–700 hPa and are analysed here. CONTRAST and CAST flights were jointly coordinated at a shared operations center in Guam.

CAST was conducted using the Facility for Airborne Atmospheric Measurements BAe-146 aircraft. NO was measured with an air quality design,

2-channel chemiluminescence instrument through reaction with O_3 with a 2σ uncertainty of 15%. O_3 was measured at 0.1 Hz with a thermo environmental 49c O_3 analyser by ultraviolet absorption with a 2σ uncertainty of ~0.8 p.p.b.v. CO was measured with an Aero-Laser 5002 instrument at 1 Hz with a 2σ uncertainty of ~1.4 p.p.b.v.. Water vapour mixing ratios were calculated from the observed dew point, measured with a General Eastern dew point hygrometer. HCN was measured by chemical ionization mass spectrometry[59].

No wingtip-to-wingtip comparisons of observations for the CAST and CONTRAST campaign instruments were acquired, due to air traffic concerns in the remote TWP. To compare HCN observations, data from TOGA and the chemical ionization mass spectrometry instrument were selected for background conditions (RH >70% and O_3 <25 p.p.b.v.) and sorted into 0.5 km bins. The mean ± 1σ values of HCN for both campaigns strongly overlap, with CAST HCN slightly lower. The mean ratio of CAST to CONTRAST HCN for all altitude bins was 0.90 ± 0.21. A similar process was used to compare O_3 observations, selecting for measurements with RH >70%. The mean CAST to CONTRAST ratio of O_3 values was 0.98 ± 0.26. Since flights were conducted in different air masses and often on different days, this agreement indicates the measurements of O_3 and HCN obtained during CAST and CONTRAST are directly comparable.

Sondes. Profiles of ozone were measured with electrochemical concentration cell ozonesondes from SHADOZ[43]. Data used here are from observations over Hanoi, Vietnam (21.03° N, 105.85° E) and Nairobi, Kenya (1.28° S, 36.82° E) and were downloaded from the SHADOZ archive (http://croc.gsfc.nasa.gov/shadoz). Nairobi data were acquired over 5 days in January and February 2014 using a 0.5% half buffer KI solution with launch times near 8 UTC. Hanoi data were taken over 4 days in January and February 2014, using a 0.5% unbuffered KI solution and launch times near 6 UTC.

Satellite data. Fire count data are from MODIS onboard the Terra satellite, with a local overpass time of ~10:30 (ref. 52). The version 1 monthly product from collection 5, available at a 1 × 1° (latitude, longitude) spatial resolution, is used. MODIS data were downloaded from ftp://neespi.gsfc.nasa.gov. The Level 3, Version 6 daily AIRS product for OLR and water vapour, at 1 × 1° horizontal resolution, is also used[60,61]. AIRS is onboard the Aqua satellite with a local overpass time of ~13:30. AIRS data were downloaded from ftp://acdisc.sci.gsfc.nasa.gov/ftp/data/s4pa/Aqua_AIRS_Level3. Water vapour data cited here are pressure layer averages.

Back trajectories. Ten day kinematic back trajectories along the flight track were calculated using the NOAA HYSPLIT model[33]. RH, temperature and pressure were output along the trajectory, and H_2O mixing ratios were calculated using the Clausius–Clapeyron relation. RH, as output by HYSPLIT, is with respect to ice for temperatures below − 20 °C and a linear blend of RH with respect to ice and water for temperatures between 0 and − 20 °C. HYSPLIT RH was post-processed to convert all points with temperatures between 0 and − 20 °C to RH with respect to ice, to render HYSPLIT RH directly comparable to *in situ* RH.

The trajectories allowed for vertical displacement, using estimates of the vertical wind from assimilated meteorological fields. Trajectories from CONTRAST were computed along the flight track at 2 min intervals for pressures between 300 and 700 hPa, corresponding with the time between TOGA observations. CAST data were averaged over 35 s (TOGA integration time) and trajectories were calculated at 2 min intervals (TOGA sampling interval) along the flight track, to make our analysis of CAST data using trajectories analogous to the CONTRAST analysis. *Global data assimilation system* meteorological fields at 1 × 1° resolution drove the HYSPLIT model. PV, at 6 h resolution from the National Center for Environmental Prediction final analysis, was interpolated to the back trajectory through a bilinear interpolation in the horizontal and a linear interpolation in the vertical and time coordinates.

Trajectories were stopped at the point of last precipitating convection. In the TWP, convection promotes mixing with MBL air, altering air parcel composition. Precipitation rates from the tropical rainfall measuring mission satellite were combined with cloud top heights calculated from geostationary satellite infrared measurements. This technique provides coverage of the entire tropics. The convective precipitation product is available at 0.25 × 0.25° (latitude, longitude) resolution with a time step of 3 h. Intersection of a trajectory with precipitating convection was defined as a point on the trajectory being within 25 km of convection in the horizontal and being at or below the cloud top height. The 25 km radius allows for the uncertainty in the back trajectory calculation.

CAM-Chem. The Community Atmosphere Model version 4.0 (CAM4) is the atmospheric component of the global chemistry-climate model Community Earth System Model[40]. When run with active chemistry, it is known as CAM-Chem. Here the model was run offline, with meteorological fields specified by the NASA GEOS5 model, with a horizontal resolution of 0.94° latitude × 1.25° longitude and 56 vertical levels. The model chemistry scheme includes a detailed representation of tropospheric and stratospheric chemistry (~180 species; ~500 chemical reactions), including very short-lived halogens. Fernandez *et al.*[62] provide details on surface emissions, wet and dry deposition, heterogeneous reactions and photochemical processes of halogens used within CAM-Chem.

Anthropogenic emissions are from the RCP 8.5 scenario and biomass burning emissions are from the Fire INventory for NCAR (FINN)[63]. FINN combines observations of biomass burning and vegetation/land cover type from MODIS and emissions factors from multiple data sets to produce a gridded global product with a 1×1 km resolution. To determine the relative contributions of biomass burning from individual regions, CO emitted from fires in Africa as well as CO emitted from fires in Southeast Asia were treated as separate variables (referred to as tagged CO in the main paper).

Box model. Net photochemical production of O_3 was calculated using equation (1) (ref. 31) for the CONTRAST profiles shown in Fig. 2, where brackets indicate concentration and k_i represents the rate constant for a given reaction:

$$dO_3/dt = k_1[NO][HO_2] + k_2[NO][CH_3O_2] - k_3[O_3][OH] - k_4[O_3][HO_2] - k_5[H_2O][O^1D] \tag{1}$$

CH_3O_2 comprised $>95\%$ of RO_2 for the majority of modelled points, so O_3 production from other RO_2 species has been ignored. HO_2, CH_3O_2 and O^1D were not measured; to determine net production of O_3, these species were calculated using the Dynamically Simple Model of Atmospheric Chemical Complexity (DSMACC) box model[64]. Model runs were only conducted for CONTRAST flights because of lack of necessary VOC data for CAST. The model uses a subset of the Master Chemical Mechanism v3.3 (ref. 65) and was initialized with observations of methyl vinyl ketone, methacrolein, acetone, isoprene, methanol and acetaldehyde. NO_2 was estimated using observations of $j(NO_2)$, O_3, NO and modelled values of HO_2 and CH_3O_2 and then used to initialize the model. The box model was constrained by Gulfstream V observations of meteorological parameters, $j(NO_2)$, $j(O^1D)$, O_3, CO, NO, HCHO, H_2O, C_3H_8 and CH_4. The abundance of NO and j-values was allowed to vary with time of day. Photolysis frequencies for all reactions were calculated with the Tropospheric Ultraviolet and Visible Radiation model version 4.2 (ref. 66) and then scaled using observations of $j(O^1D)$ and $j(NO_2)$. Data were averaged over 10 s. Data from the TOGA instrument was linearly interpolated in time to create a 10 s data set. All output has been integrated over the diel cycle to produce 24 h mean photochemical production of O_3, because of the diurnal variation of radical species. Model results were compared with output from another box model, the University of Washington Chemical Model (UWCM)[67], which was constrained and initialized with the same set of input parameters. Between 300 and 700 hPa, the mean ratio of net O_3 production found using DSMACC and UWCM was 1.01.

Mixing line analysis. Elevated O_3 in the remote TWP can have plausible sources from the polluted troposphere or the O_3 rich stratosphere, imposing enhancements of a similar magnitude. Mixing line analysis using measurements of at least two other tracers with distinctly different abundances in the troposphere and stratosphere can be used to assess the relative contribution of each potential source[48]. Here we use CO and H_2O as tracers of the polluted troposphere and the stratosphere, respectively. The fraction of stratospheric air, f_{STRAT} was calculated from equation (2), where H_2O_{OBS} is the observed H_2O mixing ratio, H_2O_{STRAT} is the stratospheric H_2O mixing ratio, and H_2O_{TROP} is the altitude dependent background tropospheric H_2O mixing ratio:

$$H_2O_{OBS} = f_{STRAT}H_2O_{STRAT} + (1 - f_{STRAT})H_2O_{TROP}(z) \tag{2}$$

A constant stratospheric H_2O mixing ratio of 3 p.p.m.v. was assumed based on the mean observed H_2O (3.1 ± 0.1 p.p.m.v.) from RF15, which probed the lower stratosphere. Background tropospheric H_2O was calculated by filtering 10 s averaged data for air parcels with RH $>70\%$ and O_3 <20 p.p.b.v., leading to a profile similar to that shown by the blue bars in Fig. 3b. These data were then averaged into 1 km altitude bins.

An inferred stratospheric CO mixing ratio, denoted $CO_{STRAT\ INFERRED}$ was calculated based on f_{STRAT} found using equation (2). The relation for $CO_{STRAT\ INFERRED}$ is:

$$CO_{STRAT\ INFERRED} = [CO_{OBS} - (1 - f_{STRAT})CO_{TROP}]/f_{STRAT} \tag{3}$$

where variables have analogous definitions to those in equation (2). A value for CO_{TROP} of 85 p.p.b.v. was used, based on the median mixing ratio of CO for air parcels with RH $>70\%$. Since CO_{TROP} showed little variation with altitude, a constant value was used.

Photochemical aging. Both benzene (C_6H_6) and ethyne (C_2H_2) are tracers of biomass burning pollution and have lifetimes much shorter than that of CO (ref. 34). All species are primarily removed by reaction with OH (see reactions 1-5), so the change in CO with respect to either C_6H_6 or C_2H_2, relative to the initial emissions ratio, allows photochemical age to be determined[39]. Rate constants for reactions R3 and R4 are from IUPAC[68] and for reactions R1, R2, and R5 from

NASA JPL 2011 (ref. 69).

$$OH + CO \rightarrow CO_2 + H \tag{R1}$$

$$OH + CO \rightarrow HOCO \tag{R2}$$

$$OH + C_6H_6 \rightarrow H_2O + C_6H_5 \tag{R3}$$

$$OH + C_6H_6 \rightarrow HOC_6H_6 \tag{R4}$$

$$OH + C_2H_2 \rightarrow HOCHCH \tag{R5}$$

All reactions have first-order kinetics, so the ratio of the CO concentration at time t, $[CO(t)]$, to that of benzene, $[C_6H_6(t)]$, is described by equation (4) (ref. 39), where k_{CO} is the sum of the rate constants for reactions R1 and R2, k_{C6H6} is the sum of the rate constants for reactions R3 and R4, $ER(t)$ is the enhancement ratio of CO to C_6H_6 at time t, and ER_0 is the ratio of $[CO]$ to $[C_6H_6]$ at the time of emission. The expression for C_2H_2 is analogous. If all the other variables are known, the expression can be solved for t, the photochemical age.

$$ER(t) = [CO(t)]/[C_6H_6(t)] = ER_0 e^{-[OH]t(k_{CO} - k_{C_6H_6})} \tag{4}$$

The enhancement ratios of CO to C_6H_6 and CO to C_2H_2 over the entire campaign were determined by using an orthogonal linear regression for parcels where $O_3 > 40$ p.p.b.v. and RH $<20\%$ (Fig. 8). The slope of these lines is the enhancement ratio. This was done over a campaign-wide basis and not for each profile because of the limited sampling frequency of C_6H_6 and C_2H_2. Separate regressions were done for the TOGA and AWAS C_6H_6 measurements. Values of r^2 for all regressions were >0.61. The value of ER_0 for the two relations was assumed to be 724 mol CO per mol C_6H_6 and 241 mol CO per mol C_2H_2, which are characteristic of a tropical forest[70]. The photochemical age varied by <0.5 days when emission ratios for other vegetation types[70] were considered. Constant values of temperature (247 K), pressure (350 hPa) and [OH] (1.7×10^6 cm^{-3}) were assumed when calculating the photochemical age. Temperature and pressure values were the average along the back trajectory for the HOLW filaments, and 1.7×10^6 cm^{-3} is the 24-h mean OH concentration for parcels between 300 and 700 hPa from our box model runs. Equation (4) also assumes that any mixing with ambient air dilutes both species equally.

Code availability. The CAM-Chem code is available for download at www2.cesm.ucar.edu. An online version of the HYSPLIT model is available at http://ready.arl.noaa.gov/HYSPLIT.php. The DSMACC photochemical box model can be downloaded at http://wiki.seas.harvard.edu/geos-chem/index.php/DSMACC_chemical_box_model and the UWCM box model can be downloaded at https://sites.google.com/site/wolfegm/models.

References

1. IPCC. *Climate Change 2013: The Physical Science Basis. Contribution of Working Group I to the Fifth Assessment Report of the Intergovernmental Panel on Climate Change* (Cambridge University Press, 2013).
2. Shindell, D. & Faluvegi, G. Climate response to regional radiative forcing during the twentieth century. *Nat. Geosci.* **2**, 294–300 (2009).
3. Stevenson, D. S. *et al.* Tropospheric ozone changes, radiative forcing and attribution to emissions in the Atmospheric Chemistry and Climate Model Intercomparison Project (ACCMIP). *Atmos. Chem. Phys.* **13**, 3063–3085 (2013).
4. Bell, M. L., Peng, R. D. & Dominici, F. The exposure-response curve for ozone and risk of mortality and the adequacy of current ozone regulations. *Environ. Health Perspect.* **114**, 532–536 (2006).
5. Avnery, S., Mauzerall, D. L., Liu, J. & Horowitz, L. W. Global crop yield reductions due to surface ozone exposure: 1. Year 2000 crop production losses and economic damage. *Atmos. Environ.* **45**, 2284–2296 (2011).
6. Parrish, D. D. Critical evaluation of US on-road vehicle emission inventories. *Atmos. Environ.* **40**, 2288–2300 (2006).
7. Vestreng, V. *et al.* Evolution of NO$_x$ emissions in Europe with focus on road transport control measures. *Atmos. Chem. Phys.* **9**, 1503–1520 (2009).
8. Cooper, O. R., Gao, R.-S., Tarasick, D., Leblanc, T. & Sweeney, C. Long-term ozone trends at rural ozone monitoring sites across the United States, 1990-2010. *J. Geophys. Res.* **117** doi:10.1029/2012jd018261 (2012).
9. Oltmans, S. J. *et al.* Long-term changes in tropospheric ozone. *Atmos. Environ.* **40**, 3156–3173 (2006).
10. Read, K. A. *et al.* Extensive halogen-mediated ozone destruction over the tropical Atlantic Ocean. *Nature* **453**, 1232–1235 (2008).
11. Dickerson, R. R. *et al.* Ozone in the remote marine boundary layer: a possible role for halogens. *J. Geophys. Res.* **104**, 21385–321395 (1999).
12. Kley, D. *et al.* Tropospheric water-vapour and ozone cross-sections in a zonal plane over the central equatorial Pacific Ocean. *Q. J. R. Meteorol. Soc.* **123**, 2009–2040 (1997).
13. Rex, M. *et al.* A tropical West Pacific OH minimum and implications for stratospheric composition. *Atmos. Chem. Phys.* **14**, 4827–4841 (2014).

14. Oltmans, S. J. *et al.* Ozone in the Pacific tropical troposphere from ozonesonde observations. *J. Geophys. Res.* **106**, 32503–32525 (2001).

15. Pan, L. L. *et al.* Bimodal distribution of tropical free tropospheric ozone over the Western Pacific revealed by airborne observations. *Geophys. Res. Lett.* **42**, 7844–7851 (2015).

16. Mapes, B. E. & Zuidema, P. Radiative-dynamical consequences of dry tongues in the tropical troposphere. *J. Atmos. Sci.* **53**, 620–638 (1995).

17. Parsons, D. B., Yoneyama, K. & Redelsperger, J. L. The evolution of the tropical western Pacific atmosphere-ocean system following the arrival of a dry intrusion. *Q. J. R. Meteorol. Soc.* **126**, 517–548 (2000).

18. Newell, R. E. *et al.* Ubiquity of quasi-horizontal layers in the troposphere. *Nature* **398**, 316–319 (1999).

19. Stoller, P. *et al.* Measurements of atmospheric layers from the NASA DC-8 and P-3B aircraft during PEM-Tropics A. *J. Geophy. Res.* **104**, 5745–5764 (1999).

20. Hayashi, H., Kita, K. & Taguchi, S. Ozone-enhanced layers in the troposphere over the equatorial Pacific Ocean and the influence of transport of midlatitude UT/LS air. *Atmos. Chem. Phys.* **8**, 2609–2621 (2008).

21. Blake, N. J. *et al.* Influence of southern hemispheric biomass burning on midtropospheric distributions of nonmethane hydrocarbons and selected halocarbons over the remote South Pacific. *J. Geophys. Res.* **104**, 16213–16232 (1999).

22. Singh, H. B. *et al.* Biomass burning influences on the composition of the remote South Pacific troposphere: analysis based on observations from PEM-Tropics-A. *Atmos. Environ.* **34**, 635–644 (2000).

23. Kondo, Y. *et al.* Impacts of biomass burning in Southeast Asia on ozone and reactive nitrogen over the western Pacific in spring. *J. Geophys. Res.* **109**, D15S12 (2004).

24. Folkins, I., Chatfield, R., Baumgardner, D. & Proffitt, M. Biomass burning and deep convection in southeastern Asia: results from ASHOE/MAESA. *J. Geophys. Res. Atmos.* **102**, 13291–13299 (1997).

25. Ridder, T. *et al.* Ship-borne FTIR measurements of CO and O_3 in the Western Pacific from 43° N to 35° S: an evaluation of the sources. *Atmos. Chem. Phys.* **12**, 815–828 (2012).

26. Yoneyama, K. & Parsons, D. B. A proposed mechanism for the intrusion of dry air into the Tropical Western Pacific region. *J. Atmos. Sci.* **56**, 1524–1546 (1999).

27. Waugh, D. W. Impact of potential vorticity intrusions on subtropical upper tropospheric humidity. *J. Geophys. Res. Atmos.* **110**, D11305 (2005).

28. Cau, P., Methven, J. & Hoskins, B. Origins of dry air in the tropics and subtropics. *J. Clim.* **20**, 2745–2759 (2007).

29. Galewsky, J., Sobel, A. & Held, I. Diagnosis of subtropical humidity dynamics using tracers of last saturation. *J. Atmos. Sci.* **62**, 3353–3367 (2005).

30. Dessler, A. E. & Minschwaner, K. An analysis of the regulation of tropical tropospheric water vapor. *J. Geophys. Res. Atmos.* **112**, D10120 (2007).

31. Jacob, D. J. *et al.* Origin of ozone and NO_x in the tropical troposphere: a photochemical analysis of aircraft observations over the South Atlantic basin. *J. Geophys. Res. Atmos.* **101**, 24235–24250 (1996).

32. Zondlo, M. A., Paige, M. E., Massick, S. M. & Silver, J. A. Vertical cavity laser hygrometer for the National Science Foundation Gulfstream-V aircraft. *J. Geophys. Res. Atmos.* **115**, D20309 (2010).

33. Stein, A. F. *et al.* NOAA's HYSPLIT atmospheric transport and dispersion modeling system. *Bull. Am. Meteorol. Sci.* doi: http://dx.doi.org/10.1175/BAMS-D-14-00110.1 (2015).

34. Mauzerall, D. L. *et al.* Photochemistry in biomass burning plumes and implications for tropospheric ozone over the tropical South Atlantic. *J. Geophys. Res. Atmos.* **103**, 8401–8423 (1998).

35. Parrington, M. *et al.* Ozone photochemistry in boreal biomass burning plumes. *Atmos. Chem. Phys.* **13**, 7321–7341 (2013).

36. Holzinger, R. *et al.* Biomass burning as a source of formaldehyde, acetaldehyde, methanol, acetone, acetonitrile, and hydrogen cyanide. *Geophys. Res. Lett.* **26**, 1161–1164 (1999).

37. Akagi, S. K. *et al.* Emission factors for open and domestic biomass burning for use in atmospheric models. *Atmos. Chem. Phys.* **11**, 4039–4072 (2011).

38. de Gouw, J. A. *et al.* Emission sources and ocean uptake of acetonitrile (CH_3CN) in the atmosphere. *J. Geophys. Res.* **108**, 4329 (2003).

39. Parrish, D. D. *et al.* Effects of mixing on evolution of hydrocarbon ratios in the troposphere. *J. Geophys. Res. Atmos.* **112**, D10S34 (2007).

40. Lamarque, J. F. *et al.* CAM-chem: description and evaluation of interactive atmospheric chemistry in the Community Earth System Model. *Geosci. Model Dev.* **5**, 369–411 (2012).

41. Pickering, K. E. *et al.* Convective transport of biomass burning emissions over Brazil during TRACE A. *J. Geophys. Res. Atmos.* **101**, 23993–24012 (1996).

42. Thompson, A. M. *et al.* Where did tropospheric ozone over southern Africa and the tropical Atlantic come from in October 1992? Insights from TOMS, GTE TRACE A, and SAFARI 1992. *J. Geophys. Res. Atmos.* **101**, 24251–24278 (1996).

43. Thompson, A. M. *et al.* Southern Hemisphere Additional Ozonesondes (SHADOZ) ozone climatology (2005–2009): Tropospheric and tropical tropopause layer (TTL) profiles with comparisons to OMI-based ozone products. *J. Geophys. Res. Atmos.* **117**, D23301 (2012).

44. Staudt, A. C. *et al.* Global chemical model analysis of biomass burning and lightning influences over the South Pacific in austral spring. *J. Geophys. Res. Atmos.* **107**, 4200 (2002).

45. Gettelman, A. *et al.* Validation of Aqua satellite data in the upper troposphere and lower stratosphere with in situ aircraft instruments. *Geophys. Res. Lett.* **31**, L22107 (2004).

46. Johanson, C. M. & Fu, Q. Hadley cell widening: model simulations versus observations. *J. Clim.* **22**, 2713–2725 (2009).

47. Ashfold, M. J. *et al.* Rapid transport of East Asian pollution to the deep tropics. *Atmos. Chem. Phys.* **15**, 3565–3573 (2015).

48. Pan, L. L. *et al.* A set of diagnostics for evaluating chemistry-climate models in the extratropical tropopause region. *J. Geophys. Res.* **112**, D09316 (2007).

49. Holton, J. R. *et al.* Stratosphere-troposphere exchange. *Rev. Geophys.* **33**, 403–439 (1995).

50. Waugh, D. W. & Polvani, L. M. Climatology of intrusions into the tropical upper troposphere. *Geophys. Res. Lett.* **27**, 3857–3860 (2000).

51. Granier, C. *et al.* Evolution of anthropogenic and biomass burning emissions of air pollutants at global and regional scales during the 1980-2010 period. *Clim. Change.* **109**, 163–190 (2011).

52. Giglio, L., Csiszar, I. & Justice, C. O. Global distribution and seasonality of active fires as observed with the Terra and Aqua Moderate Resolution Imaging Spectroradiometer (MODIS) sensors. *J. Geophys. Res. Biogeosci.* **111**, G02016 (2006).

53. Ridley, B. A. & Grahek, F. E. A Small, low-flow, high-sensitivity reaction vessel for NO chemiluminescence detectors. *J. Atmos. Ocean. Technol.* **7**, 307–311 (1990).

54. Boylan, P., Helmig, D. & Park, J. H. Characterization and mitigation of water vapor effects in the measurement of ozone by chemiluminescence with nitric oxide. *Atmos. Meas. Tech.* **7**, 1231–1244 (2014).

55. Gerbig, C. *et al.* An improved fast-response vacuum-UV resonance fluorescence CO instrument. *J. Geophys. Res. Atmos.* **104**, 1699–1704 (1999).

56. Cazorla, M. *et al.* A new airborne laser-induced fluorescence instrument for in situ detection of formaldehyde throughout the troposphere and lower stratosphere. *Atmos. Meas. Tech.* **8**, 541–552 (2015).

57. Apel, E. C. A fast-GC/MS system to measure C_2 to C_4 carbonyls and methanol aboard aircraft. *J. Geophys. Res.* **108**, 8794 (2003).

58. Shetter, R. E. & Muller, M. Photolysis frequency measurements using actinic flux spectroradiometry during the PEM-Tropics mission: Instrumentation description and some results. *J. Geophys. Res. Atmos.* **104**, 5647–5661 (1999).

59. Le Breton, M. *et al.* Airborne hydrogen cyanide measurements using a chemical ionisation mass spectrometer for the plume identification of biomass burning forest fires. *Atmos. Chem. Phys.* **13**, 9217–9232 (2013).

60. Tian, B. *et al.* Evaluating CMIP5 models using AIRS tropospheric air temperature and specific humidity climatology. *J. Geophys. Res. Atmos.* **118**, 114–134 (2013).

61. Susskind, J., Molnar, G., Iredell, L. & Loeb, N. G. Interannual variability of outgoing longwave radiation as observed by AIRS and CERES. *J. Geophys. Res. Atmos.* **117**, D23107 (2012).

62. Fernandez, R. P., Salawitch, R. J., Kinnison, D. E., Lamarque, J. F. & Saiz-Lopez, A. Bromine partitioning in the tropical tropopause layer: implications for stratospheric injection. *Atmos. Chem. Phys.* **14**, 13391–13410 (2014).

63. Wiedinmyer, C. *et al.* The Fire INventory from NCAR (FINN): a high resolution global model to estimate the emissions from open burning. *Geosci. Model Dev.* **4**, 625–641 (2011).

64. Emmerson, K. M. & Evans, M. J. Comparison of tropospheric gas-phase chemistry schemes for use within global models. *Atmos. Chem. Phys.* **9**, 1831–1845 (2009).

65. Jenkin, M. E., Young, J. C. & Rickard, A. R. The MCM v3.3.1 degradation scheme for isoprene. *Atmos. Chem. Phys.* **15**, 11433–11459 (2015).

66. Madronich, S. Implications of recent total atmospheric ozone measurements for biologically-active ultraviolet-radiation reaching the earth's surface. *Geophys. Res. Lett.* **19**, 37–40 (1992).

67. Wolfe, G. M. & Thornton, J. A. The chemistry of atmosphere-forest exchange (CAFE) model—part 1: model description and characterization. *Atmos. Chem. Phys.* **11**, 77–101 (2011).

68. Atkinson, R. *et al.* Evaluated kinetic and photochemical data for atmospheric chemistry: volume II—gas phase reactions of organic species. *Atmos. Chem. Phys.* **6**, 3625–4055 (2006).

69. Sander, S. P. *et al.* *Chemical Kinetics and Photochemical Data for Use in Atmospheric Studies, Evaluation No. 17* (Jet Propulsion Laboratory, 2011).

70. Andreae, M. O. & Merlet, P. Emission of trace gases and aerosols from biomass burning. *Global Biogeochem. Cycles* **15**, 955–966 (2001).

Acknowledgements

We thank L. Pan for coordinating the CONTRAST flights and her constructive criticism of an early version of the manuscript; S. Schauffler, V. Donets and R. Lueb for collecting and analysing AWAS samples; T. Robinson and O. Shieh for providing meteorology forecasts in the field; and the pilots and crews of the CAST BAe-146 and CONTRAST Gulfstream V aircrafts for their dedication and professionalism. CAST was funded by the Natural Environment Research Council; CONTRAST was funded by the National Science Foundation. Research at the Jet Propulsion Laboratory, California Institute of Technology, is performed under contract with the National Aeronautics and Space Administration (NASA). A number of the US-based investigators also benefitted from the support of NASA as well as the National Oceanic and Atmospheric Administration. The views, opinions, and findings contained in this report are those of the author(s) and should not be construed as an official National Oceanic and Atmospheric Administration or US Government position, policy or decision. We would like to acknowledge high-performance computing support from Yellowstone (ark:/85065/d7wd3xhc) provided by NCAR's Computational and Information Systems Laboratory. NCAR is sponsored by the National Science Foundation.

Author contributions

D.C.A., R.J.S. and J.M.N. wrote the manuscript. All authors discussed the results, commented on the manuscript and many suggested new ways to examine the data. N.R.P.H., J.D.L. and L.J.C. were CAST co-PIs; E.A. and R.J.S. were CONTRAST co-PIs. J.F.B. provided in-field meteorological forecasting and guidance on meteorological portions of the manuscript. J-.F.L., A.S-.L., D.E.K., R.P.F. and R.B.P. provided chemistry transport modelling used for flight planning in the field as well as data analysis. M.E. assisted in CAST flight planning and provided the box model used in the analysis. R.R.D. and T.P.C. assisted in the data analysis and numerous revisions of the manuscript. S.B. measured CO and O_3; M.L.B., T.B. and C.P. measured HCN; J.D.L. and A.V. measured NO, all for CAST. A.J.W. measured O_3, NO and NO_2; E.A., E.C.A., R.S.H., N.J.B. and D.D.R. measured organic trace gases; T.L.C. measured CO; T.F.H., G.M.W. and D.C.A. measured HCHO; S.R.H. and K.U. measured actinic flux and provided photolysis frequencies, all for CONTRAST. M.D.C. and B.J.B.S. provided the HYSPLIT model and guidance in its use. L.P. created the convective precipitation product. B.H.K. provided AIRS data and guidance. A.M.T. provided the ozonesonde profiles. J.M.N. and D.C.A. conducted the box modelling analysis.

Additional information

2

The timescales of global surface-ocean connectivity

Bror F. Jönsson[1] & James R. Watson[2,3]

Planktonic communities are shaped through a balance of local evolutionary adaptation and ecological succession driven in large part by migration. The timescales over which these processes operate are still largely unresolved. Here we use Lagrangian particle tracking and network theory to quantify the timescale over which surface currents connect different regions of the global ocean. We find that the fastest path between two patches—each randomly located anywhere in the surface ocean—is, on average, less than a decade. These results suggest that marine planktonic communities may keep pace with climate change—increasing temperatures, ocean acidification and changes in stratification over decadal timescales—through the advection of resilient types.

[1] Department of Geosciences, Princeton University, Princeton, New Jersey 08544, USA. [2] College of Earth, Ocean and Atmospheric Sciences, Oregon State University, Corvallis, Oregon 97331-5503, USA. [3] The Stockholm Resilience Centre, Stockholm University, 118 14 Stockholm, Sweden. Correspondence and requests for materials should be addressed to B.F.J. (email: bjonsson@princeton.edu) or to J.R.W. (email: james.watson@su.se).

Two different paradigms are used to explain the structuring of planktonic communities (bacteria, phytoplankton and zooplankton) in ocean ecosystems. One fundamental idea is that 'everything is everywhere but the environment selects'[1,2]. That is, different regions of the ocean are connected by ocean currents, resulting in potentially panmictic planktonic communities[3]. It is then the differential response of species to environmental conditions that leads to community structure[4,5]. The alternative is that regions of the oceans are not so well connected, and that this isolation leads to divergent evolution and hence differences in which species are where[6,7]. Recent work has shown that neither concept is wholly accurate, with a number of examples showing slight spatial structure on global and regional scales in marine microbial communities[8-10] and sometimes strong genetic differentiation at small spatial scales[11]. Such examples suggest that dispersal limitation is important in specific areas of the ocean. However, most studies have focused on patterns of community composition, or genetic differences of individuals within a species. The key mechanism itself—dispersal by surface ocean currents—has rarely[12] been explored and quantified.

One common method for investigating the timescales of dispersal for marine organisms is to calculate oceanographic distances by tracking virtual particles in modeled velocity fields[13,14]. Particle tracking has been used in a variety of ways to explain Lagrangian processes in the ocean, such as biological dispersal on regional scales[14-17], connectivity between different coastal habitats like coral reefs[18,19] and deep-water transport pathways[20].

Normally, oceanographical distances are defined as the expected connectivity times, or the mean time it takes for particles to travel from one location to another[15,21]. Another option is to instead use minimum connectivity times (Min-T), or distances based on the fastest times that particles can travel from one location to another. Minimum connection times have major advantages over expected connection times for problems concerning global plankton communities. First, the minimum connection time is a more appropriate metric for phytoplankton and bacterial connectivity since asexually reproducing organisms have high reproductive output that attenuates low dispersal probabilities, and only a few individuals are required to 'connect' two places, especially in terms of population genetics[22]. It is therefore possible for such organisms to exploit dispersal routes where the probability to reach a given destination is very low. Previous empirical work have also shown that minimum connection times provide better correspondence with the genetic similarity of groups[13] and that minimum connection times are more relevant to community similarity than the mean. Second, mean or median transit times in the global ocean are not well defined, as water can recirculate eternally and, hence, every particle seeded in a given patch eventually will reach all other patches if enough time is provided. Constraining the particles to a maximum advection time or using a lower percentile of connections (for example, the fastest 20%) would create results that mainly depend on on those arbitrary cutoffs. It is challenging to identify one physically motivated timescale (loop time of the the subtropical gyres, typical time spent in the gulf stream, time to circumfer the Antarctic Circumpolar Current and so on) that is generally applicable to all regions in the ocean. The estimates of minimum connection times for the global surface ocean are stored in the form of a matrix (the Min-T connectivity matrix) where each i, j element represents the shortest transit time between a given source patch i and destination patch j.

Connectivity matrices produced from Lagrangian particle tracking tend to be highly sparse with most pairs of patches unconnected. Indeed, to estimate connection times between all pairs of patches globally, an infeasible number of Lagrangian particles[23] would be needed. To circumvent this obstacle, a shortest path algorithm can be used to calculate missing connections. Here the network is the global ocean, with patches in the ocean as nodes and minimum connection times as edges connecting the nodes. Applied to this network, shortest path algorithms identify the shortest path between every global ocean patch pair, accounting for all possible multistep connections. For example, if there is no direct connection between nodes A and D, then these algorithms identify the multistep connection from $A \to B \to C \to D$. We use Dijkstra's algorithm[24], which is one of the most commonly used shortest path algorithms, and which fits our specific application.

Each step along these minimum-time routes may be unlikely, making the conditional multistep probability (of going from A to B to C to D...) very low as well. However, one can assume that the effect of these low probabilities is attenuated by the large reproductive output of microorganisms drifting with ocean currents. Over monthly to annual timescales, microorganisms moving with water masses can grow by the million[25]. Hence, there will still be planktonic organisms traveling along the potentially low probability paths identified here. Indeed, if one considers the dispersal of genetic material, then there need only be a small number of individuals traveling along these Min-T routes to make them evolutionarily relevant[17,22,25].

The outcome of our study when applying Dijkstra's algorithm to a raw Min-T matrix is a full matrix that contains estimates of minimum connection times between every region of the world's surface ocean. Both the raw and full Min-T matrices are rich with spatial information, but most importantly are the distribution of minimum connection times themselves. We find that different regions in the global surface ocean are connected on very short timescales, within ~10 years. This is in contrast to deep-water circulation, where water is thought to recirculate around the globe in roughly 1,000 years. These short surface-connection times are relevant to anyone studying dispersion in the surface ocean beyond planktonic species, including radioactive materials, plastics and other forms of pollution.

Results

Particle advection. We seeded particles in near-surface velocity fields from the ECCO2 $1/4° \times 1/4°$ state estimate[26] over 9 years and advected them for 100 years by looping fields for the years 2000–2010. The resulting paths were used to estimate the shortest time taken for water to travel from one patch in the surface ocean to another. Minimum connectivity times were then calculated by aggregating the ECCO2 grid cells (Supplementary Fig. 1) into 8×8 patches, each approximately $2° \times 2°$ in size (11,116 patches in total: Supplementary Fig. 2). On average, particles seeded in any given source patch reached 1,150 destination patches after 100 years of advection by ocean currents.

Connectivity matrices. The raw Min-T connectivity matrix, produced from the 2D Lagrangian particle simulations, is highly sparse (Fig. 1a, grey areas). Connections are made primarily within each ocean basin, reflecting the computational limits of the simulation integration period (see Methods), with values hugging the diagonal. Some cross-basin connections are made, and these typically take much longer, on the order of 20–30 years. In contrast, the Min-T matrix, modified by applying Dijkstra's algorithm (Fig. 1b), is full with minimum connection times for every ocean–patch pair. Short values still hug the diagonal, but now the cross-basin connection times are shorter, on the order of 10–20 years. For example, the largest connection time values in the full Min-T matrix occur between the Arctic and Southern

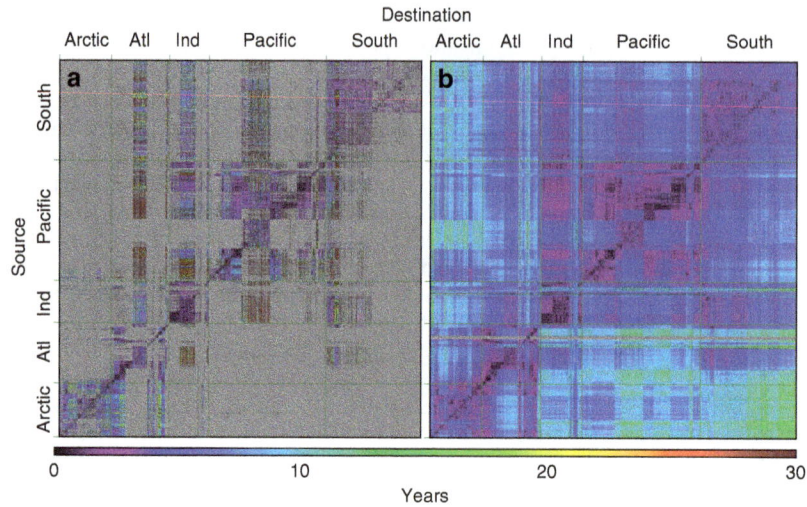

Figure 1 | Connectivity matrices. The raw minimum time connectivity matrix (**a**) and after Dijkstra's algorithm was applied (**b**). Major oceans are delimited by green lines. The number of patches (and hence the number of rows and columns) is 11,116.

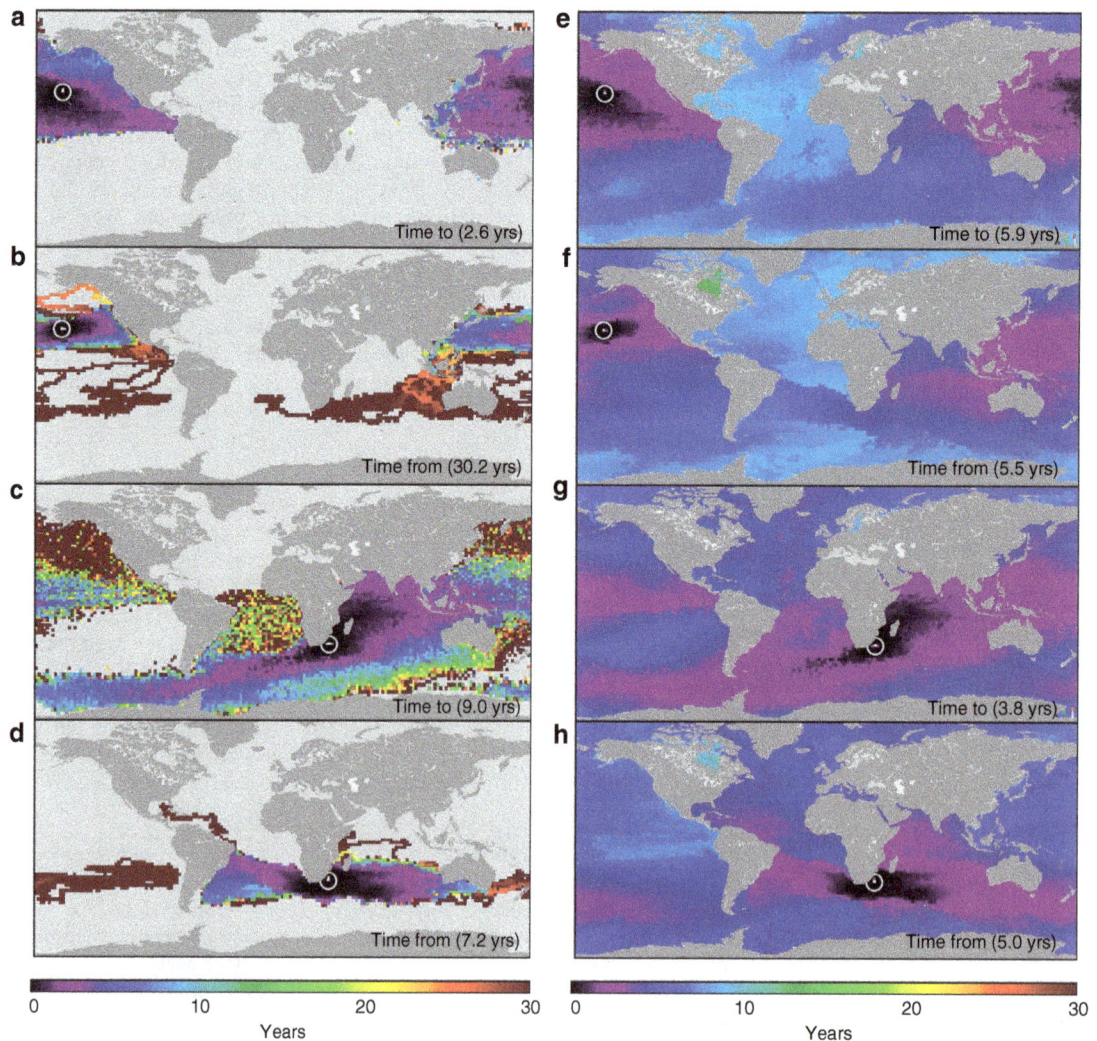

Figure 2 | Connectivity examples. Examples of minimum connection times (Min-T) to and from two locations identified by white circle-dots: off Hawaii (**a,b,e,f**) and off South Africa (**c,d,g,h**). Times 'to' are the shortest times taken for water from other patches to arrive at these locations. Times 'from' are the shortest times taken for water from these locations to go to all others. The left column shows raw minimum connection times, with the large number of no-connections noted in grey, and median Min-T in parentheses. The right column panels show Min-T values generated using Dijkstra's algorithm. Here connections occur between all areas of the ocean and median values are much lower on average than those of the raw minimum connection times.

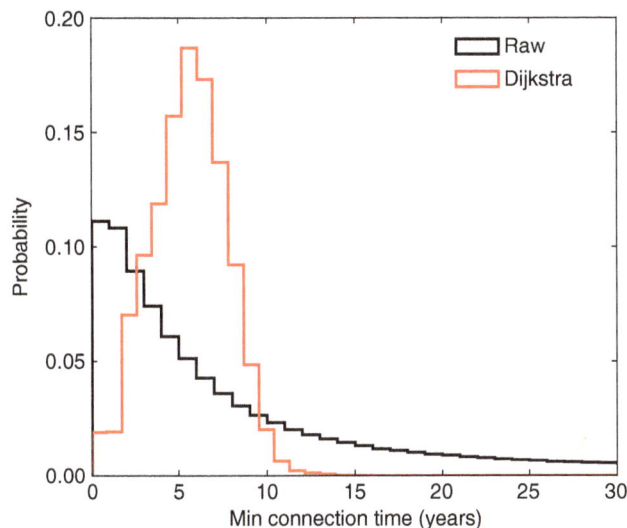

Figure 3 | Global connectivity distributions. Probability distributions of raw minimum connections times (blue) and those produced from Dijkstra's algorithm (red). Median minimum connection times (identified by the dashed vertical lines) are 6.13 years for the raw matrix, and 6.11 years for the modified. Note that connection times shorter than 1 year for Dijkstra are in fact raw connection times.

Table 1 | Median minimum connectivity time.

	Destination				
Source	Arctic	Atlantic	Indian	Pacific	Southern
Arctic	3.7 (1.9)	5.8 (2.6)	8.6 (2.0)	8.8 (1.9)	11.2 (1.6)
Atlantic	6.4 (2.2)	3.5 (2.3)	7.7 (2.6)	9.2 (1.8)	7.9 (3.0)
Indian	8.3 (2.1)	5.5 (1.9)	2.3 (3.1)	5.2 (2.0)	5.0 (2.5)
Pacific	8.1 (2.3)	7.2 (1.7)	3.9 (2.1)	3.4 (1.7)	5.9 (1.7)
Southern	9.1 (2.0)	6.4 (2.2)	5.0 (2.2)	6.2 (1.8)	4.1 (1.8)

Median minimum connectivity time between ocean basins in years. S.d. in parentheses.

Oceans. Asymmetry is present, too, revealing that there are differences in the time taken to go to, and come from, two places.

Spatial properties of connectivity. Rows of the raw and full Min-T matrices describe the minimum connection times from particular patches to all other patches in the global surface ocean. Similarly, the columns of the Min-T matrices describe the minimum time it takes for water to go from all patches to a given patch. This information is shown in Fig. 2 for two locations: Hawaii and a coastal location off of South Africa. From the raw Min-T information (Fig. 2a), the limitations of the particle tracking are evident in the large number of locations that are not connected by any particle trajectories (Fig. 2, ocean areas in grey). Of those that are connected, patches near the release point have low Min-T values relative to those locations farther away, with median connection times varying from location to location. In contrast, the full Min-T values (Fig. 2b) have connections everywhere, as expected from using Dijkstra's algorithm. Spatial structure is still seen, with some places more connected than others, but long connection times are now absent, and all median values have changed relative to their raw Min-T counterparts (Fig. 2, values in parentheses).

Timescales of global surface ocean connectivity. The most notable result from our analysis is the distribution of Min-T values themselves. The distribution of raw Min-T values (Fig. 3, blue distribution) is roughly log-normal, with a median value of 6.13 years, and a long tail extending towards 100 years when the simulations were stopped. After being modified by Dijkstra's algorithm, the global distribution of minimum connection times is changed (Fig. 3, red distribution) with a median minimum connection time of 5.61 years and with the bulk of the distribution now below 15 years, showing that the global ocean can be connected over timescales of a decade. The maximum full Min-T value is still about 100 years, relating to water traveling from the Weddell Sea to the California coast. In scaling up, the average Min-T values between different ocean basins are shown in Table 1. These aggregated metrics again highlight the short connection times between ocean regions, but also show physical consistency (that is, on average basins farther away take longer time to reach).

Discussion

Our results mirror the data from unintended, often tragic, natural experiments that quantify analogous connectivity in the Pacific Ocean. Large quantities of shoes[27] and toys[28] washed overboard from container ships en route from Asia to North America have been useful in estimating connectivity times by acting as drifters. Such results show timescales that are similar to, or often shorter than, our findings in the North Pacific. These kinds of drifters are, however, susceptible to wind-drift and could record faster times than our models. A more comparable experiment is the 2011 Fukushima disaster, in which a Japanese nuclear reactor released a large quantity of radioactive isotopes into the Pacific Ocean. Traces of radioactivity were detected on the Pacific Coast of the U.S. in November of 2014—3.6 years later (Ken Buessler WHOI, personal communication). Our estimated minimum connectivity time between the Fukushima release site and its detection site on the U.S. west coast is 3.5 years.

In summary, our results provide evidence for a highly connected global surface ocean with all regions connected to each other over decadal timescales. This suggests that plankton communities may keep pace with climate change through the immigration of new types that are better suited in changing local conditions[5,29,30]. Beyond this result, the utility of calculating global surface connectivity extends to its spatial information. For example, in many regional studies it is common to identify connectivity modules or subpopulations[31] and also the location of key stepping-stone patches, which are central to maintaining the overall connectance of the system[32,33]. These network theoretic analyses have an applied nature, such as in the design of spatial management units[34]; but they are also important for basic research, for example in generating hypotheses about genetic or taxonomic similarity across the ocean[7], or when testing models of community assembly[35]. Finally, it is important to note that we have only estimated the timescales of physical connectivity without addressing environmental factors such as nutrient availability or temperature gradients[36]. Gauging the effect of environmental barriers on global-scale dispersal will further contribute to our understanding of how marine communities adapt to their changing ocean environment.

Methods

Lagrangian particle tracking. Two-dimensional Lagrangian particle tracking was used to make our connectivity calculations[15,21,33,37]. We used velocity fields from ECCO2 (http://ecco2.org), a high-resolution (1/4) global ocean model that assimilates available satellite and the *in situ* data[26], to advect particles in the surface ocean (Supplementary Fig. 1). ECCO2 is based on a global full-depth ocean and sea-ice configuration of the Massachusetts Institute of Technology general circulation model (MITgcm) and applies an ad-joint approach to generate the physically consistent data assimilations. ECCO2's resolution is high enough to permit the formation of eddies and other narrow current systems within the ocean.

Particles were advected in the surface ocean using TRACMASS (http://tracmass.org), an off-line particle tracking code that calculates trajectories using Eulerian velocity fields. TRACMASS estimates the trajectory path through each grid cell of every Lagrangian particle, using an analytical solution to a differential equation that depends on the velocities on the grid-box walls. The scheme was originally developed for stationary velocity fields[20,38], and thereafter extended for time-dependent fields by solving a linear interpolation of the velocity field both in time and in space over each grid box[39]. This differs from the Runge-Kutta method, where trajectories are iterated forward in time with short time steps.

Particle seeding and connectivity patches. We seeded six particles in the second depth layer of each ECCO2 grid cell (a total of 4 million particles at each seeding time, or 36 million particles in total over all seeding times). When calculating connectivity, we aggregated the model's $1/4° \times 1/4°$ grid cells to 11,116 discrete $2° \times 2°$ patches. The size of these connectivity patches was selected as a balance of computational feasibility and biogeographic detail. Each connectivity patch is therefore seeded with 384 particles at each seeding event (9 in total). The second depth layer is between 5 and 20 m depth and was used to avoid potential numerical problems due to how ECCO2 implement a varying sea surface, precipitation, and evaporation. See Supplementary Fig. 2 for the spatial distribution of connectivity patches. Particles were seeded at 9 points in time: 1 January 2001, 1 February 2002, 1 March 2003, 1 April 2004, 1 May 2005, 1 June 2006; 1 July 2007, 1 August 2008; and 1 September 2009 in model years. As a consequence of the multiple seeding times, a total of 3,456 particles were used per patch to estimate connectivity. Particles were then advected using horizontal velocity fields from the second depth layer in ECCO2 so that they were locked in the surface ocean. We looped velocity fields for the years 2000–2010 continuously and advected the particles for 100 years in total. Particle positions were saved every 3 days and used to calculate minimum connection times. No extra diffusivity was added to the movement of the particles. Supplementary Figure 6 shows the relationship between advection time and number of other patches reached. It is clear from this figure that the number of connectivity patches reached saturates after about 12 years. In other words, like the number of particles released, there are diminishing returns to running simulations for longer integration times.

Estimating the timescales of connectivity. The resulting Lagrangian particle trajectories were used to estimate the shortest time taken for water to travel from one patch in the surface ocean to another. This minimum connection time is a variant on the standard measure of ocean distance, which is the expected transit time for water to travel from one patch to another[13,14]. We use the minimum and not the expected connection time for two reasons. First, the minimum connection time is a more appropriate metric for phytoplankton and bacterial connectivity since asexually reproducing organisms have high reproductive output that attenuates low dispersal probabilities, and only a few individuals are required to 'connect' two places, especially in terms of population genetics[22]. It is therefore possible for such organisms to exploit dispersal routes where the probability to reach a given destination is very low. Second, expected transit times in the global ocean are not properly defined, as water can recirculate for an infinitely long time. There is no limit, therefore, to the distribution of connectivity times over which to calculate expected connection times. Thus, the minimum connection time is a preferable alternative measure of ocean distance for this global application.

Minimum connection times for the global surface ocean are called Min-T, and they are stored in the form of a matrix—the Min-T connectivity matrix, where each i, j element represents the shortest transit time between a given source patch i and destination patch j (Fig. 1a). The *raw* Min-T matrix, produced from the Lagrangian particle tracking, is highly sparse with most pairs of patches being unconnected.

Network analysis of shortest/quickest paths. Estimation of connection times between all pairs of patches globally using Lagrangian particle simulations alone would require a currently infeasible number of particles[23] (see particle density sensitivity test described below). To circumvent this obstacle, a shortest-path algorithm was used to calculate missing values in the raw Min-T connectivity matrix. Here the network is the global ocean, with patches in the ocean as nodes, and minimum connection times as edges connecting the nodes. Applied to this network, shortest path algorithms identify the shortest path between every global ocean patch-pair, accounting for all possible multistep connections (see Supplementary Online Material for details). For example, if there is no direct connection between nodes A and D, then these algorithms identify the multistep connection from $A \to B \to C \to D$. We use Dijkstra's algorithm[24], which is one of the most commonly used shortest path algorithms, and which fits our specific application. The end result is a modified Min-T connectivity matrix (Fig. 1b), where all possible minimum connection times between patches are calculated.

Each step along these minimum-time routes may be unlikely, and so the conditional multistep probability (of going from A to B to C to D...) can have a very low probability as well. However, we assume that the effect of these low probabilities is attenuated by the large reproductive output of microorganisms drifting with ocean currents. Over the timescales that we are considering, microorganisms moving with water masses can grow by the million[25]. Hence, there will still be planktonic organisms traveling along the potentially low probability

paths identified here. Indeed, if one considers the dispersal of genetic material, then there need only be a small number of individuals traveling along these Min-T routes, to make them evolutionarily relevant[17,22,25].

While nodes in a network are usually defined as singular nodes with well-defined distances between them, our ocean patches have relatively large areas and are continuously adjacent to one another. This difference creates a problem when using Dijkstra's algorithm since a particle seeded next to the boundary of its initial patch can rapidly move to an adjacent patch. However, shortest path algorithms assumes that the travel time across each intermediate node is zero, or at least included in the edge distances. (This phenomenon is also a problem when analysing the speed of tracer transport in General Circulation Models[40].) By removing all calculated connectivity times shorter than 1 year before applying the shortest-path algorithm, we limit the effect of not including within-patch crossing times. The 365-day cutoff is based on calculated typical residence times in the patches, which are on the order of weeks. All initial minimum connectivity times are based on travel distances at least an order of magnitude longer than typical patch crossing distances.

The removed connectivity times were added back to the final connectivity matrix, allowing for connection times shorter than 365 days, as shown in Figs 2 and 3. It should be noted that the absolute number of connectivity times shorter than 1 year in Fig. 3 are identical for the Raw and Dijkstra cases. The lack of discontinuities between sub-annual and longer connection times in Fig. 2 (and all other cases we have explored) give us confidence that the resulting connectivity matrix is reasonable and that our approach works.

After applying Dijkstra's algorithm, we find that the resulting minimum connection time matrices are all connected. However, we do find some areas that are only connected in one direction (that is, there are connection time to, but not from, particular regions). These areas are mainly inland seas—the Baltic and Mediterranean, for example. However, they only account for a small fraction (2%) of the modified Min-T matrix and, consequently, do not impact the general result of the timescales of global surface ocean connectivity.

Particle seeding sensitivity test. Since the number of particles seeded per grid-cell and the seeding times are limited, we have not accounted for all possible Min-T pathways. As a result, our estimates of the timescales of global surface connectivity are conservative, since adding more particles and seeding dates could only lead to shorter Min-T pathways (that is, we look for the shortest connection times over all possibilities including seeding times). Thus, the few seeding dates—although arguably numerically incomplete—strengthen our conclusion that the global surface ocean is well connected over a few decades.

To examine the effect of particle seeding density, we performed a particle sensitivity test. Minimum connection times from a patch in the north pacific to all others were estimatedusing simulations with increasing numbers of seeded particles. Supplementary Figure 3 shows the results of these simulations. It is clear that a larger oceanic extent is reached as the number of particles released increases. However, when we examine only those patches that were reached in all seeding experiments (Supplementary Fig. 4), we can see that increasing the number of particles serves only to decrease the minimum connection times in these patches (Supplementary Fig. 4: with 84 particles some areas are reached after 100 years—the patches in gold, in contrast with 16,660 particles, these same patches are reached after around 20 years—patches now in light red).

Finally, we show the aggregated results of the sensitivity test in Supplementary Fig. 5, where we plot the fraction of patches reached (over the whole ocean) and the median minimum connection time from this study patch. The fraction of patches reached saturates at around 30%, which means that there are diminishing returns (in terms of estimating minimum connection times to new patches) to adding more particles. It also indicates that, to release enough particles to estimate minimum connection times to all patches globally, a currently impossible number of Lagrangian particles would be required. Similarly, the median minimum connection time from this patch saturates at around 8,000 particles released. In our simulations we use 3,456 particles per patch as this achieved a balance of connectivity sampling power and computational efficiency.

References

1. Baas-Becking, L. *Geobiologie of Inleiding tot de Milieukunde* (ed Van Stockum, W. P. & Zoon) (The Hague, 1934).
2. Fenchel, T. & Finlay, B. J. The ubiquity of small species: patterns of local and global diversity. *Bioscience* **54**, 777–784 (2004).
3. De Wit, R. & Bouvier, T. 'Everything is everywhere, but, the environment selects'; what did Baas Becking and Beijerinck really say? *Environ. Microbiol.* **8**, 755–758 (2006).
4. McGillicuddy, D. J. *et al.* Eddy/wind interactions stimulate extraordinary mid-ocean plankton blooms. *Science* **316**, 1021–1026 (2007).
5. Thomas, M. K., Kremer, C. T., Klausmeier, C. A. & Litchman, E. A global pattern of thermal adaptation in marine phytoplankton. *Science* **338**, 1085–1088 (2012).
6. Martiny, J. *et al.* Microbial biogeography: putting microorganisms on the map. *Nat. Rev. Microbiol.* **4**, 102–112 (2006).

7. Casteleyn, G. *et al.* Limits to gene flow in a cosmopolitan marine planktonic diatom. *Proc. Natl Acad. Sci. USA* **107,** 12952–12957 (2010).
8. Saez, A. G. *et al.* Pseudo-cryptic speciation in coccolithophores. *Proc. Natl Acad. Sci. USA* **100,** 7163–7168 (2003).
9. Rynearson, T. A. & Virginia Armbrust, E. Genetic differentiation among populations of the planktonic marine diatom Ditylum brightwellii (Bacillariophyceae). *J. Phycol.* **40,** 34–43 (2004).
10. Sul, W. J., Oliver, T. A., Ducklow, H. W., Amaral-Zettler, L. A. & Sogin, M. L. Marine bacteria exhibit a bipolar distribution. *Proc. Natl Acad. Sci. USA* **110,** 2342–2347 (2013).
11. Godhe, A. *et al.* Seascape analysis reveals regional gene flow patterns among populations of a marine planktonic diatom. *Proc. R. Soc. B Biol. Sci.* **280,** 1773 (2013).
12. Froyland, G., Stuart, R. M. & van Sebille, E. How well-connected is the surface of the global ocean? *Chaos* **24,** 3 (2014).
13. Alberton, F. *et al.* Isolation by oceanographic distance explains genetic structure for Macrocystis pyrifera in the Santa Barbara Channel. *Mol. Ecol.* **20,** 2543–2554 (2011).
14. Watson, J. R. *et al.* Currents connecting communities: nearshore community similarity and ocean circulation. *Ecology* **92,** 1193–1200 (2011).
15. Mitarai, S., Siegel, D. & Winters, K. A numerical study of stochastic larval settlement in the California Current system. *J. Mar. Syst.* **69,** 295–309 (2008).
16. Cowen, R., Paris, C. & Srinivasan, A. Scaling of connectivity in marine populations. *Science* **311,** 522–527 (2006).
17. Kool, J. T., Paris, C. B. & Andre, S. Complex migration and the development of genetic structure in subdivided populations: an example from Caribbean coral reef ecosystems. *Evolution, September* **2009,** 1–10 (2010).
18. Mora, C. *et al.* High connectivity among habitats precludes the relationship between dispersal and range size in tropical reef fishes. *Ecography* **35,** 89–96 (2012).
19. Wood, S., Paris, C. B., Ridgwell, A. & Hendy, E. J. Modelling dispersal and connectivity of broadcast spawning corals at the global scale. *Global Ecol. Biogeogr.* **23,** 1–11 (2014).
20. Döös, K. Interocean exchange of water masses. *J. Geophys. Res.* **100,** 13499–13514 (1995).
21. Cowen, R., Gawarkiewicz, G., Pineda, J., Thorrold, S. & Werner, F. E. Population connectivity in marine systems. *Oceanography* **20,** 14–20 (2007).
22. Hedgecock, D., Barber, P. H. & Edmands, S. Genetic approaches to measuring connectivity. *Oceanography* **20,** 70–79 (2007).
23. Simons, R. D., Siegel, D. A. & Brown, K. S. Model sensitivity and robustness in the estimation of larval transport: A study of particle tracking parameters. *J. Mar. Syst.* **119-120,** 19–29 (2013).
24. Dijkstra, E. W. A note on two problems in connexion with graphs. *Numerische Mathematik* **1,** 269–271 (1959).
25. Falkowski, P. G. *et al.* The evolution of modern eukaryotic phytoplankton. *Science* **305,** 354–360 (2004).
26. Wunsch, C., Heimbach, P., Ponte, R. M. & Fukumori, I. The ECCO-GODAE Consortium Members. The global general circulation of the ocean estimated by the ECCO-consortium. *Oceanography* **22,** 88–103 (2009).
27. Ebbesmeyer, C. C. & Ingraham, W. J. Shoe spill in the North Pacific. *Eos Trans. Am. Geophys. Union* **73,** 361–365 (1992).
28. Ebbesmeyer, C. C. & Ingraham, W. J. Pacific toy spill fuels ocean current pathways research. *Eos Trans. Am. Geophys. Union* **75,** 425–430 (1994).
29. Schaum, E., Rost, B., Millar, A. J. & Collins, S. Variation in plastic responses of a globally distributed picoplankton species to ocean acidification. *Nat. Clim. Change* **3,** 298–302 (2012).
30. Barton, A. D., Dutkiewicz, S., Flierl, G., Bragg, J. & Follows, M. J. Patterns of diversity in marine phytoplankton. *Science* **327,** 1509–1511 (2010).
31. Jacobi, M. N., André, C., Döös, K. & Jonsson, P. R. Identification of subpopulations from connectivity matrices. *Ecography* **35,** 1004–1016 (2012).
32. Jacobi, M. N. & Jonsson, P. R. Optimal networks of nature reserves can be found through eigenvalue perturbation theory of the connectivity matrix. *Ecol. Appl.* **21,** 1861–1870 (2011).
33. Watson, J. R. *et al.* Identifying critical regions in small-world marine metapopulations. *Proc. Natl Acad. Sci. USA* **108,** 907–913 (2011).
34. Treml, E., Halpin, P., Urban, D. & Pratson, L. Modeling population connectivity by ocean currents, a graph-theoretic approach for marine conservation. *Landscape Ecol.* **23,** 19–36 (2008).
35. Chust, G., Irigoien, X., Chave, J. & Harris, R. P. Latitudinal phytoplankton distribution and the neutral theory of biodiversity. *Global Ecol. Biogeogr.* **22,** 531–543 (2013).
36. Sarmiento, J. L. *et al.* Response of ocean ecosystems to climate warming. *Global. Biogeochem. Cycles.* **18,** GB3003 (2004).
37. Treml, E. A. & Halpin, P. N. Marine population connectivity identifies ecological neighbors for conservation planning in the Coral Triangle. *Conserv. Lett.* **5,** 441–449 (2012).
38. Blanke, B. & Raynaud, S. Kinematics of the pacific equatorial undercurrent: An Eulerian and Lagrangian approach from GCM results. *J. Phys. Oceanogr.* **27,** 1038–1053 (1997).
39. de Vries, P. & Döös, K. Calculating Lagrangian trajectories using time-dependent velocity fields. *J. Atmos. Oceanic Technol.* **18,** 1092–1101 (2001).
40. Griffies, S. M. Elements of the Modular Ocean Model (MOM): 2012 release. Geophysical Fluid Dynamics Laboratory (GFDL) Ocean Group Technical Report No. 7, 1–631 (2012).

Acknowledgements

We thank Dave Siegel and Debora Iglesias-Rodriguez for comments and discussions and Amy Ehntholt for help with the manuscript. The project was in part funded by the NSF Coupled Natural-Human Systems grant GEO-1211972, NASA ROSES NNX13AC52G and the Nippon Foundation Nereus Program.

Author contributions

The work was initiated by J.R.W. and B.F.J. equally. B.F.J. conducted the particle tracking runs and the data analysis. Figures were produced by B.F.J. with support from J.R.W. The text was written by B.F.J. and J.R.W. equally.

Additional information

Competing financial interests: The authors declare no competing financial interests.

3

Rapid intensification and the bimodal distribution of tropical cyclone intensity

Chia-Ying Lee[1], Michael K. Tippett[2,3], Adam H. Sobel[2,4] & Suzana J. Camargo[4]

The severity of a tropical cyclone (TC) is often summarized by its lifetime maximum intensity (LMI), and the climatological LMI distribution is a fundamental feature of the climate system. The distinctive bimodality of the LMI distribution means that major storms (LMI >96 kt) are not very rare compared with less intense storms. Rapid intensification (RI) is the dramatic strengthening of a TC in a short time, and is notoriously difficult to forecast or simulate. Here we show that the bimodality of the LMI distribution reflects two types of storms: those that undergo RI during their lifetime (RI storms) and those that do not (non-RI storms). The vast majority (79%) of major storms are RI storms. Few non-RI storms (6%) become major storms. While the importance of RI has been recognized in weather forecasting, our results demonstrate that RI also plays a crucial role in the TC climatology.

[1] International Research Institute of Climate and Society, Columbia University, Palisades, New York 10964, USA. [2] Department of Applied Physics and Applied Mathematics, Columbia University, New York 10027, USA. [3] Center of Excellence for Climate Change Research, Department of Meteorology, King Abdulaziz University, Jeddah 21589, Saudi Arabia. [4] Division of Ocean and Climate Physics, Lamont-Doherty Earth Observatory, Columbia University, Palisades, New York 10964, USA. Correspondence and requests for materials should be addressed to C.-Y.L. (email: clee@iri.columbia.edu).

The question of how climate change will affect tropical cyclone (TC) activity has drawn considerable attention in the past two decades[1-4]. The current expectation is that we can expect a small increase in the global frequency of intense storms along with a small reduction in the total number of storms[4]. However, details of how the TC intensity distribution may change remain uncertain. This uncertainty reflects both the difficulty of simulating the most intense storms in climate change projections[4], as well as an incomplete understanding of what determines the climatological intensity distribution in the current climate[5]. Improved understanding of the TC intensity distribution in the current climate seems necessary if we are to understand TC intensity changes in a warming climate. Among various metrics of TC intensity, lifetime maximum intensity (LMI) is an integrated statistic of TC intensification, and its distribution represents a fundamental property of the TC climatology[6-8].

Several authors have noted that the probability density function (PDF) of global LMI is bimodal. The LMI PDF from International Best-Track Archive for Climate Stewardship for the period 1975–2007 has two local maxima around 40 and 100 kt, and a local minimum at 65 kt (ref. 9). The LMI PDF from a more temporally consistent global data for the period 1982–2009 shows the first maximum at 50 kt and the second one around 120 kt (ref. 10). Considering individual basins, the LMI PDF has its first peak at 50 kt for the North Atlantic, eastern and western North Pacific storms and its second maximum at 110 and 90 kt for eastern and western North Pacific storms[11]. The LMI distribution of Atlantic storms shows no clear secondary maximum, but substantial right skewness[11]. While the precise locations of the LMI PDF peaks vary with the data set and basin, the bimodal feature is quite robust and impossible to overlook. The existence of a second mode means that major storms (sustained winds >96 kt, that is, categories 3–5 TCs in Saffir-Simpson Hurricane Wind Scale) are much less rare than would be expected from the behaviour of the LMI distribution at lower values. The nonmonotonic behaviour of the LMI distribution is in contrast to that of the intensification rate distribution, which is exponential[12].

Although no complete explanation has been offered in the literature for the bimodality of the LMI distribution, there have been suggestions and indications of what processes might be involved. Uncertainties in the best-track winds is one of them. For example, the widely used Dvorak technique for estimating TC intensity from satellite imagery has low resolution at higher intensities, with only one bin in the category 3 range of the Saffir-Simpson scale. It has been argued that this may result in an artificially low number of category 3 hurricanes (LMI between 96–112 kt) in the Atlantic[13]. A recent study, on the other hand, proposed a parameterization of the ratio of surface exchange coefficients C_k/C_d, which appears in potential intensity theory[14,15], as a function of wind speed with a local maximum around 115 kt, and speculated that such a maximum would be favourable for rapid intensification (RI) and might explain the bimodal distribution of LMI[5]. In addition to the slope change in C_k/C_d, observations have suggested an association between RI and eye formation[16]. Organized convective heating results in increased vortex efficiency[17-20], which can potentially increase RI probability and cause intense systems. Perhaps most directly relevant, storms reaching the highest intensities in the western North Pacific[21] and North Atlantic[22] typically do so after undergoing RI at least once.

Here we extend these results to all basins, and show explicitly that RI explains the bimodality of the LMI distribution. We separate storms into two groups: those that undergo RI during their lifetime (RI storms) and those that do not (non-RI storms).

We find that the bimodality of the LMI distribution reflects the mixture of these two unimodal distributions, with the higher intensity peak consisting mostly of storms which have undergone RI at some point. In other words, the LMI distribution is unimodal when RI storms are excluded—RI storms are responsible for the bimodality of the LMI distribution. Various thresholds have been used to define RI[23,24]. Here we define RI as an increase of at least 35 kt in the maximum sustained surface wind over a period of 24 h or less. We find this definition of RI to be the most effective in separating the LMI distribution into two unimodal distributions. Sensitivity of the results to other RI thresholds, such as 30 kt, is discussed later.

Results

The bimodal LMI distribution. The global distribution of LMI for the period 1981–2012 has a peak at 45 kt and an indication of a secondary maximum around 120 kt (Fig. 1). Basin distributions of LMI are bimodal with local maxima around 45 and 120–135 kt except for the Atlantic where there is only a hint of a second maximum (Fig. 2), similar to what is shown in the literature[11]. The observed LMI distribution shows that category 3 and 4 storms are more common than category 1 and 2 storms. Most types of natural hazards become more rare as they become more extreme, for example, earthquakes[25] and tornadoes[26]. TCs, as measured by LMI, are unusual in having a range over which frequency increases with intensity. The second peak in the western North Pacific occurs at a higher intensity than in the other basins, consistent with the observation that the stronger storms globally occur more often in that basin[27].

Relation to RI. The LMI distribution of the 2,303 non-RI storms in the global record is unimodal and forms the first peak of the complete distribution (blue curve in Fig. 1). About 6% (141) of the non-RI storms are major storms (LMI >96 kt, categories 3–5), and the largest LMI value in this group is 145 kt (category 5). The second peak in the LMI distribution is formed by the 766 RI storms (red curve in Fig. 1), and 79% (603) of them are major storms. The same separation by RI in individual basins yields

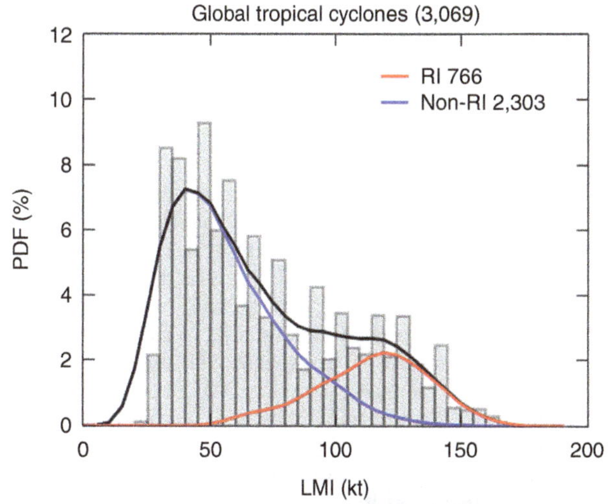

Figure 1 | Distributions of global tropical cyclone LMI. PDFs are calculated using 1981–2012 global tropical cyclone LMI. The grey bars show the raw data binned in 5 kt bins. The black, red and blue lines show the smoothed PDF for all storms, storms those undergo rapid intensification during their lifetime (RI storms), and those do not (non-RI storms), respectively. Smoothing is by moving average with window width of 15 kt. Total number of storms is listed in the title, while the numbers of RI and non-RI storms are given in the legend.

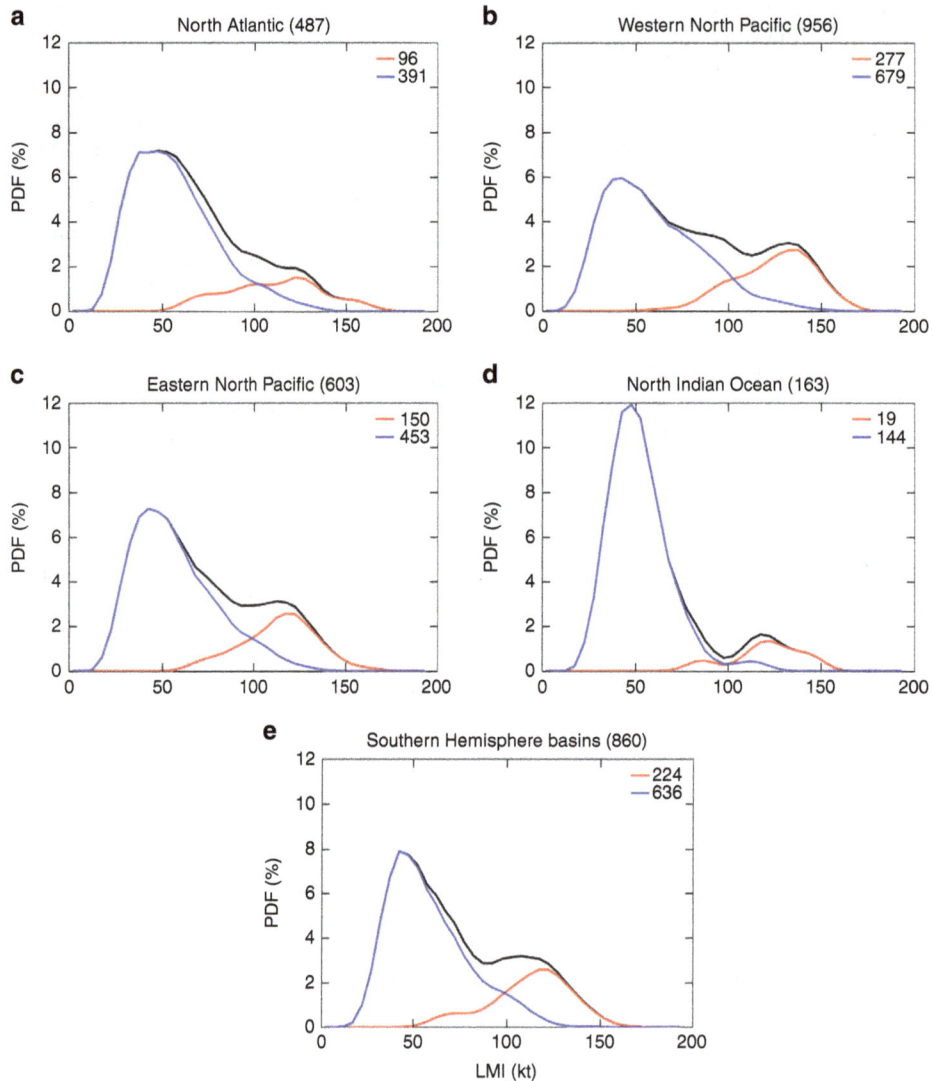

Figure 2 | Distributions of regional tropical cyclone LMI. PDFs are calculated using 1981–2012 tropical cyclone for individual basins: (**a**) North Atlantic, (**b**) Western North Pacific, (**c**) Eastern North Pacific, (**d**) North Indian Ocean, and (**e**) Southern Hemisphere basins. The black, red and blue lines show the smoothed PDFs for all, the subset of storms those undergo rapid intensification during their lifetime (RI storms), and those do not (non-RI storms), respectively. The raw data is binned in 5 kt bins and smoothing is by moving average with window width of 15 kt. The number of storms in each basin is given in the title, and the numbers of RI and non-RI storms are given in the legend.

similar results (Fig. 2). The intensification rate of 35 kt in 24 h (representing the 97th percentile of the intensification rate over that duration) is the optimal RI threshold for explaining the bimodality of the LMI distribution. Using other thresholds, such as 25, 30 or even 40 kt, does not separate the two peaks as clearly, and results a secondary maximum or a hint of it in either the RI or the non-RI LMI distributions (Supplementary Fig. 1). With the 35 kt definition, 80–85% of major storms in the eastern and western North Pacific, North Indian Ocean and southern Hemisphere basins are RI storms, but only 70% of major storms in the Atlantic are RI storms. In the Atlantic, the LMI distribution of RI storms is quite different from that in other basins, being much less peaked (red lines in Fig. 2). The probability that an RI storm in the Atlantic will become a minor hurricane (categories 1–2) is close to the probability that it will become a major hurricane. Still, RI storms are responsible for the rightmost shoulder of the LMI distribution in the Atlantic.

Relation to storm lifetime. Storm lifetime is another factor related to LMI. The distribution of lifetime itself is unimodal

(Supplementary Fig. 2). Overall, LMI is positively correlated with storm lifetime with longer lived storms tending to have higher LMI values. The correlation of storm lifetime with LMI is 0.66 globally. Here the idea that the bimodality in LMI can be explained by variations in lifetime is tested by separating storms into two groups, longer and shorter lived, and examining the LMI distributions for the two groups, analogously to what was done for RI and non-RI storms above. Thresholds of lifetime from 7 to 13 days are tested for this purpose (Supplementary Fig. 3). The classification produces less completely unimodal LMI PDFs than were achieved using RI as the criterion, especially for storms in the North Indian Ocean and southern Hemisphere basins. The best threshold of lifetime is also basin dependent, and noticeable departures from uni-modality are found in even when that best value is used. We, therefore, conclude that while lifetime has an impact on LMI, RI is a better criteria for explaining the LMI bimodality.

Discussion

The observed relation of RI with the bimodal LMI distribution is consistent with numerical simulation studies in which higher horizontal resolution results in both higher intensification rates and the appearance of the second LMI peak[9,28]. Such numerical simulations provide evidence that the observed LMI bimodality and its relationship to RI is not an artifact of the uncertainty associated with intensity estimation[13,29,30]. Our findings are also in agreement with the hypothesis of ref. 5 that RI is responsible for the bimodal distribution of LMI, although we cannot comment on the link to the ratio of exchange coefficients. On the other hand, a recent study showed that RI has no apparent signature in the overall intensification rate statistics[12]. The intensification rate distributions of RI and non-RI storms do differ, even when the RI events are excluded, but no distinguishing features result when all storms are considered (Supplementary Fig. 4). A key difference between the distribution of intensification rates and that of LMI is that the LMI distribution accounts for the intensification behaviour over the storm lifetime, and thus is a measure of the joint (across different times) intensification rate statistics.

Another measure of the TC intensity climatology is LMI normalized by local potential intensity. The distribution of normalized LMI for storms whose peak intensities are not limited by declining potential intensity has a linear cumulative distribution with two slopes[6], corresponding approximately to tropical storms (>34 kt) and hurricanes (>64 kt). This separation of the normalized LMI distribution appears to be distinct from the one here based on RI. The subset of storms not limited by declining potential intensity is a relatively small fraction of all storms, and the normalized LMI distribution over all storms (which we consider here) is not uniform (not shown; for example, Fig. 13 in ref. 6).

The basic physical mechanisms of RI are still not completely understood, and whether they are distinct from those of lower intensification rates remain an open question. We do not attempt to answer this question. The message here is that RI is relevant not only to short-term weather forecasting, but also to the relationship between TCs and climate. The most intense storms are those that undergo RI, and the storms that undergo RI are responsible for the observed bimodality of the LMI distribution. This finding suggests that a complete understanding of the most intense storms in either the current climate or future (or past) climates may need to include some understanding of RI. Our results also suggest that numerical models that do not simulate RI are likely to be incomplete in their representation of the LMI distribution and in the frequency of major storms. Therefore, an important research question is to what extent simulations and projections of the frequency of major storm occurrence can be accurate without either resolving RI or accounting for its absence.

Methods

Data. Best-track data from the National Hurricane Center (NHC)[29,31] and the Joint Typhoon Warming Centers (JTWC)[32] from 1981 to 2012 are used in this study. Best-track data include 1-min maximum sustained wind, minimum sea level pressure and location every 6 h.

Calculations. The LMI here is defined using maximum wind speed. Maximum wind speed itself is not an observed quantity, but rather estimated from *in situ* observations, remotely sensed estimates of winds or via satellite-based techniques[29]. The distribution of LMI is calculated globally, as well as for individual basins, following the definitions of the NHC and JTWC, that is, North Atlantic and eastern North Pacific (from NHC), and western North Pacific, North Indian Ocean and southern Hemisphere basins (from JTWC). Correlations between global LMI and storm lifetime are calculated using Spearman's rank correlation coefficient.

Terminology. We refer to storms with LMI >96 kt (categories 3–5 in Saffir-Simpson Hurricane Wind Scale) as major storms.

References

1. Bengtsson, L., Botzet, M. & Esch, M. Will greenhouse-induced warming over the next 50 years lead to higher frequency and greater intensity of hurricanes? *Tellus* **48A**, 57–73 (1996).
2. Emanuel, K. A. Increasing destructiveness of tropical cyclones over the past 30 years. *Nature* **436**, 686–688 (2005).
3. Webster, P. J., Holland, G., Curry, J. A. & Chang, H.-R. Changes in tropical cyclone number, duration, and intensity in a warming environment. *Science* **309**, 1844–1846 (2005).
4. Knutson, T. R. et al. Tropical cyclones and climate change. *Nat. Geosci.* **3**, 157–163 (2010).
5. Soloviev, A. V., Lukas, R., Donelan, M. A., Haus, B. K. & Ginis, I. The air-sea interface and surface stress under tropical cyclones. *Sci. Rep.* **4**, 5306 (2014).
6. Emanuel, K. A. A statistical analysis of tropical cyclone intensity. *Mon. Weather Rev.* **128**, 1139–1152 (2000).
7. Kossin, J. P., Emanuel, K. A. & Vecchi, G. A. The poleward migration of the location of tropical cyclone maximum intensity. *Nature* **509**, 349–352 (2014).
8. Park, D.-S. R., Ho, C.-H. & Kim, J.-H. Growing threat of intense tropical cyclones to East Asia over the period 1977-2010. *Environ. Res. Lett.* **9**, 014008 (2014).
9. Manganello, J. V. et al. Tropical cyclone climatology in a 10-km global atmospheric GCM: toward weather-resolving climate modeling. *J. Climate* **25**, 3867–3893 (2012).
10. Kossin, J. P., Olander, T. L. & Knapp, K. R. Trend analysis with a new global record of tropical cyclone intensity. *J. Climate* **26**, 9960–9976 (2013).
11. Zhao, M., Held, I. M., Lin, S.-J. & Vecchi, G. A. Simulations of global hurricane climatology, interannual variability, and response to global warming using a 50-km resolution GCM. *J. Climate* **22**, 6653–6678 (2009).
12. Kowch, R. & Emanuel, K. Are special processes at work in the rapid intensification of tropical cyclones? *Mon. Weather Rev.* **143**, 878–882 (2014).
13. Torn, R. D. & Snyder, C. Uncertainty of tropical cyclone best-track information. *Weather Forecast.* **27**, 715–729 (2012).
14. Bister, M. & Emanuel, K. A. Low frequency variability of tropical cyclone potential intensity 1. Interannual to interdecadal variability. *J. Geophys. Res.* **107**, 4801 (2002).
15. Emanuel, K. A. Sensitivity of tropical cyclones to surface exchange coefficients and a revised steady-state model incorporating eye dynamics. *J. Atmos. Sci.* **52**, 3969–3976 (1995).
16. Vigh, J. L., Knaff, J. A. & Schubert, W. H. A climatology of hurricane eye formation. *Mon. Weather Rev.* **140**, 1405–1426 (2012).
17. Shapiro, L. J. & Willoughby, H. E. The response of balanced hurricanes to local-sources of heat and momentum. *J. Atmos. Sci.* **39**, 378–394 (1982).
18. Hack, J. J. & Schubert, W. H. Nonlinear response of atmospheric vortices to heating by organized cumulus convection. *J. Atmos. Sci.* **43**, 1559–1573 (1986).
19. Vigh, J. L. & Schubert, W. H. Rapid development of the tropical cyclone warm core. *J. Atmos. Sci.* **66**, 3335–3350 (2009).
20. Nolan, D. S., Moon, Y. & Stern, D. P. Tropical cyclone intensification from asymmetric convection: energetics and efficiency. *J. Atmos. Sci.* **64**, 3377–3405 (2007).
21. Holliday, C. R. & Thompson, A. H. Climatological characteristics of rapidly intensifying typhoons. *Mon. Weather Rev.* **107**, 1022–1034 (1979).
22. Kaplan, J. & DeMaria, M. Large-scale characteristics of rapidly intensifying tropical cyclones in the North Atlantic basin. *Weather Forecast.* **18**, 1093–1108 (2003).
23. Knaff, J. et al. in *Sixth International Workshop on Tropical Cyclones* vol. 1.5 (WMO, Jose, Costa Rica., 2006).
24. Kaplan, J., DeMaria, M. & Knaff, J. A. A revised tropical cyclone rapid intensification index for the Atlantic and Eastern North Pacific basins. *Weather Forecast.* **25**, 220–241 (2010).
25. Hristopulos, D. T. & Mouslopoulou, V. Strength statistics and the distribution of earthquake interevent times. *Physica A* **392**, 485–496 (2013).
26. Feuerstein, B., Dotzek, N. & Grieser, J. Assessing a tornado climatology from global tornado intensity distributions. *J. Climate* **18**, 585–596 (2005).
27. Knapp, K. R., Kruk, M. C., Levinson, D. H., Diamond, H. J. & Neumann, C. J. The international best track archive for climate stewardship (IBTrACS). *Bull. Am. Meteor. Soc.* **91**, 363–376 (2010).
28. Murakami, H. et al. Future changes in tropical cyclone activity projected by the new high-resolution MRI-AGCM. *J. Climate* **25**, 3237–3260 (2012).
29. Landsea, C. W. & Franklin, J. L. Atlantic hurricane database uncertainty and presentation of a new database format. *Mon. Weather Rev.* **141**, 3576–3592 (2013).
30. Knaff, J. A., Brown, D. P., Courtney, J., Gallina, G. M. & Beven, J. L. An evaluation of Dvorak technique-based tropical cyclone intensity estimates. *Weather Forecast.* **25**, 1362–1379 (2010).
31. Jarvinen, B. R., Neumann, C. J. & Davis, M. A. S. *A Tropical Cyclone Data Tape for the North Atlantic Basin, 1886-1983: Contents, Limitations, and Uses* pp 21 (NOAA Technical Memorandum NWS NHC 22, 1984).
32. Chu, J.-H., C. R. Sampson, A. S. L. & Fukada, E. *The Joint Typhoon Warning Center tropical cyclone best-tracks, 1945-2000.* Report No. NRL/MR/7540-02-16, 22pp (Washington, D. C., USA, 2002).

Acknowledgements

The research is supported by Office of Naval Research under the research grant of MURI (N00014-12-1-0911). We thank Dr Shuyi S. Chen from University of Miami for her influential suggestion on the possible role of RI on the LMI distribution during 2014 AGU Fall Meeting. Comments and suggestions from Dr John Knaff and the other anonymous reviewer are appreciated.

Author contributions

The study was led by C.-Y.L. and M.K.T., calculations were carried out and manuscript was drafted by C.-Y.L. All authors were involved with designing the research, analysing the results and revising and editing the manuscript. The data were prepared by S.J.C.

Additional information

Competing financial interests: The authors declare no competing financial interests.

4

Humans choose representatives who enforce cooperation in social dilemmas through extortion

Manfred Milinski[1], Christian Hilbe[2,3], Dirk Semmann[1], Ralf Sommerfeld[1] & Jochem Marotzke[4]

Social dilemmas force players to balance between personal and collective gain. In many dilemmas, such as elected governments negotiating climate-change mitigation measures, the decisions are made not by individual players but by their representatives. However, the behaviour of representatives in social dilemmas has not been investigated experimentally. Here inspired by the negotiations for greenhouse-gas emissions reductions, we experimentally study a collective-risk social dilemma that involves representatives deciding on behalf of their fellow group members. Representatives can be re-elected or voted out after each consecutive collective-risk game. Selfish players are preferentially elected and are hence found most frequently in the 'representatives' treatment. Across all treatments, we identify the selfish players as extortioners. As predicted by our mathematical model, their steadfast strategies enforce cooperation from fair players who finally compensate almost completely the deficit caused by the extortionate co-players. Everybody gains, but the extortionate representatives and their groups gain the most.

[1] Department of Evolutionary Ecology, Max-Planck-Institute for Evolutionary Biology, August-Thienemann-Strasse 2, 24306 Plön, Germany. [2] Department of Organismic and Evolutionary Biology, Department of Mathematics, Program for Evolutionary Dynamics, Harvard University, One Brattle Square, Cambridge, Massachusetts 02138, USA. [3] Institute of Science and Technology Austria, Am Campus 1, Klosterneuburg 3400, Austria. [4] Max Planck Institute for Meteorology, Department "The Ocean in the Earth System", 20146 Hamburg, Germany. Correspondence and requests for materials should be addressed to M.M. (email: milinski@evolbio.mpg.de) or to C.H. (email: hilbe@fas.harvard.edu) or to J.M. (email: jochem.marotzke@mpimet.mpg.de).

Although humans are regarded as champions of cooperation[1,2], there are social dilemmas that so far have defied solution—we have not yet collaborated successfully to stop the increase of global greenhouse-gas emissions[3,4], Europe continues to overexploit its marine fish stock[5] and the European Union has so far failed to reach an equitable solution to accommodating the large number of refugees arriving from Africa and the Middle East[6]. In these and other dilemmas, essential decisions are made not by individual social actors but by representatives such as officials from elected governments. Representatives have been shown to display a more competitive mindset than 'ordinary' group members[7]. However, the behaviour of representatives in a social dilemma has, to our knowledge, not been investigated experimentally. To fill this gap is the aim of our paper.

While we believe that our results apply to the role of representatives in social dilemmas more broadly, we have drawn our main inspiration and the concrete setting of our experiments from the challenge to prevent 'dangerous anthropogenic interference with the climate system'[8]. This challenge is now usually interpreted as limiting global warming to below 2 °C compared with the pre-industrial period. To prevent temperature from exceeding this limit, greenhouse-gas emissions should be reduced from about 2020 onwards; by 2050, emissions should fall to a level of $\leq 50\%$ of the year 2000 emissions[4,9–12]. However, as representatives attend climate summits to negotiate their country's share in reducing greenhouse-gas emissions, they are eagerly watched by their voters who might not re-elect their representatives when others negotiate a lower share[13]. Though everybody profits only if dangerous climate change is averted, none of the many climate summits has achieved sustained emissions reductions, the relative success of the Paris negotiations at COP21 notwithstanding.

The global emissions-reduction problem has been simulated experimentally in the 'collective-risk social dilemma' game[14–18]. A number of volunteers can invest anonymously from their individual endowments into a climate account in each of 10 consecutive rounds. If the group collectively reaches a specified target sum, everybody receives in cash what she has not invested from her endowment. However, if the group fails to reach the target, individuals risk losing all their remaining endowment with a high probability, mimicking the drastic economic losses that result from dangerous climate change. The social dilemma arises because all players benefit only if the collective target is reached, but individual payoff is maximised by lower-than-average contributions, spurred by the hope that others will compensate to reach the target[13].

In contrast to previous work, we have here assembled 15 groups of 18 players each where the groups are sub-divided into 6 'countries' of 3 players each who elect, re-elect or vote out their representative for the 6 representatives' 'summit'. For control, we have assembled 15 groups of 6 players each (as in ref. 14) and 15 groups of 18 players each. In 3 consecutive collective-risk games with 10 rounds each, each player in the 6-players and 18-players treatments contributes from her initial endowment of €40; in the 6-representatives treatment, each representative contributes from the combined endowments (€120) of her watching country mates and on their behalf (Fig. 1; see Methods). The target sum that must be collected by each group to prevent simulated dangerous climate change is €120 in the 6-players treatment and €360 both in the 18-players and the 6-representatives treatments.

We find that selfish players are preferentially elected and are hence found more frequently in the six-representatives treatment than in the other two treatments. Across all treatments, we identify the selfish players as extortioners. We develop a mathematical model and confirm its prediction that the extortioners' steadfast strategies enforce cooperation from fair players who finally compensate almost completely the deficit caused by the extortionate co-players.

Results

Simulated dangerous climate change. In the first game of the 18-players treatment and of the 6-representatives treatment, only 33% of the groups reach the target sum. By contrast, groups in the six-players treatment are almost twice as likely to collect sufficient contributions in the first game, with 60% of the groups reaching the target sum (Fig. 2a–c), similar to a previous study[14]. The percentage of groups reaching the target sum increases towards game 3 in the six-players and the six-representatives treatment, but the increase is not statistically significant. In game 3, the groups in the 18-players treatment are the least successful (Fig. 2a–c), but again differences are not statistically significant.

The total sums contributed per group do not differ among treatments in games 2 and 3 (Fig. 2e,f). In game 1, the six representatives contribute less than the six players ($P = 0.019$,

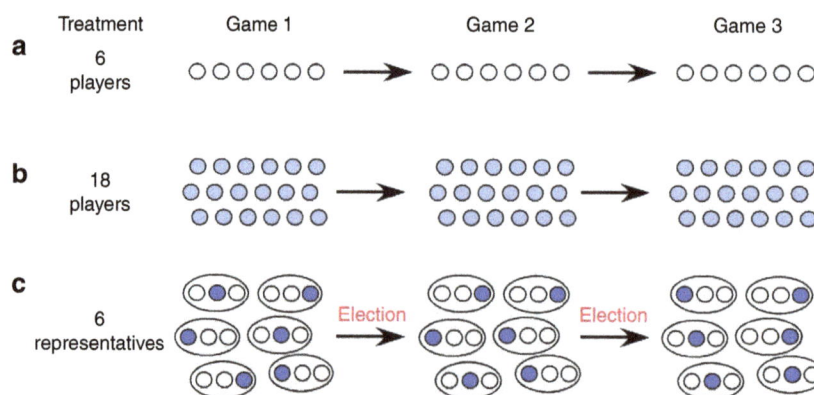

Figure 1 | Design of the three treatments. (**a**) The 6-players treatment, (**b**) the 18-players treatment, (**c**) the 6-representatives treatment. Each game consists of 10 rounds, during which players need to raise sufficient contributions to reach a specified target sum. Games 2 and 3 are replicates of game 1. The players remain the same in the 6-players and the 18-players treatment. In the 6-representatives treatment, representatives are randomly picked in game 1 and re-elected or voted out for games 2 and 3. Re-election of a representative may depend on the representatives' performance in previous games. In addition, except for the first four groups, after games 1 and 2 all players in the 6-representatives treatment are asked to write non-binding pledges about how they would contribute if elected. Players are only informed about the pledges of members of their own subgroup.

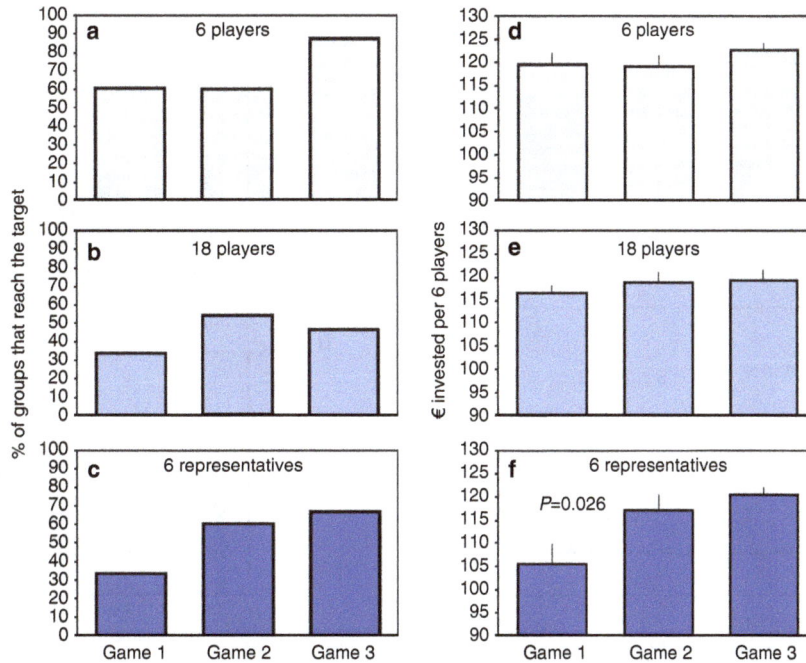

Figure 2 | Group success in reaching the target sum (left) and group investments (right). (**a,d**) Six-players treatment; (**b,e**) 18-players treatment; (**c,f**) 6-representatives treatment. In **f**, the sum invested is divided by 3 to allow comparison among treatments. Means ± s.e.m. of 15 groups per game and treatment are shown. See text for statistics.

$z = -2.341$, $n_1 = n_2 = 15$ groups, Mann–Whitney U-test, two-tailed; we use two-tailed tests throughout, with the group of six or 18 players as our statistical unit if not stated otherwise). Because in game 1, representatives are randomly picked from the group (see methods), the only difference between the two treatments is that representatives are contributing on behalf of their observing group. In such situations, representatives may have a more competitive mindset[7], which would explain why groups in the six-representatives treatment reach the target less often. Total contributions show a small increasing trend from the first to the third game in all treatments (Fig. 2e,f), but the differences are statistically significant only between games 1 and 2 in the six-representatives treatment ($P = 0.026$, $z = -2.230$, $n = 15$, Wilcoxon signed-rank matched pairs test). Summed up over all three games per group, contributions relative to the target sum are lowest in the six-representatives treatment, significantly lower than in the six-players treatment ($P = 0.0061$, $z = -2.742$, $n_1 = n_2 = 15$, Mann–Whitney U-test).

Fair and selfish players. For the group to reach the target sum, each player must on average contribute half of her total endowment—the 'fair share' of €20 (€60 per representative in the six-representatives treatment). Thus, whenever the target sum is not reached, one or several players must have contributed less than their fair share. We call these 'selfish players' to distinguish them from the 'fair players' who give at least their fair share. The percentage of selfish players is highest in the 6-representatives treatment (Fig. 3a), higher than in the 6-players treatment ($P = 0.01$, $z = -2.559$, $n_1 = n_2 = 15$, Mann–Whitney U-test) and almost significantly higher than in the 18-players treatment ($P = 0.06$, $z = -1.862$, $n_1 = n_2 = 15$, Mann–Whitney U-test). The average contribution of a selfish player (relative to the fair-share contribution) is lower in the 18-players treatment than in both the 6-players ($P = 0.02$, $z = -2.302$, $n_1 = n_2 = 15$, Mann–Whitney U-test; Fig. 3b) and the 6-representatives treatment ($P = 0.006$, $z = -2.739$, $n_1 = n_2 = 15$, Mann–Whitney

U-test) (Fig. 3b). Over all three games, the net payoff (including trials where the group fails to collect the target sum and loses all remaining money) is higher for selfish than for fair players (Fig. 3c). Selfish players achieve a higher net payoff in the 6-players treatment, compared with both the 18-players treatment ($P = 0.024$, $z = -2.261$, $n_1 = n_2 = 15$, Mann–Whitney U-test) and the 6-representatives treatment ($P = 0.020$, $z = -2.325$, $n_1 = n_2 = 15$, Mann–Whitney U-test, shown per represented player; Fig. 3c).

Using a classification of players in a social dilemma proposed by Fischbacher and Gächter[19], the selfish representatives might be 'pessimistic conditional cooperators' who dislike that others contribute less than their fair share and thus stop contributing. However, all selfish representatives contribute more in the end than in the beginning ($P = 0.0002$, linear regression of contribution per selfish representative per group on rounds 1–10, analysed for game 3) and resemble 'imperfect conditional cooperators'[19]. By increasing their contribution during the 10 rounds as do fair representatives ($P = 0.002$), the selfish players help reaching the target, though they contribute much less than fair representatives.

Voters choose selfish representatives. After both games 1 and 2, representatives can be either re-elected or voted out. After game 1, those representatives who are re-elected have contributed significantly less in game 1 than those who are voted out (Fig. 4a) ($P = 0.01$, $z = -2.587$, $n = 15$, Wilcoxon signed-rank matched pairs test). While this is not the case after game 2, we still find a tendency that selfish representatives are preferentially re-elected, based on their past contributions. In addition, before each election the players formulate election pledges specifying their contribution strategy if elected. The percentage of selfish pledges (see Methods) is higher among the 6 elected representatives than among all 18 players of that treatment (Fig. 4b), although significantly so only after game 2 ($P = 0.0071$, $z = -2.692$, $n = 11$, Wilcoxon signed-rank matched pairs test). Thus,

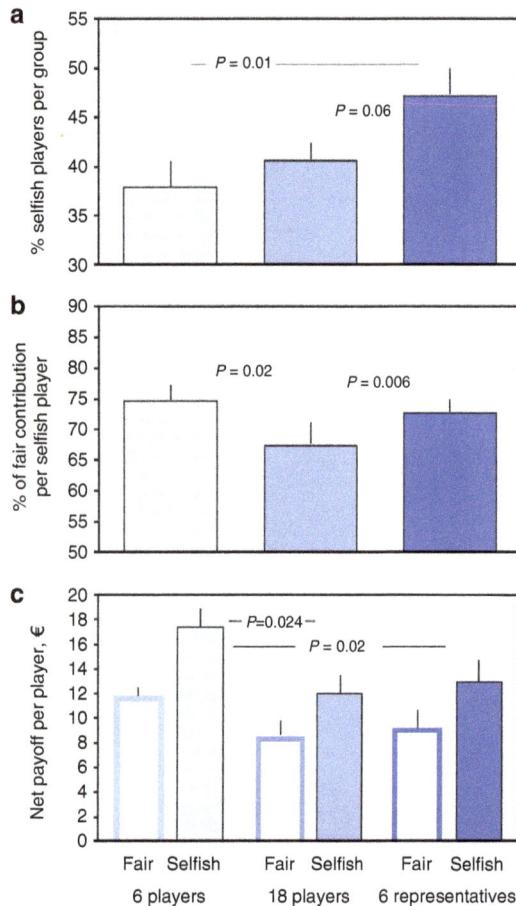

Figure 3 | Fair and selfish strategies. (**a**) The percentage of selfish players per group, (**b**) the average contribution of a selfish player (relative to the fair-share contribution), (**c**) the net payoff per fair and selfish player. Means ± s.e.m. of 15 groups per treatment are shown. See text for statistics.

representatives who act selfishly in game 1 are preferentially re-elected, and players who pledge to be selfish are preferentially elected after game 2.

Players classified as selfish according to their election pledges vote in 71.3% for classified selfish players and in 10.1% for classified fair representatives. Players classified as fair vote in 78.9% for classified fair players and in 14.6% for classified selfish players (the complement missing from 100% is due to players that could not be classified as either selfish or fair). Hence, selfish players want selfish representatives, and fair players want fair representatives.

Representatives who have pledged to be selfish contribute less in the following game than those who have pledged to be fair (Fig. 4c; after game 1: $P = 0.007$, $z = -2.692$, $n = 10$; after game 2: $P = 0.0051$, $z = -2.803$, $n = 10$, Wilcoxon signed-rank matched pairs test). Thus, players fulfil their pledges when acting as representatives.

Identification of selfish players as extortioners. Theorists have predicted for a long time that cooperative and fair strategies such as Tit-for-Tat would eventually succeed in social dilemmas[20-23]. Why then would subjects vote for representatives who mainly pursue the success of their own subgroup while disregarding the risks for the whole community? We hypothesize that the election procedure would favour representatives who motivate the other subgroups' representatives to reach the target, but at the same time ensure that the own subgroup contributes less than other

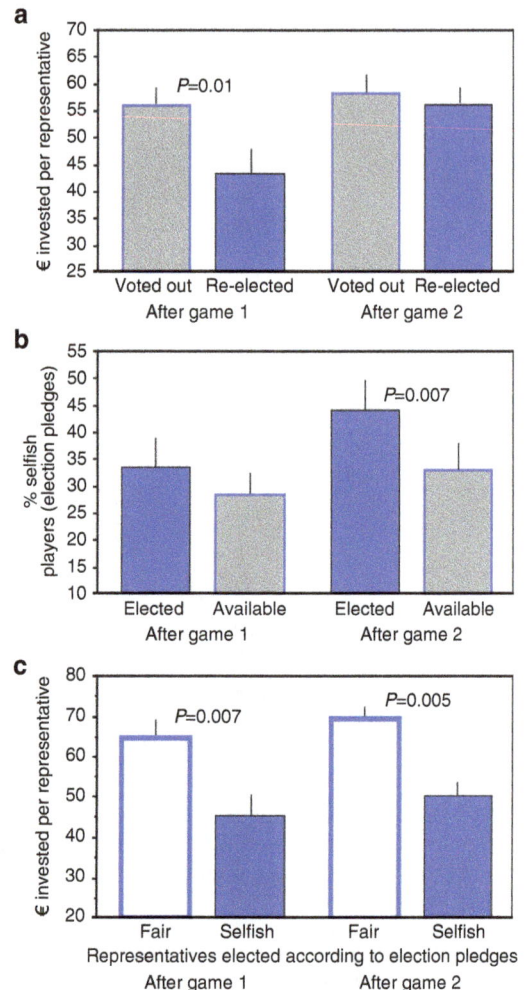

Figure 4 | Voting success and behaviour of selfish and fair representatives in the six-representatives treatment. (**a**) Previous investment of representatives who are either voted out or re-elected, (**b**) percentage of selfish players, according to their election pledges, available and elected, (**c**) future fulfilment of election pledges by selfish and fair players. Means ± s.e.m. groups are shown, for 15 groups in **a** and 11 groups in **b** and 10 groups in **c**. See text for statistics.

subgroups. Individuals would like their representatives to be steadfast and to convince the other subgroups' representatives to compensate for any missing contributions. Such behaviour is reminiscent of the recently discovered class of extortionate ZD strategies for the repeated prisoner's dilemma[24-30], where extortionate players incentivize their opponents to cooperate although they themselves are not fully cooperative. In pairwise encounters, these extortionate players cannot be beaten by any other strategy, and they are predicted to perform well among adaptive co-players[24,25,27,29]. In the Methods section, we extend the theory of ZD strategies to the collective-risk social dilemma, and we prove that also in our experiment players may adopt extortionate strategies. Such players exhibit the following three characteristics: (i) Extortioners gain higher payoffs than their co-players by contributing less towards the climate account; that is, if x_i is the total contributions of an extortioner, and if x_{-i} is the average contribution of the other group members, then

$$x_i \leq x_{-i}. \tag{1}$$

(ii) Extortioners persuade their co-players to make up for the missing contributions; that is, the collective best response for the remaining $N-1$ group members is to choose x_{-i} such that the group reaches the target sum T,

$$x_i + (N-1) \cdot x_{-i} = T \qquad (2)$$

(iii) Extortioners are consistent, meaning that the properties (i) and (ii) are not only satisfied in one particular instance of the game, but in every game the player participates in. We now test whether the selfish players in our experiment meet these three criteria.

Because we find both fair and selfish players in all three treatments, we perform a proof-of-principle with players of all treatments combined. To keep the group as statistical unit, we enter contribution averaged over all fair players of each group; contributions of representatives are divided by 3 to be comparable 'per player' to the other treatments. The contribution per fair player increases over the three games (Fig 5a; $P = 0.0057$, $F_{2,130} = 5.3788$, generalized linear model (GLM) with family = Gaussian). By contrast, the contribution per selfish player does not increase significantly ($P = 0.66$, $F_{2,131} = 0.4163$, GLM). We find a significant interaction between fair and selfish players' contributions over the three games ($P = 0.032$, $F_{2,261} = 3.4798$, GLM). Over the three games, as the contributions

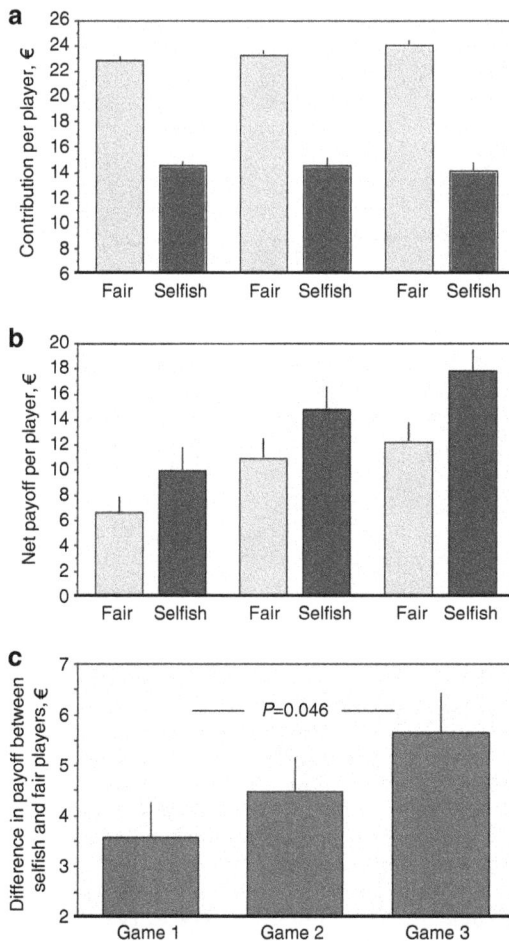

a

b

c

Figure 5 | Comparison of contributions and payoffs for fair players and selfish players across all three games. (**a**) Contribution of fair and selfish players; (**b**) net payoff of selfish and fair players; (**c**) difference in payoff between fair and selfish players. We enter contributions averaged over both all fair and all selfish players of each group. Contributions of representatives are divided by 3 to be comparable to other treatments. See text for statistics.

of fair players increase, so does the payoff of both fair players ($P = 0.010$, $F_{2,132} = 4.7574$, GLM, with family = gamma) and selfish players ($P = 0.015$, $F_{2,132} = 4.339$, GLM, with family = gamma; Fig. 5b). In each game, selfish players gain more than fair players; the difference increases from game 1 to game 3 (Fig. 5c) ($P = 0.046$, $z = -1.995$, $n = 45$, Wilcoxon matched pairs signed ranks test).

To test whether other group members are willing to compensate for missing contributions, we compare the contribution deficit of all selfish players in a group (the sum of all their negative deviations from the fair share) with the contribution surplus of all fair players (the sum of positive deviations of all the fair players; Fig. 6). For example, in game 1 in the six-players treatment, the dot most to the left (Fig. 6a) shows a group where the five selfish players contribute only €80 instead of the fair-share contribution of €100. The single fair player of that group contributes €22, €2 more than her fair share but not enough to compensate for the deficit of €20 caused by the selfish players. Hence the group misses the target sum of €120, and everybody loses the money not invested with 90% probability. As another example, the leftmost dot of those exactly on the red line depicts a group where the three selfish players invest €44 instead of €60, causing a deficit of €16, which is exactly compensated by the three remaining fair players. Thus the group meets the target of €120, but the selfish players receive a higher payoff than the fair players.

If selfish players were indeed able to persuade the remaining group members to compensate for missing contributions, we would expect the regression lines in Fig. 6 to have a significantly negative slope and to be close to the red lines marking exact (hypothetical) compensation. We see this compensation in the six-players treatment in game 2 (Fig. 6b, simple regression, F-test = 36.257, degree of freedom (DF) = 1, $P = 0.0001$) and in game 3 (Fig. 6c, simple regression, F-test = 26.204, DF = 1, $P = 0.0002$) and in the six-representatives treatment in game 3 (Fig. 6f, simple regression, F-test = 17.286, DF = 1, $P = 0.0011$). By contrast, we find no significant compensation in the 18-players treatment.

In the 6-players treatment, fair players compensate or over-compensate the selfish players' deficit in 9 groups in games 1 and 2 (Fig. 6a,b) and in 13 groups in game 3 (Fig. 6c). In the 18-players treatment, fair players compensate or overcompensate the selfish players' deficit in 5 groups in game 1 (Fig. 6g) and in 7 groups in games 2 and 3 (Fig. 6h,i). In the 6-representatives treatment, the deficit of the selfish players is only compensated in 4 groups in game 1 (Fig. 6d) but in 9 groups in game 2 (Fig. 6e) and in 10 groups in game 3 (Fig. 6f). Over all treatments and games, selfish players or selfish representatives successfully drive their fair counterparts to compensation in 73 out of 135 individual games (54%). Moreover, groups become increasingly successful in reaching the target, improving from game 1 (40%) to game 2 (56%) and game 3 (67%). Because only fair players raise their contributions over the three games but not selfish players (see Fig. 5a), these results suggest that a considerable fraction of fair players learn to become even more cooperative in response to extortioners. The learning effect is demonstrated by the observation that the contribution per fair representative has no relation to the number of selfish representatives per group in game 1 but correlates significantly in game 3 (Supplementary Fig. 2).

Players behave consistently across the 3 games in the 6-players and the 18-players treatments, as witnessed by significant positive correlation of the contributions (see Supplementary Information for detailed analysis). For the six-representatives treatment, we have analysed the behaviour of representatives after being re-elected. In 34 out of the 42 cases in which a selfish representative is re-elected, the representative remains selfish in

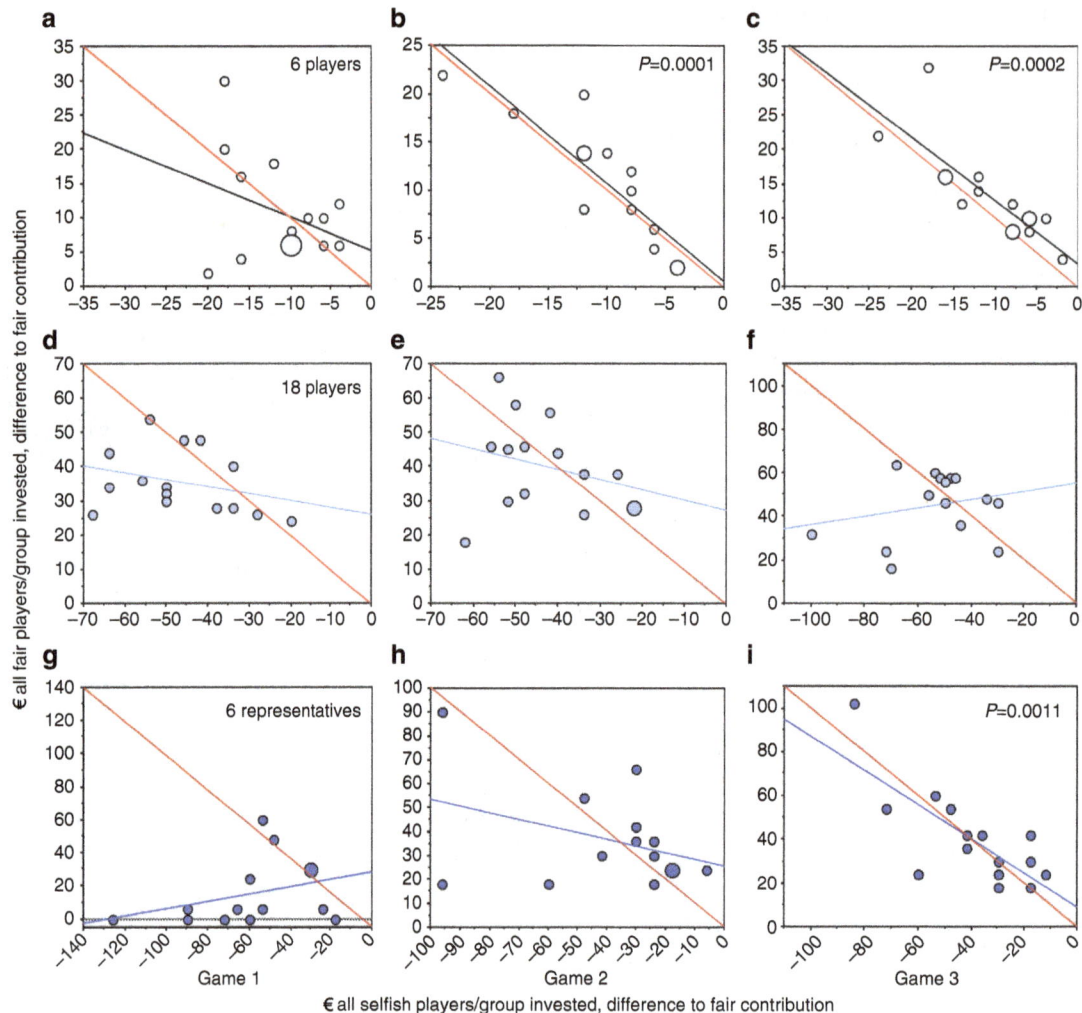

Figure 6 | Fair players' compensation of their selfish players' deficit. (**a-c**) Six-players treatment; (**d-f**) 18 players treatment; (**g-i**) 6-representatives treatment. Each dot represents a group; larger dots show overlaid results from two or three groups. Black and blue lines depict simple regressions. The red lines depict all combinations of hypothetical contributions in which fair players exactly compensate for the deficit caused by all selfish players of that group. Thus, dots on or above the red line correspond to groups that reach the target sum. See text for statistics.

the next game ($P = 0.005$, Fisher's exact test, two-tailed compared with 50%). Overall, we have thus established that selfish players gain much higher payoffs (Fig. 5); they are often successful in persuading their fair co-players to compensate for missing contributions (Fig. 6); and they are consistent across different games. Thus, selfish players show all three characteristics of extortionate behaviour.

Discussion

We have introduced into the collective-risk social dilemma the innovation that contributions into the climate account are decided on not by individual players but by representatives (six-representatives treatment). For control, we have assembled groups of 6 players and 18 players in 2 further treatments. We find selfish players in all treatments, but their concentration is highest in the 6-representatives treatment. Selfish representatives are preferentially elected or re-elected if they either contribute less than the fair share or pledge to do so. Having to cater to their electorates' preferences thus has the adverse effect that representatives risk losing the climate game to win elections. As a consequence, groups in the six-representatives treatment contribute less than groups in the six-players treatment (relative

to the required target sum), and they receive lower average payoffs. On the other hand, in games 2 and 3 the groups tend to reach the target sum more often in the 6-representatives than in the 18-players treatment. While fair representatives compensate for missing contributions in game 3, 18 players do not achieve that compensation. Thus, our representatives tend to be more successful in preventing simulated dangerous climate change than 18 players deciding themselves. We speculate that this 'representatives' advantage' is much greater with much larger groups such as real countries.

The psychological consequences of acting as a representative of a group have been characterised as evoking both more competitive interaction goals and more competitive expectations of others[7]. A representative is faced with a powerful responsibility to provide good outcomes for her constituency and may face strong pressures by being monitored and evaluated[7]. The mindset that is activated by the role of representative shows up clearly in our experiments when we compare the behaviour of six players randomly selected to decide for themselves with the behaviour of six representatives randomly selected to decide for their group. Otherwise the players find themselves in exactly the same situation in both cases. As psychology predicts[7], the six players' groups are twice as successful as the six-representatives' groups in reaching the target sum. When voting, players preferentially choose representatives who either have displayed a competitive mindset as former representatives or have pledged to do so if elected.

In our experiments, selfish behaviour pays off only if others compensate any missing contributions. Selfish subjects apply an implicit form of extortion[24]—they contribute less than is needed on average, but in a way that makes it optimal for their peers to become even more cooperative. The effect of extortion in our experiments differs from that in the repeated prisoner's dilemma, in which subjects strongly oppose exploitation[31]. Here subjects in the six-players and six-representatives treatments eventually accept extortion up to a certain degree, especially in game 3, in which subjects have already gained some experience. We speculate that the higher tolerance towards extortioners in our experiment is due to the higher stakes involved—resisting extortion comes relatively cheap in the prisoner's dilemma, but it endangers the entire payoff in the collective-risk game. Only in the 18-players treatment was extortion unsuccessful in persuading others to cooperate, presumably because in larger groups it becomes more difficult to induce individuals to behave in a desired way.

Our identification of extortionate behaviour in the collective-risk social dilemma suggests two counteracting major effects when, with all due caution, we try to interpret the social dynamics of climate summits with our results in mind. On the one hand, the competitive advantage of selfish players in getting elected or re-elected appears to work against reaching a collective target such as preventing dangerous climate change—there might not be enough fair representatives around to support the target. On the other hand, selfish players, who are ubiquitous and show up in all but 1 of the 135 individual collective-risk games, consistently act as extortioners. Their steadfast strategies enhance the already-existing willingness of our fair players to contribute towards reaching the collective target. If we compare extortionate to hypothetical non-extortionate selfish players, we conclude—with more than just a hint of Machiavellian thinking—that extortion benefits the prevention of dangerous climate change.

Methods

Experimental procedures. A total of 630 undergraduate students from the Universities of Bonn, Hamburg, Göttingen, Kiel and Münster voluntarily participated in 45 experimental sessions with either 18 or 6 subjects each in a computerized experiment (for example, ref. 32). The subjects were separated by opaque partitions and each had a computer, on which they received the instructions for the experiment and with which they communicated their decisions. Throughout the whole experiment, subjects were anonymous, and they made their decisions under a neutral pseudonym.

There are three treatments (Fig. 1). For each treatment, we had 15 groups of subjects interacting in a variant of the 'collective-risk social dilemma' game[14]: subjects received an initial endowment, and they were asked, in each of 10 rounds, to contribute money from this endowment into a 'climate account'. At the end of round 10, the game software checked whether total contributions of all group members matched (or exceeded) a previously specified target sum. If that was the case, subjects received their remaining endowment in cash (in a way that maintained the subjects' anonymity). If the collective target was not reached, subjects lost their remaining endowment with 90% probability. Each game was repeated twice, such that every group played three games with all players keeping their pseudonyms (each time with a new endowment).

In the 6-players and the 18-players treatments, groups consisted of 6 and 18 subjects, respectively, and each subject had an initial endowment of €40. In each round, subjects could choose whether to contribute €0, €2 or €4 to the climate account, and the decisions of all subjects were shown to all subjects after each round. The target was reached if on average subjects contributed half their endowment (the target sum was €120 in the 6-players treatment, and it was €360 in the 18-players treatment).

In the 6-representatives treatment, groups of 18 subjects were sub-divided in 6 'countries' of 3 players. For game 1, the computer randomly determined six representatives, one from each country. Only the representatives were able to contribute money to the group's climate account: they had 3 times €40 at their disposal, for investing €0, €6 or €12 in each of 10 rounds. The decisions of all representatives were shown to all 18 subjects after each round. The target was to collect at least €360 in donations (€60 per representative or €20 per subject). If the target was reached, subjects received a third of their country's remaining endowment.

After games 1 and 2, the three subjects of each country could re-elect the previous representative or vote her out and elect a different member of their country with a majority vote. Except for the first four groups of this treatment, subjects could compose both game 1 and game 2 election pledges of up to 500 characters on their laptop that could be seen only by the 3 subjects of a country. The pledges described how the person would decide if elected. We have blindly classified all pledges; those that promise to contribute 'less than the others', or 'less than the fair share' have been classified as 'selfish' and the others as 'fair'. When making the voting decision, each subject knew the observed decisions of her previous representative and those of the other representatives, and saw the three election pledges within her country (each with the respective pseudonym and a button for voting). In cases with no majority vote, the computer decided randomly for the next representative (in 9% of cases).

Subjects knew that the total sum of money in the climate account, accumulated from all participating groups, would be used to publish a press advertisement on climate protection in a daily German newspaper simultaneously with the publication of the present study. However, they received the 'little information' version from ref. 32 to explain the climate account, so that we could expect very weak motivation to invest in publishing the advertisement *per se*.

Theoretical model. Press and Dyson[24] describe a class of so-called ZD strategies for the repeated prisoner's dilemma, and they demonstrate that a subset of ZD strategies can be used to extort opponents. However, the collective-risk dilemma game used in our experiment is not a repeated two-player game. Herein, we thus extend the theory of ZD strategies to collective-risk dilemmas. As an application, we show the existence of extortionate strategies. Such strategies ensure that (i) a player gets at least the average payoff of the co-players; (ii) the collective best reply for the remaining group members is to reach the target; and (iii) the properties (i) and (ii) hold in any game the player participates in.

To this end, we consider a group of N individuals, with each group member having an initial endowment of E. The group engages in a collective-risk dilemma[12]: in each of R rounds, players can decide how much they want to contribute towards a common pool. We denote player i's contribution in round r by $x_i(r)$, and we assume that the minimum contribution per round is 0, whereas the maximum contribution is $x_{max} = E/R$. To calculate the total contributions x_i of player i, we sum up over all rounds, $x_i = \sum x_i(r)$. The group's total contributions x are obtained by summing up over all individual contributions, $x = \sum x_i$. Payoffs for the collective-risk dilemma are defined as follows: if total contributions x after round R exceed a threshold T, then all players receive their remaining endowment; that is, if $x \geq T$, then player i's expected payoff is $E - x_i$. Otherwise, if total contributions are below the threshold, all players risk losing their remaining endowment with some probability $p > 0$, and player i's expected payoff becomes $(1 - p)(E - x_i)$. Supplementary Table 1 gives a summary of all used variables.

In the experiment, players had to choose between three possible contribution levels in a given round, but for the model we assume for simplicity that players can contribute any amount $x_i(r) \in [0, x_{max}]$. We note that the definition of ZD strategies given below can be extended to the case of discrete contribution levels. To achieve an arbitrary contribution level $y \in [0, x_{max}]$, player i would need to randomize between the given discrete contribution levels such that the expected value satisfies $E[x_i(r)] = y$. Similar to Tit-for-Tat-like strategies in the Prisoner's Dilemma, we define ZD strategies in the collective-risk dilemma as behaviours that condition their contribution in the next round on the co-players' contributions in the previous round:

Definition (ZD strategies). Player i applies a ZD strategy for the collective-risk dilemma if i's contributions $x_i(r)$ in every round r satisfy

$$x_i(r) = s x_{-i}(r - 1) + (1 - s)\gamma E/R, \qquad (3)$$

where $x_{-i}(r - 1)$ is the average contribution of the other group members in the previous round, with $x_{-i}(0) := 0$, and s and γ are parameters that can be chosen by player i.

The parameter s is a measure for how a player reacts to the co-players' contributions of the previous round. The parameter γ, on the other hand, determines a player's baseline contribution level. These two parameters cannot be chosen arbitrarily—since player i's contribution needs to be in the interval $[0, E/R]$, the two parameters need to satisfy

$$\begin{aligned} -1 &\leq s \leq 1 \\ -s/(1-s) &\leq \gamma \leq 1/(1-s) \end{aligned} \qquad (4)$$

It is the following property that makes ZD strategies interesting.

Proposition 1 (properties of ZD strategies). Suppose player i applies a ZD strategy with parameters s and γ.

1. If x_i denotes the total contributions of player i, and if x_{-i} denotes the average total contribution of i's co-players, then

$$|x_i - s x_{-i} - (1 - s)\gamma E| \leq E/R \qquad (5)$$

2. Similarly, if π_i and π_{-i} denote the corresponding realized payoffs, then payoffs either satisfy $\pi_i = \pi_{-i} = 0$ (if the group fails to reach the threshold and dangerous climate change occurs), or

$$|\pi_i - s\pi_{-i} - (1-s)(1-\gamma)E| \leq E/R \qquad (6)$$

Proof

1. By summing up Equation (3) over all rounds $1 \leq r \leq R$, we obtain

$$x_i = s[x_{-i} - x_{-i}(R)] + (1-s)\gamma E.$$

As a consequence,

$$|x_i - sx_{-i} - (1-s)\gamma E| \leq |sx_{-i}(R)| \leq E/R.$$

2. In case players do not lose their remaining endowment, Equation (6) follows directly from Equation (5) because $\pi_i = E - x_i$ and $\pi_{-i} = E - x_{-i}$.

For a collective-risk dilemma with sufficiently many rounds R, Proposition 1 thus implies that $x_i \approx sx_{-i} + \gamma(1-s)E$. That is, there is a linear relationship between the total contributions of player i, and the total contributions of i's co-players. Similarly, it follows for the realized payoffs that either $\pi_i = \pi_{-i} = 0$ or $\pi_i \approx s\pi_{-i} + (1-\gamma)(1-s)E$. Therefore, unless payoffs are zero, there is also a linear relationship between the players' realized payoffs. This property makes strategies having the form of Equation (3) analogous to the ZD strategies described for the repeated prisoner's dilemma[24]. It is important to note that the above Proposition makes no restrictions on the strategies of i's co-players—the stated results hold no matter what the other group members do. As a particular instance of ZD strategies, let us consider the following special case.

Definition (extortionate ZD strategies). A player applies an extortionate ZD if the parameters s and γ are chosen such that

$$\gamma = 0 \text{ and } \max[0, T/(pE) - (N-1)] \leq s < 1. \qquad (7)$$

If some player i applies such an extortionate strategy, it follows from Proposition 1 that approximately i's total contribution only make up a fraction of the average contribution of the other group members, since $x_i \approx sx_{-i}$ (Supplementary Fig. 1 gives an illustration).

The following Proposition shows that the name 'extortionate ZD strategy' is justified: players with such a strategy show the typical characteristics of extortionate behaviour.

Proposition 2 (properties of extortionate ZD strategies). Suppose player i applies an extortionate ZD strategy. Then, irrespective of the strategies applied by the other group members (that is, in any game player i participates in),

1. Player i's realized payoff is never below the mean payoff of the other group members, $\pi_i \geq \pi_{-i}$.
2. The collective best reply for the remaining group members is to reach the threshold T. In that case, player i's payoff is strictly better than average, $\pi_i > \pi_{-i}$.

Proof. Because a player with an extortionate ZD strategy contributes strictly less than average, $x_i < x_{-i}$, it follows that either $\pi_i = \pi_{-i} = 0$ (if the group misses the target and players lose their remaining endowment) or $\pi_i > \pi_{-i}$ (otherwise). Moreover, for the other group members, it is collectively optimal to reach the target: by contributing nothing, their expected payoff becomes $(1-p)E$, whereas if they make the minimum contribution (in the first $R-1$ rounds) such that total contributions reach the target, then their payoff is $E - T/(N-1+s)$. Because $s \geq T/(pE) - (N-1)$, reaching the target is a collective best reply.

Proposition 2 is a proof-of-principle: there are strategies for the collective-risk dilemma that allow a player to extort the other group members. We note that the set of all extortionate strategies will typically be considerably bigger than the set of all extortionate ZD strategies. When we analyse experimental data, we therefore do not specifically look for strategies that have the functional form described in Equations (3) and (7); we rather look for all possible strategies that indicate extortionate behaviour (that is, we look whether players satisfy the conditions (i)–(iii) defined in the main text).

References

1. Fehr, E. & Fischbacher, U. The nature of human altruism. *Nature* **425**, 785–791 (2003).
2. Nowak, M. A. & Sigmund, K. Evolution of indirect reciprocity. *Nature* **437**, 1291–1298 (2005).
3. Peters, G. P. *et al.* The challenge to keep global warming below 2 degrees C. *Nat. Clim. Change* **3**, 4–6 (2013).
4. IPCC. *Climate Change 2014: Synthesis Report. Contribution of Working Groups I, II and III to the Fifth Assessment Report of the Intergovernmental Panel on Climate Change* (IPCC, 2014).
5. Froese, R. Fishery reform slips through the net. *Nature* **475**, 7–7 (2011).
6. Keep a welcome. *Nature* **525**, 157–157 (2015).
7. Reinders Folmer, C. P., Klapwijk, A., De Cremer, D. & Van Lange, P. A. M. One for all: what representing a group may do to us. *J. Exp. Soc. Psychol.* **48**, 1047–1056 (2012).
8. UNFCCC. *United Nations Framework Convention on Climate Change* (United Nations, 1992).
9. IPCC. in *Contribution of Working Group I to the Fourth Assessment Report of the Intergovernmental Panel on Climate Change.* (eds Solomon, Susan *et al.*) (Cambridge Univ. Press, 2007).
10. Meinshausen, M. *et al.* Greenhouse-gas emission targets for limiting global warming to 2°C. *Nature* **458**, 1158–1162 (2009).
11. Allen, M. R. *et al.* Warming caused by cumulative carbon emissions towards the trillionth tonne. *Nature* **458**, 1163–1166 (2009).
12. IPCC. in *Climate Change 2013: The Physical Science Basis. Contribution of Working Group I to the Fifth Assessment Report of the Intergovernmental Panel on Climate Change.* (eds Stocker, T. F. *et al.*) 3–29 (Cambridge Univ. Press, 2013).
13. Esslinger, D. There is currently no climate policy—because the voters do not want one (Es gibt derzeit keine Klimapolitik—weil die Wähler keine wollen). http://sz.de/1.1787481 (2013).
14. Milinski, M., Sommerfeld, R. D., Krambeck, H.-J., Reed, F. A. & Marotzke, J. The collective-risk social dilemma and the prevention of simulated dangerous climate change. *Proc. Natl Acad. Sci. USA* **105**, 2291–2294 (2008).
15. Milinski, M., Röhl, T. & Marotzke, J. Cooperative interaction of rich and poor can be catalyzed by intermediate climate targets. *Clim. Change* **109**, 807–814 (2011).
16. Tavoni, A., Dannenberg, A., Kallis, G. & Loeschel, A. Inequality, communication, and the avoidance of disastrous climate change in a public goods game. *Proc. Natl Acad. Sci. USA* **108**, 11825–11829 (2011).
17. Abou Chakra, M. & Traulsen, A. Evolutionary dynamics of strategic behavior in a collective-risk dilemma. *PLoS Comput. Biol.* **8**, e1002652 (2012).
18. Jacquet, J. *et al.* Intra- and intergenerational discounting in the climate game. *Nat. Clim. Change* **3**, 1025–1028 (2013).
19. Fischbacher, U. & Gächter, S. Social preferences, beliefs, and the dynamics of free riding in public goods experiments. *Am. Econ. Rev.* **100**, 541–556 (2010).
20. Axelrod, R. & Hamilton, W. D. The evolution of cooperation. *Science* **211**, 1390–1396 (1981).
21. Axelrod, R. The evolution of cooperation. (Basic Books, 1984).
22. Nowak, M. A. & Sigmund, K. Tit-for-tat in heterogeneous populations. *Nature* **355**, 250–253 (1992).
23. Nowak, M. A. & Sigmund, K. The alternating prisoner's dilemma. *J. Theor. Biol.* **168**, 219–226 (1994).
24. Press, W. H. & Dyson, F. J. Iterated Prisoner's Dilemma contains strategies that dominate any evolutionary opponent. *Proc. Natl Acad. Sci. USA* **109**, 10409–10413 (2012).
25. Hilbe, C., Nowak, M. A. & Sigmund, K. Evolution of extortion in iterated prisoner's dilemma games. *Proc. Natl Acad. Sci. USA* **110**, 6913–6918 (2013).
26. Stewart, A. J. & Plotkin, J. B. From extortion to generosity, evolution in the iterated prisoner's dilemma. *Proc. Natl Acad. Sci. USA* **110**, 15348–15353 (2013).
27. Chen, J. & Zinger, A. The robustness of zero-determinant strategies in iterated prisoner's dilemma games. *J. Theor. Biol.* **357**, 46–54 (2014).
28. Szolnoki, A. & Perc, M. Defection and extortion as unexpected catalysts of unconditional cooperation in structured populations. *Sci. Rep.* **4**, 5496 (2014).
29. Hilbe, C., Wu, B., Traulsen, A. & Nowak, M. A. Evolutionary performance of zero-determinant strategies in multiplayer games. *J. Theor. Biol.* **374**, 115–124 (2015).
30. Akin, E. What you gotta know to play good in the Iterated Prisoner's Dilemma. *Games* **6**, 175–190 (2015).
31. Hilbe, C., Roehl, T. & Milinski, M. Extortion subdues human players but is finally punished in the prisoner's dilemma. *Nat. Commun.* **5**, 3976 (2014).
32. Milinski, M., Semmann, D., Krambeck, H.-J. & Marotzke, J. Stabilizing the Earth's climate is not a losing game: supporting evidence from public goods experiments. *Proc. Natl Acad. Sci. USA* **103**, 3994–3998 (2006).

Acknowledgements

We thank the students for participation; H.-J. Krambeck for writing the software for the game; H. Arndt, T. Bakker, L. Becks, H. Brendelberger, S. Dobler and T. Reusch for support; and the Max Planck Society for the Advancement of Science for funding.

Author contributions

M.M. and J.M. designed the experiment; C.H. developed the mathematical model; M.M., D.S. and R.S. performed the experiments; M.M. analysed the data, M.M., J.M. and C.H. wrote the manuscript, and all authors revised the manuscript.

Additional information

Competing financial interests: The authors declare no competing financial interests.

5

The absence of an Atlantic imprint on the multidecadal variability of wintertime European temperature

Ayako Yamamoto[1] & Jaime B. Palter[1,2]

Northern Hemisphere climate responds sensitively to multidecadal variability in North Atlantic sea surface temperature (SST). It is therefore surprising that an imprint of such variability is conspicuously absent in wintertime western European temperature, despite that Europe's climate is strongly influenced by its neighbouring ocean, where multidecadal variability in basin-average SST persists in all seasons. Here we trace the cause of this missing imprint to a dynamic anomaly of the atmospheric circulation that masks its thermodynamic response to SST anomalies. Specifically, differences in the pathways Lagrangian particles take to Europe during anomalous SST winters suppress the expected fluctuations in air–sea heat exchange accumulated along those trajectories. Because decadal variability in North Atlantic-average SST may be driven partly by the Atlantic Meridional Overturning Circulation (AMOC), the atmosphere's dynamical adjustment to this mode of variability may have important implications for the European wintertime temperature response to a projected twenty-first century AMOC decline.

[1] Department of Atmospheric and Oceanic Sciences, McGill University, 805 Sherbrooke Street West, Montreal, Quebec, Canada H3A 2K6. [2] Graduate School of Oceanography, University of Rhode Island, Narragansett Bay Campus, Narragansett, Rhode Island 02882, USA. Correspondence and requests for materials should be addressed to A.Y. (email: ayako.yamamoto@mail.mcgill.ca).

Large-scale, multidecadal variability in North Atlantic sea surface temperature (SST), frequently referred to as the Atlantic Multidecadal Oscillation (AMO), is a prominent feature of Northern Hemisphere climate[1,2]: Sahel drought[3], Atlantic hurricanes[4], large-scale atmospheric circulation[2,5–7] and summertime European temperature and precipitation[8,9] all respond sensitively to this low-frequency variability in North Atlantic SST. A number of studies suggest that the cause of this SST oscillation is internal variation in ocean heat transport, possibly related to the Atlantic Meridional Overturning Circulation (AMOC) variability[10–12], with the role of external and/or atmospheric stochastic forcing provoking recent controversy[13–16]. Evidence in support of the AMO variability being driven internally comes in the form of proxy evidence of a persistent oscillation throughout the past 8,000 years[17] and the reconstruction of this mode of variability in a number of modelling studies, even in the absence of external forcing[18–20].

It is well known that the North Atlantic strongly influences western European climate, with the most obvious manifestation being the anomalous wintertime warmth of the region relative to the zonal mean at equivalent latitudes[21]. Moreover, a recent study showed that temporal variability in western European wintertime temperature is set largely by the size of the air–sea turbulent fluxes along the trajectories of Lagrangian air parcels en route to Europe[22]. Coupled with evidence that variability in air–sea heat fluxes over the Atlantic is controlled by the ocean on decadal and longer time scales[23], it is natural to expect that decadal, basin-scale SST fluctuations should translate to variability in European temperature. Indeed, in all seasons besides winter, the imprint of the AMO is evident in European temperature[8,9]. The SST anomaly associated with the AMO persists throughout the year[5,10], making the absence of a wintertime AMO signal in western Europe all the more puzzling (Fig. 1a).

We propose here that the AMO is closely associated with variability in the position and strength of the storm track, which suppresses the influence of the anomalous SST on the heat fluxes seen by Lagrangian parcels transiting to Europe. To evaluate this possibility, we examine the National Centers for Environmental Prediction/National Center for Atmospheric Research 20th-Century Reanalysis (20CR)[24] from complementary Eulerian and Lagrangian perspectives. The 20CR is one of the longest reanalysis products currently available, and is faithful with independent observations of northeast Atlantic storminess from 1940 onward[25]. Thus, we analyse storm track and Lagrangian pathway variability in the period from 1940 to 2011, which encompasses one full AMO cycle.

Results

Eulerian perspective. The AMO index, computed by taking an area-weighted mean of the linearly detrended SST field over the North Atlantic[9,26], is generally positive from 1940 to 1963 and 1996 to 2011 and negative from 1966 to 1994 (Fig. 1a). There are various approaches to defining an AMO index[27,28], yet the main features remain almost identical to those identified here regardless of the method chosen[18,20,27]. In particular, both modelling[6,10,19] and observational studies[2,6,9,11], utilizing different methodologies to isolate only the internal mode of variability, show spatial patterns of SST anomalies similar to that shown in Fig. 2a, and the approximate timing of transitions between AMO phases is not sensitive to its definition[28].

During each multidecadal period characterized by a given phase of the AMO index, short-term variability gives rise to months in which the basin-averaged SST anomaly is near zero or of the opposite sign relative to the decade in which it is embedded (Fig. 1a). To expose the association of the atmospheric anomaly pattern with SST anomalies, we made composite periods using only the January months with the most extreme AMO index. The extreme AMO months are chosen such that they meet the criteria, $|AMO index| > 0.15$, which is nearly one standard deviation beyond zero (Fig. 1a). All of these extreme months fall within a longer period where the 10-year running mean AMO index has the appropriate sign. In this manner, 17 positive and 18 negative AMO January months are selected (Fig. 1a). We repeated the analysis using all years within the corresponding AMO phase and present the results in the Supplementary Materials, where it is apparent that the key results and interpretation are essentially unchanged, although their statistical power is slightly weaker.

The large-scale atmospheric flow varies with the AMO index (Fig. 2b). The difference in the 500-hPa geopotential height (Z500) field, which is analogous to streamlines, shows that the direction of winds arriving in western Europe changes between the two AMO phases: winds are more northerly during the anomalous AMO-positive years, whereas they are more zonal during the AMO-negative years (Fig. 2b). The more tightly spaced isohypses during the AMO-negative years indicate a swifter flow relative to the AMO-positive years. Accordingly, the AMO-negative years see an elongated and more zonal January storm track (Supplementary Fig. 1), which is consistent with results from a free-running climate model[7]. Composite Z500 maps constructed with more complete sampling of the longer decadal periods associated with the AMO show similar, albeit weaker, anomaly patterns (Supplementary Fig. 2a).

Lagrangian perspective. The impact of the modulation of the large-scale atmospheric flow on temperature in western Europe is best evaluated in a Lagrangian framework, where the dynamic variability of the atmosphere and the variability in the air–sea turbulent exchange can be assessed simultaneously for the atmospheric particles that influence Europe. Therefore, we launch virtual Lagrangian particles from the surface of forty-one uniformly distributed points over land in western Europe (Fig. 1b) and track them backward in time for 10 days using the atmospheric dispersion model, FLEXPART[29] (see Methods for details). Ten atmospheric particles are released twice a day in January from 1940 to 2011 from each of the forty-one release points. A climatological two-dimensional histogram of the positions of the resulting Lagrangian particles is shown in Fig. 3a, in which their trajectories are seen spreading out over the North Atlantic, many stretching back to the Labrador Sea and northern Canada. The statistical significance of these results is strengthened when separating the particle launch locations into northern and southern sub-regions of western Europe (see Supplementary Fig. 3 for details).

Our previous work showed that air–sea turbulent fluxes at the base of the atmospheric planetary boundary layer (PBL) govern variability in the potential temperature change along these particle trajectories in January almost entirely: turbulent fluxes alone explain more than 80% of the variability in the potential temperature change along 10-day back trajectories from western Europe[22]. Although the fluxes through the top of the planetary boundary layer are important for closing the heat budget of the layer, they are not crucial for understanding the low-frequency variability of the temperature tendency along Lagrangian trajectories. Therefore, we track the ocean-atmosphere turbulent fluxes along each particle's trajectory by interpolating these fluxes from 20CR to each particle's hourly position when the particle is within PBL. The turbulent fluxes are a function of the temperature and moisture gradients at the air–sea interface and the surface wind speed (see Methods for details).

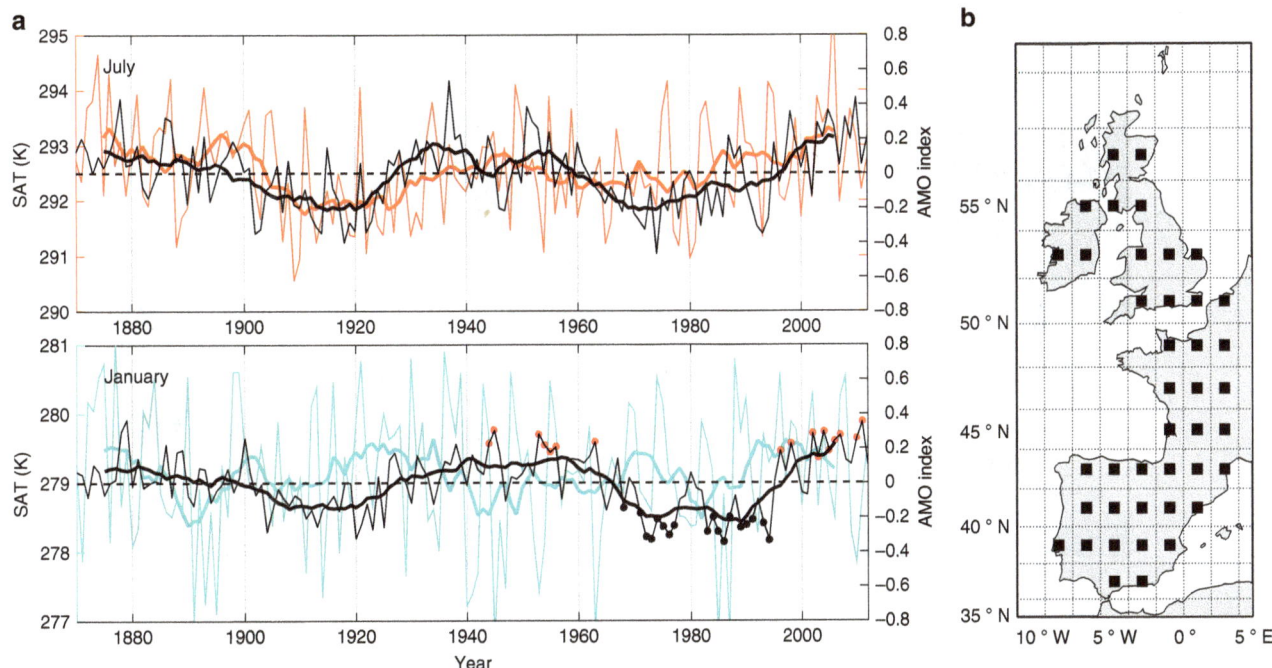

Figure 1 | Decadal variability in North Atlantic SST and western European SAT. (a) Time series of the linearly detrended North Atlantic SST (black lines, referred to as the AMO index) and SAT averaged over western Europe ([36N 60N] × [10W 3E]; shown in coloured lines) in July (top panel) and January (bottom panel). Bold lines show 10-year running means. The correlation coefficient between the 10-year running mean of the detrended SAT and AMO index is 0.61 in July (statistically significant at 10% confidence level even after accounting for the reduced effective degrees of freedom due to autocorrelation of the time series) and − 0.02 in January; these correlations are insensitive to the averaging region chosen for western Europe. The red circles on January plot indicate the AMO-positive years chosen for the composite analysis, whereas the blue circles indicate the AMO-negative years chosen. (**b**) Study region encompassing western Europe ([36N 60N] × [10W 3E]) and locations for the backtracked Lagrangian particle release (black squares).

The turbulent fluxes are calculated in 20CR with the product's wind speeds and temperature and moisture gradients, and the size of the fluxes is influenced by correlations between the wind speed and temperatures[22].

In winter in the North Atlantic, SST is almost always warmer than the surface air temperature (SAT), so the ocean loses heat rapidly to the atmosphere over the entirety of the basin (that is, positive fluxes in our convention; Fig. 3b and Supplementary Fig. 3b). The fluxes over the warm Gulf Stream and its North Atlantic Current extension are generally a factor of five higher than found elsewhere. However, a view of the fluxes weighted by the fraction of time the particles spend in each location on their journey to western Europe (Fig. 3c and Supplementary Fig. 3c) suggests a reduced role of these strong flux regions in establishing western European wintertime temperature.

The difference in the number density of the particle positions between the composite AMO periods (Fig. 3d) shows a significant distinction in the preferred pathways, with the statistical significance increasing when results are separated by particles launched from northern and southern sub-regions of western Europe (Supplementary Fig. 3d). In the AMO-positive years, particles spend more of their 10-day trajectory recirculating locally to the southwest of Iceland. During the AMO-negative years, the pathways are anomalously long, and a greater number of trajectories originate from North America and the Arctic, before transiting over the Labrador Sea and mid-latitude North Atlantic. These differences in the atmospheric trajectories are explained mechanistically by the difference in the Z500 anomalies associated with the AMO, which shows swifter, more zonal winds during AMO-negative years (Fig. 2b). This largely barotropic anomaly pattern has been noted in a number of modelling and observational studies[5,10,30,31], and is somewhat similar

to the atmospheric circulation patterns associated with the North Atlantic Oscillation (NAO)[5]. To explore whether this atmospheric anomaly pattern is linked with anomalous SST conditions of the AMO regardless of the NAO phase, we performed an additional analysis excluding the strong NAO years. This exclusion only amplifies the signal of intensified zonal flow during negative AMO years relative to positive years (c.f. Fig. 3d and Supplementary Fig. 4a). The cause of the linkage between AMO and NAO has been the subject of debate, with several papers arguing that North Atlantic SST anomalies force an atmosperic NAO response[32–34], and others arguing the reverse[12,35]. Regardless of what drives the relationship, the association between the atmospheric circulation and the AMO index is clear in the Lagrangian trajectory composites (Fig. 3d).

The difference map of turbulent fluxes along these Lagrangian trajectories points towards a leading cause of the missing AMO imprint on European wintertime temperatures. During AMO-positive years, the shorter trajectories arriving from the north and southwest (Fig. 3d and Supplementary Fig. 3d) are accompanied by high fluxes (Fig. 3e and Supplementary Fig. 3e). However, the long and zonal trajectories associated with the AMO-negative years are accompanied by even stronger turbulent fluxes over much of the mid-latitude North Atlantic. Therefore, there are partially compensating regions of elevated and depressed flux during both phases of the AMO. The same pattern is found when the difference maps are constructed from composites of the full decadal periods associated with the AMO, but with slightly weaker statistical strength (Supplementary Fig. 2b,c).

Time series constructed by averaging along Lagrangian back trajectories (Fig. 4) further reveal the net effect of the combined

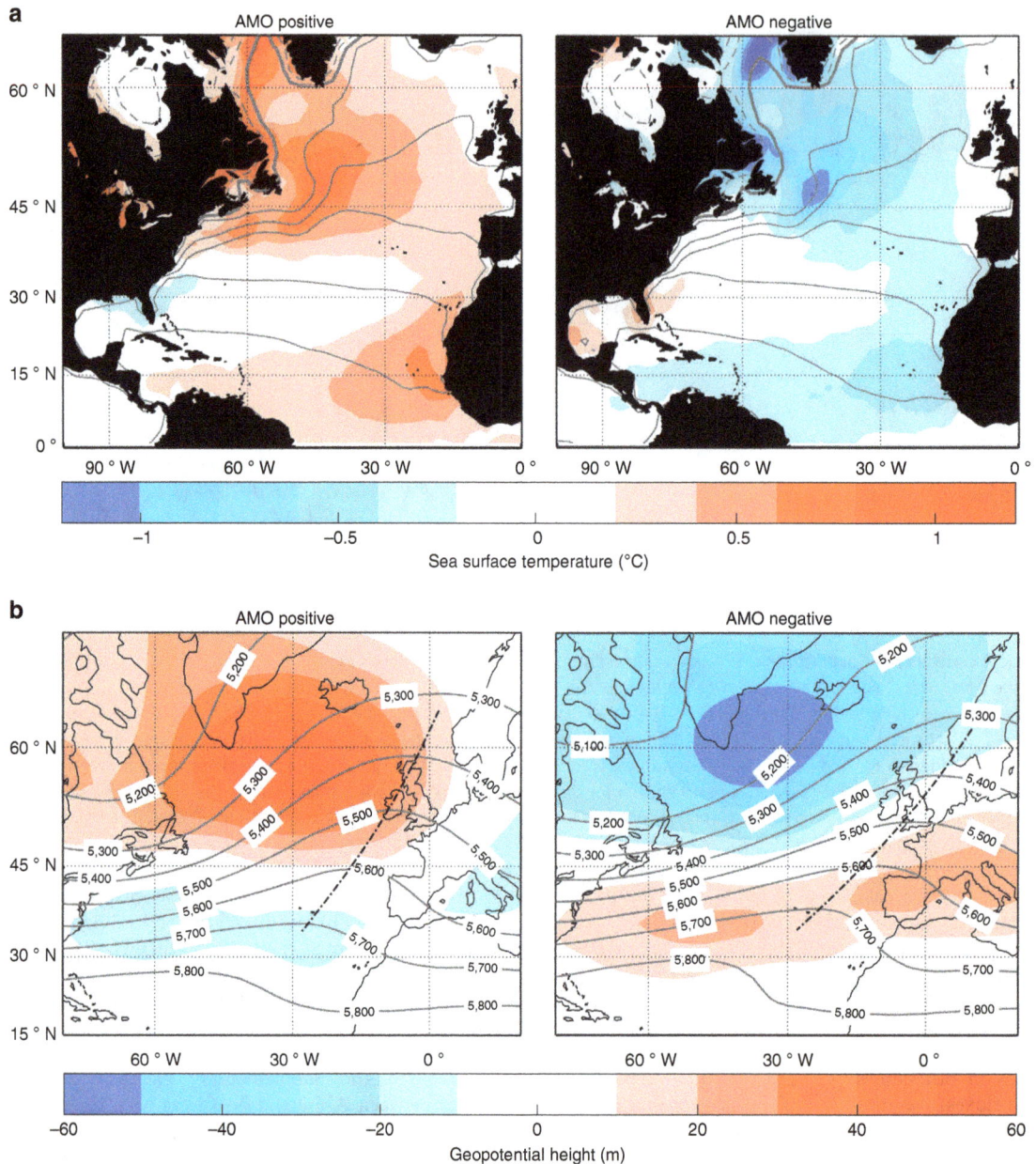

Figure 2 | The spatial pattern of the AMO index and its relationship with the atmospheric flow in January. Composite maps of (**a**) sea surface temperature (SST) field and (**b**) 500 hPa geopotential height field (Z500) for AMO anomalously positive years (left panel) and negative years (right panel). The January mean field is shown in contours, and its departure from the 72-year climatology is represented by colour shading. The thick grey contour line in **a** denotes 0 °C, whereas thin (dashed) lines denote positive (negative) SST every 5 °C. The black dashed lines in **b** are drawn through the local maxima of the geopotential height field at each latitude, which is the point where the wind changes direction from south-westerly to north-westerly.

changes to Lagrangian pathways and the properties along them. Notably, the SST sampled along the Lagrangian trajectories lacks a clear AMO signal (Fig. 4a), because decadal variability in atmospheric trajectories, which travel over a spatially variable SST field, swamps the temporal variability of the North Atlantic average SST. The SAT is highly correlated with the SST (not shown), as turbulent fluxes work to bring the surface boundary layers of the atmosphere and ocean towards equilibrium. Yet, during the negative AMO years from the mid-1970s to 1990, the air advected along Lagrangian trajectories is more anomalously cold than the SST, producing strengthened air–sea temperature gradients (Fig. 4b). Further heightened by stronger winds (Fig. 4c), the largest turbulent fluxes are achieved during these

AMO-negative years (Fig. 4d). We acknowledge that at the onset of the AMO-negative period around 1968, both SST and SAT averaged along the trajectories were elevated and fluxes were approximately average. Nevertheless, the overall effect is that the 10-year running mean turbulent fluxes sampled along Lagrangian trajectories are weakly anticorrelated with the AMO index ($r = -0.39$, not significant given the few effective degrees of freedom in the smoothed time series; Fig. 4d). We conclude that, in winter, the dynamic modulation of Lagrangian pathways and the atmospheric properties transported with them oppose the influence of basin-scale SST fluctuations on turbulent air–sea fluxes, thereby concealing the temperature expression of the AMO in atmospheric particles arriving in Europe.

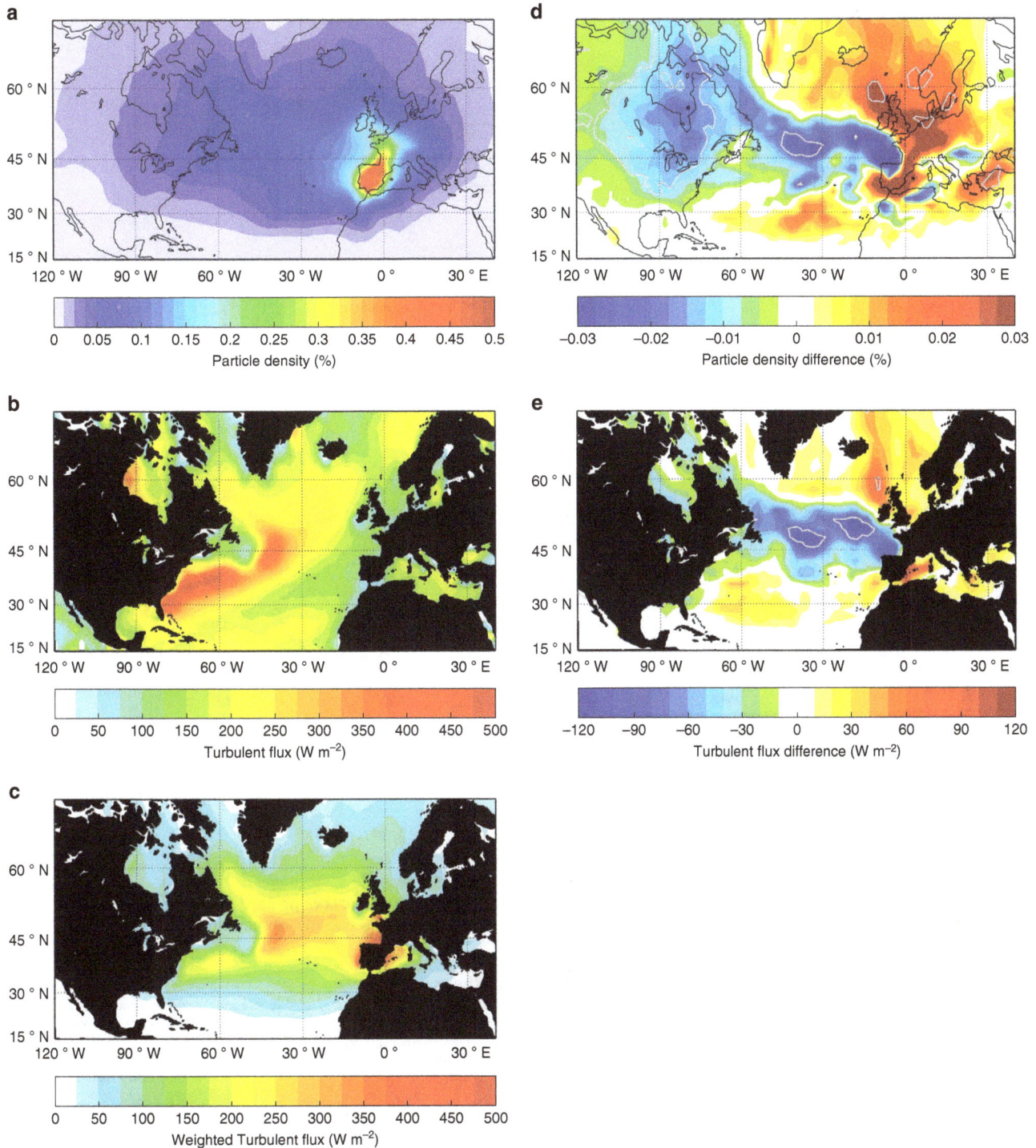

Figure 3 | Climatological mean and AMO influence on backward Lagrangian trajectories and the properties along them. (**a**) Lagrangian particle climatological number density, given as the percentage of all hourly positions that were spent in any 2° × 2° grid cell. (**b**) Climatological turbulent fluxes (sensible + latent; W m^{-2}) calculated by averaging the fluxes along the Lagrangian trajectories. (**c**) Turbulent fluxes as in **b** but weighted by the fraction of hourly particle positions spent in each grid cell, and normalized to have an equal spatial mean as the unweighted fluxes (see Methods for details; W m^{-2}). (**d**) The difference in number density for AMO-positive state minus AMO-negative state (% particle's hourly positions). (**e**) The difference in turbulent fluxes (W m^{-2}) for AMO-positive state minus AMO-negative state, weighted and normalized as in **c**. In **d**, statistical significance at 10 and 15% is shown, whereas in **e**, the 15% significance level is shown in grey contours. All the significance levels were obtained using a bootstrapping method (see Methods for details).

To further assess the degree to which the dynamic modulation of the trajectories is responsible for suppressing the AMO imprint on the fluxes, we re-ran our Lagrangian simulations with randomly selected, unvarying trajectories (see Methods for

details). The time series of the 10-year running mean conditions along these random trajectories is plotted in blue in Fig. 4. The SST (Fig. 4a) averaged along these randomly selected trajectories vary in phase with the AMO index, and the anticorrelation of the

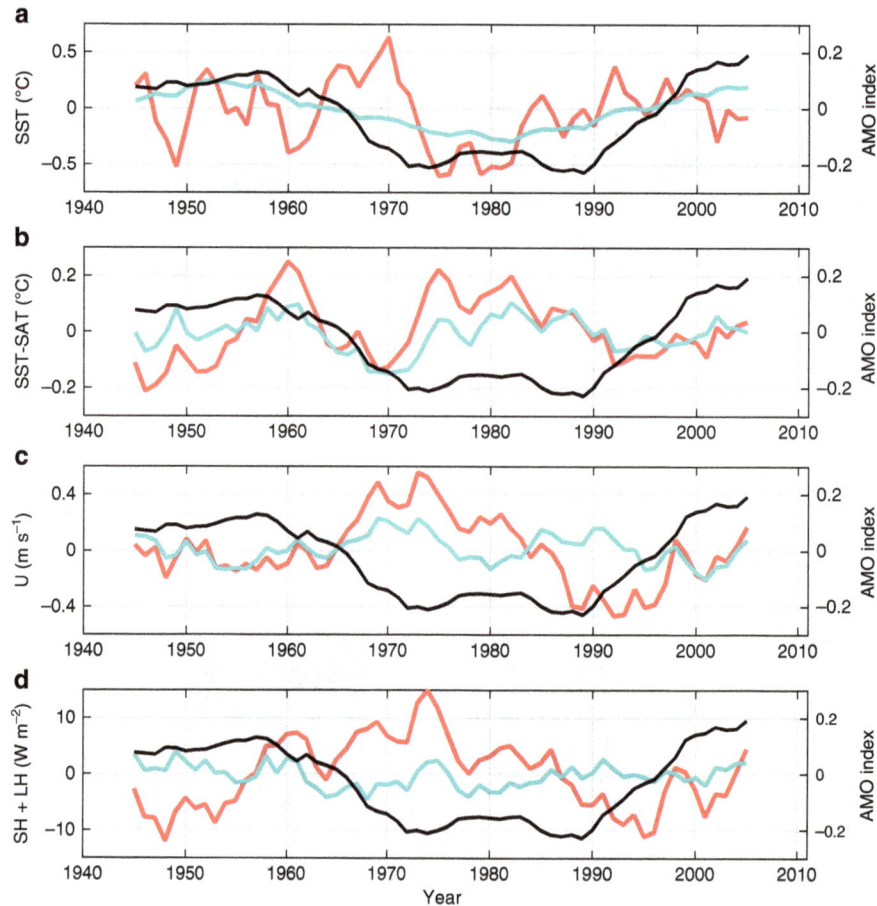

Figure 4 | Time series of properties and heat fluxes averaged along Lagrangian trajectories. 10-Year running mean of each linearly detrended variable along true trajectories (red), and along 10 sets of 10 random trajectories (blue), averaged across all 41 release locations. The January 10-year running mean AMO index is overlaid in black in each panel. (**a**) SST (°C), (**b**) SST − SAT (°C), (**c**) wind speed (m s^{-1}) and (**d**) turbulent fluxes (W m^{-2}).

fluxes with the AMO is eliminated (Fig. 4d). Finally, we confirm that, in summer, when the European temperature reflects AMO variability, the turbulent fluxes vary in phase with the AMO. In July, there is a minimal difference between the preferred pathways by AMO phase (Supplementary Fig. 5). Hence, the dominant signature of the AMO-positive phase appears to be due to higher basin-scale SST, which allows for a broad region of enhanced fluxes, consistent with an extended analysis of air–sea fluxes over the past century[23].

Discussion

The strengthening and lengthening of the storm track in sync with anomalously cooler North Atlantic SSTs has important implications for future climate. Given that decadal variability in North Atlantic SSTs may be driven partly by fluctuations in the strength of the AMOC[10-12], our result suggests the possibility of a stabilizing feedback for ocean circulation: Cooler SSTs associated with a sluggish AMOC is linked with an atmospheric adjustment that enhances turbulent heat fluxes over oceanic convective regions in winter. These larger fluxes could possibly reinvigorate convection, deep water formation and the AMOC. Moreover, the observed link of the atmospheric circulation with the cool SST anomalies of the late 1970s to early 1990s is much like the predicted change of the storm track in response to a decline of the AMOC under global warming[36]. A weakened AMOC has long been thought to cause anomalous cooling in western Europe via a decline in oceanic heat transport and associated atmospheric feedbacks[21]. However, the changes we describe here in

atmospheric Lagrangian trajectories and the heat fluxes along them could provide a mechanism that reduces the magnitude of European wintertime cooling on decadal time scales, even as they might stabilize the oceanic circulation.

Methods

FLEXPART. We adapted the Lagrangian atmospheric dispersion model FLEXPART version 9.02 (ref. 29) for use with the Twentieth Century Atmospheric Reanalysis product (20CR)[24] in order to simulate the atmospheric particles released from 41 equally spaced western European locations (Fig. 1b). These release points are chosen from an evenly spaced 2° × 2° grid over the study region [36N 60N] × [10W 3E], when these points fall on land. Every January from 1940 to 2011, ten particles are released from the surface of each of these points twice daily at 0 coordinated universal time (UTC) and 12 UTC and advected backward in time for the duration of 10 days, following the three-dimensional wind field. There are three components to this wind field: (i) resolved wind, (ii) turbulent wind fluctuations and (iii) mesoscale wind fluctuations. FLEXPART accounts for the turbulent wind fluctuations by adding a perturbation to the velocity field for atmospheric particles in the PBL, where these random motions are calculated by solving Langevin equations for Gaussian turbulence. Mesoscale velocity, whose spectral interval falls between the resolved flow and the turbulent flow, is included by solving an independent Langevin equation. The PBL height is diagnosed at each particle's hourly position. The duration of 10 days for the back trajectories was chosen based on the fact that the Lagrangian decorrelation time scale is ∼3 days[22]; thus, the choice of 10 days is long enough for the memory of each particle's initial temperature to be erased under the effect of diabatic processes along the trajectory.

20CR is one of the longest reanalysis products currently available. It has 6-h temporal resolution and 2° × 2° spatial resolution. The product assimilates only observations of surface pressure, monthly SST and sea-ice distributions, and we only use the ensemble mean fields. To assess the reliability of our FLEXPART results with use of 20CR, the trajectories computed using 20CR was compared with those with a default input for FLEXPART, Climate Forecast System Reanalysis

(CFSR) forecast and reanalysis data sets from National Centers for Environmental Prediction[37], which has hourly temporal and $0.5° \times 0.5°$ spatial resolution, for the period of 1981–2009 under the same set up as Yamamoto et al.[22] We found that the trajectory paths computed with 20CR are generally very close to those computed with CFSR especially over the ocean, with particles in 20CR taking slightly more northern paths relative to CFSR (Supplementary Fig. 6). We note that 20CR assimilates monthly mean SST data, whereas CFSR assimilates SST every 6 h. The agreement of the amplitude and variability of the turbulent fluxes along Lagrangian trajectories constructed from the two reanalysis products (Supplementary Fig. 7) suggests that the missing sub-monthly SST variability in 20CR has a minimal impact on these fluxes on interannual and longer time scales.

Bootstrapping. Bootstrapping was used in order to gauge the statistical significance of the difference in the spatial patterns of Lagrangian trajectory pathways and the fluxes along the trajectories for the two AMO phases (Fig. 3d,e). We sample the Lagrangian particle trajectories or the fluxes along them from randomly selected January months ('pseudo-periods') for the same number of years as each AMO phase (17 years for AMO positive and 18 years for AMO negative, respectively). We then take a difference between the composite pseudo-periods. This operation was repeated 500 times. We consider differences of the Lagrangian particle density and the fluxes along them from the true AMO composites significant at the 10% (15%) level when this true difference exceeds the 90th (85th) percentile of the pseudo-period differences. We repeated the same procedure to make the composites with entire AMO periods (40 AMO positive years and 29 AMO negative years) and show these results in the Supplementary Materials.

Surface fluxes along the trajectories. Along the particle trajectories simulated using FLEXPART, the surface turbulent fluxes are interpolated from 20CR sensible heat (SH) and latent heat flux (LH) fields, whenever the particle's hourly position falls within PBL, under the assumption that the turbulent fluxes influence the entire air mass within the PBL. The turbulent fluxes in 20CR are computed using bulk formulae with a typical formulation[38]:

$$SH = \rho_a c_p C_h U(T_s - T_a) \qquad (1)$$

$$LH = \rho_a L_e C_e U(q_s - q_a) \qquad (2)$$

where ρ_a is the atmospheric density, c_p is the atmospheric heat capacity, C_h and C_e are transfer coefficients, U is the mean value of wind speed relative to the surface ocean current, T_s is sea surface temperature, T_a is the atmospheric potential temperature at a reference height, L_e is latent heat of evaporation, q_s is interfacial value of water vapour mixing ratio, and q_a is the atmospheric water vapour mixing ratio at a reference height.

Time series of the mean accumulated surface heat fluxes along the trajectories using 20CR are highly correlated with those using CFSR (Supplementary Fig. 7), with the average correlation coefficient being $r = 0.92$.

Weighting of composite fluxes. In the composite figures of weighted surface fluxes along the trajectories (Fig. 3c,e), weights are proportional to the fraction of all particle positions that pass over a each $2° \times 2°$ grid cell (that is, the number density, given as a percentage in Fig. 3a). These weights are scaled so that the mean value of the climatological map (Fig. 3c) is equal to the mean of the unweighted climatology (Fig. 3b). The spatial mean of both mapped fields used in this scaling includes only those grid cells visited by at least 0.01% of the climatological particle positions; these collectively contain 90.1% of all hourly particle positions.

Randomly selected trajectories. The unvarying particle trajectory pathways used to produce Fig. 4 (blue lines) were chosen by randomly selecting ten particles from the set of all possible pathways generated from the full 72-year Lagrangian simulation. The truly varying surface fluxes are interpolated along these random trajectories. We then repeat this process ten times, each time by picking a different random set of ten trajectories from each release point, thereby creating a spread of particle positions. In total, 100 particles (10 particles $\times$ 10 realizations) are selected for each release location.

References

1. Schlesinger, M. E. & Ramankutty, N. An oscillation in the global climate system of period 65–70 years. *Nature* **367**, 723–726 (1994).
2. Kushnir, Y. Interdecadal variations in North Atlantic sea surface temperature and associated atmospheric conditions. *J. Clim.* **7**, 141–157 (1994).
3. Rowell, B. D. P., Folland, C. K., Maskell, K. & Ward, M. N. Variability of summer rainfall over tropical north Africa (1906-92): Observations and modelling. *Quat. J. R. Meteorol. Soc.* **121**, 669–704 (1995).
4. Goldenberg, S. B., Landsea, C. W., Mestas-Nunez, A. M. & Gray, W. M. The recent increase in Atlantic hurricane activity: causes and implications. *Science (New York, N.Y.)* **293**, 474–479 (2001).
5. Sutton, R. T. & Hodson, D. L. R. Influence of the ocean on North Atlantic climate variability 1871-1999. *J. Clim.* **16**, 3296–3313 (2003).
6. Gastineau, G., D'Andrea, F. & Frankignoul, C. Atmospheric response to the North Atlantic Ocean variability on seasonal to decadal time scales. *Clim. Dyn.* **40**, 2311–2330 (2012).
7. Zhang, R. & Delworth, T. L. Impact of the Atlantic Multidecadal Oscillation on North Pacific climate variability. *Geophys. Res. Lett.* **34**, 2–7 (2007).
8. Arguez, A., O'Brien, J. J. & Smith, S. R. Air temperature impacts over Eastern North America and Europe associated with low-frequency North Atlantic SST variability. *Int. J. Climatol.* **10**, 1–10 (2009).
9. Sutton, R. T. & Dong, B. Atlantic Ocean influence on a shift in European climate in the 1990s. *Nat. Geosci.* **5**, 788–792 (2012).
10. Delworth, T. L. & Mann, M. E. Observed and simulated multidecadal variability in the Northern Hemisphere. *Clim. Dyn.* **16**, 661–676 (2000).
11. Latif, M., Roeckner, E., Botzet, M. & Esch, M. Reconstructing, monitoring, and predicting multidecadal-scale changes in the North Atlantic thermohaline circulation with sea surface temperature. *J. Clim.* **17**, 1605–1614 (2004).
12. McCarthy, G. D., Haigh, I. D., Hirschi, J. J.-M., Grist, J. P. & Smeed, D. A. Ocean impact on decadal Atlantic climate variability revealed by sea-level observations. *Nature* **521**, 508–510 (2015).
13. Otterå, O. H., Bentsen, M., Drange, H. & Suo, L. External forcing as a metronome for Atlantic multidecadal variability. *Nat. Geosci.* **3**, 688–694 (2010).
14. Booth, B. B. B., Dunstone, N. J., Halloran, P. R., Andrews, T. & Bellouin, N. Aerosols implicated as a prime driver of twentieth-century North Atlantic climate variability. *Nature* **484**, 228–232 (2012).
15. Zhang, R. et al. Have Aerosols Caused the Observed Atlantic Multidecadal Variability? *J. Atmos. Sci.* **70**, 1135–1144 (2013).
16. Clement, A. et al. The Atlantic Multidecadal Oscillation without a role for ocean circulation. *Science* **350**, 320–324 (2015).
17. Knudsen, M. F., Seidenkrantz, M.-S., Jacobsen, B. H. & Kuijpers, A. Tracking the Atlantic Multidecadal Oscillation through the last 8,000 years. *Nat. Commun.* **2**, 178 (2011).
18. Ting, M., Kushnir, Y., Seager, R. & Li, C. Forced and Internal Twentieth-Century SST Trends in the North Atlantic. *J. Clim.* **22**, 1469–1481 (2009).
19. Ting, M., Kushnir, Y., Seager, R. & Li, C. Robust features of Atlantic multidecadal variability and its climate impacts. *Geophys. Res. Lett.* **38**, L17705 (2011).
20. DelSole, T., Tippett, M. K. & Shukla, J. A Significant Component of Unforced Multidecadal Variability in the Recent Acceleration of Global Warming. *J. Clim.* **24**, 909–926 (2011).
21. Palter, J. B. The Role of the Gulf Stream in European Climate. *Annu. Rev. Marine Sci.* **7**, 113–137 (2015).
22. Yamamoto, A., Palter, J. B., Lozier, M. S., Bourqui, M. S. & Leadbetter, S. J. Ocean versus atmosphere control on western European wintertime temperature variability. *Clim. Dyn.* **45**, 3593–3607 (2015).
23. Gulev, S. K., Latif, M., Keenlyside, N., Park, W. & Koltermann, K. P. North Atlantic Ocean control on surface heat flux on multidecadal timescales. *Nature* **499**, 464–467 (2013).
24. Compo, G. P. et al. The Twentieth Century Reanalysis Project. *Quat. J. R. Meteorol. Soc.* **137**, 1–28 (2011).
25. Krueger, O., Schenk, F., Feser, F. & Weisse, R. Inconsistencies between Long-Term Trends in Storminess Derived from the 20CR Reanalysis and Observations. *J. Clim.* **26**, 868–874 (2013).
26. Enfield, D. B., Mestas-Nunez, A. M. & Trimble, P. J. The Atlantic multidecadal oscillation and its relation to rainfall and river flows in the continental U.S. *Geophys. Res. Lett.* **28**, 2077–2080 (2001).
27. Trenberth, K. E. & Shea, D. J. Atlantic hurricanes and natural variability in 2005. *Geophys. Res. Lett.* **33**, L12704 (2006).
28. Nigam, S., Guan, B. & Ruiz-Barradas, A. Key role of the Atlantic Multidecadal Oscillation in 20th century drought and wet periods over the Great Plains. *Geophys. Res. Lett.* **38**, L16713 (2011).
29. Stohl, A., Forster, C., Frank, A., Seibert, P. & Wotawa, G. Technical note: The Lagrangian particle dispersion model FLEXPART version 6.2. *Atmos. Chem. Phys.* **5**, 2461–2474 (2005).
30. Ting, M., Kushnir, Y. & Li, C. North Atlantic Multidecadal SST Oscillation: External forcing versus internal variability. *J. Marine Sys.* **133**, 27–38 (2014).
31. Gastineau, G. & Frankignoul, C. Influence of the North Atlantic SST on the atmospheric circulation during the twentieth century. *J. Clim.* **28**, 1396–1416 (2015).
32. Frankignoul, C., Chouaib, N. & Liu, Z. Estimating the observed atmospheric response to SST anomalies: maximum covariance analysis, generalized equilibrium feedback assessment, and maximum response estimation. *J. Clim.* **24**, 2523–2539 (2011).
33. Peings, Y. & Magnusdottir, G. Forcing of the wintertime atmospheric circulation by the multidecadal fluctuations of the North Atlantic ocean. *Environ. Res. Lett.* **9**, 034018 (2014).
34. Omrani, N.-E., Keenlyside, N. S., Bader, J. & Manzini, E. Stratosphere key for wintertime atmospheric response to warm Atlantic decadal conditions. *Clim. Dyn.* **42**, 649–663 (2014).

35. Mecking, J. V., Keenlyside, N. S. & Greatbatch, R. J. Stochastically-forced multidecadal variability in the North Atlantic: a model study. *Clim. Dyn.* **43**, 271–288 (2014).

36. Woollings, T., Gregory, J. M., Pinto, J. G., Reyers, M. & Brayshaw, D. J. Response of the North Atlantic storm track to climate change shaped by ocean—atmosphere coupling. *Nat. Geosci.* **5**, 313–317 (2012).

37. Saha, S. *et al.* The NCEP climate forecast system reanalysis. *Bull. Amer. Meteor. Soc.* **91**, 1015–1057 (2010).

38. Fairall, C. W., Bradley, E. F., Rogers, D. P., Edson, J. B. & Young, G. S. Bulk parameterization of air-sea fluxes for Tropical Ocean-Global Atmosphere Coupled-Ocean Atmosphere Response Experiment difference relative analysis. *J. Geophys. Res.* **101**, 3747–3764 (1996).

Acknowledgements

This work was supported by a funding from the McGill University, the NSERC Discovery Program, FQRNT's Programme Etablissement de Nouveaux Chercheurs Universitaires and Quebec-Ocean. We thank three anonymous reviewers and E. Galbraith and M. Gervais for their constructive comments that greatly improved earlier versions of this manuscript.

Author contributions

A.Y. and J.B.P. designed the experiments, analysed the results and wrote the manuscript together. A.Y. adapted FLEXPART for 20CR input and performed the experiments.

Additional information

Competing financial interests: The authors declare no competing financial interests.

6

Ice-sheet-driven methane storage and release in the Arctic

Alexey Portnov[1], Sunil Vadakkepuliyambatta[1], Jürgen Mienert[1] & Alun Hubbard[1]

It is established that late-twentieth and twenty-first century ocean warming has forced dissociation of gas hydrates with concomitant seabed methane release. However, recent dating of methane expulsion sites suggests that gas release has been ongoing over many millennia. Here we synthesize observations of ∼1,900 fluid escape features—pockmarks and active gas flares—across a previously glaciated Arctic margin with ice-sheet thermomechanical and gas hydrate stability zone modelling. Our results indicate that even under conservative estimates of ice thickness with temperate subglacial conditions, a 500-m thick gas hydrate stability zone—which could serve as a methane sink—existed beneath the ice sheet. Moreover, we reveal that in water depths 150–520 m methane release also persisted through a 20-km-wide window between the subsea and subglacial gas hydrate stability zone. This window expanded in response to post-glacial climate warming and deglaciation thereby opening the Arctic shelf for methane release.

[1]CAGE—Centre for Arctic Gas Hydrate, Environment and Climate, Department of Geology, UiT The Arctic University of Norway, 9037 Tromsø, Norway. Correspondence and requests for materials should be addressed to A.P. (email: portnovalexey@gmail.com).

atural gas can exist in solid form of crystalline ice-like structures known as gas hydrates that are stable within the subsurface under high-pressure and low-temperature conditions bounded by the gas hydrate stability zone (GHSZ). The kinetics of hydrate formation and dissociation also critically depends on the supply and composition of gas and liquid water within available pore space of sediments, hence even under an appropriate envelope of GHSZ pressure and temperature conditions, gas hydrates are not, *per se*, guaranteed[1]. However, wherever persistent subsurface methane (or heavier fractions of natural gas) and water coexist within available pore space, then the GHSZ is a robust indication of the conditions under which gas hydrate is likely to form. The present distribution and stability of gas hydrates beneath oceans and permafrost, along with their potential to release large fluxes of methane and other potent greenhouse gases, are fundamental to determining long-term atmospheric composition and its impact on climate change. Previous research reveals that subglacial soils, lakes, peatlands and marine sediments can store significant reserves of carbon within a GHSZ beneath the palaeo-ice sheets that covered North America and also beneath the Antarctic ice sheet today[2,3]. It has been argued that as a result of active methanogenesis, this significant carbon pool could provide a major contribution to global atmospheric methane emissions and composition following deglaciation of Antarctica[4]. However, to date, few studies have investigated how gas hydrates responded to past climate change—specifically—the impact of extensive ice-sheet expansion on gas hydrate stability and dissociation during the last glaciation. In particular, three leading research questions warrant attention: how did the subglacial footprint of the former ice-sheet affect the GHSZ? How could this GHSZ govern methane storage and release across the glaciated margin? How could post-glacial ice-sheet retreat impact on this former subglacial gas hydrate reservoir?

Persistent and extensive discharge of methane gas flares into the water column offshore of Prins Karls Forland (PKF), western Svalbard has been reported since they were first observed in 2008 (refs 5,6). More than 1,000 individual gas flares, predominantly ejecting methane (C_1 ~98.9–99.9%) from known hydrocarbon sources[7] cluster across a broad zone of the seabed between 80 and 420 m depth[8] (Fig. 1). Gas flares can be grouped in two distinct sets based on depth: a deep zone that spans 380–420 m below sea level (m.b.s.l.) and a shallow zone between 80 and 130 m.b.s.l. Deeper gas flares can theoretically be associated with the base of the present-day GHSZ, which pinches out on the seabed at 396 m water depth[6]. It has been argued that a temperature increase of 1 °C within the West Spitsbergen Current bottom water and/or annual fluctuations of 0.6–4.9 °C has been sufficient to force a downslope migration of the GHSZ in this area from 360 to 396 m.b.s.l. over the past three decades[5,6]. Despite a lack of records of vertical fluid flow in the study area, U/Th isotope analyses on authigenic carbonate records reveal that there has been significant methane flow-induced precipitation at the deeper gas flares sites since at least 3 ka (ref. 5). In contrast, shallow gas flares cluster across the main ridge of the Forlandet moraine complex[9] (Fig. 1). The region is characterized by a sediment blanket that diminishes from several hundred metres over the continental slope to a minimum of a few tens of metres on the shelf[10,11]. The tectonically induced Forlandsundet graben (Fig. 1) is infilled with several kilometres thick sediment section[12,13], and hence has major potential to host free gas. Sediment thickness to the southeast at Isfjorden varies up to ~100 m (ref. 14). Hydrocarbon source rocks are extensive in the Svalbard region, including Triassic and Early Jurassic formations[7], as well as organic-rich Miocene deposits[15]. The major hydrocarbon reservoirs within our study area are concentrated within Early and Middle Triassic source sequences[7]. Natural gas including thermogenic methane—the lightest hydrocarbon fraction—is generated under high pressure and

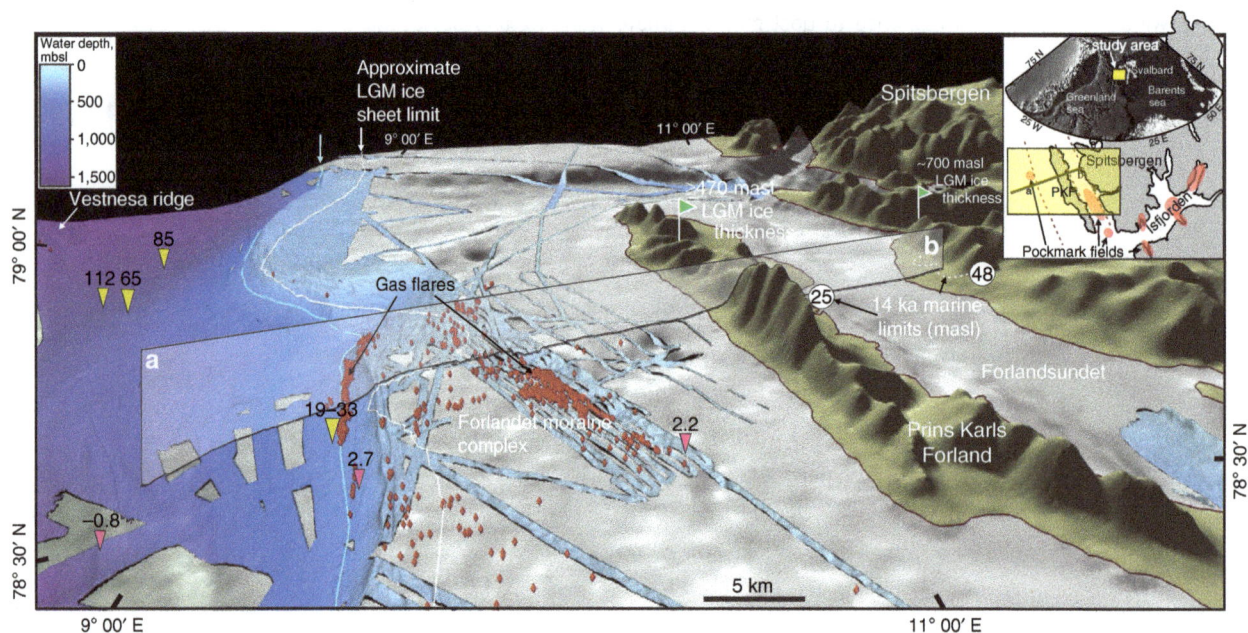

Figure 1 | The West Svalbard shelf at present. The western Svalbard margin (IBCAO v.3 (ref. 51) in grey and high-resolution multibeam data in the blue scale—see Methods) showing the observational compilation used in the modelling experiments along the transect a to b (semi-transparent vertical curtain). Yellow and red triangles show geothermal temperature gradients (19–112 °C km^{-1})[5,35,36] and average long-term bottom water temperatures (−0.8 to 2.2 °C)[32], respectively, which were used to constrain the LGM boundary conditions along the transect. Also indicated are the LGM marine limits[25,28], used to estimate isostatic loading (25 and 48 m in white rounds), minimum ice-surface elevation[21] (green flags), location of modern GHSZ, approximate LGM ice-sheet limit and modern gas flare locations (see Methods). Inset is the location of the study area in respect to pockmark fields (red ovals) and major tectonic lineaments of Hornsund fracture zone (dashed brown lines) across the western Svalbard margin.

temperatures up to 200 °C at depth within sedimentary basins from a mixture of insoluble organic compounds known as kerogen[16]. In 1992, during exploratory drilling in Svalbard, gas blow outs from depths of 630 m beneath today's sediment surface brought operations to a complete halt[17].

It is well established that the ultimate stable ice-sheet stand across the West Spitsbergen shelf was concurrent with maximum thickness and horizontal extent (representing the Last Glacial Maximum (LGM) stage) and persisted for at least 5 kyr (20–15 ka)[18,19]. Onshore and offshore radiocarbon dating[20], cosmogenic [10]Be surface exposure dating[20,21], numerical modelling[22], offshore high-resolution multibeam[23] and seismic surveys[9] reveal a LGM sequence that extended across the continental margin to the shelf break. By inference, the ice sheet in this sector was cold based and covered the Spitsbergen continental shelf to a distance of ∼45 km offshore (∼150 m.b.s.l.; Fig. 1). Field evidence further reveals that the ice surface was at least ≥473 m above the present sea level (m.a.s.l.) as determined from the [10]Be exposure age of the boulders over the PKF, and ∼700–900 m.a.s.l. over the west Spitsbergen margin[21,22] (Fig. 1). Wet-based, fast-flowing ice streams discharged across the shelf from Kongsfjorden and Isfjorden, and bounded the cold-based ice lobe that flowed across our study area and west of PKF[24].

Radiocarbon (^{14}C) dating of whale bones and mollusk shells sampled on the raised beaches define the earliest post-LGM marine limits on the northern PKF (25 m.a.s.l.) and western Spitsbergen (45–48 m.a.s.l.)[25,26] (Fig. 1). The resulting sea level curves define a local glacio-isostatic loading scenario that was at least 100 m at PKF and western Spitsbergen. The time scale for ice-sheet stabilization at its LGM stand and subsequent retreat was 20–15 ka in this region of western Svalbard, followed by complete deglaciation of the continental margin that was complete by 12–10 ka (ref. 24). This glacial chronology is consistent with an initial phase of relatively slow post-glacial isostatic rebound of 1.5–5 m ka[−1] for West and North Svalbard that commenced at ∼13–12 ka, followed by an episode of accelerated uplift (15–30 m ka[−1]) between 10.5 and 9 ka (ref. 27).

In this study, we model the impact of the paleo-ice sheet on the GHSZ offshore of western Svalbard by integrating geophysical mapping with the glacial geology. Our results reveal a potentially large subglacial gas hydrate reservoir that accumulated under high-pressure/low-temperature LGM conditions and that would have subsequently dissociated, releasing methane during deglaciation and marine incursion.

Results

Modelling predicts thick GHSZ for 20 ka. The GHSZ model applied here uses a reconstructed LGM ice-sheet configuration, constrained by the available empirical evidence as described above[9,23,28]. Field data provide minimum constraints on ice-sheet thickness, isostatic loading and offshore extent to provide boundary conditions for two-dimensional, steady-state ice-sheet modelling under a perfect-plastic ice-flow assumption[29,30] (Figs 1 and 2; see Methods). This LGM, cold-based ice-sheet reconstruction attained ∼700-m thickness over Spitsbergen and thinned westward where it terminated into the ocean. We set the sea level datum to conform to the global eustatic sea level curve[31], with zero-elevation aligned to its 20 ka level (120 m below the present; Fig. 2). Observations indicate significant subsidence of subglacial ground surface, accruing eastward and attaining ∼128 m under the West Spitsbergen margin (see Methods). LGM bottom water temperatures range from +2.3 to +3.8 °C from West to East along the transect (Fig. 2; Supplementary Fig. 1). These temperatures are ∼1 °C higher than the present-day average bottom water temperatures[32] due to episodes of intensive Atlantic Water inflow and cold freshwater discharge from glaciated margins[33,34]. The juncture of the ocean–ice sheet interface critically controls the ground surface temperature change between the submarine and subglacial environments. It abruptly decreases from +3.5 °C within the ocean to −4.5 °C within the subglacial environment and consequently impacts significantly on the temperature distribution in the lower subsurface (Fig. 2; Supplementary Fig. 1).

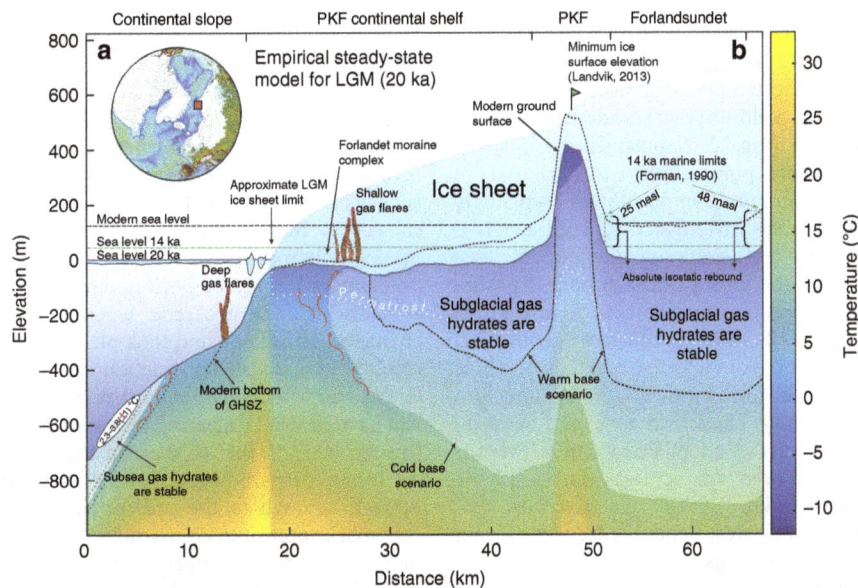

Figure 2 | Empirical steady-state model for the Last Glacial Maximum. Modelled for LGM (20 ka) ice-sheet and gas hydrate reconstruction showing both subsea and subglacial GHSZ in a minimum cold-based and warm-based ice-sheet scenario. The elevation datum is set at the LGM sea level (−120 m compared with present). Black dashed line shows the base of GHSZ in a warm-based ice-sheet scenario (pressure melting point temperature along the ice-sheet bed surface). Red and blue dashed lines show the shift of subsea GHSZ under ±1 °C bottom water temperature deviation. Gas-hydrate-free window existed between subsea and subglacial GHSZ during the LGM within the upper margin, leaving the potential for methane release (red arrows). Present-day seabed gas flares are confined to modern and ancient subsea and subglacial GHSZ pinch-out areas.

These LGM conditions, combined with present-day heat-flow measurements[5,35,36], were coupled with a thermal diffusion model to yield the two-dimensional distribution of subsurface temperature during the LGM. This is subsequently used to model an extensive subsea and subglacial GHSZ with a thickness in excess of 800 m below Forlandsundet that declines westward to where it abruptly tapers-out under the Forlandet moraine complex (Fig. 2). In this configuration, the subglacial GHSZ system resembles its deep offshore analogue, but with the ice sheet providing the high-pressure- and low-temperature loading conditions. Approximately, 20 km westward from where the subglacial GHSZ terminates, the subsea GHSZ pinches out across the seabed at ∼400 m water depth (∼520 m below the present sea level; Fig. 2). Here and in contrast to the subglacial situation, the subsea GHSZ attained a maximum thickness of ∼160 m beneath the western and deepest submerged part of the transect (Fig. 2).

Modelling also reveals an upper margin window for potential methane release since the subsea and subglacial GHSZ configurations leave the ice-sheet margin with a GHSZ-free region. Hence, continuous gas release was possible, indeed, likely across the upper continental slope during the LGM, and, potentially, throughout the last glacial cycle. A cold-based ice-sheet scenario that also agrees with available field data predicts at least 100 m of thick subglacial permafrost above a 0 °C isotherm, abruptly terminating under the shelf break due to the thermal diffusion beneath the subglacial/subsea interface (Fig. 2).

GHSZ is sensitive to environmental changes. The GHSZ modelling presented here is bounded by the subglacial and subsea temperature distributions associated with an optimal LGM ice-sheet reconstruction constrained by the glacial geological record and cosmogenic dating. We investigate the sensitivity of the modelled GHSZ to alteration of these boundary conditions by, first, changing the subglacial thermal conditions from a cold to a warm-based scenario and, second, by increasing the ice-sheet reconstruction from a minimum thickness in accordance to available field data, to a maximum profile.

The transition from a cold to temperate subglacial conditions (defined as being at pressure melting point) leads to a twofold decrease in GHSZ thickness (Fig. 2; Supplementary Fig. 1). Moreover, its surface exposure point shifts eastward ∼3 km across the modern Forlandet moraine complex. Model sensitivity to oceanic thermal conditions was investigated through a ±1 °C deviation of the bottom water temperature profile and reveals a reciprocal ∼±1.5 km shift of the subsea GHSZ surface exposure point along the continental slope, fixing it between 390 and 450 m below LGM sea level (Fig. 2). The subsurface geothermal heat-flow pattern is interpolated from a sparse array of borehole measurements from across the region (see Methods) and also represents a source of uncertainty. The sensitivity experiments do though confirm that the approach we adopt along with our results are robust and indicate that even an increase in lithospheric temperature gradients by 10 °C km^{-1} shifts the base of GHSZ upwards by some 100 m at most and hence, does not substantively detract from our findings.

The ice-sheet surface profile that optimally concurs with the available field evidence is achieved with a yield strength for perfect-plastic ice-sheet flow of 40 kPa (ref. 29). An ice-sheet reconstruction with such a low-yield strength is glaciologically valid[30] and within bounds of published values under temperate subglacial conditions provides conservative, end-member, boundary conditions for pressure and thermal loading of the subglacial GHSZ during the LGM (that is, it yields the thinnest ice-sheet profile with warmest basal temperatures). Yield stress, if increased to a more typical value of 80 kPa for a cold-based ice sheet[29,30] yields an increase both in ice thickness to a maximum of 950 m and associated subglacial GHSZ to ∼900 m beneath Forlandsundet (Supplementary Fig. 1). Under this maximum ice thickness reconstruction, the surface exposure point of subglacial GHSZ shifts ∼2.5 km westward, reflecting the steeper ice profile to the margin.

Discussion

Post-LGM eustatic sea level rise shifted the subsea GHSZ ∼6.5 km laterally up the continental slope, where it attained its present location at 396 m.b.s.l. and is, today, marked by abundant gas flares that align parallel to the slope (Figs 1 and 2). We hypothesize that the lateral displacement of subsea GHSZ developed in parallel with the migration of gas-saturated fluids along the base of GHSZ (Fig. 2). This would impact on the formation of multiple fluid flow features within the glacio-marine sedimentary sequence as observed below the modern subsea GHSZ in recent high-resolution seismic data[11,37].

We collate over 1,000 pockmarks and active gas flares across the shallow West Spitsbergen shelf and fjord environments (Fig. 3)[14,23,38]. Gas from deeper hydrocarbon sources is likely to migrate along the tectonic lineaments. Alternatively, release from thawing relic permafrost also should be considered[8]. However, 546 of 1,304 individual pockmarks studied in Isfjorden are confined to thrust faults, indicating focused fluid flow migration[38]. The present-day gas discharge across the shelf west of PKF is coincident with the exposure point of subglacial GHSZ modelled here (Figs 1 and 2). This apparent match between our modelled GHSZ using a range of LGM ice-sheet scenarios and the incidence of observed gas flares is a robust yet paradoxical coincidence, since the subglacial GHSZ might have been expected to follow the post-LGM pattern of deglaciation. However, at present, there is no convincing observational evidence for pockmark and related gas flare activity that synchronously shifted eastward with on-going ice retreat and deglaciation. During its LGM stand, the West Spitsbergen ice sheet provided continuous high-pressure and thermal subglacial conditions for a period of at least 5 kyr (ref. 18) —sufficient time for the establishment of an extensive and persistent GHSZ that hence had the potential for the formation of gas hydrates in regions with favourable geological setting[39]. On deglaciation, the former subglacial environment associated with the ice sheet was subject to marine incursion and became exposed to shallow marine environments, accompanied by decreased hydrostatic pressure and increased bottom temperatures related to the inflow of ∼2–3 °C warm Arctic water. This major environmental change was on-going and took place within 7 kyr (refs 19,24) and resulted in a transient reduction of the previous thick subglacial GHSZ and associated permafrost layer. As a result, this sequence will have enabled triggering of seabed release of methane formerly trapped beneath the ice sheet in the form of gas hydrate, fuelled by deep hydrocarbon reservoirs that continued to drive methane upward through to the present day[40,41]. Active fluid flow and pockmark footprints across the vast area of western Spitsbergen margin have been mapped outside and inside the fjords (Fig. 3), coincident with known deeper petroleum reservoirs[7]. Our post-glacial scenario for natural greenhouse gas release is supported by age constrains for Isfjorden pockmarks, which have been dated as post-glacial, and agree with the age estimation for Grønfjorden (southernmost branch of Isfjorden) pockmarks of ∼11.3 ka (ref. 14) (Fig. 3). Such a geological setting appears to be characteristic for extensive post-LGM seabed gas escape across the Arctic

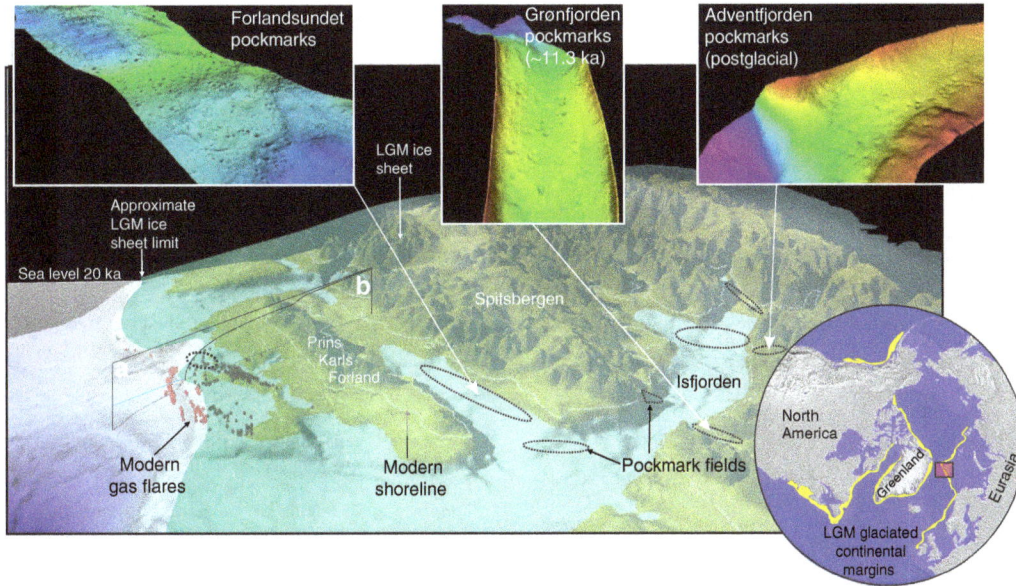

Figure 3 | Conceptual reconstruction of the glaciated western Svalbard margin. Dotted ellipses reveal the location of discovered pockmark fields that we infer to be due to the post-LGM retreat of the ice sheet. Upper insets demonstrate pockmark fields in high-resolution multibeam data. Lower inset schematically shows the widespread distribution of the formerly glaciated margins at the LGM across the Arctic, for which the western Svalbard margin is a potential GHSZ analogue.

margin and potentially across the US east coast continental margins where thousands of gas flares have been observed[42].

High pressures to at least 50 MPa coupled with temperatures −5 °C or cooler under the LGM ice sheet likely promoted expansion of the gas hydrate reservoir beneath the western Svalbard margin, but also left a ∼20-km-wide gas-hydrate-free zone, which served as a corridor for upward migration of fluids and release of methane throughout the LGM. The North West Svalbard shelf and upper slope provides a robust test bed for understanding gas hydrate formation and dissociation across a formerly glaciated continental margin and reference point for further natural greenhouse gas release studies in Polar Regions. Former continental margins with a similar paleo-subglacial footprint and legacy extend some 20,000 km in the northern hemisphere where major ice sheets advanced across western Eurasia shallow shelves, Greenland and eastern and western American shelves (Fig. 3, inset). We infer that the ongoing deglaciation of these continental margins between 18 and 12 ka (ref. 43), many of which are underlain by deep hydrocarbon reservoirs, would have likely been accompanied by widespread dissociation of gas hydrates and associated increase of methane flux from the seabed. In the outer continental shelves where the eustatic signal outpaced isostatic rebound, methane emissions from recently inundated shallow shelves (first tens of metres) would have been expelled into the atmosphere, similar to present-day process of methane transport across the shallow East Siberian Arctic Shelf[44]. This study not only implies the potential for significant gas hydrate storage and release capacity during past glacial/inter-glacial conditions but is also significant in its implication for current and future greenhouse gas release under the ongoing thinning and retreat of contemporary ice sheets and glaciers[45].

Methods

Ice-sheet configuration and bed surface temperature modelling. The equilibrium profile of the West Svalbard ice sheet was obtained using a steady-state model based on the refined perfect-plastic assumption for ice flow[29,30,46]. Under these conditions, plastic-flow yielding a steady-state ice-sheet thickness distribution

(H) occurs when basal shear stress (τ_b) is equal to the yield stress (τ_0) at all points across the bed:

$$\tau_b = -\rho g H \left(\frac{\partial h}{\partial x} \right) = \tau_0$$

where ρ is the density of ice, g is the gravitation constant and dh/dx is the ice-surface slope. This equation can be rearranged and applied to reconstruct former ice sheets using observational constraints that indicate its maximum extent, such as offshore moraines sequences, and its former surface from, for example, cosmogenically dated erratics and trimlines. To reconstruct the ice-sheet surface, the above equation is integrated numerically[20], from the margin (H = 0) under a given subglacial topographic profile with an appropriate value of yield strength (in our case 40 and 80 kPa), which for grounded ice masses can fall between 40 and 100 kPa (refs 30,47).

Reconstruction of the LGM isostatic loading involved adjustment of past marine limits to account for relative sea level rise. Oldest post-LGM marine limits discovered on the PKF (25 m.a.s.l.) and Spitsbergen margin (48 m.a.s.l.) were dated as 14 ka (refs 25,28; Fig. 2). Thus, given the ∼80 m sea level rise since 14 ka, we estimate the absolute isostatic rebound as ∼105 and ∼128 m for PKF and Spitsbergen margin, respectively (Fig. 2).

We determine the steady-state cold-subglacial temperature distribution based on conservation of energy at the bed from vertical diffusion, advection and frictional heating as implemented and validated for various glaciers and the Antarctic ice sheet[30,48,49]. As implemented at Taylor Glacier, Dry valleys[50], for its application to the LGM ice sheet in western Svalbard (Supplementary Fig. 1), we use a lapse rate of −0.007 °C m^{-1}, an accumulation rate of 0.3 m a^{-1} and mean annual temperature at the margin of −14 °C. For the warm-bed situation, basal temperatures are assumed to be at pressure meting point calculated according to H × 8.70 × 10^{-4} (ref. 30).

Subsurface heat-flow model. We sampled the seafloor depths along the transect (Fig. 1) at every 100 m from IBCAO v.3 gridded bathymetric data[51] and adjusted the depths to sea level at 20 ka (ref. 31). Then, a 671 × 1,000 cell temperature grid (100 × 1-m cell dimensions) was generated, with its upper boundary at the seafloor and basal boundary 1 km below seafloor. To define initial temperature conditions, existing heat-flow measurements[5,35,36] (Fig. 1) located near the selected transect were utilized. A spin-up model temperature grid was generated by assuming a one-dimensional linear temperature gradient, with ocean bottom water temperatures[32] adjusted to 20 ka (refs 33,34) and ice-bottom temperatures (where ice exists) as the top boundary condition (Supplementary Fig. 1). To constrain the thermal diffusivity, we assumed a constant thermal conductivity of 2 W m^{-1} K^{-1}, average density of 1,900 kg m^{-3} for the bulk sediments and an average specific heat capacity of 2,000 J kg^{-1} K^{-15}.

Assuming no significant *in situ* generation of heat, we run the two-dimensional finite difference heat-flow model[52] for 5 kyr (assuming no significant variation of ice extent, sea level and ocean bottom temperatures in the study area during this period). The model is run for two different ice-bottom temperatures (Supplementary Fig. 1) and two different ice-sheet thicknesses, corresponding to 40 and 80 kPa yield stress.

Gas hydrate stability modelling. To identify the base of methane hydrate stability, we integrate results from our heat-flow model with theoretical hydrate stability phase diagrams generated using the CSMHYD program[1], which uses an algorithm based on Gibbs energy minimization and account for different pressure and temperature conditions, the composition of gas forming hydrates and the presence of inhibitors of hydrate formation (for example, salt). Assuming pure methane gas and a constant pore water salinity of 35‰, CSMHYD program estimates the pressure at which hydrates are stable for any given temperature. With temperature constrained from our heat-flow model, we estimated the hydrostatic pressure at each cell location using standard values for density of seawater $(1,027\,kg\,m^{-3})$ and acceleration due to gravity $(9.8\,m\,s^{-2})$. This pressure grid is then compared with theoretical predictions from CSMHYD program to determine the hydrate stability at each cell location.

Hydroacoustic data. High-resolution bathymetry data have been acquired in 2004–2014 onboard RV 'Helmer Hanssen' by the Arctic University of Norway (UiT) using Kongsberg-Simrad EM300 multibeam echosounder system. Multibeam data were gridded with the cell 7–10 m in the water depths ≤200 and 15–20 m in the water depths >200 m, which provided sufficient resolution for pockmark detections. Locations of gas flares offshore PKF were derived from previously published study[8]. New water column data, acquired in 2014 with Simrad ER-60 echosounder during UiT cruise onboard RV 'Helmer Hanssen' has been processed using Fledermaus software and included in the current study.

References

1. Sloan, E. D. & Koh, C. A. *Clathrate Hydrates of Natural Gases* 3rd edn (CRC Press, 2008).
2. Weitemeyer, K. A. & Buffett, B. A. Accumulation and release of methane from clathrates below the Laurentide and Cordilleran ice sheets. *Global Planet. Change* **53**, 176–187 (2006).
3. Wadham, J. L. *et al.* Potential methane reservoirs beneath Antarctica. *Nature* **488**, 633–637 (2012).
4. Wadham, J. L., Tranter, M., Tulaczyk, S. & Sharp, M. Subglacial methanogenesis: A potential climatic amplifier? *Global Biogeochem. Cycles* **22**, 1–16 (2008).
5. Berndt, C. *et al.* Temporal Constraints on Hydrate-Controlled Methane Seepage off Svalbard. *Science* **343**, 284–287 (2014).
6. Westbrook, G. K. *et al.* Escape of methane gas from the seabed along the West Spitsbergen continental margin. *Geophys. Res. Lett.* **36**, L15608 (2009).
7. Mørk, A. & Bjorøy, M. *Mesozoic Source Rocks on Svalbard* (Springer, 1984).
8. Sahling, H. *et al.* Gas emissions at the continental margin west of Svalbard: mapping, sampling, and quantification. *Biogeosciences* **11**, 6029–6046 (2014).
9. Landvik, J. Y. *et al.* Rethinking Late Weichselian ice-sheet dynamics in coastal NW Svalbard. *Boreas* **34**, 7–24 (2005).
10. Amundsen, I., Blinova, M., Hjelstuen, B., Mjelde, R. & Haflidason, H. The Cenozoic western Svalbard margin: sediment geometry and sedimentary processes in an area of ultraslow oceanic spreading. *Mar. Geophys. Res.* **32**, 441–453 (2011).
11. Rajan, A., Mienert, J. & Bünz, S. Acoustic evidence for a gas migration and release system in Arctic glaciated continental margins offshore NW-Svalbard. *Mar. Pet. Geol.* **32**, 36–49 (2012).
12. Blinova, M., Thorsen, R., Mjelde, R. & Faleide, J. I. Structure and evolution of the Bellsund Graben between Forlandsundet and Bellsund (Spitsbergen) based on marine seismic data. *Norw. J. Geol.* **89**, 215–228 (2009).
13. Gabrielsen, R. H. *et al.* A Structural outline of Forlandsundet Graben, Prins Karls Forland, Svalbard. *Nor. Geol. Tidsskr.* **72**, 105–120 (1992).
14. Forwick, M., Baeten, N. J. & Vorren, T. O. Pockmarks in Spitsbergen fjords. *Nor. J. Geol.* **89**, 65–77 (2009).
15. Knies, J. & Mann, U. Depositional environment and source rock potential of Miocene strata from the central Fram Strait: introduction of a new computing tool for simulating organic facies variations. *Mar. Pet. Geol.* **19**, 811–828 (2002).
16. Tissot, B. P. & Welte, D. H. *Petroleum Formation and Occurrence* 2 edn (Springer-Verlag, 1984).
17. Elvevold, S., Dallmann, W. & Blomeier, D. *Geology of Svalbard* (Norwegian Polar Institute, 2007).
18. Jakobsson, M. *et al.* Arctic Ocean glacial history. *Quat. Sci. Rev.* **92**, 40–67 (2014).
19. Patton, H. *et al.* Geophysical constraints on the dynamics and retreat of the Barents Sea Ice Sheet as a palaeo-benchmark for models of marine ice-sheet deglaciation. *Rev. Geophys.* doi: 10.1002/2015RG000495 (2015).
20. Landvik, J. Y. *et al.* Northwest Svalbard during the last glaciation: Ice-free areas existed. *Geology* **31**, 905–908 (2003).
21. Landvik, J. Y. *et al.* 10Be exposure age constraints on the Late Weichselian ice-sheet geometry and dynamics in inter-ice-stream areas, western Svalbard. *Boreas* **42**, 43–56 (2013).
22. Landvik, J. Y. *et al.* The Last Glacial Maximum of Svalbard and the Barents Sea Area: Ice Sheet Extent and Configuration. *Quat. Sci. Rev.* **17**, 43–75 (1998).
23. Ottesen, D. A. G., Dowdeswell, J. A., Landvik, J. Y. & Mienert, J. Dynamics of the Late Weichselian ice sheet on Svalbard inferred from high-resolution sea-floor morphology. *Boreas* **36**, 286–306 (2007).
24. Ingólfsson, Ó. & Landvik, J. Y. The Svalbard–Barents Sea ice-sheet – Historical, current and future perspectives. *Quat. Sci. Rev.* **64**, 33–60 (2013).
25. Andersson, T., Forman, S. L., Ingolfsson, O. & Manley, W. F. Late Quaternary environmental history of central Prins Karls Forland, western Svalbard. *Boreas* **28**, 292–307 (1999).
26. Forman, S. L. Post-glacial relative sea-level history of northwestern Spitsbergen, Svalbard. *Geol. Soc. Am. Bull.* **102**, 1580–1590 (1990).
27. Forman, S. L. *et al.* A review of postglacial emergence on Svalbard, Franz Josef Land and Novaya Zemlya, northern Eurasia. *Quat. Sci. Rev.* **23**, 1391–1434 (2004).
28. Forman, S. L. Post-glacial relative sea-level history of northwestern Spitsbergen, Svalbard. *Geol. Soc. Am. Bull.* **102**, 1580–1590 (1990).
29. Nye, J. F. The flow of glaciers and ice-sheets as a problem in plasticity. *Proc. R. Soc. Lond. A* **207**, 554–572 (1951).
30. van der Veen, C. J. *Fundamentals of Glacier Dynamics* (Taylor & Francis, 1999).
31. Fleming, K. *et al.* Refining the eustatic sea-level curve since the Last Glacial Maximum using far- and intermediate-field sites. *Earth Planet. Sci. Lett.* **163**, 327–342 (1998).
32. Boyer, T. P. *et al.* in *NOAA Atlas NESDIS* 72 (ed. Center NOD) (National Oceanic And Atmospheric Administration, 2013).
33. Hebbeln, D., Dokken, T., Andersen, E. S., Hald, M. & Elverhoi, A. Moisture supply for northern ice-sheet growth during the Last Glacial Maximum. *Nature* **370**, 357–360 (1994).
34. Rasmussen, T. L., Thomsen, E. & Nielsen, T. Water mass exchange between the Nordic seas and the Arctic Ocean on millennial timescale during MIS 4–MIS 2. *Geochem. Geophys. Geosyst.* **15**, 530–544 (2014).
35. Crane, K., Sundvor, E., Buck, R. & Martinez, F. Rifting in the northern Norwegian-Greenland Sea: Thermal tests of asymmetric spreading. *J. Geophys. Res. Solid Earth* **96**, 14529–14550 (1991).
36. Isaksen, K., Holmlund, P., Sollid, J. L. & Harris, C. Three deep Alpine-permafrost boreholes in Svalbard and Scandinavia. *Permafrost Periglacial Processes* **12**, 13–25 (2001).
37. Sarkar, S. *et al.* Seismic evidence for shallow gas-escape features associated with a retreating gas hydrate zone offshore west Svalbard. *J. Geophys. Res. Solid Earth* **117**, B09102 (2012).
38. Roy, S., Hovland, M., Noormets, R. & Olaussen, S. Seepage in Isfjorden and its tributary fjords, West Spitsbergen. *Mar. Geol.* **363**, 146–159 (2015).
39. Sloan, E. D. & Koh, C. A. *Clathrate Hydrates of Natural Gases* 3rd edn (CRC Press, 2007).
40. Damm, E., Mackensen, A., Budéus, G., Faber, E. & Hanfland, C. Pathways of methane in seawater: Plume spreading in an Arctic shelf environment (SW-Spitsbergen). *Cont. Shelf Res.* **25**, 1453–1472 (2005).
41. Knies, J., Damm, E., Gutt, J., Mann, U. & Pinturier, L. Near-surface hydrocarbon anomalies in shelf sediments off Spitsbergen: evidences for past seepages. *Geochem. Geophys. Geosyst.* **5**, CiteID Q06003 (2004).
42. Skarke, A., Ruppel, C., Kodis, M., Brothers, D. & Lobecker, E. Widespread methane leakage from the sea floor on the northern US Atlantic margin. *Nat. Geosci.* **7**, 657–661 (2014).
43. Bennett, M. in *Boreas*. (eds Benn, D. I. & Evans, D. J.) **40**, 555–555 (Blackwell Publishing Ltd, 2011).
44. Shakhova, N *et al.* Ebullition and storm-induced methane release from the East Siberian Arctic Shelf. *Nat. Geosci.* **7**, 64–70 (2014).
45. Hanna, E *et al.* Ice-sheet mass balance and climate change. *Nature* **498**, 51–59 (2013).
46. Reeh, N. A plasticity theory approach to the steady-state shape of a three-dimensional ice sheet. *J. Glaciol.* **28**, 431–455 (1982).
47. Paterson, W. S. B. *The Physics of Glaciers* (Pergamon Press, Elsevier Science Ltd, 1994).
48. Budd, W. F., Jenssen, D. & Radok, U. *Derived Physical Characteristics of the Antarctic Ice Sheet* (University of Melbourne, 1971).
49. Robin, G. deQ. Ice movement and temperature distribution in glaciers and ice sheets. *J. Glaciol.* **3**, 589–606 (1955).
50. Hubbard, A., Lawson, W., Anderson, B., Hubbard, B. & Blatter, H. Evidence for subglacial ponding across Taylor Glacier, Dry Valleys, Antarctica. *Ann. Glaciol.* **39**, 79–84 (2004).
51. Jakobsson, M. *et al.* The International Bathymetric Chart of the Arctic Ocean (IBCAO) Version 3.0. *Geophys. Res. Lett.* **39**, L12609 (2012).

52. Phrampus, B. J. & Hornbach, M. J. Recent changes to the Gulf Stream causing widespread gas hydrate destabilization. *Nature* **490**, 527–530 (2012).

Acknowledgements

The research is part of the Centre for Arctic Gas Hydrate, Environment and Climate and was supported by the Research Council of Norway through its Centres of Excellence funding scheme grant No. 223259. A.P. was supported by a Statoil fellowship grant through the UiT The Arctic University of Norway. We are thankful to Ellen Damm and two anonymous reviewers for very constructive comments, which helped to improve the manuscript.

Author contributions

J.M. and A.P. developed the ideas that led to the paper. A.P. processed and interpreted hydroacoustic data and designed the figures; S.V. applied heat-flow and GHSZ modelling. J.M., A.P. and S.V. wrote the original manuscript. A.H. modelled the ice-sheet thickness and temperature reconstruction and advised on the study's scope and conception. All authors contributed to the final editing of the manuscript.

Additional information

Competing financial interests: The authors declare no competing financial interests.

7

No inter-gyre pathway for sea-surface temperature anomalies in the North Atlantic

Nicholas P. Foukal[1] & M. Susan Lozier[1]

Recent Lagrangian analyses of surface drifters have questioned the existence of a surface current connecting the Gulf Stream (GS) to the subpolar gyre (SPG) and have cast doubt on the mechanism underlying an apparent pathway for sea-surface temperature (SST) anomalies between the two regions. Here we use modelled Lagrangian trajectories to determine the fate of surface GS water and satellite SST data to analyse pathways of GS SST anomalies. Our results show that only a small fraction of the surface GS water reaches the SPG, the water that does so mainly travels below the surface mixed layer, and GS SST anomalies do not propagate into the SPG on interannual timescales. Instead, the inter-gyre heat transport as part of the Atlantic Meridional Overturning Circulation must be accomplished via subsurface pathways. We conclude that the SST in the SPG cannot be predicted by tracking SST anomalies along the GS.

[1] Nicholas School of the Environment, Duke University, Durham, North Carolina 27708, USA. Correspondence and requests for materials should be addressed to N.P.F. (email: Nicholas.Foukal@Duke.edu).

The surface waters of the eastern subpolar gyre (50° N–60° N and 30° W–15° W) in the North Atlantic are exceptionally warm and salty compared with the surface waters at similar latitudes, leading to the supposition that the subpolar gyre (SPG) is primarily supplied by the similarly warm and salty waters of the subtropical gyre (STG)[1–3]. This connection has been of particular interest because the sea-surface characteristics of the North Atlantic SPG provide the environment for density-driven overturning—the warm surface water loses heat to the atmosphere, becomes denser and sinks to depth as it flows around the SPG. This water mass transformation is one component of the Atlantic Meridional Overturning Circulation (AMOC), which moves warm water northwards in its upper layers and colder water southwards at depth, affecting regional and global climate by redistributing heat[4]. Another impetus for understanding the structure and pathways of the upper-layer AMOC is the potential for predicting climate on timescales of months to years in advance[5–7].

Eulerian-based analyses of surface drifter velocities—in which drifter velocities within a spatially defined bin are averaged over time—indicate that surface waters in the Cape Hatteras region are advected north-eastward into the eastern SPG[3,8]. In contrast, Lagrangian analyses of the pathways of individual trajectories from the same surface drifter data show little to no surface connection between the Gulf Stream (GS) and the eastern SPG[9,10]. Only 1 out of 273 (0.37%) surface drifters deployed south of 45° N in the North Atlantic between 1990 and 2002 made it to the SPG[9]. Though this surface throughput has increased since 2002, the largest exchange observed during any given time period is only 3.3% (ref. 10). A subsequent modelling study demonstrates that Lagrangian trajectories initialized in the STG at depth shoal as they move northward along isopycnal surfaces and are more likely to reach the SPG than those initialized at the surface in the STG[11]. When coupled with recent findings on the formation and fate of subtropical mode water (STMW)[12,13], these Lagrangian results yield a new circulation pattern of North Atlantic upper-layer currents: surface GS water recirculates in the STG, becomes part of the STMW, re-enters the GS at depth and is then exported to the eastern SPG[14].

This proposed paradigm, with subsurface STG waters supplying the eastern SPG, begs the question as to whether winter sea-surface temperature (SST) anomalies are transmitted to the SPG via the advective pathway described in previous studies[5,15,16] in which SST anomalies re-emerge during subsequent winters as they progress along the GS/North Atlantic Current (GS/NAC). The goal of this paper is to answer that question by first carefully analysing and quantifying the fate of surface GS waters and second, determining whether winter SST anomalies in the modern satellite SST record propagate from the STG into the SPG.

Here we find that only a small fraction of surface GS water reaches the SPG, and that the water that does reach the SPG does so at depths below the winter-mixed layer and is thus out of contact with the SST. We also find that the SST anomaly pathway previously reported is most likely an artefact of a smoothing filter on the same timescale as SST changes associated with the North Atlantic Oscillation (NAO), and not an actual physical pathway for SST anomalies. We conclude that interannual changes in the SST of the SPG cannot, as once thought, be predicted from information on the GS SST and that inter-gyre heat transport of the AMOC must occur below the surface layers.

Results

Modelled Lagrangian pathways. To trace the pathways and fate of surface GS waters, we initialize Lagrangian particles in an

eddy-resolving ocean circulation model (FLAME)[17,18]. FLAME is an ocean-only, fully dynamic model with 1/12° horizontal resolution and 45 depth layers varying from 5-m resolution at the surface to 250-m resolution at depth. The model is forced by NCEP-NCAR reanalysis wind and buoyancy forcing for 15 years (1990–2004). We find Lagrangian particle positions using the model's 3-day velocity fields, and record those positions every 15 days. Particles are initialized at the surface in the vicinity of Cape Hatteras (VCH) in a spatial box defined as in a previous study[5] (31.5° N–38.5° N and 80° W–60° W). We also require that the particles' initial positions be located south of the GS north wall—defined to be the 15 °C isotherm at 200-m depth[19]—so that all trajectories are of subtropical origin. To calculate particle pathways from the model's velocity field, we initially use only horizontal velocities—that is, we restrict the simulated floats to the surface to mimic the constraint placed on the observational surface drifters (Fig. 1a). In a subsequent Lagrangian experiment, we use the three-dimensional flow field (Fig. 1b) for the trajectory calculations. In both experiments, water particles are tagged at the surface in the VCH box every 15 days for the first 10 years of the model run (1990–1999) and tracked for 5 years. In the surface-constrained run, the trajectories are heavily influenced by the winds, accumulating in the Ekman convergence zone at 30° N.

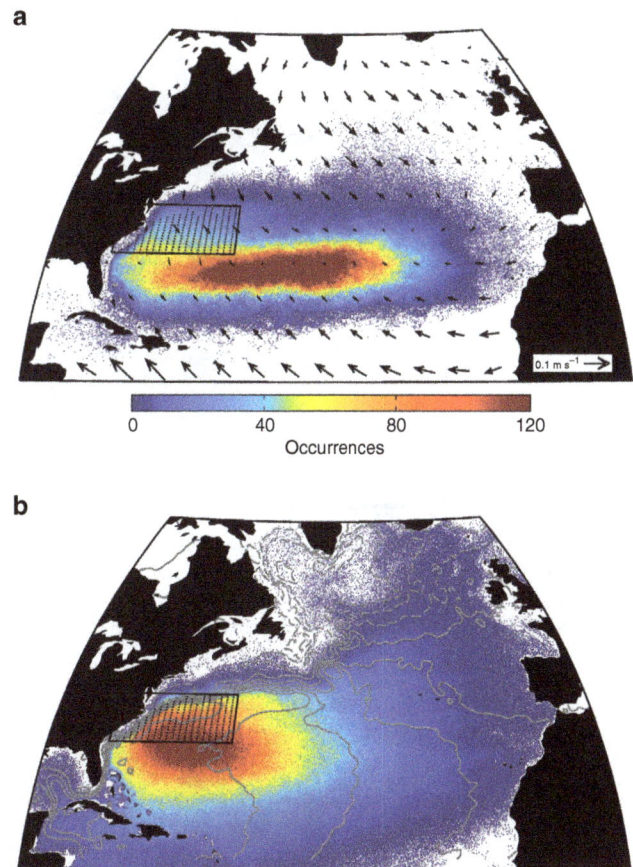

Figure 1 | Distributions of modelled particle positions. Floats are initialized from the surface of the VCH box (black rectangle) in FLAME and are constrained to follow either the two-dimensional surface flow field (**a**) or the three-dimensional flow field (**b**) for 5 years. The colour bar refers to the total number of trajectories that pass through a given location (that is, occurrences) over the 15 years. Arrows in **a** are the time-mean Ekman velocities from the Center for Topographic Studies of the Ocean and Hydrosphere[28] and contours in **b** are the time-mean SSH values from AVISO[29] (dashed contours indicate negative values).

Only 5 out of 60,906 modelled trajectories (<0.01%) cross north of 53° N, the latitude at which we consider the floats to be subpolar. When allowed to follow the three-dimensional flow field, the vast majority of the trajectories (97.5%) trace the wind-driven, western-intensified geostrophic STG, accumulating in its convergent centre. Only 1,526 out of the 60,906 total floats (2.51%) reach the SPG in this experiment. Along with vertical mixing, Ekman pumping from negative wind-stress curl forcing yields a net downward transport in the STG[20], removing particles from the surface layer and creating the disparity between a and b in Fig. 1. These results agree with the observed and modelled Lagrangian experiments on the fate of surface water in the GS[9–11,21], and we elaborate on those previous studies by exploring the pathways of the Lagrangian particles from the GS region to the SPG.

An examination of those trajectories that reach the SPG within their 5-year lifetime reveals the canonical GS/NAC pathway (Fig. 2). However, only 0.5% of the floats remain above the 125-m climatological winter-mixed layer in the GS/NAC[22] on their way to the SPG (Fig. 2, inset). In addition, the small number of particles that travel to the eastern SPG vary in their transit time (median = 3.4 years, s.d. = 1.2 year); the spread demonstrates that

waters from the VCH do not travel coherently to the eastern SPG and instead mix with waters of other ages (Fig. 3). In comparison, the Lagrangian integral timescale of temperature anomalies for the trajectories that reach the SPG (~48 days) is much shorter than the advective timescale, thus individual trajectories are not expected to carry their temperature anomaly to the SPG.

Satellite SST anomaly pathways. Given the lack of a surface advective pathway from the STG to the SPG, we revisit the question of whether winter SST anomalies propagate along the GS/NAC[5,15,16]. Our investigation uses a lagged correlation analysis with the modern satellite SST record—31 years (1981–2012) of AVHRR Pathfinder SST data[23]. We switch to observational data for this analysis because a reliable data set exists with an observational period long enough for this analysis. A number of steps are followed to provide consistency with previous studies: (1) we use winter (November–April mean) SST anomalies at 1° × 1° spatial resolution; (2) the VCH box is defined as 80° W–60° W and 31.5° N–38.5° N; and (3) we apply a 5-year running mean filter to the SST anomalies before calculating correlations. A time series of winter SST anomalies averaged over the VCH box is extracted and then correlated with

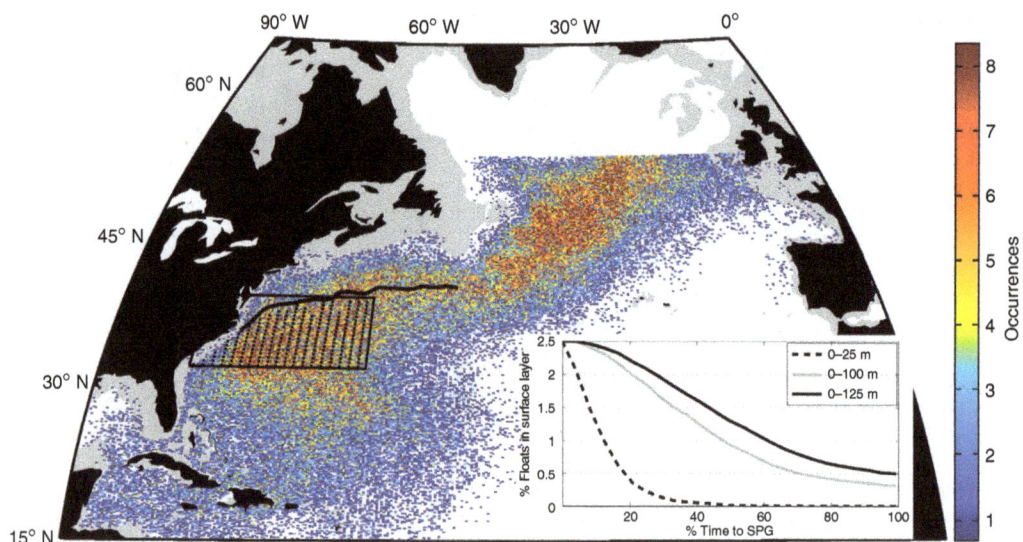

Figure 2 | Distribution of modelled particle positions that are SPG-bound. Particles that reach 53° N within 5 years in the three-dimensional flow field (2.51% of total). The colour bar refers to the total number of trajectories that passed through a given location (that is, occurrences) over the 15 years. Particle locations north of 53° N are not displayed. Solid black line indicates the average position of the GS north wall in FLAME. Inset panel shows the percent of trajectories that remain in the surface layer as a function of time, normalized by the total time required to reach the SPG. Three definitions of the surface layer are provided (0–25, 0–100 and 0–125 m).

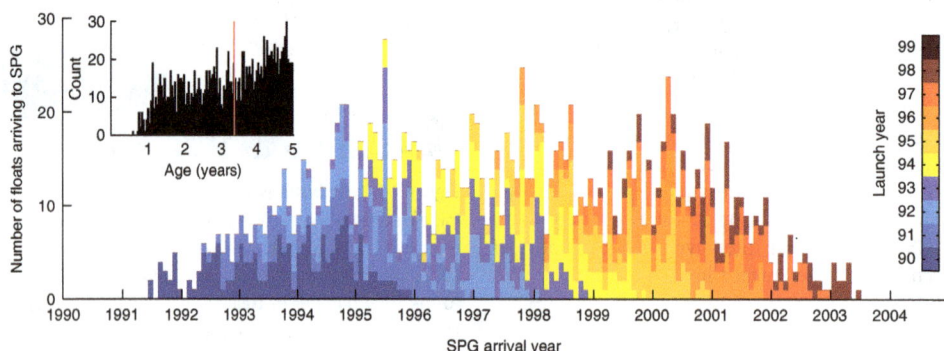

Figure 3 | Number of modelled particles crossing 53° N every 30 days. Colours denote the year of initialization in the VCH box. Inset panel shows the distribution of float ages when crossing 53° N, with a median (red line) at 3.37 years.

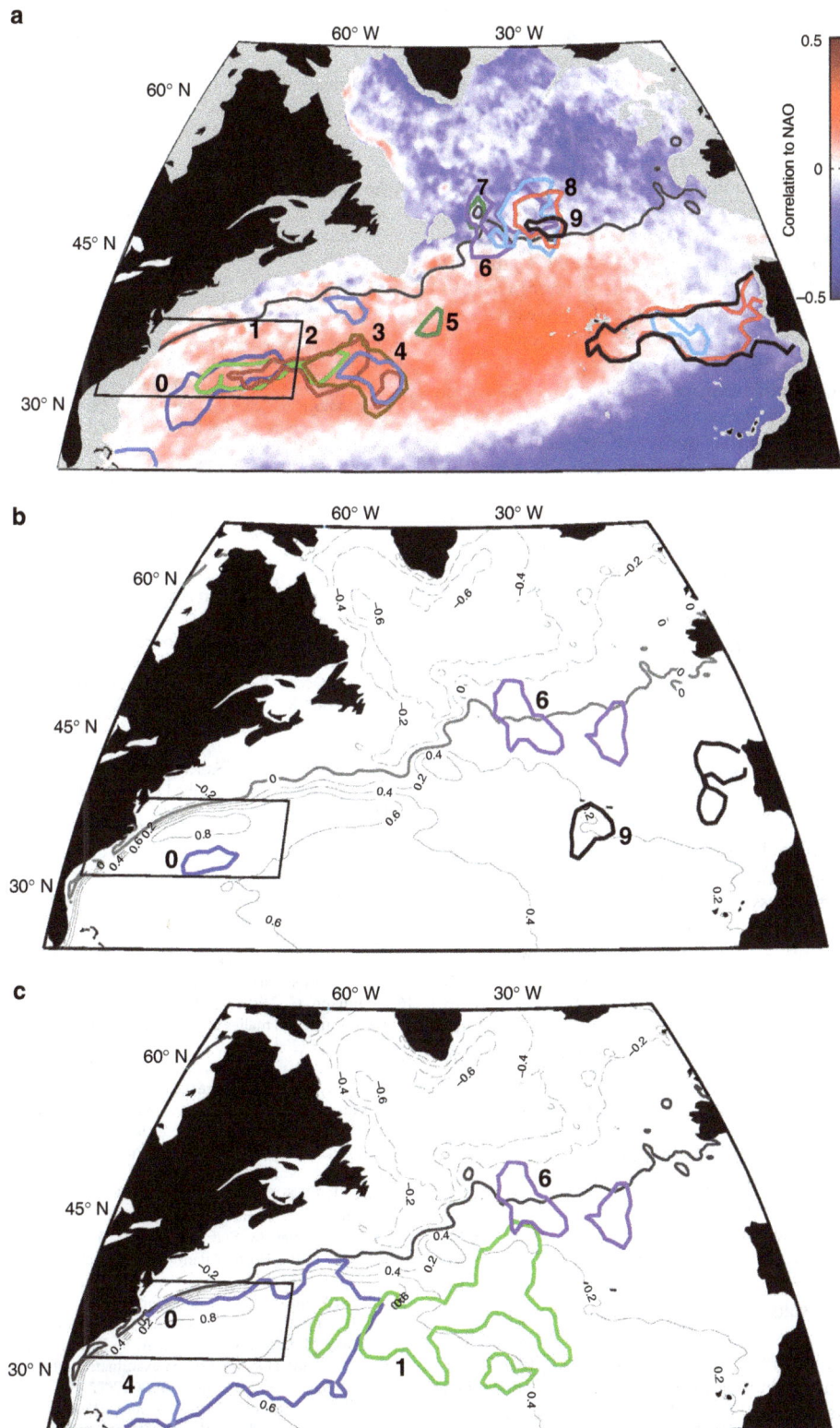

Figure 4 | Lagged correlation analysis with the modern satellite SST record. Correlations are calculated between the local SST and the time series from the spatially averaged VCH box (black rectangle). (**a**) Correlations that exceed subjectively chosen thresholds that decrease as the lag increases (lag 0 = 0.8, lag 1 = 0.75, lag 2 = 0.65, lag 3 = 0.55, lag 4 = 0.50, lag 5 = 0.40, lag 6 = 0.40, lag 7 = 0.40, lag 8 = 0.35 and lag 9 = 0.35). The thresholds used in lags 7 through 9 are not statistically significant ($P > 0.05$). Winter SST anomalies are smoothed with a 5-year running mean before calculating correlations. Correlations between the NAO and North Atlantic SST from 1981–2012 are shown in the background. Bathymetry above 1,000 m is shaded light grey, and the dark grey line represents the zero SSH contour. (**b**) Correlation contours are calculated as in **a** but the 5-year running mean filter is removed. (**c**) Objectively chosen, statistically significant ($P < 0.05$) correlation contour thresholds with no running mean filter. In **b** and **c**, time-mean AVISO SSH (1993–2014) contours (metres) are shown in grey. An identical analysis with March SST yields qualitatively similar results.

winter SST anomalies for each $1° \times 1°$ grid in the North Atlantic. Correlations are calculated for zero to nine-year lags. Areas that exceed subjectively chosen threshold correlations that decrease as the lag increases so as to maximize a propagation pattern are outlined for each lag in Fig. 4a. The resulting contours reveal a propagation pattern similar to that in Fig. 2 of Sutton and Allen[5]. However, if the 5-year running mean filter is removed before calculating the lagged correlations (Fig. 4b), the propagation pattern disappears. This implies that the apparent pathway in Fig. 4a results solely from the use of temporal smoothing. When we instead outline objectively chosen, statistically significant correlation thresholds ($P < 0.05$) and do not use a running mean filter (Fig. 4c), years 0 and 1 encompass most of the western and central STG (STMW formation regions). STMW temperature anomalies have been hypothesized to resurface in the following winter due to summer stratification capping the anomalies and protecting them from air–sea exchanges[13,24]. Our results support this hypothesis with the caveat that our cutoff for significance at 0- and 1-year lag was relatively low ($r \geq 0.36$). Though the individual particles have Lagrangian integral timescales of < 50 days, the STMW as a whole exhibits a 1-year SST memory due to particles entering and exiting the STMW during their lifetimes[12,13]. We also note that the STMW re-emergence region (Fig. 4c, green) is south of the GS/NAC indicated by either the GS north wall (Fig. 2) or the zero SSH contour (Fig. 4), both of which can be used to track the northern extent of the STG[25]. Thus, not only are the thermal anomalies short-lived, they are also primarily constrained to the STG. We conclude that SST anomalies are not advected or propagated along the GS/NAC from the STG to the SPG.

To determine why an inter-gyre pathway for SST anomalies appears with the 5-year running mean filter, we correlate the monthly time series of the NAO, an index of the atmospheric pressure gradient between Iceland and the Azores, with the AVHRR satellite SST (Fig. 4a). The spatial expression of the NAO forms a tri-pole pattern across the North Atlantic, with the SPG in phase with the tropical Atlantic and out of phase with the STG. This spatial pattern, combined with NAO timescales of roughly 2–7 years, gives context to the lagged correlation contours in Fig. 4a. With the 5-year running mean filter, the correlation contours of lags 0 through 5 years occur in the STG, where NAO-induced SST variability is in phase with the SST of the VCH box. For lags of 6 through 9 years, the correlations shift into the SPG where the NAO forcing is out of phase with the VCH SST. This shift into the SPG results not from the advection of water from the STG, but from the changing phase of the NAO and its opposing effect on the two gyres[26]. In addition, the 5-year running mean filter artificially increases the magnitude and prolongs the duration of the correlations that appear without the smoothing; the original 1-year memory of STMW SST in Fig. 4c is extended to 5 years by the temporal smoothing in Fig. 4a. Taken together, these results imply that the appearance of a SST anomaly pathway from the GS to the SPG in Fig. 4a is an artefact of the 5-year running mean filter introducing memory into NAO-induced SST oscillations rather than a signal of SST anomalies propagating along the GS/NAC[16,27]. Though other studies have pointed to the divergences and convergences in oceanic heat transport due to NAO forcing as a means to generate the SST anomalies in the GS region[16], we propose that the apparent propagation pathway results from a smoothing window (5 years) that is on the same timescale as the NAO forcing (2–7 years).

Though for decades it has been assumed that waters from the surface GS flow northward as part of the upper limb of the AMOC, building on a recent study[11] we demonstrate here that this throughput is not achieved by a surface connection.

Importantly, we show that surface thermal anomalies generated in the subtropics are not transmitted to the surface waters at higher latitudes on interannual timescales. This work has implications for our understanding of how the AMOC achieves its poleward transport of heat.

References

1. Maury, M. F. *The Physical Geography of the Sea and its Meteorology* 13th edn (Sampson Low, 1868).
2. Broecker, W. S. The Great Ocean Conveyor. *Oceanography* **4**, 79–89 (1991).
3. Fratantoni, D. M. North Atlantic surface circulation during the 1990s observed with satellite-tracked drifters. *J. Geophys. Res.* **106**, 22067–22093 (2001).
4. Kostov, Y., Armour, K. C. & Marshall, J. Impact of the Atlantic meridional overturning circulation on ocean heat storage and transient climate change. *Geophys. Res. Lett.* **41**, 2108–2116 (2014).
5. Sutton, R. T. & Allen, M. R. Decadal predictability of North Atlantic sea surface temperature and climate. *Nature* **388**, 563–567 (1997).
6. Latif, M., Collins, M., Pohlmann, H. & Keenlyside, N. A review of predictability studies of Atlantic sector climate on decadal time scales. *J. Clim.* **19**, 5971–5987 (2006).
7. Msadek, R., Dixon, K. W., Delworth, T. L. & Hurlin, W. Assessing the predictability of the Atlantic meridional overturning circulation and associated fingerprints. *Geophys. Res. Lett.* **37**, L19608 (2010).
8. Lumpkin, R. & Johnson, G. C. Global ocean surface velocities from drifters: mean, variance, El-Niño-Southern Oscillation response, and seasonal cycle. *J. Geophys. Res.* **118**, 2992–3006 (2013).
9. Brambilla, E. & Talley, L. D. Surface drifter exchange between the North Atlantic subtropical and subpolar gyres. *J. Geophys. Res.* **111**, C07026 (2006).
10. Häkkinen, S. & Rhines, P. B. Shifting surface currents in the northern North Atlantic Ocean. *J. Geophys. Res.* **114**, C0405 (2009).
11. Burkholder, K. C. & Lozier, M. S. Subtropical to subpolar pathways in the North Atlantic: deductions from Lagrangian trajectories. *J. Geophys. Res.* **116**, C07017 (2011).
12. Gary, S. F., Lozier, M. S., Kwon, Y.-O. & Park, J. J. The fate of North Atlantic subtropical mode water in the FLAME model. *J. Phys. Oceanogr.* **44**, 1354–1371 (2014).
13. Kwon, Y.-O., Park, J. J., Gary, S. F. & Lozier, M. S. Year-to-year reoutcropping of eighteen degree water in an eddy-resolving ocean simulation. *J. Phys. Oceanogr.* **45**, 1189–1204 (2015).
14. Burkholder, K. C. & Lozier, M. S. Tracing the pathways of the upper limb of the North Atlantic Meridional Overturning Circulation. *Geophys. Res. Lett.* **41**, 4254–4260 (2014).
15. Hansen, D. V. & Bezdek, H. F. On the nature of decadal anomalies in the North Atlantic sea surface temperature. *J. Geophys. Res.* **101**, 8749–8758 (1996).
16. Krahmann, G., Visbeck, M. & Reverdin, G. Formation and propagation of temperature anomalies along the North Atlantic current. *J. Phys. Oceanogr.* **31**, 1287–1303 (2001).
17. Böning, C. W., Scheinert, M., Dengg, J., Biastoch, A. & Funk, A. Decadal variability of subpolar gyre transport and its reverberation in the North Atlantic overturning. *Geophys. Res. Lett.* **33**, L21S01 (2006).
18. Biastoch, A., Böning, C. W., Getzlaff, J., Molines, J.-J. & Madec, G. Causes of interannual-decadal variability in the meridional overturning circulation of the midlatitude North Atlantic Ocean. *J. Clim.* **21**, 6599–6615 (2008).
19. Joyce, T. M. & Zhang, R. On the path of the Gulf Stream and the Atlantic Meridional overturning circulation. *J. Clim.* **23**, 3146–3154 (2010).
20. Palter, J., Lozier, M. S. & Barber, R. T. The effect of advection on the nutrient reservoir in the North Atlantic subtropical gyre. *Nature* **437**, 687–692 (2005).
21. Rypina, I. I., Pratt, L. J. & Lozier, M. S. Near-surface transport pathways in the North Atlantic Ocean: looking for throughput from the subtropical to the subpolar gyre. *J. Phys. Oceanogr.* **41**, 911–925 (2011).
22. de Boyer Montégut, C., Fischer, A. S., Lazar, A. & Iudicone, D. Mixed layer depth over the global ocean: an examination of profile data and a profile-based climatology. *J. Geophys. Res.* **109**, C12003 (2004).
23. Casey, K. S., Brandon, T. B., Cornillon, P. & Evans, R. in *Oceanography from Space: Revisited* (eds Barale, V., Gower, J. F. R. & Alberotanza, L.) 273–287 (Springer, 2010).
24. Alexander, M. A. & Deser, C. A mechanism for the recurrence of wintertime midlatitude SST anomalies. *J. Phys. Oceanogr.* **25**, 122–137 (1995).
25. Frankignoul, C., de Coëtlogon, G., Joyce, T. & Dong, S. Gulf Stream variability and ocean-atmosphere interactions. *J. Phys. Oceanogr.* **31**, 3516–3529 (2001).
26. Lozier, M. S. *et al.* The spatial pattern and mechanisms of heat content change in the North Atlantic. *Science* **319**, 800–803 (2008).
27. de Coëtlogon, G. & Frankignoul, C. The persistence of winter sea surface temperature in the North Atlantic. *J. Clim.* **16**, 1364–1377 (2003).
28. Sudre, J. & Morrow, R. A. Global surface currents: a high-resolution product for investigating ocean dynamics. *Ocean Dyn.* **58**, 101–118 (2008).

29. Ducet, N., Traon, P.-Y. & Reverdin, G. Global high resolution mapping of ocean circulation from TOPEX/Poseidon and ERS-1 and -2. *J. Geophys. Res.* **105,** 19477–19498 (2000).

Acknowledgements

We thank Stefan Gary (SAMS) for developing the platform to run the Lagrangian trajectories in the FLAME model, Apurva Dave (EPA) for his code to process the Ekman velocities and Claus Böning and Arne Biastoch (IFM-GEOMAR) for access to the FLAME model output. We are also grateful for the publicly available data from NOAA (AVHRR SST), CNES (AVISO SSH) and CTOH (Ekman velocities). This work was funded by the NSF (OCE-1259102) and the NASA Earth and Space Science Fellowship (NNX13AO21H).

Author contributions

N.P.F. and M.S.L. devised the study and wrote the manuscript.

Additional information

Competing financial interests: The authors declare no competing financial interests.

8

Radiocarbon constraints on the extent and evolution of the South Pacific glacial carbon pool

T.A. Ronge[1], R. Tiedemann[1], F. Lamy[1], P. Köhler[1], B.V. Alloway[2], R. De Pol-Holz[3], K. Pahnke[4], J. Southon[5] & L. Wacker[6]

During the last deglaciation, the opposing patterns of atmospheric CO_2 and radiocarbon activities ($\Delta^{14}C$) suggest the release of ^{14}C-depleted CO_2 from old carbon reservoirs. Although evidences point to the deep Pacific as a major reservoir of this ^{14}C-depleted carbon, its extent and evolution still need to be constrained. Here we use sediment cores retrieved along a South Pacific transect to reconstruct the spatio-temporal evolution of $\Delta^{14}C$ over the last 30,000 years. In ~2,500–3,600 m water depth, we find ^{14}C-depleted deep waters with a maximum glacial offset to atmospheric ^{14}C ($\Delta\Delta^{14}C = -1,000‰$). Using a box model, we test the hypothesis that these low values might have been caused by an interaction of aging and hydrothermal CO_2 influx. We observe a rejuvenation of circumpolar deep waters synchronous and potentially contributing to the initial deglacial rise in atmospheric CO_2. These findings constrain parts of the glacial carbon pool to the deep South Pacific.

[1] Alfred-Wegener-Institut Helmholtz-Zentrum für Polar- und Meeresforschung, Department for Marine Geology, PO Box 120161, Bremerhaven 27515, Germany. [2] School of Geography, Environment and Earth Sciences, Victoria University of Wellington, PO Box 600, 6012 Wellington, New Zealand. [3] GAIA-Antárctica Universidad de Magellanes, Department of Paleclimatology, Oceanography, Punta Arenas 01855, Chile. [4] Max Planck Research Group—Marine Isotope Geochemistry, Institute for Chemistry and Biology of the Marine Environment, Department of Marine Isotope Geochemistry, Carl von Ossietzky University, PO Box 2503, Oldenburg 26111, Germany. [5] School of Physical Science, Department of Earth Science, University of California, Irvine, California 92697-4675, USA. [6] Laboratory of Ion Beam Physics (HPK), Eidgenössische Technische Hochschule, Schafmattstrasse 20, Zürich 8093, Switzerland. Correspondence and requests for materials should be addressed to T.A.R. (email: Thomas.Ronge@awi.de).

The deep ocean contains the largest carbon reservoir within the global carbon cycle that might interact with the atmosphere on glacial/interglacial timescales. Therefore, the deglacial rise in atmospheric CO_2 by ~ 90 p.p.m.v. (ref. 1) was probably linked to significant modifications in oceanic circulation that resulted in increasing rates of CO_2 outgassing[2–4]. Thus, in order to sequester large amounts of atmospheric CO_2, the deep glacial ocean must have been effectively cut off from gas exchange with the atmosphere. Throughout a glacial period, the isolation of deep waters from the surface, and hence the atmosphere, leads to an accumulation of carbon (CO_2) and nutrients in the deep ocean, which is accompanied by a progressive depletion of radiocarbon (^{14}C). Consequently, the older a water mass gets, the more enriched in ^{14}C-depleted CO_2 it becomes.

So far, only isolated occurrences of old glacial water masses have been identified in the North and South Pacific, as well as in the South Atlantic, which suggests that the storage of CO_2 occurred in the deep glacial ocean[4–8] (Supplementary Fig. 1 and Supplementary Table 1). In particular, the overturning circulation of the Southern Ocean (SO), where nowadays $\sim 65\%$ of all deep waters make first contact with the atmosphere[9], controls the ventilation of the oceans interior. However, the surface residence time of upwelled waters before re-subduction is an important factor controlling the efficiency of air–sea gas exchange. Changes in the climate system of the SO, such as the intensification or weakening of stratification or westerly winds have the potential to significantly alter the oceanic uptake or release of CO_2 (refs 2, 3 and 10) and likewise the radiocarbon budget of deep waters. Therefore, the circum-Antarctic upwelling region is considered the most likely deglacial pathway of stored old carbon from the abyss to the atmosphere. In this oceanic window, carbon-rich deep waters like Pacific Deep Water (PDW) are mixed and upwelled and provide a major source for Antarctic Intermediate Water (AAIW), formed close to the Subantarctic Front (SAF)[11]. Hence, AAIW is able to propagate the circulation- and outgassing signals into the major ocean basins. In this context, numerous intermediate-water records have been analysed to track the timing and pathways of SO deep water upwelling[12–16]. The spatial and temporal dimension of the glacial reservoir itself as well as the pathway, magnitude and process of the deglacial CO_2 release remain elusive, although evidence for carbon storage in the deep glacial southwest Pacific is increasing[5,6].

Here, to better constrain the glacial carbon pool, its vertical extent and evolution, we use $\Delta^{14}C$-records from six sediment cores at the New Zealand Margin (NZM; Fig. 1a and Supplementary Fig. 2), covering the major South Pacific water masses AAIW and Upper Circumpolar Deep Water/Lower Circumpolar Deep Water between ~ 830 and $\sim 4,300$ m water depth (Fig. 1b). To assess the lateral extent of the glacial carbon pool, we have additionally analysed an open-ocean sediment core from the East Pacific Rise (EPR; PS75/059-2; 3,613 m) located more than 4,000 km east of the NZM (Fig. 1a). Our NZM depth transect is well suited for the analysis of SO water mass ventilation as we can record ^{14}C-depleted deep waters on their way to the upwelling region further south, as well as recently subducted intermediate waters moving towards the north (Fig. 1b).

We show that throughout the water column, a wide range of radiocarbon activities ($\Delta^{14}C$) indicates a highly stratified South Pacific during the last glacial. This stratification implicates pronounced sequestration of CO_2 in circumpolar deep waters below a water depth of $\sim 2,000$ m. Building on the hypothesis of increased glacial outgassing of volcanic CO_2 along mid-ocean ridges (MORs)[17–19], we use a simple box model to highlight that

the most extreme ^{14}C-depletion between $\sim 2,500$ and $\sim 3,600$ m water depth might be explained by a combination of aged ^{14}C-depleted waters and the additional admixture of ^{14}C-dead hydrothermal CO_2. At the end of the glacial period, our deep water $\Delta^{14}C$-values increase throughout the water column in unison to rising atmospheric CO_2-values[1]. On the basis of these patterns, we conclude that the deep South Pacific was an important contributor to the deglacial rise in atmospheric CO_2.

Results

Radiocarbon. To assess both glacial and deglacial ventilation changes, we used paired samples of *Globigerina bulloides* and mixed benthic foraminifers from seven new sediment cores, located south of the present Subtropical Front (STF) (Fig. 1a). The core locations have sedimentation rates between 2.5 cm per kyr and 22 cm per kyr (Supplementary Table 2). Before the Last Glacial Maximum (LGM), ~ 29 cal. ka, all deep-water masses (below $\sim 2,000$ m) show $\Delta^{14}C$-values ranging from $+200‰$ to $-100‰$ (Fig. 2). At the same time, AAIW $\Delta^{14}C$ is clearly elevated with values of $\sim 360‰$. The most obvious feature of our reconstructed $\Delta^{14}C$-values over the LGM is the large glacial range of $\Delta^{14}C$ between ~ 830 and $\sim 4,300$ m (400‰ to $-550‰$; Fig. 2). About 21 cal. ka, deep-water $\Delta^{14}C$ at 4,300 and 2,066 m increases, followed by increasing radiocarbon values at 2,500 m at the onset of the last deglaciation. Parallel to increasing deep-water radiocarbon concentrations, AAIW-$\Delta^{14}C$ slightly decreases. At ~ 14.7 cal. ka, the $\Delta^{14}C$-records of all water depths converge and continue to evolve parallel to each other throughout the deglaciation and into the Holocene (Fig. 2). In the discussion, we used $\Delta\Delta^{14}C$-records for our interpretations. These records represent the $\Delta^{14}C$ offset of our data to the $\Delta^{14}C$-value of the past atmosphere[20]. We also calculated the $\Delta\Delta^{14}C_{adj}$ according to Cook and Keigwin[21] (Supplementary Fig. 3). This method corrects the initial $\Delta^{14}C$ values to the modern, pre-industrial ^{14}C profile (ref. 21). As the trend in our data remains the same, regardless of the method used, we use $\Delta\Delta^{14}C$ for our discussion, in order to improve the comparability to other studies.

Box modelling. We used a 1-box model to investigate the hypothesis that hydrothermal CO_2-fluxes might have contributed to our reconstructed maximum depletion in $\Delta^{14}C$ in the glacial deep ocean (Methods). We simulated the single effect of hydrothermal CO_2 inflow on oceanic $\Delta^{14}C$, as well as two sensitivity runs in which hydrothermal CO_2 inflow is combined with either carbonate compensation, leading to sediment dissolution[22] or with CO_2 sequestration, potentially connected with deep-ocean volcanism[23]. Our model simulates for the single effect of different CO_2 flux rates F (in µmol per kg per year; Supplementary Fig. 4) a drop in $\Delta^{14}C$ by $-240‰$ (F = 0.3), $-380‰$ (F = 0.6), $-480‰$ (F = 0.9) and $-550‰$ (F = 1.2). Only if the response of the marine carbonate system to this CO_2 flux (by carbonate compensation and the dissolution of sediments) is considered[22] (Supplementary Fig. 4), we calculate $\Delta^{14}C$ amplitudes in agreement with our maximally depleted data (approximately $-500‰$ to $-600‰$) for a hydrothermal flux of 0.6 µmol kg per year or larger. However, as hot rocks interact with seawater, MOR volcanism is also discussed as a potential CO_2 sink[23]. If we implement this process of similar size of the hydrothermal CO_2 flux (no net oceanic carbon change and therefore no carbonate compensation) we need a hydrothermal CO_2 flux F of 1.2 µmol kg per year to meet the maximum $\Delta^{14}C$ depletion of $\sim -500‰$, as observed in our data (Fig. 3a). To convert the CO_2 fluxes to gross carbon fluxes (PgC per year) we estimated the minimum area covered by our sediment cores as a $\sim 1,000$-m-thick water mass (2,500–3,600 m as covered

a

b

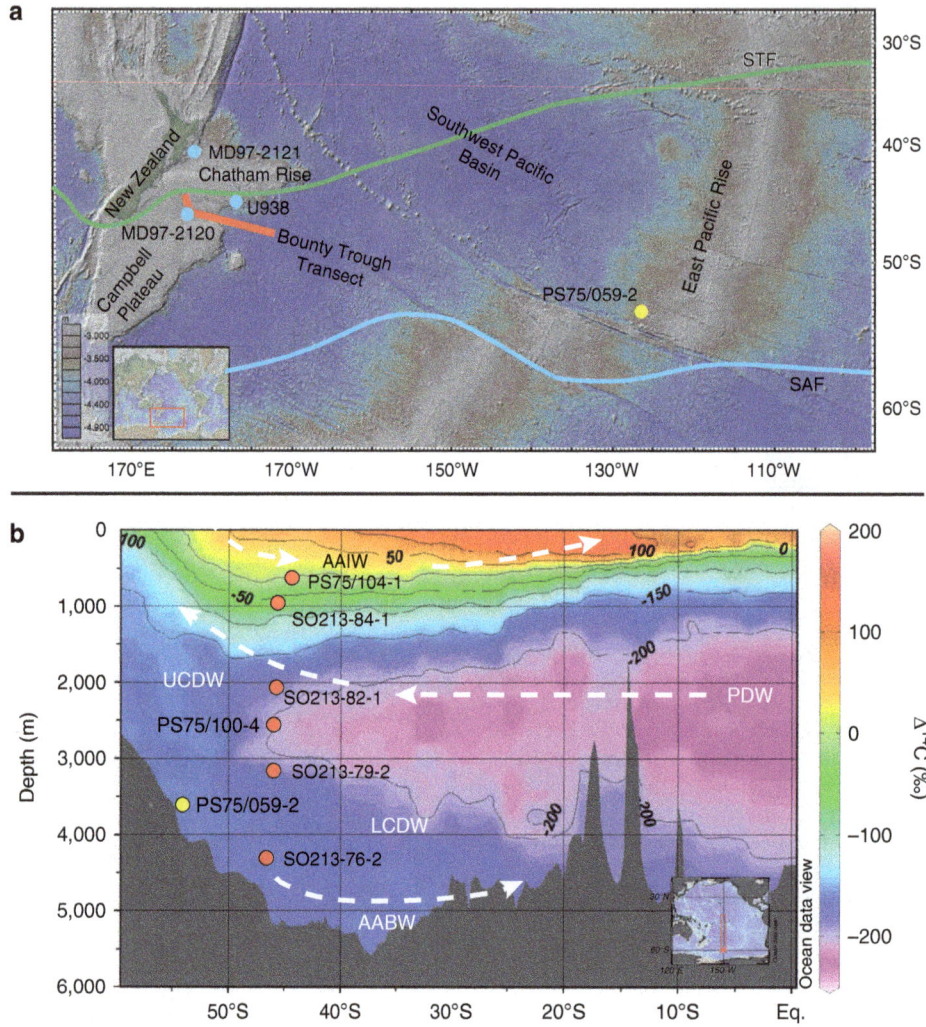

Figure 1 | Overview of the NZM showing core locations and water masses. (**a**) Map of the Southwest Pacific. The red bar represents the lateral area covered by our sediment core transect at the NZM. Yellow circle—position of our sediment core from the EPR. Blue circles—previous studies. U938 (ref. 5), MD97-2120 (ref. 15), MD97-2121 (ref. 6). Green line—Subtropical Front. Blue line—Subantarctic Front[66]. Map created using GeoMapApp. (**b**) Water mass section of modern South Pacific $\Delta^{14}C$-concentrations. Sediment cores (red and yellow dots) projected into WOCE line P16 (ref. 67). AABW, Antarctic Bottom Water; WOCE, World Ocean Circulation Experiment. Section generated using ODV 4.7.2 (ref. 68).

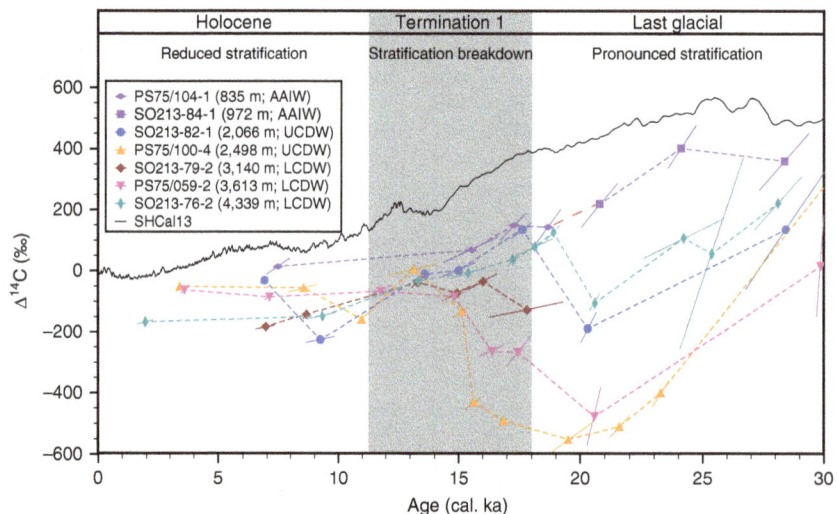

Figure 2 | $\Delta^{14}C$ changes of the major South Pacific water masses. The large glacial range in $\Delta^{14}C$ indicates a pronounced water mass stratification, followed by a progressive stratification breakdown during Termination 1 (grey shaded area). The broken red line indicates the area, where we spliced PS75/104-1 and SO213-84-1.

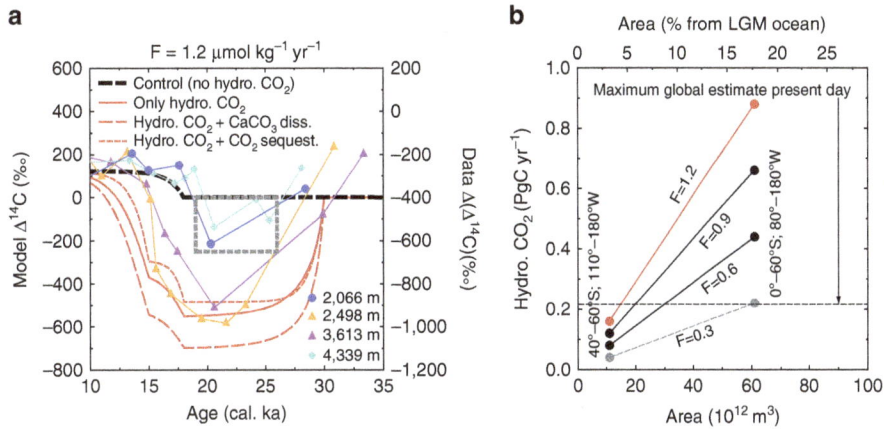

Figure 3 | Simulation results of a 1-box model for the effect of hydrothermal CO$_2$-outgassing on oceanic Δ^{14}C. (**a**) Comparison of model-based Δ^{14}C with data-based $\Delta\Delta^{14}$C. SO213-82-1 (2,066 m; blue line); PS75/100-2 (2,498 m; orange line); PS75/059-2 (3,613 m; pink line); and SO213-76-2 (4,339 m; green line). The impact of our best-guess hydrothermal CO$_2$ flux (F) of 1.2 µmol kg^{-1} yr^{-1} between 30 and 15 cal. ka (red solid line) is compared with a control run (black broken line). In the control run, we only decreased the turnover time at 18 cal. ka according to Skinner et al.[6] from 2,700 to 1,500 years. In two sensitivity runs, we estimated the influence of CaCO$_3$ dissolution or CO$_2$ sequestration (red broken lines). The grey box indicates the $\Delta\Delta^{14}$C-area covered by previous Southern Ocean studies[4,6,8]. (**b**) Upscaling of our localized results for different hydrothermal CO$_2$ fluxes (F) to regional carbon fluxes as a function of area. Present day maximum global estimate of hydrothermal CO$_2$ fluxes (0.22 PgC yr^{-1}) (ref. 24) indicated by the black broken line.

by PS75/100-4 and PS75/059-2) ranging from 40°S to 60°S and 110°W to 180°W. The fluxes necessary to influence such a water mass would lead to a hydrothermal injection of CO$_2$ of 0.08–0.16 PgC per year (modern global flux is estimated to up to ~0.22 PgC per year)[24]. Upscaling to a larger water mass, spanning most of the South Pacific (1,000 m; 0°–60°S; 80°–180°W) would imply that the hydrothermal CO$_2$ flux might be as large as 0.44–0.88 PgC per year (Fig. 3b).

Discussion

Throughout the water column, the observed glacial Δ^{14}C-range in our cores exceeds the modern and Holocene values by a factor of ~5 and indicates strong age differences and therefore enhanced stratification of the intermediate and deep glacial South Pacific. Along its pathway in the global thermohaline circulation PDW is fed into circumpolar waters and constitutes today's oldest water mass. The ^{14}C-depleted PDW presently extends to 2,000–2,500 m north of the Chatham Rise close to New Zealand[25] (Fig. 1b). From our transect, we are able to show that during the LGM, significantly ^{14}C-depleted and aged water masses occupied depths between ~2,000 and ~4,300 m in the Southern Westerly (SW) Pacific (Fig. 2). Our data locate the core of the ^{14}C-depleted water mass at the NZM in a of ~2,500 m (PS75/100-4; modern depth of Upper Circumpolar Deep Water; Fig. 4), yielding a maximum deep water to atmosphere offset in radiocarbon activities ($\Delta\Delta^{14}$C) of approximately −1,000‰. Analysing $\Delta\Delta^{14}$C corrects for any impacts of changes in ^{14}C production[26] as well as for variable ocean–atmosphere exchange rates. A corresponding apparent ventilation age, based on benthic minus reservoir-corrected planktic ^{14}C ages would equate to ~8,000 years (Supplementary Fig. 3d). A similar glacial $\Delta\Delta^{14}$C depletion of about −870‰ was reported from sediment core U938, which was recovered at the NZM in a water depth of 2,700 m (ref. 5) (Figs 1a and 4). We hypothesize that these extremely low $\Delta\Delta^{14}$C-values might be the result of the admixture of ^{14}C-dead hydrothermal CO$_2$ into a water mass with an initial high ventilation age, which was estimated to at least 2,700 years[6]. The upper and lower boundary of this old water mass are marked by higher $\Delta\Delta^{14}$C-values of −550‰ to −600‰ indicating a highly stratified water column (Fig. 4b). Similar $\Delta\Delta^{14}$C-values were

reported at the NZM north of Chatham Rise at 2,314 m (ref. 6). This confines the most radiocarbon-depleted waters to a depth below ~2,300 m. The observed trend of ^{14}C between 830 and 4,300 m parallels the highest glacial nutrient concentrations off New Zealand, between 2,000 and 3,000 m (ref. 27) (Fig. 5), likewise indicative for the presence of aged, nutrient rich (low δ^{13}C) and radiocarbon-depleted waters. Yet, the δ^{13}C reconstructions might yield a certain bias, as endobenthic (Uvigerina) instead epibenthic (Cibicidoides) foraminifera were used[27].

We traced the ^{14}C-depleted glacial carbon reservoir off New Zealand to the central South Pacific (EPR) 4,000 km east of the NZM. At this location, $\Delta\Delta^{14}$C-values are as negative as −900‰ at 3,600 m (PS75/059-2; Figs 1a and 4). Hence, we are confident that this water mass was not only restricted to the NZM but seems to have occupied large parts of the South Pacific. Further off, in the Drake Passage and the South Atlantic, glacial water masses have been identified in CDWs with $\Delta\Delta^{14}$C-values as low as −330‰ (ref. 8) and −540‰ (ref. 6), respectively (Fig. 4). In their timing and amplitudes, these records are similar to our Pacific radiocarbon signature characterizing the upper and lower boundary of the old carbon pool (Fig. 4b). Our intermediate-water record (SO213-84-1; modern depth of AAIW) shows the highest glacial $\Delta\Delta^{14}$C-values of our transect (approximately −90‰). We suggest that the ^{14}C-depleted deep waters represent the very old return flow from the North Pacific (PDW), similar to the modern circulation pattern (Fig. 1b). The distribution of radiocarbon in our reconstruction might indicate a floating carbon pool instead of a stagnant bottom layer. Several records from the North Pacific might corroborate this assumption. In the Gulf of Alaska[28] (MD02-2489) and off Kamchatka[29] (MD01-2416), as well, the glacial mid-depth water mass (Kamchatka) shows a considerable higher benthic to planktic ^{14}C offset than the deeper water mass off Alaska. Additional data from the northwest Pacific suggest the lowest glacial ^{14}C values in a water depth of ~2,300 m with better ventilated waters above and below[21]. In the Atlantic Ocean as well, Ferrari et al.[30] and Burke et al.[31] observed a mid-depth (floating) radiocarbon anomaly. A floating carbon pool might furthermore explain why no sign of old carbon was found in the deep equatorial Pacific below 4,000 m (ref. 32). Therefore, this record might have 'missed' the old, mid-depth carbon pool above.

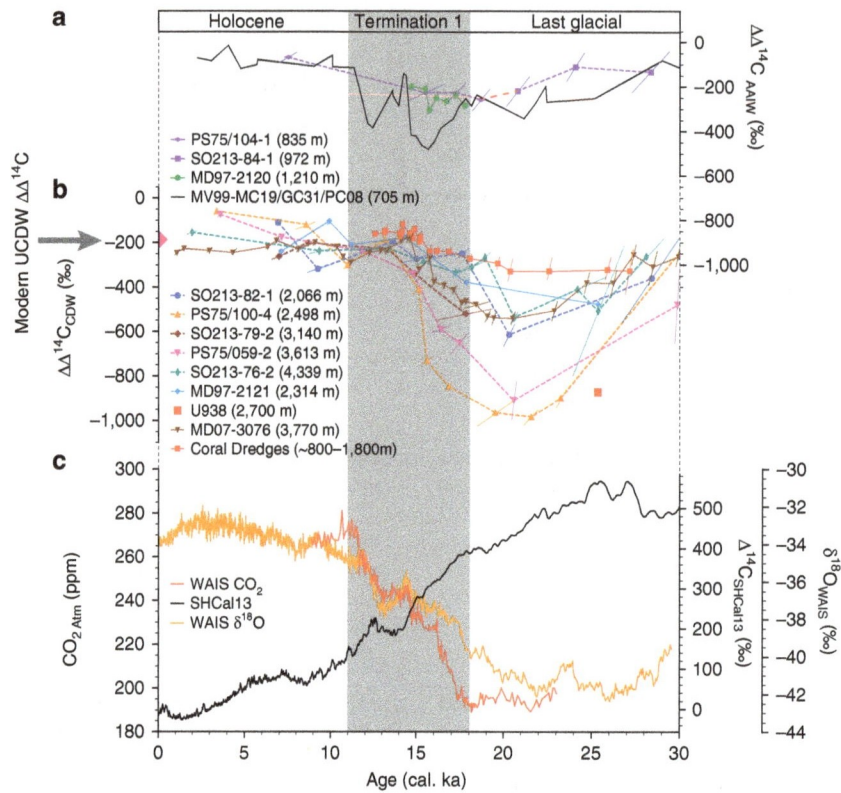

Figure 4 | Comparison of oceanic and atmospheric proxy records. $\Delta\Delta^{14}$C-values (ocean–atmosphere) of (**a**) the intermediate Pacific region; PS75/104-1 and SO213-84-1 (this study); MD97-2120 (ref. 15) (green line: Bounty Trough); MV99-MC19/GC31/PC08 (ref. 12) (black line: Northeast Pacific); (**b**) CDW $\Delta\Delta^{14}$C; PS75 and SO213 records (this study); MD97-2121 (ref. 6) (light blue line: north of Chatham Rise); U938[5] (red square: Bounty Trough); Coral dredges[8] (red line: Drake Passage); MD07-3076 (ref. 4 (brown line: South Atlantic); and Modern SW-Pacific UCDW $\Delta\Delta^{14}$C (ref. 67) (pink triangle). (**c**) Atmospheric CO_2 concentrations[1] (red line); WAIS δ^{18}O record[69] (orange line) and atmospheric Δ^{14}C-values[20] (black line). UCDW, Upper Circumpolar Deep Water.

Any explanation for the pronounced glacial radiocarbon-depletion of the deep SO has to involve a limited ocean–atmosphere exchange due to strengthened ocean stratification under glacial boundary conditions[3] (Fig. 6a). Northward-expanded Antarctic sea ice and SW Winds[33] contributed to reduced air–sea gas exchange and upwelling of deep-waters[30]. Surface freshening by melting sea-ice in the source regions of intermediate waters[34] and enhanced formation of highly saline Antarctic Bottom Water[35] may have set a density structure that led to reduced mixing and the encasement of old PDW (Fig. 6a). In addition, the shoaling of North Atlantic Deep Water[30] might have reduced the contribution of freshly ventilated waters into South Pacific CDW below ~2,000 m. These interacting key processes may have significantly contributed to the low radiocarbon values and are consistent with an enhanced glacial storage of carbon in the deep ocean.

Old water masses of 5,000–8,000 years are expected to be strongly oxygen depleted[36]. According to Sarnthein *et al.*[7], water masses with Δ^{14}C values lower than −350‰ would be completely anoxic. However, pronounced anoxia have not been documented in the deep South Pacific between 2,500 and 3,600 m. Therefore, an admixture of ^{14}C-dead carbon via submarine tectonic activity along MOR[18] into a an old water mass in the deep South Pacific might have contributed to the extremely low radiocarbon values of the water mass at ~2,500–3,600 m. During the LGM, sea floor eruption rates along tectonically active plate boundaries may have intensified due to the lower glacial sea level[17-19]. This process might have released significant amounts of ^{14}C-dead CO_2 into the water column. Using a simple 1-box model (Methods), we tested our hypothesis and calculated if

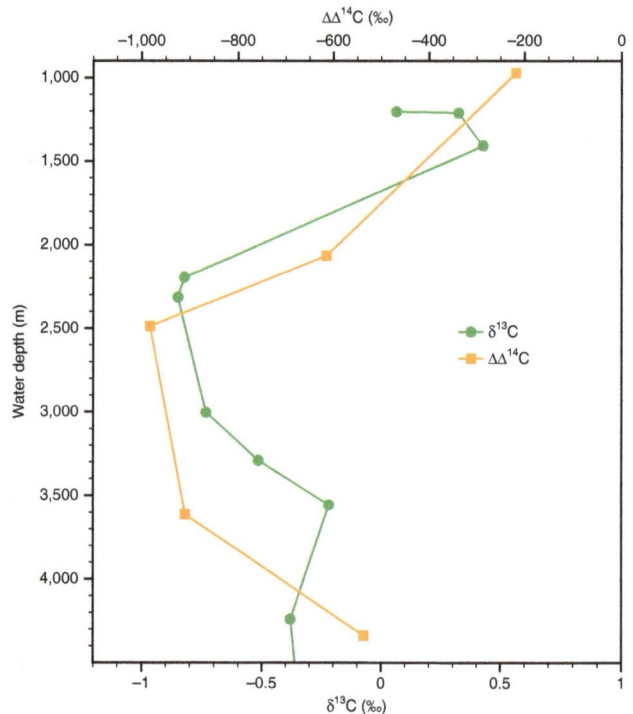

Figure 5 | Vertical distribution of carbon isotopes off New Zealand throughout the LGM. Orange line—$\Delta\Delta^{14}$C (this study). Green line $\delta^{13}C_{Uvigerina}$[27].

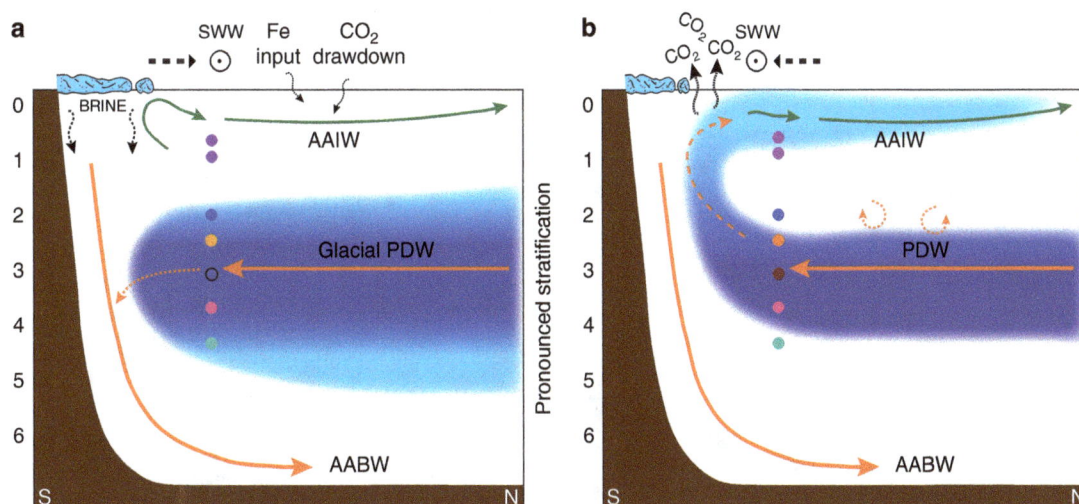

Figure 6 | Schematic representation of South Pacific overturning circulation. (a) Glacial pattern: northernmost extent of sea ice and SWW. Increased AABW-salinity by brine rejection favours stratification. Increased dust input promotes primary production and drawdown of CO_2. **(b)** Deglacial pattern: upwelling induced by southward shift of Antarctic sea ice and SWW. The erosion of the deep-water carbon pool releases ^{14}C-depleted CO_2 towards the atmosphere. Following air–sea gas exchange, the outgassing signal is incorporated into newly formed AAIW (light blue shading). Blue shading: poorly ventilated old and CO_2-rich waters; Darkest shading 2,500–3,600 m: water level influenced by hydrothermal CO_2. Green arrows: intermediate water; orange arrows: deep-water; light-blue areas: sea ice; SWW: Southern Westerly Winds; coloured circles: sediment cores (colour coding according to Fig. 2); black circle: SO213-79-2—no glacial data; and circular arrows: diffusional and diapycnal mixing.

the injection of hydrothermal CO_2 into the deep Pacific has the potential to amplify the $\Delta\Delta^{14}C$ minimum throughout the LGM (Fig. 3). To overcome the influence of the variable atmospheric ^{14}C-levels[26,37], we compared our simulated $\Delta^{14}C$ to our reconstructed $\Delta\Delta^{14}C$ values (deep ocean-to-atmosphere offset). The probably time-delayed response of submarine volcanism to changes in sea level complicate our flux calculations[19]. A crucial prerequisite for our hypothesis of the admixture of hydrothermal CO_2 is the presence of an already aged water mass with high nutrient concentrations and low $\Delta^{14}C$ levels (Fig. 7). According to the record of MD97–2121 (ref. 6) (2,314 m), the glacial ventilation age off New Zealand is at least 2,700 years. However, radiocarbon values might have been even lower as the MD97–2121 record lacks any data points between ~25 kyr and ~18 cal. ka (Fig. 4b). Our $\Delta\Delta^{14}C$ record of SO213-82-1 (2,066 m; Fig. 4b) is −600‰ at ~20kyr, ~100‰ lower than the minimum observed in MD97–2121 ~25 cal. ka, potentially indicating even higher turnover times. When we combine the radioactive decay, caused by an estimated water mass age of 2,700 years for the time of 35–18 cal. ka, with submarine ^{14}C-free volcanic CO_2 influx, our model calculates a decrease in $\Delta^{14}C$ for the corresponding water mass by additional −500‰ to −600‰ (Fig. 3a). This hydrothermal CO_2 outgassing (potentially accompanied by carbonate compensation and/or CO_2 sequestration) would lead to a maximum atmosphere-to-deep ocean offset of −800‰ to −1,000‰ $\Delta\Delta^{14}C$ (Supplementary Fig. 4e–g), comparable to the maximum depletion observed in PS75/059-2 and PS75/100-4. Today, in a water depth between ~2,500 and ~3,500 m, pronounced volcanic outgassing occurs along the southern EPR[38,39]. The resulting hydrothermal plume spreads towards the west and can be traced by the [3]He-signal in the broader western Pacific (Fig. 8)[40] and off northern New Zealand, right in the water depth under debate of ~2,500 m (ref. 41). Therefore, we argue that increased glacial outgassing of ^{14}C-dead volcanic CO_2 into a stratified ocean has the potential to significantly lower the $\Delta^{14}C$-content of an old (at least 2,700 years) water mass. The prominent Chatham Rise (Fig. 1a) might have acted as a physical barrier, blocking MD97–2121 (ref. 6)

(~2,300 m) from the volcanic plume. As MD97–2121 lacks data for most of the last glacial (~18–25 cal. ka), we cannot fully exclude that this core might have been affected by volcanic CO_2 to some extent. Nevertheless, the ^{14}C-data of MD97–2121 are already significantly higher at ~18 cal. ka compared to the values of PS75/100-4. Therefore, we argue that the influence (if any) of hydrothermal activity must have been lower to the north of the Chatham Rise and/or at 2,300 m water depth. Despite the in detail unknown processes accompanying such a hydrothermal carbon flux, its admixture might add additional carbon to the ocean–atmosphere–biosphere system (0.08–0.16 PgC per year; Fig. 3b). However, the net carbon injection depends in detail on the strength of the additional processes carbonate compensation and CO_2 sequestration and might also be zero. Once the glacial processes, favouring stratification, are reversed, any net injected carbon might eventually be released to the atmosphere along with the carbon already stored within the deep ocean (Fig. 5b). A further quantification of the net carbon injection and its contribution to atmospheric CO_2 is not yet possible, since future investigations with process-based models are necessary. Furthermore, as the distribution of MOR is inhomogeneous in the world ocean, it is difficult to compare our local results to global CO_2 flux estimates[24]. In Fig. 3b, we illustrate that the water mass affected by hydrothermal CO_2 might span an area of between 3 and 17% of the global glacial ocean. While the results for the minimum area, covered by our sediment cores, are below the maximum estimate of present day global estimate of hydrothermal outgassing, the results for an area representative for most of the South Pacific are a factor of 2–4 times higher (Fig. 3b). This suggests that if the admixture of hydrothermal CO_2 is the process that can explain the minimum $\Delta\Delta^{14}C$ values recorded in the mid-depth South Pacific (PS75/100-4; PS75/059-2; U938 (ref. 5)) the global CO_2-fluxes from MOR throughout the LGM might have been much larger than today. However, if the initial water mass was older than the 2,700 years assumed in our model, the resulting fluxes might also have been smaller than stated here. Although mantle-CO_2 is depleted in both, $\Delta^{14}C$ and also in $\delta^{13}C$ (−5±3‰ (refs 42 and 43)), its

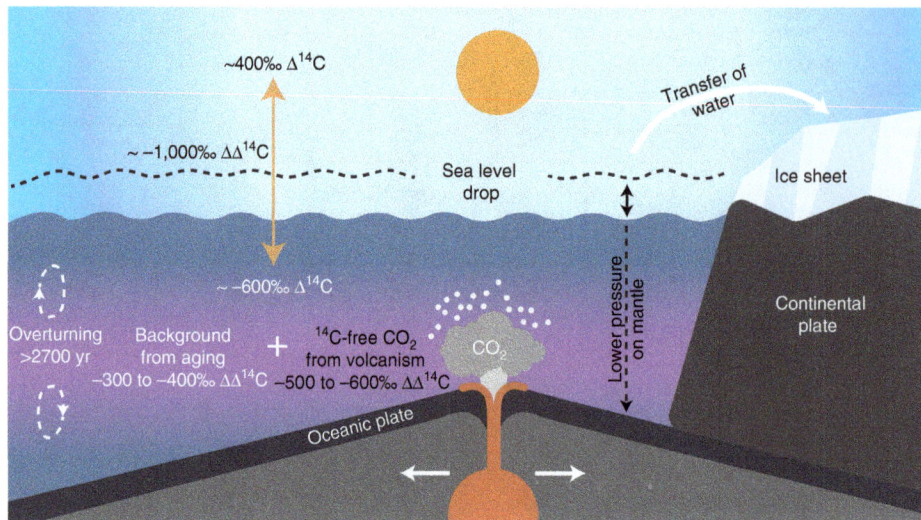

Figure 7 | Processes linking the glacial release of hydrothermal CO₂ and water mass Δ¹⁴C. The drop in global sea level triggers increased volcanic activity at MORs. The plume of ¹⁴C-dead hydrothermal CO_2 is mixed into an aged water mass, in which the combined effects of surface reservoir age and deep-ocean turnover time, of $\sim 2,700$ years, led already to a background ocean-to-atmosphere offset in $\Delta\Delta^{14}C$ of $-300‰$ to $-400‰$. This admixture of hydrothermal CO_2 further lowers the water masses $\Delta\Delta^{14}C$ by another $-500‰$ to $-600‰$, yielding a total $\Delta\Delta^{14}C$ of about $-1,000‰$ (purple layer). Modified after Hand[70]. Reprinted with permission from AAAS.

influence might be stronger on radiocarbon. As ^{14}C is by far less common in the ocean than ^{13}C, it can be diluted (lowered) more easily than its non-radiogenic counterpart.

We assume that the previously outlined glacial/interglacial changes in the SO climate system (position of sea ice and westerlies; changes in water mass densities; changes in upwelling and circulation) are the major factors influencing the spatio-temporal evolution of the oceanic carbon pool. However, the extreme minima in $\Delta\Delta^{14}C$ between 2,500 and 3,600 m water depth in the South Pacific are most plausibly explained by the hypothesized admixture of hydrothermal CO_2 into an old and already ^{14}C-depleted water mass. Admittedly, this process would complicate the use of ^{14}C as a ventilation proxy for the water masses affected by hydrothermal CO_2. However, as the presence of an already existing ^{14}C-depleted glacial water mass is a crucial prerequisite for our model, our theory does not interfere with the concept of stratification, decreased ventilation and the presence of a glacial oceanic carbon pool.

In the Pacific, other processes affect the radiocarbon inventory of water masses as well. Stott and Timmermann[44] already suggested the release of ^{14}C-depleted carbon from gas clathrates. As the stability of such clathrates is located in shallow waters ~ 400 m (ref. 44), this process is not applicable for our mid-depth anomaly below $\sim 2,500$ m. Therefore, we reject the possibility of any large influence of ^{14}C-depleted CO_2 and CH_4 clathrates.

At the end of the LGM and during the transition into the Holocene (~ 20–11.5 cal. ka), converging $\Delta\Delta^{14}C$-values argue for a progressive destratification (Fig. 4). During this interval, the deep-water-to-atmosphere offset in radiocarbon between $\sim 2,000$ and $\sim 4,300$ m decreases significantly (Fig. 4b). Although the resolution of our deep-water sediment cores is rather low, their $\Delta\Delta^{14}C$-values increase within error synchronous to the rise in atmospheric CO_2 rise (Fig. 4).

During Termination 1, when the most radiocarbon-depleted deep waters rejuvenate, no pronounced depletion in AAIW $\Delta\Delta^{14}C$ is recorded (Fig. 4a). However, the intermediate-water $\Delta\Delta^{14}C$-values remain low from ~ 18–15 cal. ka. The decrease in the deep water-to-atmosphere $\Delta\Delta^{14}C$-offset and the abrupt drop in $\delta^{13}C$ of atmospheric CO_2 (ref. 45) suggests increased air–sea gas exchange and the oceanic release of upwelled old CO_2

(Fig. 5b). The intermediate-waters from the NZM (this study) and the Chile Margin[14] significantly deviate from the (sub)tropical East Pacific, which shows two prominent drops in $\Delta\Delta^{14}C$ at the intermediate-water level during Termination 1 (refs 12, 13) (Fig. 4a). Therefore, it seems unlikely that southern sourced AAIW represents the source for the deglacial radiocarbon signals in the (sub)tropical East Pacific.

As we mentioned before, because of the close proximity, similar reservoir ages for all sediment cores are a requirement for our interpretations. However, the Holocene reservoir ages of two of our cores differ by $\sim 1,000$ years (Supplementary Fig. 5). Although this offset does not change the overall story, it prevents us from discussing the data of this time interval in more detail.

Our reconstructions provide new insights into the evolution and dynamic of the marine carbon inventory, its aging and the process of CO_2 release to the atmosphere. Growing evidence throughout large parts of the glacial South Pacific (this study; Skinner et al.[6]), the North Pacific[21,28,29], the Drake Passage[8] and the South Atlantic[4] suggest the existence of a floating body of very old, ^{14}C-depleted water between 2,000 and 4,300 m water depth, particularly in the tectonically active Pacific Ocean, where the glacial admixture of volcanic CO_2 might have influenced the ^{14}C-signature of this carbon pool from $\sim 2,500$ to $\sim 3,600$ m water depth. The effect of such volcanic CO_2 flux on the global carbon cycle as a whole and on atmospheric CO_2 in particular needs to be assessed in future studies, using more sophisticated models. During Termination 1, the atmosphere to deep-water $\Delta\Delta^{14}C$ of the carbon reservoir was reduced from about $-1,000‰$ to about $-200‰$ (Fig. 4b), indicating the erosion of the glacial carbon pool. This erosion is in accordance with the concept of a deglacial breakdown of SO stratification[3,45,46] and intensified deglacial wind-driven SO upwelling[2]. These processes ultimately culminated in the release of ^{14}C-depleted CO_2 from the deep ocean reservoir to the atmosphere (Fig. 5b), although in detail, the interaction of processes affecting the biological and physical pumps—and thus atmospheric CO_2— was probably more complex[47].

Methods
Sediment core details and sample treatment. The water mass transect that forms the backbone of this study consists of seven sediment cores retrieved during

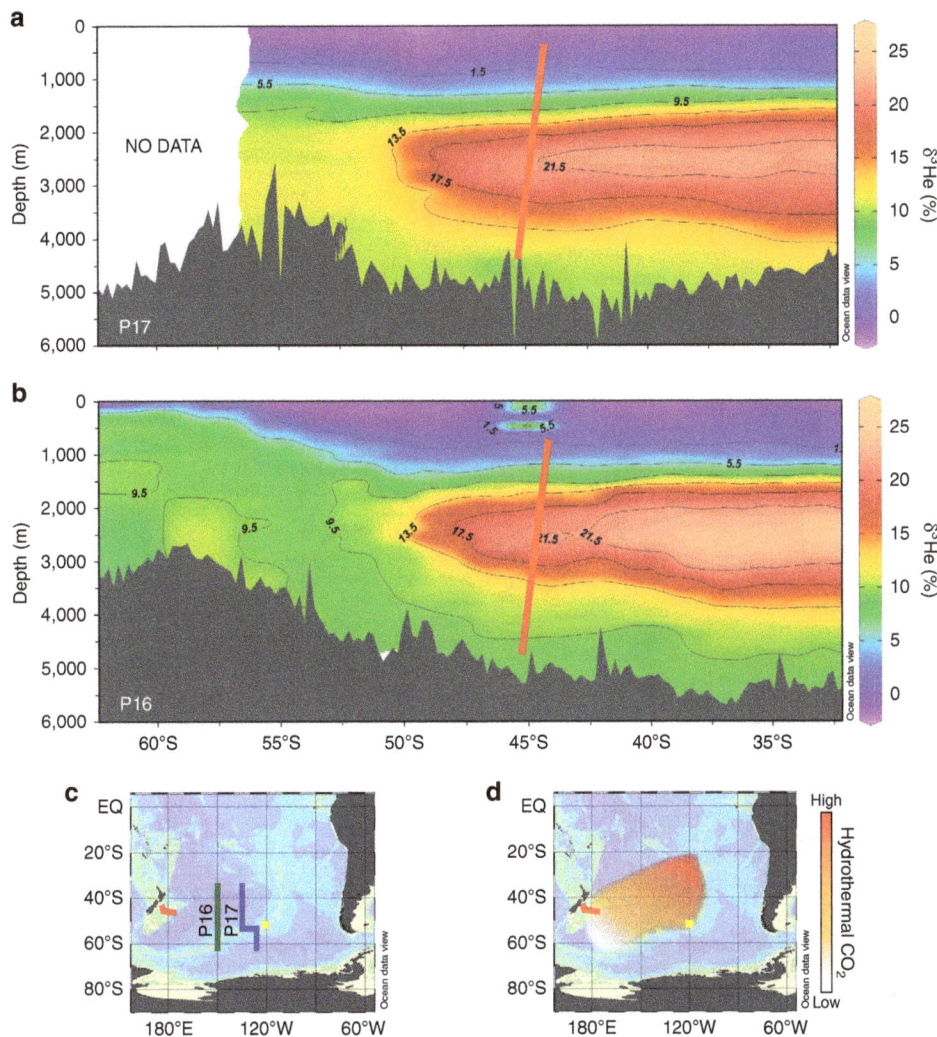

Figure 8 | Dispersal of hydrothermal ³He in the Southwest Pacific. The modern hydrothermal ³He plume, emanating at the southern EPR in a water depth of ~2,500 m (refs 38 and 39), can be traced throughout the southern Pacific towards New Zealand[41]. ³He distribution along (**a**) WOCE line P17 (ref. 67) (~135° W) and (**b**) WOCE line P16 (ref. 67) (150° W). (**c**) Locations of ³He sections P16 and P17. (**d**) Hypothesized dispersal of the glacial hydrothermal plume emanating from the EPR. Core locations and depths indicated by red bars and yellow squares. Panels generated using ODV 4.7.2 (ref. 68). WOCE, World Ocean Circulation Experiment.

the ANTXXVI/2 and SO213/2 cruises from the Bounty Trough off New Zealand (NZM) and from the EPR (Supplementary Figs 1 and 2). Collectively, these sediment cores record all water masses between 835 and 4,339 m, thus the modern water depths of the AAIW down to the Lower Circumpolar Deep Water with an imprint of Antarctic Bottom Water[27]. Positions, water depth, modern water masses and average sedimentation rates are reported in Supplementary Table 2.

An advantage of the NZM core transect is that it was not affected by potential glacial/interglacial shifts of the STF or the SAF[48,49]. The STF is bathymetrically fixed by the Chatham Rise (Supplementary Fig. 2), while the SAF is topographically steered by the submerged Campbell–Bounty Plateau[48,49]. As all sediment cores were located south of the STF and because of their close proximity to each other, we assume that changes in surface reservoir ages would affect all locations in a similar way. An exception is core PS75/059-2, which was retrieved ~4,200 km east of the Bounty Trough, at the western flank of the EPR (Supplementary Fig. 1) ~2°N of the SAF.

All sediment cores were split to form working and archival halves. The working half was sampled at 2 cm intervals. Depending on their water content, all samples were freeze dried for 2–3 days. Subsequently, the samples were wet sieved, using a 63 μm mesh sieve and dried at 50 °C for 2 days. As a last step, all samples were subdivided into the size fractions >400 μm, 315–400 μm, 250–315 μm, 125–250 μm and <125 μm. Benthic and planktic foraminifera were picked from 250 to 315 μm and 315 to 400 μm size fractions, taking great care to group similar sized individuals into samples for isotope analyses.

For the analysis of AAIW ventilation, we spliced sediment cores PS75/104-1 (835 m) and SO213-84-1 (972 m). We were forced, to combine both records as

PS75/104-1 did not yield a sufficient amount of benthic foraminiferal fauna below a core depth of ~100 cm (LGM and older), while SO213-84-1 was significantly disturbed above ~50 cm core depth (Termination 1 and younger).

Radiocarbon measurements. For the reconstruction of radiocarbon activities, corresponding pairs of planktic (monospecific *G. bulloides*) and benthic (mix of *Cibicidoides wuellerstorfi* and *Uvigerina peregrina*) foraminifera were picked. To minimize the effect of contamination on our analyses, we paid special attention not to pick any broken, discoloured or filled tests. Radiocarbon measurements were performed at the National Ocean Science Accelerator Mass Spectrometer (NOSAMS) facility in Woods Hole, USA, the W. M. Keck Carbon Cycle AMS Laboratory at the University of California in Irvine, USA and at the Laboratory for Ion Beam Physics at the Eidgenössische Technische Hochschule in Zurich, Switzerland.

We calculated the difference in benthic and reservoir-corrected planktic (B-P) ¹⁴C-ages to reconstruct the apparent ventilation ages of different water masses and compared our reconstructions with the method proposed by Cook and Keigwin[21] (Supplementary Fig. 4).

The equation of Adkins and Boyle[50] was used to determine the initial (paleo) radiocarbon activity (Δ^{14}C) of the benthic samples.

The difference of deep-water Δ^{14}C to the contemporaneous past atmospheric Δ^{14}C (called $\Delta\Delta^{14}$C) is the offset of our data from the IntCal13 reference curve[20].

Despite a potential bias by uncertainties in our age models and the IntCal13 (ref. 20) curve, we show in Supplementary Fig. 6 that the trend in $\Delta\Delta^{14}$C remains the same, regardless of the method used.

The radiocarbon dates for all cores are stored in the PANGAEA-database. Ventilation age errors were calculated from combined errors in ^{14}C ages and calibrated calendar ages. The error in $\Delta^{14}C$ was calculated from ^{14}C and calibrated age errors. All ^{14}C-data can be found in Supplementary Table 3 and on the PANGAEA-database.

Age control. For all sediment cores, an initial radiocarbon chronology was obtained from planktic ^{14}C-datings. Planktic radiocarbon ages were calibrated to calendar ages, using the calibration software Calib 7.0 (refs 51 and 52) with the embedded SHCal13 calibration curve[20]. To account for surface reservoir effects, we corrected all ^{14}C-ages according to reservoir age estimates by Skinner et al.[6]. However, to allow the use of ^{14}C as an unbiased proxy for deep-water ventilation, we fine-tuned our records to the nearby reference core MD97-2120 (ref. 53) (Supplementary Figs 7–9). We chose MD97-2120 (1,210 m water depth) as a reference core because of its position close to our sediments cores in the Bounty Trough. The original stratigraphy of MD97-2120 (refs 53 and 54) is based on nine ^{14}C measurements in the time interval between 0–35 kyr. Following the method applied by Rose et al.[15], we correlated the planktic $\delta^{18}O$ and Mg/Ca-derived SST records of MD97-2120 (ref. 54) with the EDC ice core δD record[55,56] (Supplementary Fig. 7). Correlating MD97-2120 via surface temperatures to Antarctic ice cores enables us to obtain a ^{14}C-independent stratigraphy that we can use as the baseline for the X-ray fluorescence (XRF) core-to-core correlation.

At the Alfred Wegener Institute in Bremerhaven, Germany, all sediment cores were analysed for their specific element abundances using an Avaatech XRF core-scanner with a preset step width of 1 cm. The resulting element content records (Sr, Ca, Fe and Sr/Fe) were then correlated to respective age-scaled XRF records of the sediment core MD97-2120 (refs 53 and 54) from the southern Chatham Rise (Supplementary Fig. 8) using the computer program AnalySeries 2.0.4.2. In particular, the Sr-counts were useful for the core-to-core correlation, as Sr represents CaCO₃, but is barley affected by differences in grain sizes or the sediments water-content[57]. Furthermore, we were able to utilize a distinctive vitric-rich tephra layer as a radiocarbon-independent stratigraphic marker within two sediment cores (SO213-76-2 and SO213-79-2). The tephra layer in both cores was geochemically characterized and identified as the widespread Kawakawa/ Oruanui tephra (KOT; for analytical details see the section on tephra analyses). The KOT is a widespread silicic tephra erupted from Taupo Volcanic Centre in the central North Island of New Zealand with an age of ~25.36 cal. ka (ref. 58) and is the most important isochronous tephra marker erupted in the SW-Pacific region during the past 30,000 years[59]. This tephra was likewise found in reference core MD97-2120 (ref. 54). We adjusted the KOT age used by Pahnke et al.[54] to the revised age by Vandergoes et al.[58].

The age model of the EPR-core PS75/059-2 is based on the original approach of Lamy et al.[60]. However, as this age model lacks detailed tie-points between 0 and 30 cal. ka, we fine-tuned PS75/059-2 to the age-scaled record PS75/100-4, using their Sr XRF element counts. Despite a certain lack in the LGM variability of most XRF-records, we consider our age models as relatively robust. In particular the comparison of our records SO213-82-1 and SO213-76-2 to records from similar water depths in the South Pacific[6] (MD97-2121), the Drake Passage[8] (coral dredges) and the South Atlantic[4] (MD07-3076), reveals a similar timing as well as comparable $\Delta\Delta^{14}C$-values (Fig. 4) for all records.

The errors for the correlated age models were estimated from the offset to the ^{14}C-derived age model plus ^{14}C-errors.

Our XRF-based correlation method creates new surface reservoir ages for our sediment records. These differ from the surface reservoir age record of MD97-2121 (ref. 6), but are still in good agreement with these reconstructions (Supplementary Fig. 5). However, owing the XRF-correlation method, our reservoir ages show a slightly higher scatter than the records of Sikes et al.[5] and Skinner et al.[6]

Tephra analyses. In cores SO213-76-2 (507–508 cm) and SO213-79-2 (150–151 cm) a distinctive vitric-rich tephra layer was identified. Morphological expression, grain size and thickness of this tephra were indistinguishable and hence, most likely to represent the same eruptive event. Glass shard major element chemistry of this tephra layer was then determined by electron microprobe analysis and the results were compared with selected onshore and offshore KOT correlatives (Supplementary Fig. 10). The glass shard geochemistries were consequently indistinguishable which (a) affirms correlation to KOT and (b) augments the overall chronology of this study. All major element determinations were made on a JEOL Superprobe (JXA-8230) housed at Victoria University of Wellington, using the ZAF correction method. Analyses were performed using an accelerating voltage of 15 kV and a static electron beam operating at 8 nA. The electron beam was defocused between 10 and 20 μm. All elements calculated on a water-free basis, with H₂O by difference from 100%. Total Fe expressed as FeO_t. Mean and ±1 s.d. (Supplementary Table 4; in parentheses), based on n analyses. All samples normalized against glass standards VG-568 and ATHO-G.

Box modelling. Here we make some first-order estimates investigating the hypothesis of an impact of a potential hydrothermal CO₂ flux on the marine carbonate system and on $\Delta^{14}C$ in the South Pacific. We simulate carbon cycle changes in a 1-box model water mass, which might represent the mid-depth South

Pacific waters ~2,500–3,500 m (Supplementary Fig. 4a). We start our carbon cycle simulations from an LGM state that was previously simulated with the BICYCLE model[46]. We thus perturb a water mass with dissolved inorganic carbon (DIC) concentration of ~2,600 μmol kg⁻¹, with a total alkalinity of 2,667 μmol kg⁻¹. For LGM conditions (temperature ~0 °C; salinity = 35.8 PSU), these factors would lead to a pH of 7.7 and a carbonate ion concentration of ~70 μmol kg⁻¹.

We base our calculations on changes in concentrations for an undefined volume of the water mass and increase the amplitude of hydrothermal CO₂ until the ^{14}C-anomaly, seen in our data, is reproduced by our most simplistic model.

Similar to our data from 2,066 m (SO213-82-1), Skinner et al.[6] provide an independent glacial ventilation age of ~2,700 years, which decreases to ~1,500 years at 18 cal. ka. To obtain a stable $\Delta^{14}C$ of 0‰ in our water mass ~30 cal. ka (Fig. 3a, black broken line), the incoming carbon flux from X_i from oceanic mixing processes has a $\Delta^{14}C$ signature of +328‰. Corresponding to the decrease in ventilation age from 2,700 to 1,500 years, this flux increases at 18 cal. ka from 1 to 1.7 μmol per kg per year (Supplementary Fig. 4c). The outgoing carbon flux X_o, leaving our simulated 1-box water mass via oceanic transport processes, is determined by the turnover/ventilation time τ. X_o might change over time in size due to hydrothermal CO₂ injection, causing a rise in DIC within the water parcel. For the time window of minimum sea level[61] (Supplementary Fig. 4b) between 30 and 15 cal. ka, we assume a hydrothermal CO₂ flux (F) of 0, 0.3, 0.6, 0.9 or 1.2 μmol per kg per year (Supplementary Fig. 4d–g). Please note that this approach assumes an instantaneous feedback of the hydrothermal CO₂ outgassing rate to the removed load from the drop in global sea level and neglects any potential time delay in the solid earth response. Time delay of this process were briefly discussed previously[18] but might also be more complex[19].

On millennial timescales, the injection of hydrothermal CO₂ might lead to a partly dissolution of CaCO₃ in oceanic sediments[22,62]. Via this process of carbonate compensation, carbonate ions are brought into solution and added to the DIC pool. In this dissolution flux (D), the ratio of alkalinity and DIC is 2:1. If enough CaCO₃ is available for dissolution, the additional dissolution flux from carbonate compensation is on the order of 88% of the initial CO₂ injection[63]. In case of insufficient amounts of available CaCO₃, carbonate ion concentration and pH would fall. So far, it is estimated that the amount of dissolvable sediments is restricted to ~1,600 PgC (ref. 64), although simulation studies show that the actual dissolution for large CO₂-injections is smaller than this[22]. While the actual ^{14}C signature of dissolved CaCO₃ is not readily known, we assume the dissolution of ^{14}C-free sediment as the upper limit. This assumption is supported by the existence of very old surface sediments in the pelagic Southeast Pacific off Chile[65]. For simplicity, we here calculate an instantaneous additional impact of carbonate compensation. However, since this process is in reality time delayed with an e-folding time of a few thousand years[22], the most likely solution lies between both scenarios, with and without carbonate compensation.

In addition, we estimate the impact on $\Delta^{14}C$, if CO₂, of similar size (S) as the hydrothermal injection flux (F), is directly sequestered via magmatic processes[23], leading to stable DIC concentrations.

Please note that the three processes considered here (hydrothermal CO₂ flux F; carbonate dissolution D; CO₂ sequestration S) are roughly estimated due to their potential impact on oceanic $\Delta^{14}C$. In detail, these processes might be time-delayed or offset from each other.

Code availability. The computer code for the 1-box model used to calculate the simulation results shown in Fig. 3 and Supplementary Fig. 4 is available upon request from one of the co-authors (PK; peter.koehler@awi.de).

References

1. Marcott, S. A. et al. Centennial-scale changes in the global carbon cycle during the last deglaciation. Nature 514, 616–619 (2014).
2. Anderson, R. F. et al. Wind-driven upwelling in the southern ocean and the deglacial rise in atmospheric CO₂. Science 323, 1443–1448 (2009).
3. Sigman, D. M., Hain, M. P. & Haug, G. H. The polar ocean and glacial cycles in atmospheric CO₂ concentration. Nature 466, 47–55 (2010).
4. Skinner, L. C., Fallon, S., Waelbroeck, C., Michel, E. & Barker, S. Ventilation of the deep Southern Ocean and deglacial CO₂ rise. Science 328, 1147–1151 (2010).
5. Sikes, E. L., Samson, C. R., Guilderson, T. P. & Howard, W. R. Old radiocarbon ages in the Southwest Pacifc Ocean during the last glacial period and deglaciation. Nature 405, 555–559 (2000).
6. Skinner, L. C. et al. Reduced ventilation and enhanced magnitude of the deep Pacific carbon pool during the last glacial period. Earth Planet. Sci. Lett. 411, 45–52 (2015).
7. Sarnthein, M., Schneider, B. & Grootes, P. M. Peak glacial ^{14}C ventilation ages suggest major draw-down of carbon into the abyssal ocean. Clim. Past 9, 929–965 (2013).
8. Chen, T. et al. Synchronous centennial abrupt events in the ocean and atmosphere during the last deglaciation. Science 349, 1537–1541 (2015).
9. DeVries, T. & Primeau, F. An improved method for estimating water-mass ventilation age from radiocarbon data. Earth Planet. Sci. Lett. 295, 367–378 (2010).

10. Toggweiler, J. R., Russell, J. L. & Carson, S. R. Midlatitude westerlies, atmospheric CO_2, and climate change during the ice ages. *Paleoceanography* **21**, PA2005 (2006).

11. Bostock, H. C., Sutton, P. J., Williams, M. J. M. & Opdyke, B. N. Reviewing the circulation and mixing of Antarctic intermediate water in the South Pacific using evidence from geochemical tracers and Argo float trajectories. *Deep-Sea Res. I* **73**, 84–98 (2013).

12. Marchitto, T. M., Lehman, S. J., Ortiz, J. D., Flückinger, J. & van Geen, A. Marine radiocarbon evidence for the mechanism of deglacial atmospheric CO_2 rise. *Science* **316**, 1456–1459 (2007).

13. Stott, L. D., Southon, J., Timmermann, A. & Koutavas, A. Radiocarbon age anomaly at intermediate water depth in the Pacific Ocean during the last deglaciation. *Paleoceanography* **24**, PA2223 (2009).

14. De Pol-Holz, R., Keigwin, L. D., Southon, J., Hebbeln, D. & Mohtadi, M. No signature of abyssal carbon in intermediate waters off Chile during deglaciation. *Nat. Geosci.* **3**, 192–195 (2010).

15. Rose, K. A. *et al.* Upper-ocean-to-atmosphere radiocarbon offsets imply fast deglacial carbon dioxide release. *Nature* **466**, 1093–1097 (2010).

16. Lindsay, C. M., Lehman, S. J., Marchitto, T. M. & Ortiz, J. D. The surface expression of radiocarbon anomalies near Baja California during deglaciation. *Earth Planet. Sci. Lett.* **422**, 67–74 (2015).

17. Lund, D. C. & Asimow, P. D. Does sea level influence mid-ocean ridge magmatism on Milankovitch timescales? *Geochem. Geophys. Geosyst.* **12**, Q12009 (2011).

18. Tolstoy, M. Mid-ocean ridge eruptions as a climate valve. *Geophys. Res. Lett.* **42**, 1346–1351 (2015).

19. Crowley, J. W., Katz, R. F., Huybers, P., Langmuir, C. H. & Park, S.-H. Glacial cycles drive variations in the production of oceanic crust. *Science* **347**, 1237–1240 (2015).

20. Reimer, P. J. *et al.* IntCal13 and Marine13 radiocarbon age calibration curves 0–50,000 years Cal BP. *Radiocarbon* **55**, 1869–1887 (2013).

21. Cook, M. S. & Keigwin, L. D. Radiocarbon profiles of the NW Pacific from the LGM and deglaciation: evaluating ventilation metrics and the effect of uncertain surface reservoir ages. *Paleoceanography* **30**, 174–195 (2015).

22. Archer, D., Kheshgi, H. & Maier-Reimer, E. Multiple timescales for neutralization of fossi fuel CO_2. *Geophys. Res. Lett.* **24**, 405–408 (1997).

23. Alt, J. C. & Teagle, D. A. H. The uptake of carbon during alteration of ocean crust. *Geochim. Cosmochim. Acta* **63**, 1527–1535 (1999).

24. Burton, M. R., Sawyer, G. M. & Granieri, D. Deep carbon emissions from volcanoes. *Rev. Mineral. Geochem.* **75**, 323–354 (2013).

25. Bostock, H. C., Hayward, B. W., Neil, H. L., Currie, K. I. & Dunbar, G. B. Deep-water carbonate concentrations in the Southwest Pacific. *Deep-Sea Res. I* **58**, 72–85 (2011).

26. Köhler, P., Muscheler, R. & Fischer, H. A model-based interpretation of low-frequency changes in the carbon cycle during the last 120,000 years and its implications for the reconstruction of atmospheric $\Delta^{14}C$. *Geochem. Geophys. Geosyst.* **7**, Q11N06 (2006).

27. McCave, I. N., Carter, L. & Hall, I. R. Glacial-interglacial changes in water mass structure and flow in the SW Pacific Ocean. *Quat. Sci. Rev.* **27**, 1886–1908 (2008).

28. Rae, J. W. B. *et al.* Deep water formation in the North Pacific and deglacial CO_2 rise. *Paleoceanography* **29**, 645–667 (2014).

29. Sarnthein, M., Grootes, P. M., Kennet, J. P. & Nadeau, M.-J. in *Ocean Circulation: Mechanisms and Impacts—Past and Future Changes of Meridional Overturning* Vol. 173 (eds Schmittner, A., Chiang, J. C. H. & Hemming, S. R.) 175–196 (AGU, 2007).

30. Ferrari, R. *et al.* Antarctic sea ice control on ocean circulation in present and glacial climates. *Proc. Natl Acad. Sci. USA* **111**, 8753–8758 (2014).

31. Burke, A. *et al.* The glacial mid-depth radiocarbon bulge and its implications for the overturning circulation. *Paleoceanography* **30**, 1021–1039 (2015).

32. Broecker, W. & Clark, E. Search for a glacial-age ^{14}C-depleted ocean reservoir. *Geophys. Res. Lett.* **37**, L13606 (2010).

33. Kohfeld, K. E. *et al.* Southern Hemisphere westerly wind changes during the Last Glacial Maximum: paleo-data synthesis. *Quat. Sci. Rev.* **68**, 76–95 (2013).

34. Ronge, T. A. *et al.* Pushing the boundaries: Glacial/Interglacial variability of intermediate- and deep-waters in the southwest Pacific over the last 350,000 years. *Paleoceanography* **30**, 23–38 (2015).

35. Adkins, J. F. The role of deep ocean circulation in setting glacial climates. *Paleoceanography* **28**, 539–561 (2013).

36. Hain, M. P., Sigman, D. M. & Haug, G. H. Shortcomings of the isolated abyssal reservoir model for deglacial radiocarbon changes in the mid-depth Indo-Pacific Ocean. *Geophys. Res. Lett.* **38**, L04604 (2011).

37. Hain, M. P., Sigman, D. M. & Haug, G. H. Distinct roles of the Southern Ocean and North Atlantic in the deglacial atmospheric radiocarbon decline. *Earth Planet. Sci. Lett.* **394**, 198–208 (2014).

38. Lupton, J. Hydrothermal helium plumes in the Pacific Ocean. *J. Geophys. Res.* **103**, 15853–15868 (1998).

39. Lupton, J. & Craig, H. A major Helium-3 source at 15°S on the East Pacific Rise. *Science* **214**, 13–18 (1981).

40. Talley, L. D. *Hydrographic Atlas of the World Ocean Circulation Experiment (WOCE). Vol. 2: Pacific Ocean* (eds Sparrow, M., Chapman, P. & Gould, J.) (International WOCE Project Office, Southampton, UK, 2007).

41. de Ronde, C. E. J. *et al.* Intra-oceaniic subduction-related hydrothermal venting, Kermadec volcanic arc, New Zealand. *Earth. Planet. Sci. Lett.* **193**, 359–369 (2001).

42. Sano, Y. & Marty, B. Origin of carbon in fumarolic gas from island arcs. *Chem. Geol.* **119**, 265–274 (1995).

43. Deines, P. The carbon isotope geochemistry of mantle xenoliths. *Earth Sci. Rev.* 247–278 (2002).

44. Stott, L. D. & Timmermann, A. in *Abrupt Climate Change: Mechanisms, Patterns, and Impacts* (eds Harunur, R., Polyak, L. & Mosley-Thompson, E.) 123–138 (American Geophysical Union, 2011).

45. Schmitt, J. *et al.* Carbon isotope constraints on the deglacial CO_2 rise from ice cores. *Science* **336**, 711–714 (2012).

46. Köhler, P., Fischer, H., Munhoven, G. & Zeebe, R. E. Quantitative interpretation of atmospheric carbon records over the last glacial termination. *Global Biochem. Cycles* **19**, GB4020 (2005).

47. Völker, C. & Köhler, P. Responses of ocean circulation and carbon cycle to changes in the position of the Southern Hemisphere westerlies at Last Glacial Maximum. *Paleoceanography* **28**, 726–739 (2013).

48. Sikes, E. L., Howard, W. R., Neil, H. L. & Volkman, J. K. Glacial-interglacial sea surface temperature changes across the subtropical front east of New Zealand based on alkenone unsaturation ratios and foraminiferal assemblages. *Paleoceanography* **17**, PA000640 (2002).

49. Hayward, B. W. *et al.* The effect of submerged plateaux on Pleistocene gyral circulation and sea-surface temperatures in the Southwest Pacific. *Global Planet. Change* **63**, 309–316 (2008).

50. Adkins, J. F. & Boyle, E. A. Changing atmospheric $\Delta^{14}C$ and the record of deepwater paleoventilation ages. *Paleoceanography* **12**, 337–344 (1997).

51. Stuiver, M. & Reimer, P. J. Extended 14C Data Base and Revised Calib 3.0 ^{14}C Age Calibration Program. *Radiocarbon* **35**, 215–230 (1993).

52. Calib 7.0 v. 7.0. Available at: http://calib.qub.ac.uk (2014).

53. Pahnke, K. & Zahn, R. Southern Hemisphere water mass conversion linked with North Atlantic climate variability. *Science* **307**, 1741–1746 (2005).

54. Pahnke, K., Zahn, R., Elderfield, H. & Schulz, M. 340,000-Year centennial-scale marine record of Southern Hemisphere climatic oscillation. *Science* **301**, 948–952 (2003).

55. EPICA Comminity Members. One-to-one coupling of glacial climate variability in Greenland and Antarctica. *Nature* **444**, 195–198 (2006).

56. Veres, D. *et al.* The Antarctic ice core chronology (AICC2012): an optimized thousand years multi-parameter and multi-site dating approach for the last 120. *Clim. Past* **9**, 1773–1748 (2013).

57. Tjallingii, R., Kölling, M. & Bickert, T. Influence of the water content on X-ray fluorescence core- scanning measurements in soft marine sediments. *Geochem. Geophys. Geosyst.* **8**, Q02004 (2007).

58. Vandergoes, M. J. *et al.* A revised age for the Kawakawa/Oruanui tephra, a key marker for the Last Glacial Maximum in New Zealand. *Quat. Sci. Rev.* **74**, 195–201 (2013).

59. Alloway, B. *et al.* Towards a climate event stratigraphy for New Zealand over the past 30 000 years (NZ-INTIMATE project). *J. Quat. Sci.* **22**, 9–35 (2007).

60. Lamy, F. *et al.* Increased dust deposition in the Pacific Southern Ocean during glacial periods. *Science* **343**, 403–407 (2014).

61. Bintanja, R., van de Wal, R. S. W. & Oerlemans, J. Modelled atmospheric temperatures and global sea levels over the past million years. *Nature* **437**, 125–128 (2005).

62. Zeebe, R. E. & Caldeira, K. Close mass balance of long-term carbon fluxes from ice-core CO_2 and ocean chemistry records. *Nat. Geosci.* **1**, 312–315 (2008).

63. Zeebe, R. E. & Wolf-Gladrow, D. in *CO2 in Seawater: Equilibrium, Kinetics, Isotopes Vol. 65 Elsevier Oceanography Series.* (eds Zeebe, R. E. & Wolf-Gladrow, D.Ch. 3, 141–250 (Elsevier, 2001).

64. Archer, D. An Atlas of the distribution of calcium carbonate in sediments of the deep sea. *Global Biochem. Cycles* **10**, 159–174 (1996).

65. Tiedemann, R. *FS Sonne Fahrtbericht/Cruise Report SO213* (Alfred Wegener Institute, 2012).

66. Orsi, A. H., Whitworth, III T. & Nowlin, Jr W. D. On the meridional extent and fronts of the Antarctic circumpolar current. *Deep Sea Res. I* **42**, 641–673 (1995).

67. Key, R. M. *et al.* A global ocean carbon climatology: results from Global Data Analysis Project (GLODAP). *Global Biochem. Cycles* **18**, GB4031 (2004).

68. Ocean Data View v. 4.7.2. Available at: http://owa.awi.de (2015).

69. WAIS Divide Project Members. Onset of deglacial warming in West Antarctica driven by local orbital forcing. *Nature* **500**, 440–444 (2013).

70. Hand, E. Seafloor grooves record the beat of the ice ages. *Science* **347**, 593–594 (2015).

Acknowledgements

We thank the crews of *R/V Sonne* and *R/V Polarstern*; M. Sarnthein for helpful comments; R. Fröhlking, B. Glückselig, N. Lensch, L. Ritzenhofen, L. Schönborn, M. Seebeck, R. Sieger, M. Warnkroß and S. Wiebe for technical support. T.A.R. was funded by the Federal Ministry of Education and Research (BMBF; Germany) project 03G0213A—SOPATRA. BVA received funded support from an internal Victoria University of Wellington Science Faculty Research Grant. R.D.P.-H. was funded by Chilean FONDAP 15110009 and Iniciativa Científica Milenio NC120066. Data are available at http://doi.pangaea.de/10.1594/PANGAEA.833663.

Author contributions

T.A.R., R.T. and F.L. designed the study and wrote most of the manuscript; T.A.R. performed XRF measurements; T.A.R., R.D.P.-H., L.W. and J.S. performed accelerator mass spectrometry measurements; P.K. constructed the box model to calculate the effect of hydrothermal CO_2-injection on $\Delta^{14}C$; K.P. provided XRF measurements on reference core MD97-2120; B.V.A. performed tephra analyses; all authors contributed to the written manuscript.

Additional information

Competing financial interests: The authors declare no competing financial interests.

9

Transition to a Moist Greenhouse with CO_2 and solar forcing

Max Popp[1,2,†], Hauke Schmidt[1] & Jochem Marotzke[1]

Water-rich planets such as Earth are expected to become eventually uninhabitable, because liquid water turns unstable at the surface as temperatures increase with solar luminosity. Whether a large increase of atmospheric concentrations of greenhouse gases such as CO_2 could also destroy the habitability of water-rich planets has remained unclear. Here we show with three-dimensional aqua-planet simulations that CO_2-induced forcing as readily destabilizes the climate as does solar forcing. The climate instability is caused by a positive cloud feedback and leads to a new steady state with global-mean sea-surface temperatures above 330 K. The upper atmosphere is considerably moister in this warm state than in the reference climate, implying that the planet would be subject to substantial loss of water to space. For some elevated CO_2 or solar forcings, we find both cold and warm equilibrium states, implying that the climate transition cannot be reversed by removing the additional forcing.

[1] Max Planck Institute for Meteorology, Bundesstrasse 53, Hamburg 20146, Germany. [2] Program in Atmospheric and Oceanic Sciences, Princeton University, 300 Forrestal Road, Sayre Hall, Princeton, New Jersey 08544, USA. † Present address: NOAA's Geophysical Fluid Dynamics Laboratory, Princeton, New Jersey, USA. Correspondence and requests for materials should be addressed to M.P. (email: mpopp@princeton.edu).

Water-rich planets such as Earth lose water by photo-dissociation of water vapour in the upper atmosphere and the subsequent escape of hydrogen. On present-day Earth, the loss occurs very slowly, because the mixing ratio of water vapour in the upper atmosphere is very low. But significant loss of water could occur over geological timescales if the surface temperature were around 70 K warmer than it is today[1-3]. For these high surface temperatures, the tropopause is expected to climb to high altitudes. As a consequence, the cold trapping of water vapour at the tropopause becomes ineffective, because the mixing ratio of water vapour increases with the rising tropopause. Steady states in which the mixing ratio in the upper atmosphere is sufficiently high for a water-rich planet to lose most of its water inventory in its lifetime are known as Moist-Greenhouse states[4]. A planet in this state would eventually become uninhabitable as all water is lost to space. For an Earth-like planet around a Sun-like star, a Moist Greenhouse would be attained if the mixing ratio in the upper atmosphere exceeds ~ 0.1 % (ref. 1). For comparison, the mixing ratio in Earth's stratosphere is presently around two orders of magnitude smaller.

Moist-Greenhouse states were found and described in several studies with one-dimensional models[1-3,5,6] and have recently been found for terrestrial planets with three-dimensional models in different setups[7-9]. However, not all three-dimensional studies found stable Moist-Greenhouse states[10-12]. Instead the climate of these models would destabilize into a Runaway Greenhouse, a self-reinforcing water-vapour feedback-loop, before the Moist Greenhouse is attained. A few studies applied large forcing but the employed models became numerically unstable before the Moist-Greenhouse regime was attained[13-15]. Therefore, it remains unclear whether planets would attain a Moist-Greenhouse state before a Runaway Greenhouse occurs, especially for planets on an Earth-like orbit, where the only two previous studies with state-of-the-art general circulation models (GCM) gave contradicting results[12,9]. Moreover, all three-dimensional studies investigating Moist-Greenhouse states only applied solar forcing without considering greenhouse-gas forcing. Several studies have applied strong greenhouse-gas forcing, but either did not run their simulations to sufficiently high temperatures[16-19] or did not investigate the emergence of a Moist Greenhouse[20]. Greenhouse-gas forcing has long been assumed to be ineffective at causing Moist-Greenhouse states, because the greenhouse effect of any additional greenhouse gas would eventually be rendered ineffective by the increasing greenhouse effect of water vapour with increasing temperatures. Furthermore, large greenhouse-gas forcing would lead to a cooling of the upper atmosphere, which would push the Moist-Greenhouse limit to much higher surface temperatures[5]. However, if clouds are considered, these arguments may not apply, because clouds themselves can contribute to the climate becoming unstable[21].

Here we compare for the first time with a state-of-the-art GCM, namely ECHAM6 (ref. 22), how effective solar and CO_2 forcing are at causing a transition to a Moist Greenhouse. We couple the atmosphere to a slab ocean and choose an aqua-planet setup (fully water-covered planet) in perpetual equinox. This idealized framework is better suited than a present-day Earth setting to understand the involved dynamics while preserving the major feedback mechanisms of the Earth[23]. It also avoids conceptual problems with the representation of land-surface processes at high temperatures. We turn off sea ice in order to investigate the possibility of solely cloud-induced multiple steady states that were recently found in a one-dimensional study[21]. We modify the model such that it can deal with surface temperatures of up to 350 K (see Methods).

Thus we show that cloud-radiative effects (CRE) destabilize a present-day Earth climate as readily with CO_2 as with solar forcing. The changes in CRE are a consequence of the weakening of the large-scale circulation with increasing global-mean surface temperature (gST). However, the resulting climate transition does not lead to a Runaway Greenhouse, but instead a new regime of warm steady state with gST above 330 K is attained. This warm regime differs substantially in its dynamics from a present-day Earth-like climate and, most importantly, the upper atmosphere exceeds the Moist-Greenhouse limit in this regime. Hence a planet in such a state would lose water at a fast rate to space. Furthermore, there is hysteresis in the warm regime and removing the imposed forcing does therefore not necessarily cause a transition back to an Earth-like climate.

Results

Simulations with increased TSI. To assess the dependence of the climate state on total solar irradiance (TSI) for fixed CO_2 levels (at 354 p.p.m. volume mixing ratio), we apply a total of five different TSI-values that range from the present-day value on Earth (S_0) to 1.15 times that value. We find two regimes of steady states that are separated by a range of gST for which stable steady states are not found (Fig. 1). The regime of steady states with gST of up to ~ 298 K exhibits similar features as present-day Earth climate, such as a large pole-to-equator surface-temperature contrast (Fig. 2a) and a similar meridional distribution of cloud cover (Fig. 2b). Hence this regime of steady state can be considered to be Earth-like. In contrast, the warm regime of steady states with gST above 334 K is characterized by a considerably smaller pole-to-equator surface-temperature difference and a substantially different meridional distribution of cloud cover (Fig. 2a,b). This illustrates that the dynamics in the warm regime are quite different from the present-day Earth regime. Most importantly, the mixing ratio of water vapour at the uppermost level at 0.01 hPa is considerably higher in the warm regime and exceeds the Moist-Greenhouse limit (Fig. 2c). Therefore, a planet in such a state would be losing water to space at a fast rate. The minimum TSI required to cause a climate transition from the Earth-like to the warm regime lies between 1.03 S_0 and 1.05 S_0, whereas the maximum TSI to cause a climate transition from the warm back to the Earth-like regime lies

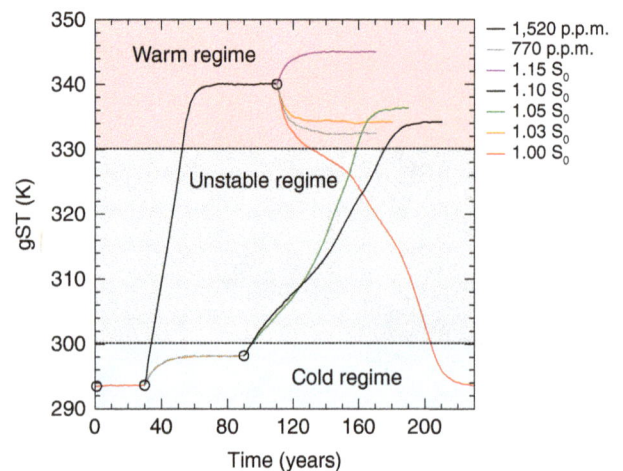

Figure 1 | Temporal evolution of gST. The circles denote the four states from which new simulations are started. For both a TSI of 1.00 S_0 and 1.03 S_0 as well as for a CO_2 concentration of 770 p.p.m., two simulations with different initial conditions are performed. The lines are interpolated from annual means.

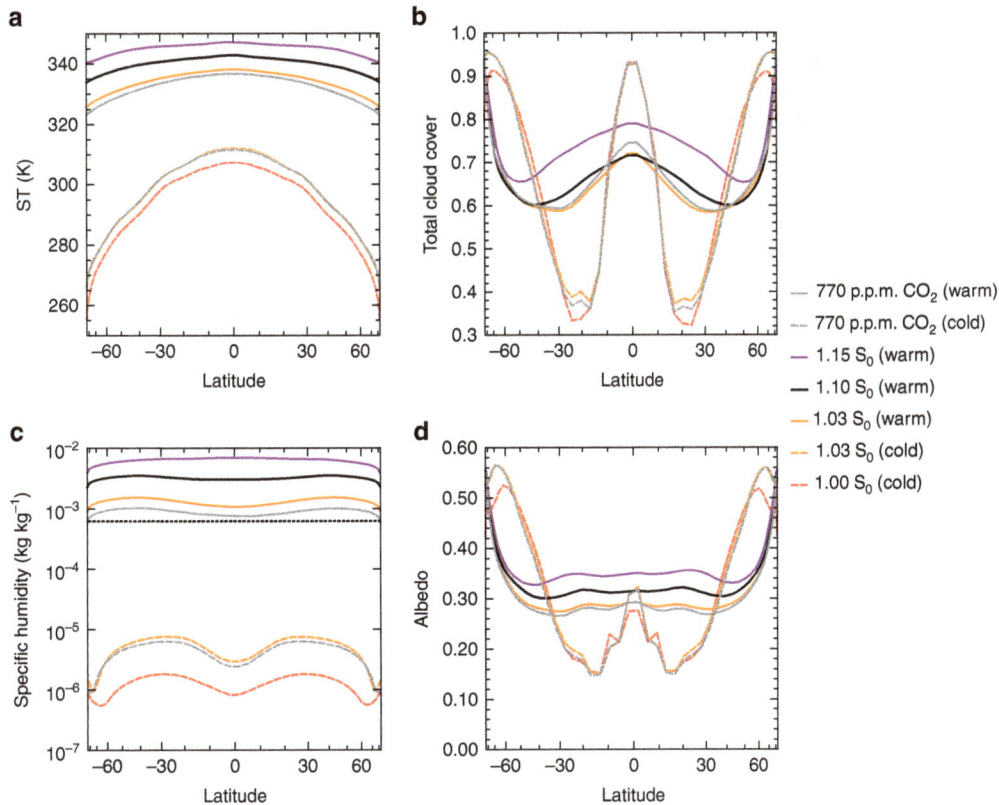

Figure 2 | Zonal means in steady state. (**a**) Surface temperature (ST); (**b**) the total cloud cover; (**c**) the specific humidity at the top level; and (**d**) the effective albedo. The effective albedo is defined as the ratio of the zonal and temporal means of the reflected solar radiation divided by the zonal and temporal means of the incoming solar radiation. The black horizontal line in **c** indicates the Moist-Greenhouse limit[1]. The temporal mean is taken over a period of 30 years. The horizontal axes are scaled with the sine of the latitude.

between 1.00 S_0 and 1.03 S_0 (Fig. 1). Consequently, there are two different stable steady states for a TSI of 1.03 S_0. Since sea ice is turned off in our model, this double steady state is entirely a consequence of atmospheric processes. In the warm regime, the cloud albedo increases at all latitudes with TSI, thus providing an efficient way to stabilize the climate against increased radiative forcing (Fig. 2d).

Energetics of the climate transition. To understand the processes governing the climate transition from the cold to the warm regime, we focus now on the transient simulation with a TSI of 1.05 S_0. The climate instability is evidenced by an increase in the total (shortwave plus longwave) net (downward minus upward component of the) global-mean top-of-the-atmosphere (TOA) radiative flux with increasing gST for gST between 300 and 330 K (Fig. 3). Note that a positive TOA radiative flux is net downward and a negative radiative flux is net upward. Therefore, the aqua-planet takes up energy for a positive TOA radiative flux and loses energy for a negative flux. The instability is caused by the cloud-radiative contribution to the total radiative flux that increases with increasing gST for gSTs below 330 K. The clear-sky contribution to the total radiative flux is decreasing with increasing gST and does thus not contribute to the climate instability. This decrease in clear-sky contribution is caused by an increase in clear-sky contribution to the outgoing longwave radiation, which is upward and thus decreases the net downward flux. At gST above 330 K, the cloud-radiative contribution decreases again with increasing gST. This allows together with the clear-sky contribution to attain a new steady state. Thus, clouds

destabilize the climate at lower gST and then stabilize again at higher gST. This change in sign of the cloud feedback is also responsible for the existence of the bistability.

Dynamics of the climate transition. The changes in cloud-radiative contribution to the total net TOA radiative flux (henceforth simply referred to as CRE) are caused by the weakening of the large-scale circulation (Fig. 4a) and the increase of water vapour in the atmosphere with increasing gST. The weakening of the circulation causes tropical convection to spread more evenly around the tropics, with less convection occurring around the equator and more convection occurring in the subsidence region. As a consequence, deep convective clouds with low cloud-top temperatures become more frequent in the subsidence region of the Hadley circulation (Fig. 4b). This in turn leads to a very strong increase in longwave CRE in this region, which dominates the increase in shortwave CRE (Fig. 5a,b). However, as the gST increases further and the specific humidity in the atmosphere increases, the clouds become thicker and thus more reflective, whereas the longwave CRE does not increase as fast anymore leading to a decrease in total CRE in the tropics for gST above 320 K. This decrease in tropical CRE with increasing gST contributes to the stabilization of the climate at gST above 330 K (Fig. 5c). In general, the changes in total CRE dominate the changes in clear-sky radiative effect in the tropics (Fig. 5c,d). In the extra-tropics, the weakening of the large-scale circulation leads to a steady decrease in cloud cover everywhere except at very high latitude (Fig. 4a,b). Therefore, the shortwave CRE increases in the extra-tropics (Fig. 5a). Since the tropopause deepens with increasing surface temperatures (not shown), the difference between the temperature at the surface and at the cloud tops increases as well and leads also to an increase in longwave CRE despite the decrease in cloud cover (Fig. 5b). Whereas the changes in the clear-sky radiative effect dominate the changes in CRE in most of the extra-tropics for gST up to 315 K, the changes in CRE increasingly dominate the extra-tropical response at gST above. This supports the idea that at high gST changes in CRE are more important than changes in clear-sky radiative effect and dominate the climate response. In general, the weak large-scale circulation in the warm regime leads to a much more uniform meridional distribution of cloud condensate at all levels than in the cold regime (Fig. 6). Note that there is a decrease in global-mean convective precipitation but a slight increase in total precipitation from the cold to the warm regime (not shown). This may indicate that in the warm regime convection is overall less frequent but more intense, such that a significant fraction of condensate is not converted to convective precipitation but detrained to form large-scale precipitation. This trend continues as surface temperatures increase further in the warm regime.

Simulations with increased CO_2 concentrations. We start our comparison of CO_2-induced to solar forcing by increasing CO_2 concentrations to 770 p.p.m., while keeping the TSI fixed to 1.00 S_0. This corresponds to an equivalent adjusted forcing as is caused by an increase of TSI from 1.00 S_0 to 1.03 S_0. The adjusted forcing is defined to be the temporal and global mean of the energy uptake over the first year of simulation. The results suggest that the increase of the CO_2 concentrations leads to an equivalent warming and a similar meridional distribution of surface temperatures and clouds as the increase in TSI does (Fig. 2a,d). Since the aqua-planet warms by 4.57 K for an increase in CO_2 concentrations from 354 to 770 p.p.m., the climate sensitivity of the aqua-planet for a doubling of CO_2 concentrations is 4.08 K, if a $\log_2$ scaling is assumed. Starting from the final state of the simulation with a TSI of 1.03 S_0, we then increase the CO_2 concentrations to 1,520 p.p.m. and set the TSI back to 1.00 S_0 (Fig. 1). The combined effect leads to an adjusted forcing that is equivalent to increasing the TSI to 1.05 S_0. In this case the aqua-planet undergoes a climate transition into the warm regime (Fig. 1). Thus, the aqua-planet can as readily be forced to transition from the Earth-like to the warm regime by increasing CO_2 concentrations as by increasing the TSI. When starting in the warm regime, a reduction of CO_2 concentrations to 770 p.p.m. does not cause the planet to fall back into the Earth-like regime, but the aqua-planet remains in the warm regime. Therefore, the aqua-planet also exhibits a bistability of the climate for a TSI of 1.00 S_0 and a CO_2 concentration of 770 p.p.m.

Overall, the results suggest that the aqua-planet behaves similarly for solar forcing and CO_2-induced forcing (Figs 1,2 and 6). The most notable difference is that the steady-state gST in the warm regime is ~2 K lower for CO_2-induced forcing. The

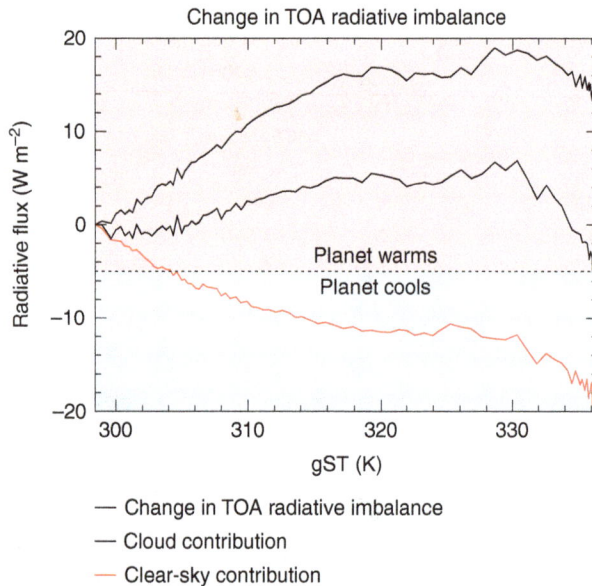

Figure 3 | Change of TOA radiative flux as a function of gST. The black line shows the change in global and annual mean of the total (shortwave plus longwave) net (downward minus upward component of the) TOA radiative flux, the blue line shows the cloud-radiative contribution and the red line the clear-sky contribution from the transient simulation with a TSI of 1.05 S_0. Note that a positive slope indicates an unstable state, because any warming/cooling would lead to an increase/decrease in energy uptake by the aqua-planet and thus to an additional warming/cooling. The cloud-radiative and clear-sky contributions sum up to the net TOA radiative flux. The changes are calculated by subtracting the global and annual means over the first year of the respective quantities. The horizontal line corresponds to the negative value of the initial total net TOA radiative flux. Hence, for a steady state to be attained, the total net TOA radiative flux must touch or intersect the horizontal line.

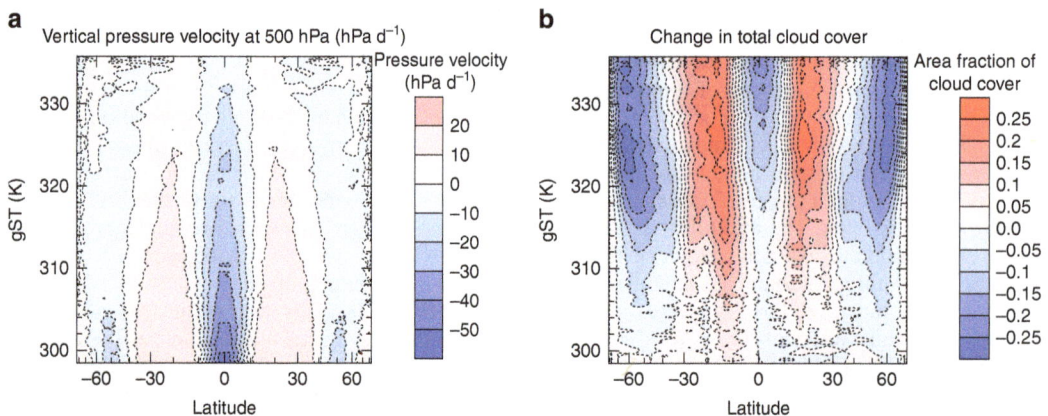

Figure 4 | Large-scale circulation and cloud cover during the climate transition. (**a**) Zonal and annual mean of the vertical pressure velocity as a function of latitude and gST for the transient period of the simulation with a TSI of 1.05 S_0. (**b**) Same as in **a** but for the change in zonal and annual mean of cloud cover from the first year of simulation.

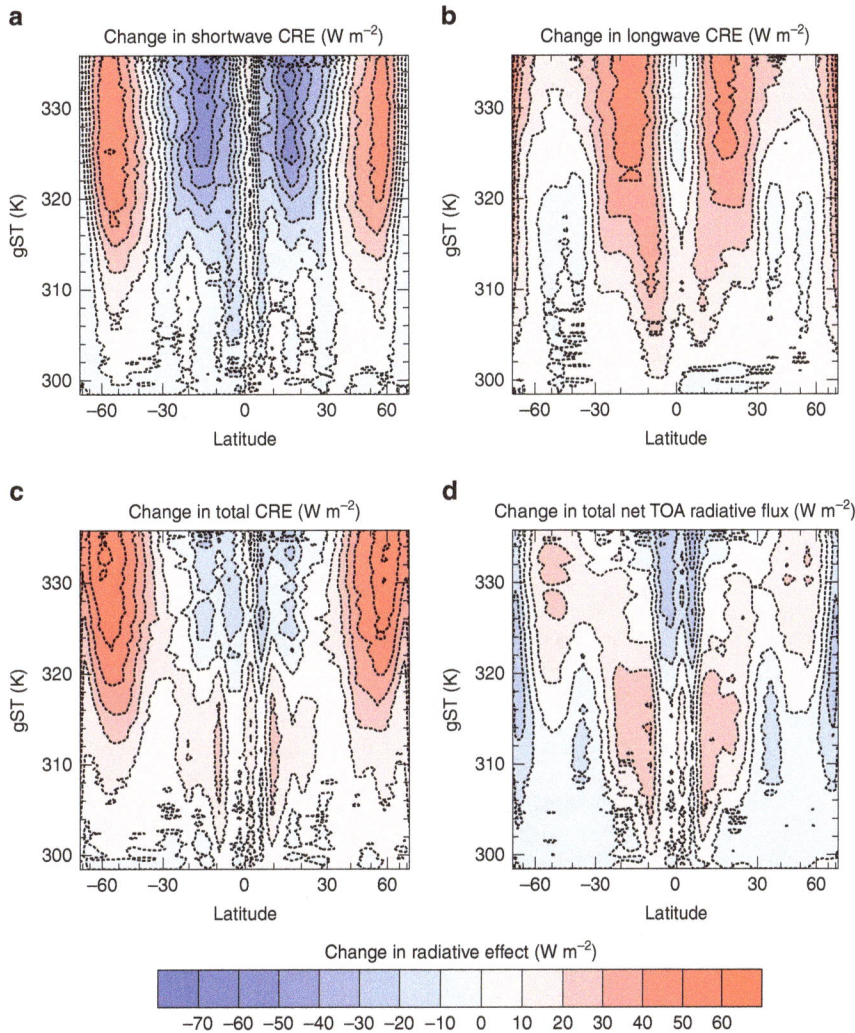

Figure 5 | Zonal means of changes in radiative effects during the climate transition. (**a**) Change of zonal and annual mean of the shortwave CRE; (**b**) change of the longwave CRE; (**c**) change of the total CRE; and (**d**) the change of the total net TOA radiative flux. All panels use the same colour bar. The horizontal axes are scaled with the sine of the latitude.

likely reason for this is that the thermal absorption by water vapour overlaps with the thermal absorption by CO_2 in the warm moist atmosphere, which renders the greenhouse effect of CO_2 less effective. However, since the climate instability in our simulations is caused by CRE at a gST at which the atmosphere is not yet sufficiently opaque to cancel the greenhouse effect of CO_2, CO_2-induced forcing can as easily cause a climate transition to the Moist Greenhouse as solar forcing does.

Sensitivity experiments. Two of the assumptions made in this study could potentially have a large influence on the results and may contribute to the differences in the results between our study and previous ones[17,12,9,20]. These assumptions concern the treatment of ozone and oceanic heat transport of the model. For these two cases, we show with sensitivity experiments that the qualitative nature of the results is not changed by the assumptions (Supplementary Figs 1–3). The neglect of sea ice should have no influence on the qualitative results, because there is no sea ice in the warm regime and because the sea-ice albedo feedback is positive and would thus favour a climate instability in the cold regime. However, the absence of sea ice may explain why our control simulation is warmer than present-day Earth.

Discussion
A recent study using the same model but in a different version found that the Earth's climate remains stable for CO_2 concentrations of at least 4,480 p.p.m. (ref. 17), whereas our study suggests that such concentrations would lead to a climate transition. Studies of Earth with other GCMs also found the climate to remain stable for higher CO_2 concentrations than we do[16,18]. However, the initial climate of our aqua-planet is ~6 K warmer than the one of present-day Earth. Such a warming would be attained by a quadrupling of CO_2 in the different version of our model used in ref. 17. By a simple estimate, this other study would thus have explored CO_2 concentrations of up to a fourth of 4,480 p.p.m.; hence, 1,120 p.p.m., if the simulations were started from a climate similar to ours. Therefore, if we account for the difference in the initial climates, the results of the two studies are not in contradiction. Indeed, the climate of the model version used in ref. 17 was recently shown to become unstable when the CO_2 concentrations were increased from 4,480 to 8,960 p.p.m. (eventually leading to numerical failure of their model)[19]. Nonetheless, the forcing required to cause a climate transition would certainly be higher on present-day Earth than on our aqua-planet, even with our version of the model. Several other studies of Earth have found lower climate sensitivities to

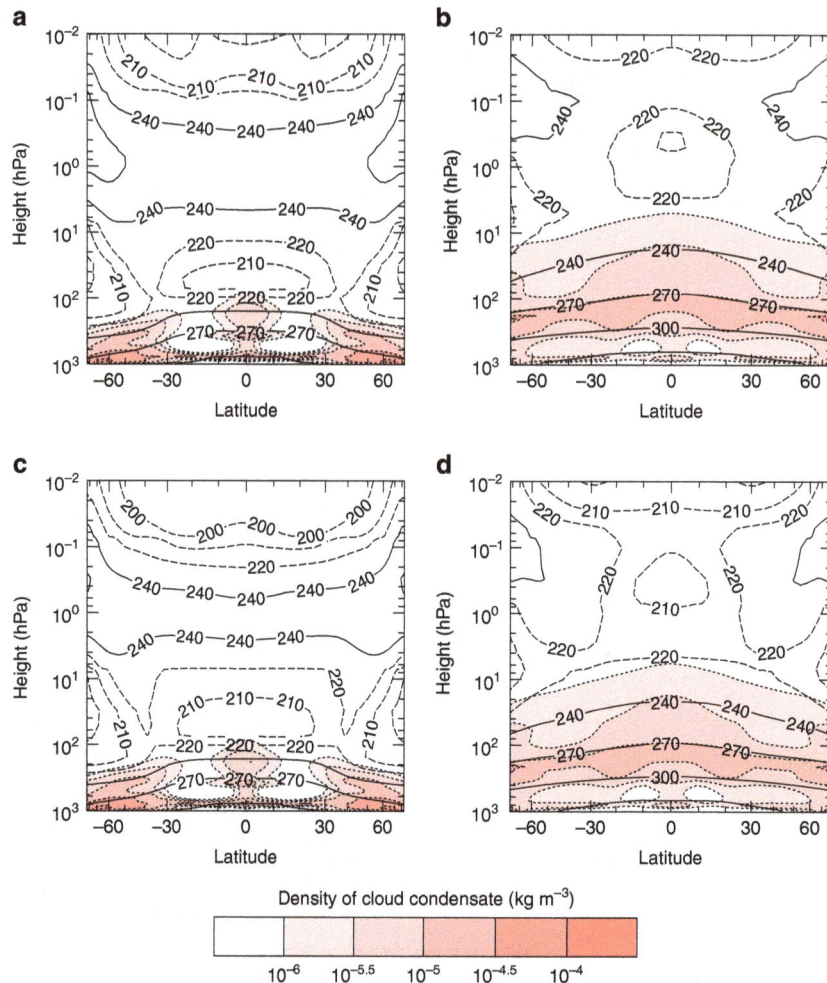

Figure 6 | Zonal means of cloud condensate in steady state. (**a**) Temporal mean over the last 30 years of simulation of the cloud condensate for a TSI of 1.03 S_0 in the cold regime and (**b**) same quantity for the same TSI but in the warm regime. (**c**) Same quantity obtained with a TSI of 1.00 S_0 but with atmospheric CO_2 concentrations of 770 p.p.m. in the cold regime (**d**) with the same CO_2 concentration and TSI in the warm regime. The contours denote temperatures in Kelvin, with the solid lines denoting the contours for 240, 270, 300 and 330 K and the dashed lines for the contours of 200, 210 and 220 K. The vertical axes are the height in terms of pressure of dry air and the horizontal axes are the latitudes scaled with their sines. The pressure of dry air is defined to be the pressure the atmosphere would have at a given level if no water vapour was present. Since the total mass of dry air does not change between simulations, the mean of the dry surface pressure is constant across simulations.

relatively large CO_2 forcing than we do which supports this notion[16,18,20].

Two studies recently investigated climates at gST above 330 K with state-of-the-art GCMs for Earth-like planets[12,9]. Wolf and Toon[9] used a modified version of the Community Atmosphere Model version 4 (CAM4) and Leconte *et al.*[12] a modified version of the Laboratoire de Météorologie Dynamique Generic (LMDG) climate model to investigate the climate response to strong solar forcing and both studies also found a region of increased climate sensitivity[12,9]. Therefore, a region of gST with increased climate sensitivity surrounded by regions of lower climate sensitivity appears to be a robust result, despite some differences in the magnitude of the region (Fig. 7c,e). We calculated the climate sensitivity here following[9], which yields a small climate sensitivity in the cold and in the warm regimes because the method uses the instantaneous forcing. If we use the adjusted forcing, the climate sensitivity is considerably larger, because the fast atmospheric adjustments reduce the initial TOA radiative imbalance quickly. Our model encounters the region of high climate sensitivity for smaller values of TSI (Fig. 7a), because our control climate is the

warmest and because the climate instability is encountered at $\sim$ 300 K, whereas the region of increased climate sensitivity starts at $\sim$ 310–315 K in the two other studies. Our steady-state albedo is a monotonically increasing function of both TSI and gST, in contrast to the two other studies where the albedo decreases with both increasing TSI and gST until either the Moist-Greenhouse state is attained or a Runaway Greenhouse occurs (Fig. 7b,d). However, in our transition from the cold to the warm regime the albedo decreases as well with increasing gST between 300 and 320 K. Only one of the two studies found a stabilizing cloud feedback similar to ours and a moist stratosphere at high gST (ref. 9), whereas the other one found that the cloud feedback is rather destabilizing and that the stratosphere remains dry[12]. In general, our results of the simulations with increased solar forcing are qualitatively similar to the ones found in ref. 9 in that we find two regimes of steady states, in that the warm regimes have a similar temperature structure (Fig. 8a), in that the cloud albedo increases with gST and most importantly in the existence of a stable Moist-Greenhouse regime. Similarly to ref. 9, the troposphere in the warm regime is characterized in our model by a particular

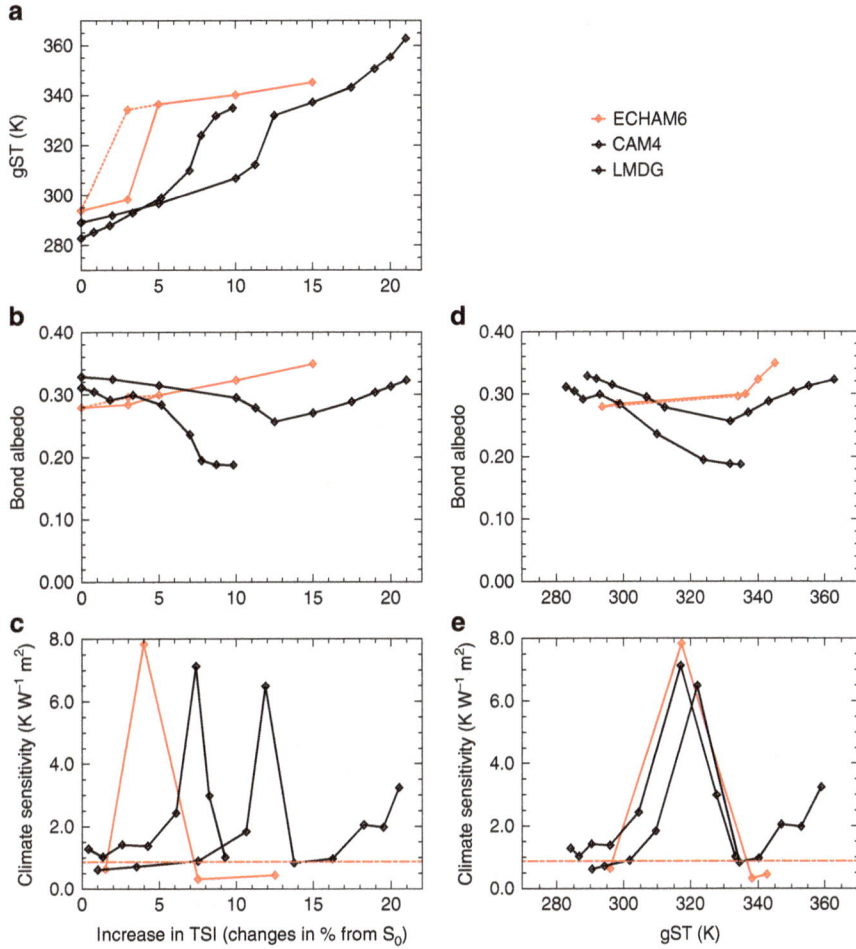

Figure 7 | Model inter-comparison. (**a**) gST; (**b**) the Bond albedo; and (**c**) the climate sensitivity as a function of the increase in TSI in per cents of S_0. The red solid lines and marks denote the results obtained with ECHAM6 when increasing TSI, whereas the dashed lines and mark denote the results obtained with ECHAM6 but when decreasing TSI. The black lines and marks denote the results obtained with CAM4 in ref. 9 and the blue lines and marks denote the results obtained with LMDG in ref. 12. (**d**) Bond albedo and (**e**) the climate sensitivity as a function of the gST. The red dashed-dotted line denotes the climate-sensitivity parameter for doubling CO_2 and is calculated from the simulation in which CO_2 concentrations are increased from 354 to 770 p.p.m. using $\log_2$ scaling. Note that the Bond albedo is equal to the global-mean of the effective albedo.

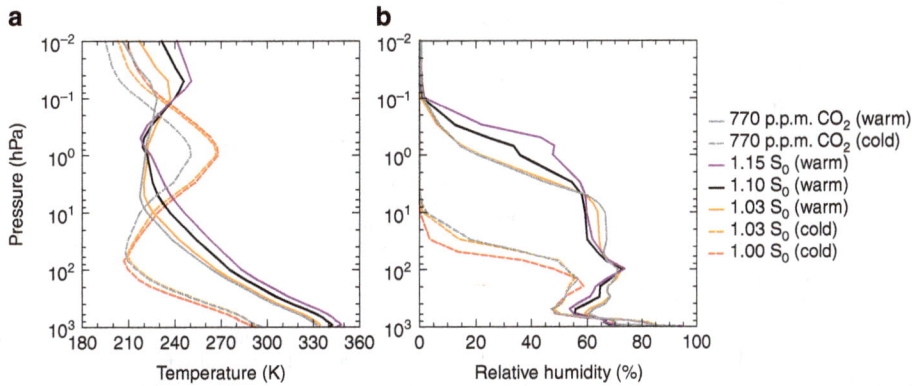

Figure 8 | Atmospheric profiles of temperature and relative humidity. (**a**) Global-mean vertical profiles of gST and (**b**) of relative humidity for the different steady states. The vertical axes are the height in terms of pressure of dry air.

radiative-convective equilibrium with a temperature inversion close to the surface, a somewhat drier region above and a more humid region up to the tropopause. A similar structure has also been found and discussed in two one-dimensional studies[5,21].

Wolf and Toon[9] argues that the change to the aforementioned radiative-convective equilibrium is crucial for the emergence of stable Moist-Greenhouse states, but our results suggest that the weakening of the large-scale circulation is equally important by allowing the radiative-convective regime to spread over the entire tropics. This spread of the convective region over a large fraction of the planet, namely the tropics, also explains some of the similarities between the three- and the one-dimensional models in the warm regime. Compared with ref. 9, the dry region is more humid in our warm regime and in general more humid than in our cold regime (Fig. 8b, Supplementary Fig. 4). Given the different dynamical cores, radiative transfer schemes, convection schemes and cloud schemes, it is, however, remarkable how many similarities the three models share.

In the two aforementioned studies of large solar forcing, ozone was entirely removed from the whole atmosphere in all simulations[12,9], whereas we do include ozone in our calculations. Since ozone has a strong warming influence in the middle atmosphere due to its strong absorption of solar radiation, the inclusion of ozone in our model may explain why our atmosphere is moister in the upper levels than theirs for similar values of gSST. To test this hypothesis, we perform a simulation with a TSI of 1.10 S_0 without atmospheric ozone. Both the upper atmospheric temperatures and specific humidity are lower but still above the Moist-Greenhouse limit in the experiments where ozone is removed (Supplementary Fig. 1). Thus, the specific humidity is still considerably higher than in our cold regime and especially than in the simulations of Wolf and Toon[9] and Leconte et al.[12] at similar gSST. In general, removing ozone does not appear to alter the results substantially.

One study investigated strong CO_2 forcing and gST above 330 K with the Fast Atmosphere-Ocean Model (FAOM) developed at the Goddard Institute for Space Studies and did not find any region of strongly increased climate sensitivity[20]. The climate sensitivity increases somewhat but not nearly as much as with our or with the other two state-of-the-art models used for solar forcing. As a consequence, considerably higher values of CO_2 are required in that study compared with ours to attain a gST of 330 K. Furthermore, the control climate is colder in that study, and it takes around one doubling of CO_2 to attain the gST of our control simulation. The humidity at the top level of FAOM remains roughly one order of magnitude below the Moist-Greenhouse limit in the warmest steady states found in ref. 20. This may be partly due to the CO_2 cooling of the upper atmosphere, but could also be a consequence of not running the model to sufficiently high temperatures. We perform a sensitivity experiment, where we increase CO_2 concentrations to 9,000 p.p.m. in order to assess whether increased CO_2 concentrations could cause a substantial drying of our upper atmosphere (not shown). The upper atmosphere-specific humidity stays; however, well above the Moist-Greenhouse limit also in that case. Some of the differences between FAOM and ECHAM6 may simply be caused by the use of different setups and parameterizations. But FAOM is a simplified model designed for fast computation and uses simplified cloud physics, which may be the cause for the absence of a region of strongly increased climate sensitivity. So, whereas our version of ECHAM6 is rather on the low side of CO_2 concentrations required to cause the gST to rise above 330 K, FAOM likely is on the high side of the concentrations.

To conclude, we have demonstrated with a state-of-the-art climate model that a water-rich planet might lose its habitability as readily by CO_2 forcing as by increased solar forcing through a transition to a Moist Greenhouse and the implied long-term loss of hydrogen. We confirm previous results that a region of increased climate sensitivity exists and show that the climate is unstable in our model in that region due to positive cloud feedbacks caused by a weakening of the large-scale circulation. We also demonstrate that there is hysteresis and that once a transition to the Moist-Greenhouse regime has occurred, the

process may not simply be reversed by removing the additional forcing.

Methods

General setup. We employ a modified version of the GCM ECHAM6 (ref. 22) in an aqua-planet setting in which the whole surface is covered by a 50-m-deep mixed-layer ocean. We run the model with a spectral truncation of T31, which corresponds to a Gaussian grid with a grid-point spacing of 3.75°. The atmosphere is resolved vertically by 47 layers up to a pressure (of dry air) of 0.01 hPa. The oceanic heat transport is prescribed by a sinusoidal function of latitude. There is no representation of sea ice in our model and as a consequence water may be colder than the freezing temperature. The orbit of the aqua-planet is perfectly spherical with a radius of 1 AU. The obliquity of the aqua-planet is 0°. For simplicity, a year is set to be 360 days. The rotation velocity of the aqua-planet corresponds to present-day Earth.

Parameterized oceanic heat transport. To account for the meridional oceanic heat transport, we introduce an additional energy flux from the mixed-layer ocean to the atmosphere, which mimics the divergence of a prescribed meridional oceanic heat flux. This energy flux is commonly called q-flux. Imposing a q-flux is necessary to prevent the atmosphere from having to take over all the meridional energy transport, which would result in an amplified atmospheric large-scale circulation. However, since the oceanic circulation depends also on atmospheric properties, the imposed q-flux may be inaccurate in a warming climate. The weak atmospheric large-scale circulation suggests that the oceanic heat transport would be weaker in the warm regime and that thus the absolute values of the q-flux are too large. To assess whether the imposed q-flux affects the results in the warm regime, a number of simulations without q-flux are performed. Despite small differences in the large-scale circulation (Supplementary Fig. 2), the warm steady states exist and are stable even without q-flux (Supplementary Fig. 3).

Treatment of ozone. If the tropopause climbs, regions with high ozone concentrations may come to lie in the troposphere of the model, because ozone concentrations are prescribed to climatological values. High ozone concentrations could, however, not occur in the presence of tropospheric water vapour concentrations. Therefore, we limit the tropospheric ozone concentrations to a volume mixing ratio of 1.5×10^{-7}. As a consequence, ozone is taken out of the atmosphere, if the tropopause rises to levels where the climatological values would exceed this limit. This process is reversible if the tropopause descends again.

Modifications to the grid-point physics. Our version of the model incorporates several changes to the grid-point physics, such that we obtain a more accurate representation of several physical processes in warm climates[21]. The grid-point physics include representation of surface exchange, turbulence and vertical diffusion[24,25], gravity-wave drag[26,27], radiative transfer and radiative heating[28,29], convection[30,31], cloud cover[32] and cloud microphysics[33]. In summary, these changes are the inclusion of the mass of water vapour when calculating the total pressure and the omission of all approximations where small specific humidities are assumed (as for example in the calculation of density). The pressure effects of water vapour are not considered for the horizontal transport. So, the model is in sorts a hybrid model, with water vapour adding to the total pressure for local effects but not so for the large-scale transport. A detailed description of the modified model thermodynamics can be found in the appendix of ref. 34. We will give here a short overview of the modified radiative transfer scheme as well as of the convection, cloud-cover and cloud-microphysical schemes.

Radiative transfer. The radiative transfer scheme has recently been described and evaluated extensively in ref. 21, but as a courtesy to the reader we will repeat some of the major features here. It is based on the Rapid Radiative Transfer Model[28,29], but includes some small modifications. It uses the correlated-k method to solve the radiative transfer equations in the two-stream approximation. The k-coefficients are calculated from the HITRAN (1996 and 2000) database using a line-by-line radiative transfer model[28,29]. The water vapour continuum is based on CKD_v2.4. The shortwave radiation spectrum is divided into 14 bands, and the longwave radiation spectrum is divided into 16 bands. Since the lookup-tables of the molecular absorption coefficients are designed for a limited range of temperatures only, an exponential extrapolation for temperatures up to 400 K for the longwave radiation scheme is performed. The same extrapolation scheme is also applied to the lookup-tables for the absorption coefficients of the water vapour self-broadened continuum in the shortwave radiation scheme, but the original linear extrapolation scheme is kept for all the other absorption coefficients. The lookup-tables for the bandwise spectrally integrated Planck function and the derivative thereof with respect to temperature have been extended to 400 K. Furthermore the water vapour self-broadened continuum is introduced in the upper atmosphere radiation calculations, to account for the increase in water vapour with increasing gST. The effect of pressure broadening by water vapour on the molecular absorption coefficients is neglected, as is scattering by water vapour. The thus modified radiation scheme is not as accurate as line-by-line radiative transfer models or models using recalculated k-coefficients for the higher temperatures, but still sufficiently accurate for the task at hand as has recently been demonstrated[21].

Convective scheme. ECHAM6 uses a mass-flux scheme for cumulus convection[30], with modifications for penetrative convection to the original scheme[31]. The contribution of cumulus convection to the large-scale budgets of heat, moisture and momentum is represented by an ensemble of clouds consisting of updrafts and downdrafts in a steady state. Depending on moisture convergence at the surface and depth of the convection cell, the model will either run in penetrative, mid-level, or shallow convection mode. The scheme allows for the formation of precipitation, but not for radiatively active convective clouds. Instead, detrained water is passed to the cloud-microphysical scheme which creates or destroys cloud condensate in a further step.

Cloud-cover scheme. ECHAM6 uses the Sundqvist scheme for fractional cloud cover[32]. This cloud scheme has been tuned for the use with present-day Earth's climate[33,35]. However, the scheme is well suited for simulations of cloud cover in warm climates as it diagnoses cloud cover directly from relative humidity, which should be crucial to cloud formation irrespective of the temperature.

Cloud-microphysical scheme. The cloud-microphysical scheme is described in detail in ref. 33. The scheme consists of prognostic equations for the vapour, liquid, and ice phases. There are explicit microphysics for warm-phase, mixed-phase and ice clouds. The cloud-condensation-nuclei concentration follows a prescribed vertical profile which is typical for the present-day Earth's maritime conditions. Since we have no estimate of the aerosol load in a hypothetical warm climate, we assume that this profile is also a reasonable choice for warmer climates. The microphysics do not require changes for the use of the scheme in warm climates, since cloud formation in a warm-phase cloud, in which potential changes may occur, is not directly temperature-dependent (at least not in the range of temperatures we consider).

Cloud-radiative interactions. Clouds are represented in the radiative transfer calculations, assuming the so-called maximum-random-overlap assumption. Under this assumption cloud layers are assumed to be maximally overlapping if they are adjacent to one another, and randomly overlapping if they are separated by a clear layer. The absorptivity of clouds depends on their combined optical depths, the gas in which they are embedded, and the interstitial aerosol. The microphysics to determine the optical properties of the cloud particles involve the liquid water and ice paths, cloud-drop radii, as well as liquid water and ice content. Cloud scattering is represented as a single-scattering albedo by assuming Mie scattering from cloud droplets in the shortwave calculations, but is neglected in the longwave calculations. Clouds are not considered in the radiative transfer routines if the specific mass of the cloud condensate does not exceed 10^{-7} kg per kg of air.

Special settings for high-temperature simulations. To run the model at high temperatures, a few special settings are necessary. The time step is reduced from 2,400 to 600 s, and the radiation time step is reduced from 7,200 to 2,400 s, except for the simulation with a TSI of 1.15 S_0, where the time step is reduced to 360 and the radiation time step to 1,440 s. Nonetheless, we sometimes encounter problems with resolved waves propagating to the top levels of the model, where their amplitude grows and where they may be reflected. To avoid frequent model failure due to these effects, we introduce Rayleigh friction to the vorticity and the divergence as well as increased horizontal diffusion in the top six layers (above ~ 0.75 hPa). The time constant of the Rayleigh friction is (10800 s) at the sixth layer and is increased by a factor of 3.2 per layer towards the top and is hence increased by a factor of $\sim 1,000$ in the top layer. The horizontal diffusion is increased by a factor of 3.2 per layer. The values of the time constant of the Rayleigh friction and the magnitude of horizontal diffusion are determined by trial and error. Since we investigate a large range of climates, it is difficult to find suitable values for these time constants, and despite these modifications the model fails occasionally. In these cases, however, the runs can be continued by slightly changing their trajectory, which is achieved by reducing the factor of multiplication per level for the Rayleigh friction from 3.2 to 3.199 for 30 days.

References

1. Kasting, J. F. Runaway and moist greenhouse atmospheres and the evolution of earth and venus. *Icarus* **74**, 472–494 (1988).
2. Kasting, J. F., Whitmire, D. P. & Reynolds, R. T. Habitable zones around main-sequence stars. *Icarus* **101**, 108–128 (1993).
3. Kopparapu, R. K. *et al.* Habitable zones around main-sequence stars: new estimates. *Astrophys. J.* **765**, 131 (2013).
4. Kasting, J. F., Pollack, J. B. & Ackerman, T. P. Response of Earth's atmosphere to increases in solar flux and implications for loss of water from venus. *Icarus* **57**, 335–355 (1984).
5. Wordsworth, R. D. & Pierrehumbert, R. T. Water loss from terrestrial planets with CO2-rich atmospheres. *Astrophys. J.* **778**, 154 (2013).
6. Ramirez, R. M., Kopparapu, R. K., Lindner, V. & Kasting, J. F. Can increased atmospheric CO2 levels trigger a runaway greenhouse? *Astrobiology* **14**, 714–731 (2014).
7. Abe, Y., Abe-Ouchi, A., Sleep, N. H. & Zahnle, K. J. Habitable zone limits for dry planets. *Astrobiology* **11**, 443–460 (2011).
8. Yang, J., Cowan, N. B. & Abbot, D. S. Stabilizing cloud feedback dramatically expands the habitable zone of tidally locked planets. *Astrophys. J. Lett.* **771**, L45 (2013).
9. Wolf, E. & Toon, O. The evolution of habitable climates under the brightening sun. *J. Geophys. Res. Atmos.* **120**, 5775–5794 (2015).
10. Ishiwatari, M., Takehiro, S., Nakajima, K. & Hayashi, Y. Y. A numerical study on appearance of the runaway greenhouse state of a three-dimensional gray atmosphere. *J. Atmos. Sci.* **59**, 3223–3238 (2002).
11. Ishiwatari, M., Nakajima, K., Takehiro, S. & Hayashi, Y.-Y. Dependence of climate states of gray atmosphere on solar constant: From the runaway greenhouse to the snowball states. *J. Geophys. Res.* **112**, D13120 (2007).
12. Leconte, J., Forget, F., Charnay, B., Wordsworth, R. & Pottier, A. Increased insolation threshold for runaway greenhouse processes on earth-like planets. *Nature* **504**, 268–271 (2013).
13. Boer, G. J., Hamilton, K. & Zhu, W. Climate sensitivity and climate change under strong forcing. *Clim. Dyn.* **24**, 685–700 (2005).
14. Heinemann, M. Warm and sensitive paleocene-eocene climate. *Reports on Earth System Science 70/2009* (Max-Planck-Institute for Meteorology, 2009).
15. Wolf, E. T. & Toon, O. Delayed onset of runaway and moist greenhouse climates for earth. *Geophys. Res. Lett.* **41**, 167–172 (2014).
16. Hansen, J. *et al.* Efficacy of climate forcings. *J. Geophys. Res.* **110**, D18104 (2005).
17. Meraner, K., Mauritsen, T. & Voigt, A. Robust increase in equilibrium climate sensitivity under global warming. *Geophys. Res. Lett.* **40**, 5944–5948 (2013).
18. Wolf, E. T. & Toon, O. B. Hospitable archean climates simulated by a general circulation model. *Astrobiology* **13**, 656–673 (2013).
19. Bloch-Johnson, J., Pierrehumbert, R. T. & Abbot, D. S. Feedback temperature dependence determines the risk of high warming. *Geophys. Res. Lett.* **42**, 4973–4980 (2015).
20. Russell, G. L., Lacis, A. A., Rind, D. H. & Colose, C. Fast atmosphere-ocean model runs with large changes in CO2. *Geophys. Res. Lett.* **40**, 5787–5792 (2013).
21. Popp, M., Schmidt, H. & Marotzke, J. Initiation of a runaway greenhouse in a cloudy column. *J. Atmos. Sci.* **72**, 452–471 (2015).
22. Stevens, B. *et al.* The atmospheric component of the MPI-M earth system model: ECHAM6. *J. Adv. Model. Earth Syst.* **5**, 146–172 (2013).
23. Medeiros, B., Stevens, B. & Bony, S. Using aquaplanets to understand the robust responses of comprehensive climate models to forcing. *Clim. Dyn.* **44**, 1957–1977 (2015).
24. Brinkop, S. & Roeckner, E. Sensitivity of a general circulation model to parameterizations of cloudturbulence interactions in the atmospheric boundary layer. *Tellus A* **47**, 197–220 (1995).
25. Giorgetta, M. A. *et al.* The atmospheric general circulation model ECHAM6 - model description. *Technical Report 135* (Max-Planck-Institute for Meteorology, 2013).
26. Hines, C. O. Doppler-spread parameterization of gravity-wave momentum deposition in the middle atmosphere. Part 1: basic formulation. *J. Atmos. Sol.-Terr. Phys.* **59**, 371–386 (1997).
27. Hines, C. O. Doppler-spread parameterization of gravity-wave momentum deposition in the middle atmosphere. Part 2: Broad and quasi monochromatic spectra, and implementation. *J. Atmos. Sol.-Terr. Phys.* **59**, 387–400 (1997).
28. Mlawer, E. J., Taubman, S. J., Brown, P. D., Iacono, M. J. & Clough, S. A. Radiative transfer for inhomogeneous atmospheres: RRTM, a validated correlated-k model for the longwave. *J. Geophys. Res.* **102**, 16663–16682 (1997).
29. Iacono, M. J. *et al.* Radiative forcing by long-lived greenhouse gases: calculations with the AER radiative transfer models. *J. Geophys. Res.* **113**, D13103 (2008).
30. Tiedtke, M. A comprehensive mass flux scheme for cumulus parameterization in large-scale models. *Mon. Wea. Rev.* **117**, 1779–1800 (1989).
31. Nordeng, T.-E. Extended versions of the convective parametrization scheme at ECMWF and their impact on the mean and transient activity of the model in the tropics. *Technical Memorandum 206, ECMWF* (European Centre for Medium-Range Weather Forecasts, Shinfield Park, Reading, RG2 9AX, UK, 1994).
32. Sundqvist, H., Berge, E. & Kristjansson, J. E. Condensation and cloud parameterization studies with a mesoscale numerical weather prediction model. *Mon. Weather Rev.* **117**, 1641–1657 (1989).
33. Lohmann, U. & Roeckner, E. Design and performance of a new cloud microphysics scheme developed for the ECHAM general circulation model. *Clim. Dyn.* **12**, 557–572 (1996).

34. Popp, M. Climate instabilities under strong solar forcing. *Reports on Earth System Science 152/2014* (Max-Planck-Institute for Meteorology, 2014).

35. Mauritsen, T. *et al.* Tuning the climate of a global model. *J. Adv. Model. Earth Syst.* **4**, M00A01 (2012).

Acknowledgements

We thank Thorsten Mauritsen for a thorough internal review. We thank Eric Wolf for an insightful discussion and for making the data from CAM4 available to us. We thank Jérémy Leconte for making the data from LMDG available to us. We thank Dorian Abbot and two anonymous reviewers for very constructive reviews of this manuscript. We thank the Max Planck Society for the Advancement of Science for financial support.

Author contributions

J.M., H.S. and M.P. conceived the study. M.P. performed the changes to the model and conducted the modelling work. J.M., H.S. and M.P. all contributed substantially to the analysis and the discussion. M.P. wrote the manuscript with comments from J.M. and H.S.

Additional information

Competing financial interests: The authors declare no competing financial interests.

10

Decrease in coccolithophore calcification and CO_2 since the middle Miocene

Clara T. Bolton[1,2], María T. Hernández-Sánchez[1], Miguel-Ángel Fuertes[3], Saúl González-Lemos[1], Lorena Abrevaya[1], Ana Mendez-Vicente[1], José-Abel Flores[3], Ian Probert[4], Liviu Giosan[5], Joel Johnson[6] & Heather M. Stoll[1]

Marine algae are instrumental in carbon cycling and atmospheric carbon dioxide (CO_2) regulation. One group, coccolithophores, uses carbon to photosynthesize and to calcify, covering their cells with chalk platelets (coccoliths). How ocean acidification influences coccolithophore calcification is strongly debated, and the effects of carbonate chemistry changes in the geological past are poorly understood. This paper relates degree of coccolith calcification to cellular calcification, and presents the first records of size-normalized coccolith thickness spanning the last 14 Myr from tropical oceans. Degree of calcification was highest in the low-pH, high-CO_2 Miocene ocean, but decreased significantly between 6 and 4 Myr ago. Based on this and concurrent trends in a new alkenone ε_p record, we propose that decreasing CO_2 partly drove the observed trend via reduced cellular bicarbonate allocation to calcification. This trend reversed in the late Pleistocene despite low CO_2, suggesting an additional regulator of calcification such as alkalinity.

[1] Geology Department, Oviedo University, Arias de Velasco s/n, 33005 Oviedo, Asturias, Spain. [2] Aix-Marseille University, CNRS, IRD, CEREGE UM34, 13545 Aix en Provence, France. [3] Grupo de Geociencias Oceánicas, Geology Department, University of Salamanca, Salamanca 37008, Spain. [4] CNRS, Sorbonne Universités-Université Pierre et Marie Curie (UPMC) Paris 06, FR2424, Roscoff Culture Collection, Station Biologique de Roscoff, Place Georges Teissier, 29680 Roscoff, France. [5] Department of Geology and Geophysics, Woods Hole Oceanographic Institution, 266 Woods Hole Road, MS# 22, Woods Hole, Massachusetts 02543-1050, USA. [6] University of New Hampshire, Department of Earth Sciences, 56 College Road, James Hall, Durham, New Hampshire 03824-3589, USA. Correspondence and requests for materials should be addressed to C.T.B. (email: bolton@cerege.fr) or to H.M.S. (email: hstoll@geol.uniovi.es).

C occolithophores, a group of unicellular marine phytoplankton, are the only primary producers of biogenic calcite in the open ocean. During their diploid life-cycle stage, calcifying coccolithophores intracellularly produce calcite plates called heterococcoliths. These circular to elliptical coccoliths are extruded through the cell wall to form an exoskeleton, usually composed of a single layer of calcite plates, called a coccosphere. Coccolithophores play an important role in the carbon cycle because they promote the sinking of particulate organic carbon to the deep ocean[1]. Changes in their production of organic carbon and calcification can alter the balance between the organic and inorganic carbon pumps, with strong feedbacks on climate and atmospheric carbon dioxide concentrations (pCO_2) on seasonal to geological timescales[2]. Despite the importance of coccolithophore calcification to biogeochemical cycles and the large range in degree of cell-calcification (defined here as the amount of calcite per unit surface area of the cell) observed both among and within modern species, it is unclear whether specific factors drive changes in cell-calcification state of the ocean's coccolithophore populations on evolutionary timescales. Rapid changes in ocean dissolved CO_2 concentration ($[CO_{2aq}]$), pH, temperature and surface-water stratification in the coming centuries may exert selective pressure on coccolithophore calcification[3,4]. Short-term experiments reveal an array of species- and strain-specific physiological responses to elevated $[CO_{2aq}]$[4–9]. However, selection experiments lasting around a year show that the negative effects of short term (<10 generations) high pCO_2 exposure on coccolithophore calcification and growth are partly reversed for populations that have been exposed to long term (about 500 generations) high pCO_2 conditions[10,11]. Such adaptability is consistent with the geological data indicating that coccolithophores were more ubiquitous and common in the warm, high-CO_2 ocean of the earlier Cenozoic, with larger coccoliths and cells[12,13]. Recent work has shown that calcification competes with photosynthesis for intracellular bicarbonate (HCO_3^-), and that multiple species of coccolithophores reallocate HCO_3^- transport from calcification to photosynthesis at low $[CO_{2aq}]$[14].

Here we explore the long-term response of coccolithophore calcification and HCO_3^- allocation to the evolution of ocean conditions and $[CO_{2aq}]$ over the past 14 million years (Myr) in a key coccolithophore family, the Noëlaerhabdaceae. This family, which includes the genera *Emiliania*, *Gephyrocapsa*, *Pseudoemiliania* and *Reticulofenestra*, dominates most modern ocean coccolithophore communities as well as fossil assemblages since the Miocene. As an indicator of coccolithophore calcification, we show that coccolith thickness correlates strongly with cellular calcification per unit surface area across the range of modern Noëlaerhabdaceae. We then document changes in the size and degree of calcification of Noëlaerhabdaceae coccoliths since 14 Myr ago in two sediment sequences from the tropical Atlantic and Indian Oceans containing well-preserved coccoliths (Ocean Drilling Program, ODP, Site 925 and Indian National Gas Hydrate Program Site NGHP-01-01A, respectively; Fig. 1). In contrast to recent studies of coccolith mass[5,15–18], we present our data as coccolith thickness within narrow size classes to focus on changes in degree of calcification of coccoliths, allowing us to better capture the potential coccolithophore calcification response to changes in the palaeo-carbonate system[19]. These records are the first to document past long-term changes in coccolithophore calcification for given cell size classes in the Miocene and Pliocene, when pCO_2 was higher than pre-industrial levels. Using the geochemical signature of coccoliths, we then assess if changes in degree of calcification correspond to changes in the allocation of intracellular HCO_3^- resources to calcification. Finally, we evaluate the potential role of changing upper ocean stratification

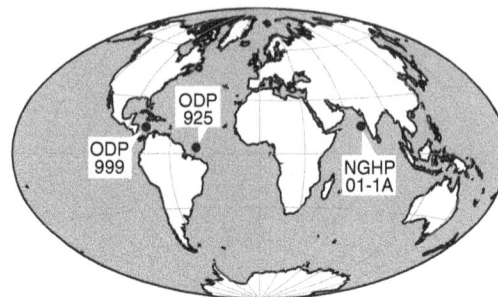

Figure 1 | Map showing site locations.

and $[CO_{2aq}]$ on both degree of calcification and HCO_3^- resource allocation, using new proxy records of foraminiferal stable isotopes ($\delta^{18}O$ and $\delta^{13}C$) and carbon isotopic fractionation by alkenone-producing haptophyte algae during photosynthesis (ε_p).

Results

Coccolith thickness and cellular calcification. New culture experiments sampling the diversity of modern Noëlaerhabdaceae coccolithophores show variation in coccolith thickness both among the different species and among different strains of the same species. This variation in thickness correlates strongly with variation in cellular calcification per surface area as well as with changes in calcite/organic carbon, a measure of calcification per cell volume (Fig. 2 and Supplementary Fig. 1). The thickness of an individual coccolith is intimately linked to the degree of calcification of a cell, because it represents a key mechanism by which cells can regulate the amount of biomineral for a given cell volume. Various factors may drive cells to adjust calcite per cell surface area. In this study, we focus on changes that are occurring within narrow size classes. In addition, calcite per cell surface area varies with cell size across the modern diversity of placolith-bearing coccolithophores, where small cells are characterized by thinner coccoliths (Supplementary Fig. 2). This latter effect may be an adaptation to compensate for the higher surface area to volume ratio of small cells that, if calcification per cell surface area were constant across all cell sizes, would impose a much higher biomineral requirement relative to cell volume in small cells. While coccolith mass has been used as an indicator of cellular calcification in Pleistocene and recent sediments[15–18], coccolith mass is driven by changes in cell size as well as degree of calcification. On the other hand, coccolith thickness within narrow size classes, or size-normalized coccolith thickness, represent degree of calcification and are indicators better suited to reconstructing coccolithophore calcification on long timescales over which significant coccolith and cell size changes occur. The range of coccolith thickness variation among cultured Noëlaerhabdaceae strains (Fig. 2) is consistent with previous observations that phenotypic differences in the degree of calcification between species and between strains of the same species tend to be much larger than the phenotypic plasticity of a single strain cultured under varying environmental conditions[8,9,20]. This may arise if coccolith morphotype or thickness is genetically regulated[9]. The potential for large intraspecific diversity may reflect the genetic architecture, in that the dominant modern Noëlaerhabdaceae *Emiliania huxleyi* has a pan-genome composed of core genes plus genes distributed variably amongst strains[21].

Decreasing cellular calcification since the late Miocene. Over the past 14 Myr, the Noëlaerhabdaceae have undergone large variations in coccolith size (Fig. 3) and degree of calcification, represented by thickness (Fig. 4). Changes in coccolith thickness

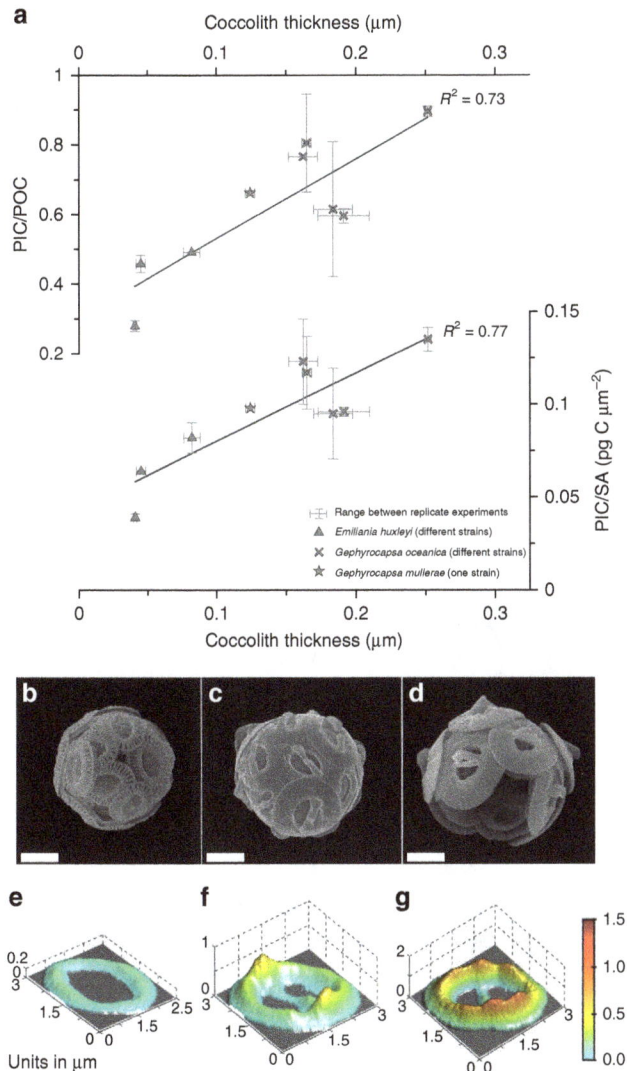

Figure 2 | Noëlaerhabdaceae coccolith morphology in culture.
(**a**) Relationship between coccolith thickness and cellular PIC/POC (particulate inorganic carbon/particulate organic carbon) and cellular PIC/cell SA (surface area) for modern strains of *Emiliania huxleyi* and *Gephyrocapsa* grown in laboratory culture (Supplementary Table 1). Symbols are averages for each experiment and lines show the range of values between replicate culture bottles for each experiment. Scanning electron microscope images of coccospheres from the strains with lowest (RCC 1257, **b**), intermediate (RCC 3370, **c**) and highest (RCC 1292, **d**) coccolith thickness. Scale bar, 2 μm (in all images). (**e-g**) Three-dimensional representations of coccolith thickness for the same strains as coccospheres. The vertical scale shows cumulative thickness from zero at the base; therefore in the central area of *Gephyrocapsa* coccoliths, the bridge (central area bar) is displaced towards the base of the plane of illustration.

are evident in both narrowly restricted size classes, as well as in measurements of size-normalized (SN) thickness and calculated 'shape factor', confirming that they are not a direct result of temporal changes in coccolith and cell size (isometric scaling, that is, changes related to proportional changes in size) (Fig. 4). The quantification of thickness was not biased by variable coccolith fragmentation (Supplementary Fig. 3). Scanning electron microscope (SEM) observations confirm that in all samples the original crystal structure of the coccolith remains well defined. Only on some older coccoliths did we identify a small amount of diagenetic overgrowth (small abiogenic crystals formed on the

surface of the collar in the central area; Supplementary Data 1) However, the presence of this minor overgrowth does not correspond to an increase in coccolith thickness, except in the oldest 14 Myr old sample at the Indian Ocean Site. Thus, with this exception, the preservation visible under SEM makes it unlikely that middle Miocene Noëlaerhabdaceae coccoliths of a given size were originally thinner and more delicate than those present in our samples. This suggests that either (1) overgrowth was minor enough not to significantly impact mean coccolith thickness, or (2) the calcite that recrystallized on the surface of coccoliths was originally derived from dissolution of primary calcite of these same coccoliths. Between 8 and 3 Myr ago at both sites, *Spenolithus* and *Discoaster* nannoliths are abundant. These are typically more susceptible than placolith coccoliths to overgrowth due to their crystal structure, yet SEM images show that these susceptible forms exhibit excellent and constant preservation, providing supporting evidence that diagenetic overgrowth was not more significant when Noëlaerhabdaceae coccoliths showed a higher degree of calcification at 6–8 Myr ago relative to at 3–4 Myr ago (Supplementary Data 1; Supplementary Figs 4 and 5).

The measured coccolith populations exhibit large variability in the morphology and degree of calcification of small coccoliths within and between each sample (Fig. 4; Supplementary Figs 4 and 5; Supplementary Data 1). For example, *Gephyrocapsa protohuxleyi*, a form close to *E. huxleyi* but with a central area bridge characteristic of *Gephyrocapsa*, was present in Pleistocene samples at both sites alongside much more heavily calcified *Gephyrocapsa* coccoliths. Despite this large diversity in morphology and thickness, there are significant changes in the dominance of more heavily calcified versus more lightly calcified forms over time, as well as the emergence during the early Pliocene of coccoliths thinner and/or with larger central area openings than those found in previous intervals. Coccolith degree of calcification was on average highest between 14 and 6 Myr ago and decreased abruptly in the late Miocene to early Pliocene (6–4 Myr ago) to low values that were maintained during the Pliocene and early Pleistocene (4–1 Myr ago). For the few sample points of the last 1 Myr ago, degree of calcification increased both in the Indian and Atlantic Ocean records relative to this Pliocene minimum (Fig. 4). However, we note that assemblages in our samples <1 Myr are dominated by *Gephyrocapsa* coccoliths and pre-date the emergence of the less heavily calcified *E. huxleyi* (see k_s values, Fig. 4d,h), which is significant especially in modern high and mid-latitude regions. Large changes in degree of coccolith calcification, including the decrease from 6 to 4 Myr ago and the increase around 1 Myr ago at both sites, occurred within the dominant genus at a given time, and do not coincide with major shifts in the contribution of different genera to the Noëlaerhabdaceae (Fig. 4).

Geochemical records of carbon isotopic fractionation into coccolith calcite ($\varepsilon_{\text{coccolith}}$) can be used to elucidate the relationship between the observed changes in degree of calcification and the resource allocation of carbon to calcification. Models of cellular carbon fluxes have shown that $\varepsilon_{\text{coccolith}}$ becomes increasingly depleted if the rate of supply of HCO_3^- to the site of calcification (coccolith vesicle) is reduced relative to calcification rate[14]. Our new records of $\varepsilon_{\text{coccolith}}$ from ODP Site 925 show that large cells begin to decrease the HCO_3^- allocation to calcification at about 8 Myr ago, evidenced by decreasing $\varepsilon_{\text{coccolith}}$ (Fig. 5a,b). This trend occurs shortly after a decrease in mean Noëlaerhabdaceae coccolith size (interpreted as a reduction in mean cell size[13]) at both sites (Fig. 3) that is also observed in other low-latitude records[12,22,23]. Reduced HCO_3^- allocation to

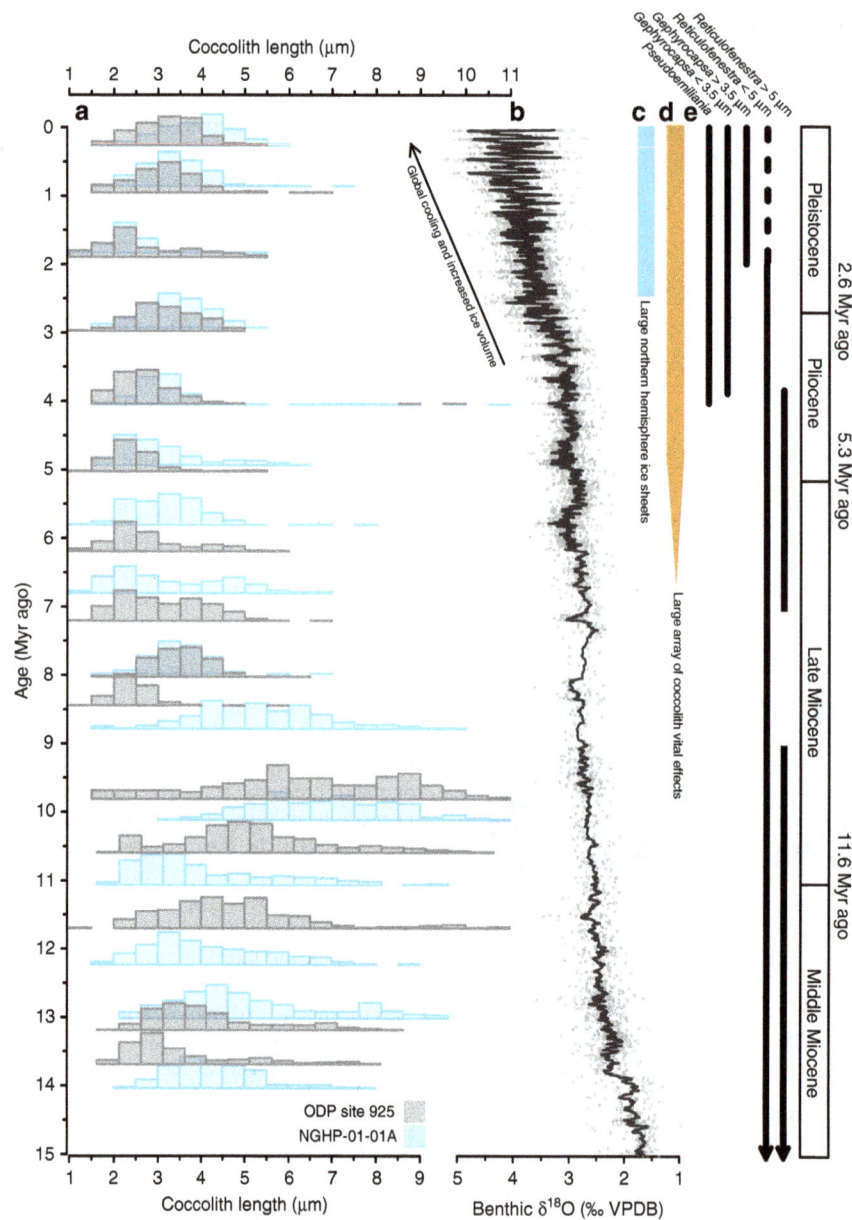

Figure 3 | Long-term evolution of Noëlaerhabdaceae coccolith size and stable isotope vital effects with climate over the last 14 Myr.
(**a**) Noëlaerhabdaceae coccolith size distributions over time at Sites ODP 925 (grey) and NGHP-01-01A (blue). (**b**) Climate evolution over the Neogene represented by a benthic foraminiferal $\delta^{18}O$ stack (data compiled by ref. 86). (**c**) Onset of major northern hemisphere glaciation at $\sim$ 2.6 Myr ago. (**d**) The emergence of large scale vital effects in the $\delta^{18}O$ and $\delta^{13}C$ of coccolith calcite around 7–5 Myr ago (ref. 14; this study). (**e**) Approximate age ranges of Neogene genera belonging to the Noëlaerhabdaceae family[76].

calcification continues in large cells from 6 to 4 Myr ago, as indicated by decreasing $\varepsilon_{coccolith}$ during this interval, despite a stable trend in mean coccolith size. Although we cannot resolve changes in the degree of calcification of large coccoliths in this study (see Methods), the $\varepsilon_{coccolith}$ trend suggests that in large cells, the change in HCO_3^- allocation to calcification was of greater magnitude than any concurrent decrease in calcification that may have occurred. This significant reduced allocation to calcification in large cells drove a divergence in the range of vital effects among small and large coccoliths after 8 Myr ago (Fig. 5a), similar to the results from Caribbean ODP Site 999 (ref. 14).

Small coccoliths show evidence for decreased HCO_3^- allocation to calcification only since 6 Myr ago. From 11 to 6 Myr ago, $\varepsilon_{coccolith}$ and SN coccolith thickness are relatively stable (Fig. 5b,c), suggesting minimal changes in HCO_3^-

allocation to calcification. In contrast, between 6 and 1 Myr ago, a near-constant $\varepsilon_{coccolith}$ indicates a stable ratio of HCO_3^- allocation to the coccolith vesicle relative to calcification rate, despite a large decrease in degree of calcification (Figs 4 and 5c). This implies a decrease in HCO_3^- allocation to calcification of comparable magnitude to the decrease in cellular calcification. In the last 1 Myr, an increase in degree of calcification in the small coccoliths with no change in $\varepsilon_{coccolith}$ suggests that allocation of HCO_3^- to calcification also increased in parallel.

Relationship between calcification and ocean stratification. Water column stratification influences productivity and production depth in the tropics. Stratification can be inferred from foraminiferal $\delta^{18}O$ gradients between the upper mixed

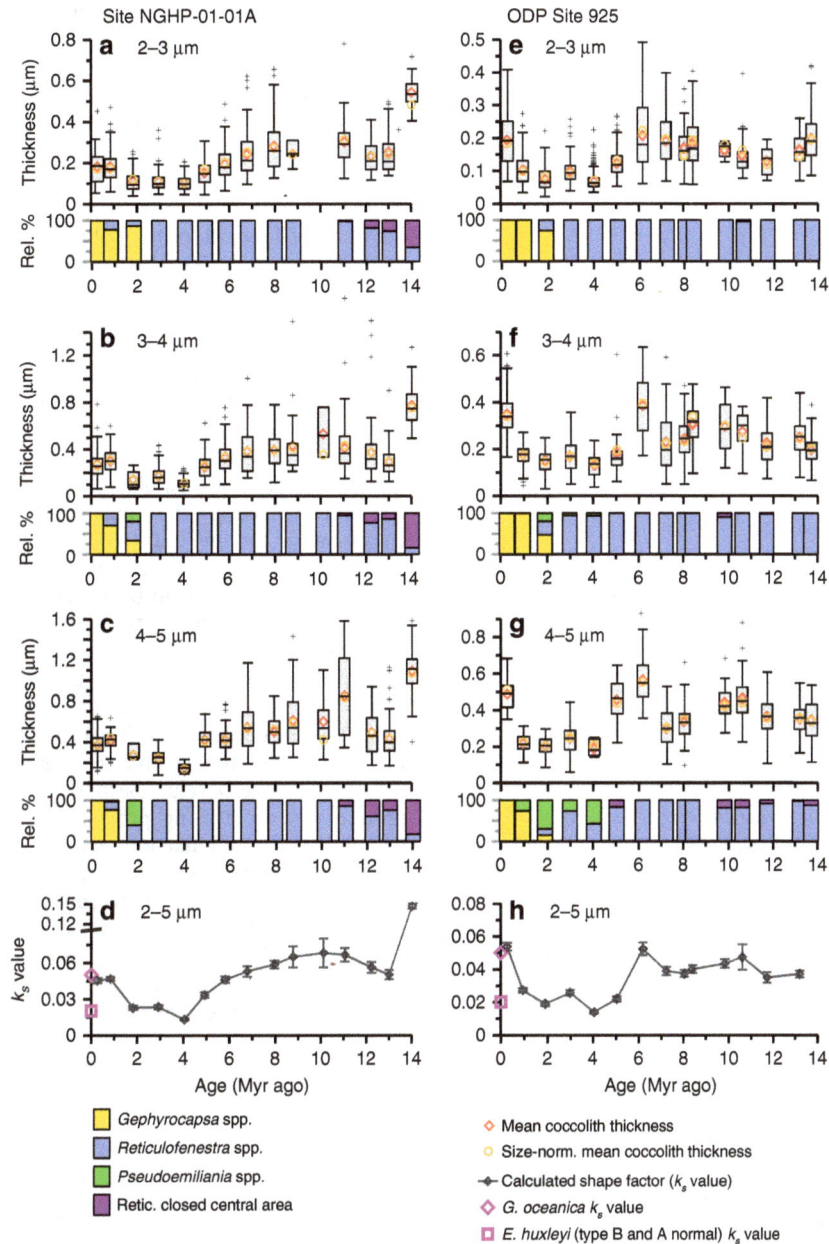

Figure 4 | Changes in Noëlaerhabdaceae coccolith thickness and k_s value at two tropical sites since 14 Myr ago. (**a–d**) Site NGHP-01-01A, and (**e–h**) ODP Site 925. (**a–c,e–g**) Thickness data for coccoliths of 2–3, 3–4 and 4–5 μm length. Box–Whisker plots illustrate coccolith thickness data for each sample and size class (box shows median value and upper/lower quartiles, whiskers show maximum and minimum values, outliers >1.5 × the interquartile range are shown as crosses). Also shown are mean values of raw (circles) and SN (diamonds) thickness (Supplementary Data 1). Bar graphs show the relative contribution of different genera to the Noëlaerhabdaceae population in each size class and sample. (**d,h**) k_s values (error bars are ±2 s.e.m.). The shape factor k_s, which expresses the fraction of the volume of a cube defined by the length of a coccolith that is composed of biomineral[77], was originally proposed to estimate coccolith mass from coccolith length and is similar to coccolith thickness. However, unlike thickness, k_s does not account for variations in coccolith circularity. Pink symbols are k_s for extant Noëlaerhabdaceae species[77].

layer (*Globigerinoides sacculifer*) and thermocline (*Globorotalia menardii*), because these reflect the upper photic zone temperature and salinity gradients[24]. The temporal evolution of planktic foraminiferal $\delta^{18}O$ at Sites ODP 925 and NGHP-01-01A is shown in Fig. 6. Between 3.5 and 2 Myr ago, a deep thermocline at Site 925 is inferred from independent foraminiferal assemblage indicators[25], potentially suggesting a deeper coccolithophore depth habitat and lower light levels. Decreased light has been shown to reduce cellular calcification (PIC/SA) twofold by a reduction in photon flux density from 80 to 15 μmol m^{-2} s^{-1} in culture[26], and low light levels have been

proposed to decrease cellular HCO_3^- transport[27]. However, neither site shows a clear decrease in $\delta^{18}O$ gradients at this time (Fig. 6a,b), as would be expected if reduced coccolith calcification from 4 to 1 Myr ago were due to a deepening of the thermocline, resulting in a reduced temperature gradient between the two foraminifer species' depth habitats. Proxy records suggest high productivity from 10 to 8 Myr ago in the Indian Ocean[28] and from 6.6 to 6 Myr ago at ODP Site 925 (ref. 29). Thus, reconstructed changes in water column structure and paleoproductivity do not consistently co-vary with changes in degree of coccolith calcification.

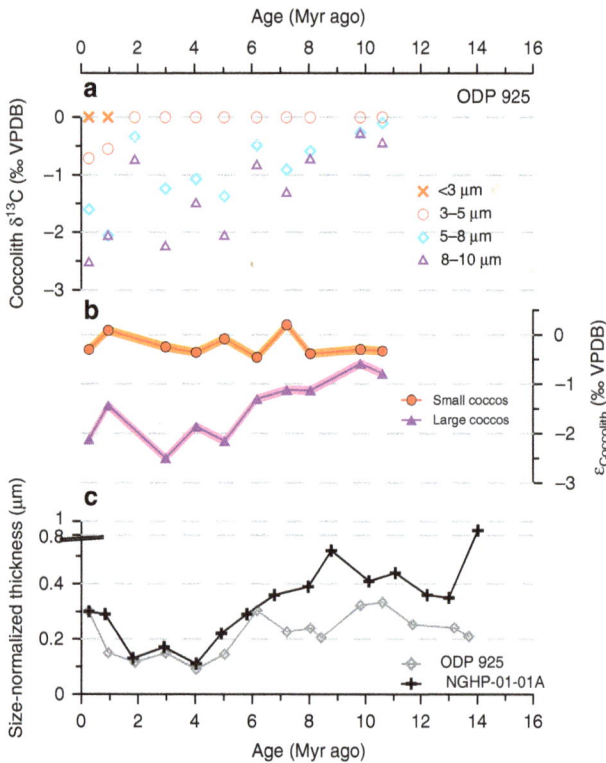

Figure 5 | Coccolith geochemistry and SN thickness trends since the Miocene. (**a**) $\delta^{13}C$ values for size-separated coccoliths from ODP Site 925, normalized to the smallest coccolith size fraction. (**b**) $\varepsilon_{coccolith}$ for small (3–5 μm) and large (8–10 μm) coccoliths from Site 925. Shading indicates propagated analytical uncertainty on $\delta^{13}C$ measurements. (**c**) mean SN coccolith thickness for Noëlaerhabdaceae coccoliths of 2–5 μm lengths at ODP Site 925 and Site NGHP-01-01A (coccolith thicknesses are normalized to mean coccolith length within the 2–5-μm size fraction over the whole time series at each site, that is, 3.52 and 3.93 μm, respectively).

Calcification and [CO$_{2aq}$] in the Miocene–Pliocene. Carbon isotopic fractionation in phytoplankton during photosynthesis (ε_p) varies directly with [CO$_{2aq}$] and has been widely applied as a CO$_2$ proxy in the Cenozoic. However, limited data exist for the interval of major changes in calcification and HCO$_3^-$ allocation between 14 and 5 Myr ago. In addition, the interpretation of any data is complex because of the expected influence of active HCO$_3^-$ allocation on ε_p (ref. 14). Our new record of ε_p extends the published record from ODP Site 999 for the last 5 Myr[30] back to 16 Myr ago (Fig. 7a). This extended record reveals a decrease in ε_p from 16 to 8 Myr ago, an excursion to higher ε_p values at 7 Myr ago, and then a continued decrease towards the present. The decline in ε_p could be driven by decreasing [CO$_{2aq}$], increasing cellular growth rates that increase carbon demand relative to supply, or increasing cell sizes that reduce surface area to volume and thus diffusive supply (see ref. 31 and references therein). Following previous workers, [CO$_{2aq}$] is estimated with the formula [CO$_{2aq}$] $= b/(\varepsilon_f - \varepsilon_p)$, where ε_f is a constant reflecting the maximum effective photosynthetic fractionation by the cell (25‰), and 'b' encompasses factors such as growth rate and cell geometry that modulate the ratio of carbon supply to demand by the cell. First, to estimate temporal variations in b due to cell size, we use previous formulations of the relationship between cell size and b[32], together with our record of tropical Noëlaerhabdaceae coccolith size evolution (Fig. 7b), which shows trends similar to those at other tropical sites[22,23]. The decrease in cell size after 9 Myr ago, compared with the average between 11 and 16 Myr ago, corresponds to a 25% reduction in the b value. Second, we estimate the influence of productivity on b using proxy records from ODP Site 999 of coccolith Sr/Ca and alkenone mass accumulation rates (Fig. 7c). These records confirm that there is no long-term productivity increase, and suggest maxima from 13 to 10 Myr ago and at 8 Myr ago. Calculated b values are shown in Fig. 7d. The resulting estimates of [CO$_{2aq}$] (Fig. 7e) show a trend of continued decline over the past 16 Myr, with the exception of a local maximum at 9.3–10.3 Myr ago resulting from the unusually large cell sizes in

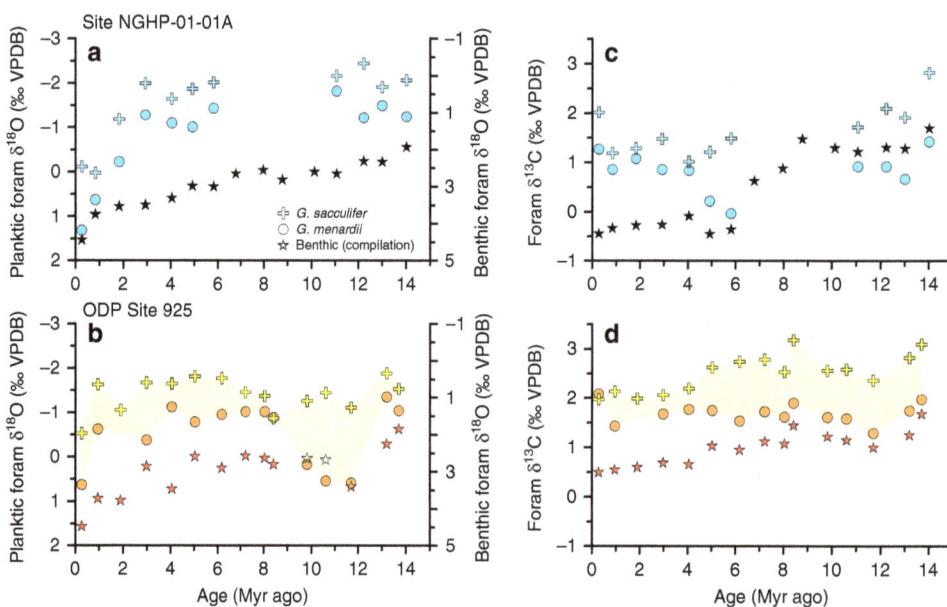

Figure 6 | Foraminiferal stable isotope records as indicators of water column structure. $\delta^{18}O$ (**a,b**) and $\delta^{13}C$ (**c,d**) records for surface (*G. sacculifer*), thermocline (*G. menardii*) and benthic species at Sites ODP 925 and NGHP-01-01A. Benthic isotope data for our sites are not available, therefore values were interpolated using a global compilation[86] separated into ocean basins. Because Neogene Indian Ocean data are sparse and trends are very similar to those in the Pacific, a compilation of Indian and Pacific Ocean data were used to interpolate the benthic values in **a** and **c**. For **b** and **d**, a compilation of all Atlantic data was used. Note: benthic $\delta^{18}O$ is plotted on a different y axis. Shading indicates the gradient between surface and thermocline-dwelling planktic foraminiferal species.

the geometry correction. Assuming equilibrium with the atmosphere, these results are similar in trend and magnitude to $[CO_{2aq}]$ predicted from the atmospheric pCO_2 curve of ref. 33 derived from inverse modelling of climate data (Fig. 7e). The absolute values of $[CO_{2aq}]$ are subject to greater uncertainty than the trend.

As in previous studies[30], our calculations would not account for the likely increase in active carbon uptake for photosynthesis as $[CO_{2aq}]$ declined[14,34], especially after 8 Myr ago. Because active carbon transport increases the chloroplast uptake of inorganic carbon relative to fixation, it can result in higher ε_p values than would be predicted from passive diffusive CO_2 uptake alone[35].

Laboratory culture experiments suggest that active HCO_3^- transport to the chloroplast becomes more significant at low $[CO_{2aq}]$. Simulations with the ACTI-CO model of HCO_3^- transport in coccolithophores[14] were used to evaluate the potential impact of changes in active carbon uptake on ε_p and calculated $[CO_{2aq}]$ (Supplementary Methods; Fig. 8). In one set of simulations, we specify a logarithmic dependence of chloroplast HCO_3^- transport/diffusive CO_2 uptake on $[CO_{2aq}]$ as observed in culture experiments[14,27]. Alternatively, if enhancement of HCO_3^- transport to the chloroplast is coupled, in part, to reallocation of HCO_3^- from the coccolith vesicle, as inferred from modelling of cultures[14], our new $\varepsilon_{coccolith}$ and SN coccolith thickness data put additional constraints on the timing of this reallocation. Therefore in a second set of simulations, we specify chloroplast HCO_3^- transport based on HCO_3^- spared from the coccolith vesicle by the reduction in cellular calcite in the last 8 Myr. We then derive the $[CO_{2aq}]$ implied by measured ε_p for the specified parameterization of active HCO_3^- uptake to the chloroplast. The results in both cases indicate a greater amplitude of decline in $[CO_{2aq}]$ compared with that reconstructed with standard cell size and growth rate considerations only, from around 17 to 6 µM (Fig. 8).

Calcification in relation to CO2 and alkalinity since 1 Myr ago. In the last 1 Myr, climate, the carbon cycle and ocean chemistry

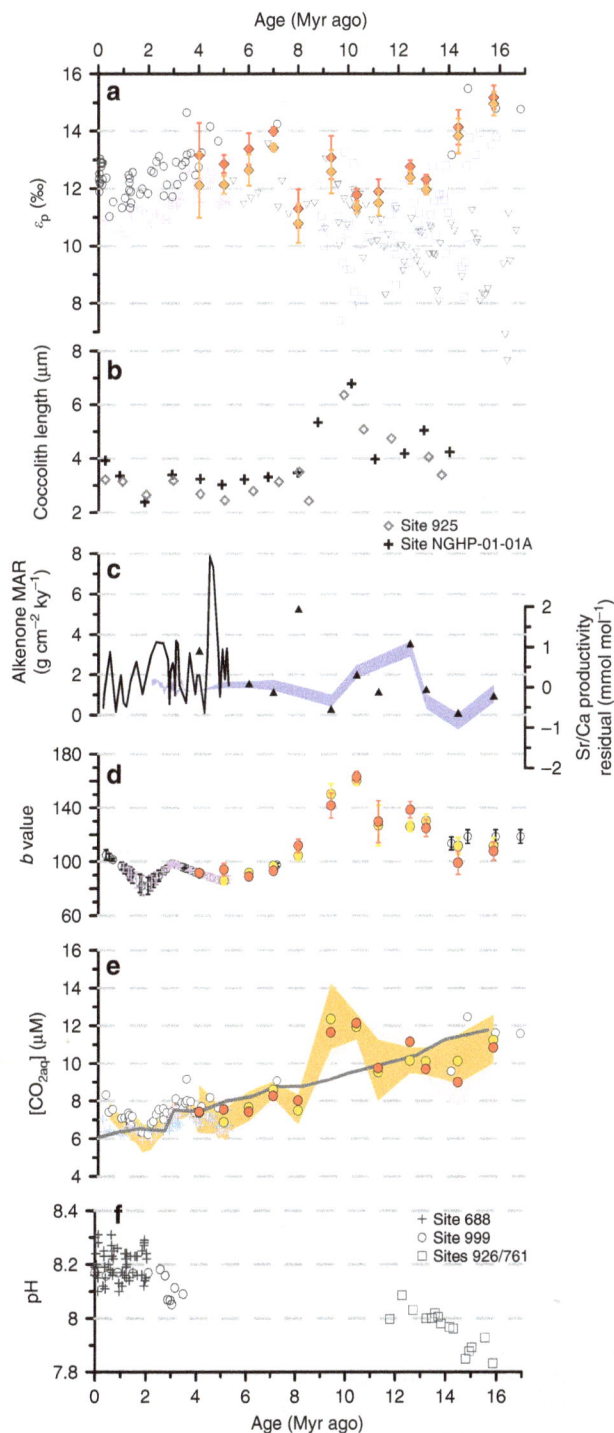

Figure 7 | ε_p values and estimates of b and $[CO_{2aq}]$ at Caribbean ODP Site 999 and other sites for the last 17 Myr. (a) New ε_p data (Site 999, red diamonds: SST max, orange diamonds: SST min; error bars show propagated analytical uncertainty on $\delta^{13}C$ measurements). Published ε_p records: Site 999 (ref. 30) (purple crosses), ODP Site 925 (ref. 37) (grey circles), DSDP Sites 588 (ref. 84) (grey triangles, maximum ε_p) and 608 (ref. 84) (blue squares). **(d)** Variations in b (Site 999) inferred to arise from changing cell size **(b)** and growth rate **(c)** (see Methods). In **c**, triangles (this study) and line[30] show alkenone MARs and blue shading shows Sr/Ca productivity estimates for small coccolithophores[14] (all Site 999). In **d**, purple crosses (Site 999 (ref. 30)), grey circles (Site 925 (ref. 37)), and orange circles (Site 999, this study) show b values recalculated using our new cell size correction. Red circles (Site 999, this study) show b values calculated with cell size and growth rate corrections. For error calculations, see Methods. **(e)** $[CO_{2aq}]$ calculated using cell size (orange circles), or cell size plus growth rate (red circles), correction and ε_p values (Site 999, this study). $[CO_{2aq}]$ was also recalculated using our cell size correction for the Plio-Pleistocene at Site 999 (ref. 30) (purple crosses) and Site 925 (ref. 37) (grey circles). For all sites, reference $b = 150$. $[CO_{2aq}]$ assuming constant b for Site 999 (ref. 30) is also shown (blue crosses). Shading indicates maximum and minimum $[CO_{2aq}]$ estimates for all data from Site 999 (see Supplementary Methods). We do not apply our size correction to DSDP Sites 608 and 588 ε_p data because these sites are at significantly higher latitudes; therefore cell size history may be different compared with the tropical sites studied here. Also shown in **e** is the $[CO_{2aq}]$ expected for the Caribbean site if it were in equilibrium with the atmospheric pCO_2 modelled by ref. 33 (grey line). **(f)** pH derived from $\delta^{11}B$ of planktic foraminifers, for the Plio-Pleistocene[30,36] and Miocene[53]. During the Miocene, ODP Site 999 ε_p values are similar to values at ODP Site 925 (ref. 37) and higher than values from DSDP Sites 588 and 608 (ref. 85). From 16 to 9 Myr ago, the maximum ε_p at DSDP Site 608 shows a similar trend to ε_p at ODP Site 999, albeit with slightly lower absolute values, suggesting that either both sites experienced similar changes in growth rates, or that a global CO_2 component exerted a dominant forcing on both ε_p records. The temporally variable scatter to low ε_p values seen in the Site 608 record may result from higher frequency oscillations in growth rates at this site[83]. The much lower average ε_p at Site 588 suggest that this site experienced on average higher phytoplankton growth rates and productivity compared to Sites 925, 999 and 608.

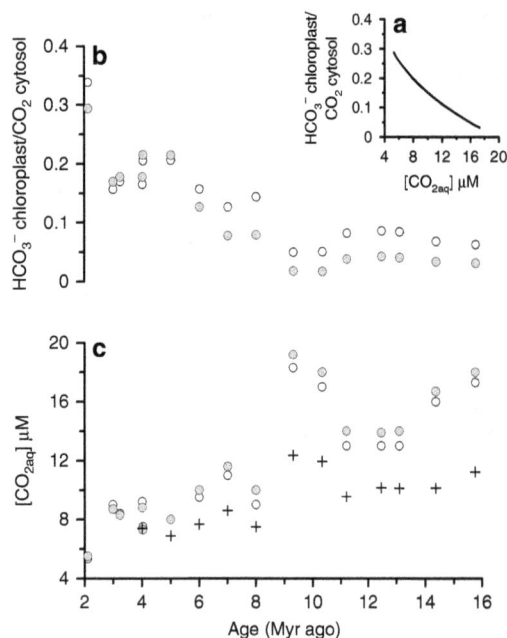

Figure 8 | Simulations of the effect of active HCO₃⁻ uptake on reconstructed [CO₂ₐq]. ACTI-CO model simulation for two potential scenarios of active HCO_3^- uptake to the chloroplast (**b**), and consequences for [CO₂ₐq] implied by alkenone ε_p measurements (**c**). A first simulation (unfilled circles) employs a logarithmic dependence of HCO_3^- transport to the chloroplast on [CO₂ₐq], similar to that observed in cultures (**a**). A second simulation (filled circles) supplements HCO_3^- supply to the chloroplast as a function of HCO_3^- spared from the coccolith vesicle by the production of thinner coccoliths (that is, a reduced PIC per cell surface area). Crosses in **c** show [CO₂ₐq] estimated from standard regressions between ε_p and [CO₂ₐq] as shown in Fig. 7e (orange circles).

underwent significantly higher amplitude variations on glacial–interglacial timescales compared with the preceding 15 Myr. Although our sampling resolution does not capture this higher frequency variability, and does not sample very recent major evolutionary events such as the emergence of *E. huxleyi*, our results nonetheless suggest in the Pleistocene a reversal of the late Miocene–Pliocene trend of more lightly calcified coccoliths and decreasing HCO_3^- allocation to the coccolith vesicle. Within the last 1 Myr, both of these factors rebound to values typical of the late Miocene (Fig. 5). Records based on boron isotopes[36], alkenone $\delta^{13}C$ (refs 30,37), and ice cores[38] suggest pCO_2 values below around 280 p.p.m. over the last 2 Myr, so this increase in degree of calcification contrasts with the generally positive correlation of [CO₂ₐq] and degree of calcification observed over the preceding interval.

The change in relationship between degree of coccolith calcification and [CO₂ₐq] is even more salient when we examine which samples fall in glacial or interglacial ocean states. Planktic foraminiferal $\delta^{18}O$ values from samples at Indian Ocean Site NGHP-01-01A and the orbital age model for Atlantic ODP Site 925 indicate that our youngest samples at both sites (about 0.27 Myr ago), with high SN coccolith thickness, coincide with glacial periods (Supplementary Fig. 6). The sample at 0.84 Myr ago from Site NGHP-01-01A with high SN coccolith thickness also falls during a glacial period, whereas the sample at 0.95 Myr ago from the Site 925, with lower SN coccolith thickness, falls in an interglacial. These particular sampling points therefore underscore the nature of a change in the relationship between degree of calcification and pCO_2, as the samples with thicker coccoliths in a

given size class are from glacial periods that coincide with pCO_2 minima in the last 800 kyr (Supplementary Fig. 6).

In the absence of a coherent relationship with [CO₂ₐq], we consider whether a change in ocean alkalinity may have increased cellular HCO_3^- uptake and reduced competition for intracellular HCO_3^-, promoting the recovery of degree of coccolith calcification and HCO_3^- allocation to calcification. No proxy record of alkalinity change has yet been produced for this time interval. Multiple lines of evidence based on geochemistry, sedimentology and modelling suggest that the rate of silicate weathering, which adds alkalinity to the ocean, accelerated around 1.5 Myr ago as the North American Precambrian basement shed regolith and experienced more intense subglacial erosion[39]. At the same time the first large amplitude sea level cycles accelerated erosion of shelf sediments[40]. Estimates of alkalinity from carbon system proxies are subject to multiple uncertainties. To explore the magnitude of alkalinity change that might be possible, we compared pH estimates from boron isotopes in planktic foraminifers with estimates of ocean [CO₂ₐq] calculated from the cycles of atmospheric pCO_2 recorded in ice cores. While not diagnostic, this analysis suggests the potential for an increase in alkalinity by up to 30% during successive glacials of the last 1.5 Myr (Supplementary Methods; Supplementary Fig. 6). Such an increase would contrast with relatively stable alkalinity inferred for the previous 14 Myr from analysis of the carbonate compensation depth[41], although such estimates are also subject to multiple uncertainties (Supplementary Fig. 7).

Discussion

Over the past 14 Myr, selective pressure has acted on the large diversity of different degrees of calcification and morphotypes found in natural coccolithophore populations[15,42–46], modifying in a similar way in the tropical Atlantic and Indian Oceans the composition of the population towards better-adapted forms. A similar selective pressure has been suggested for natural populations on seasonal timescales, modulating the relative contribution of different *E. huxleyi* morphotypes with specific degrees of calcification[46]. Long-term mono- and multi-clonal experiments also reveal genotypic selection, as well as beneficial new mutations, as a mechanism for adaptive evolution[11]. Such coccolithophore species or morphotype shifts as a result of ocean changes in the future will arguably have a greater impact on carbon cycle feedbacks than direct physiological responses, highlighting the importance of studying integrated community calcification as well as species- or clone-specific responses[47,48].

The similar long-term decreases in degree of coccolith calcification at our Atlantic and Indian Ocean sites suggest a common selective pressure. This trend in calcification occurs alongside reduced HCO_3^- allocation to the coccolith vesicle. The lack of a coherent relationship between stratification and productivity and SN coccolith thickness at both sites suggest that these factors are not strong candidates to force the common trends in degree of calcification at both sites. In contrast, while the relationship between [CO₂ₐq] and cellular calcification has been ambiguous in clonal cell cultures[49–51], decreasing [CO₂ₐq] is one factor shown to reduce HCO_3^- allocation to calcification in modern cells[14]. Changes in [CO₂ₐq] are expected to be globally synchronous across the stratified tropical oceans. While the magnitude of [CO₂ₐq] decline is sensitive to the inferences about active carbon uptake by algae and detailed steps in [CO₂ₐq] cannot be reliably identified given the resolution of our record, a progressive [CO₂ₐq] decline since the middle Miocene is evident and correlates with a succession of adaptations in coccolithophore calcification and cell size. A decline in cell size and the first evidence for reduced allocation of HCO_3^- to calcification (decreasing $\varepsilon_{coccolith}$) in the larger coccolithophores occurs several million years before the reduced allocation of HCO_3^- to calcification and reduced cellular calcification in small cells.

The factors driving this differential timing and type of response between the smaller and larger coccolithophores are at this time uncertain but might include lesser plasticity of coccolith thickness in larger genera (*Helicosphaera*, *Coccolithus* and *Calcidiscus*), or a much stronger pressure for HCO_3^- reallocation by larger cells whose diffusive CO_2 supply was more limited by their low surface area to volume ratio. While decreasing coccolith size has been suggested as one adaptation to decreasing CO_2 availability[12,52], changes in coccolith size, like changes in degree of calcification or changes in the allocation of available carbon to calcification, appear to be part of an array of possible adaptations that may be used simultaneously or sequentially. These strategies appear to be used to varying degrees in different cell size classes, potentially with different thresholds, as each adaptation may come with its own trade-offs. In addition to varying resource availability documented by geochemical indicators, 'top down' ecological pressures may contribute to changes in coccolithophore calcification, but unfortunately no proxies are yet available to evaluate their significance in the geological past.

Here, we show a new approach to exploiting independent geochemical and morphological records of coccolithophores to explore the effect of changing cellular HCO_3^- allocation on the magnitude of $[CO_{2aq}]$ change inferred from ε_p. This approach significantly increases the magnitude of inferred $[CO_{2aq}]$ decline over the last 16 Myr, a result that if substantiated by future high-resolution work, would have important implications for our understanding of climate sensitivity.

The coincidence of greatest SN coccolith thickness and inferred degree of cell-calcification with the period of highest $[CO_{2aq}]$ in the Miocene is at odds with hypotheses for less-calcified cells under future ocean acidification[15,20,48]. Similar to future scenarios, proxy records of past pH derived from boron isotopes in planktic foraminifer shells suggest that high Miocene $[CO_{2aq}]$ coincided with lower surface ocean pH compared with the Pleistocene (Fig. 7f)[30,53]. Although extracellular pH may influence the ease with which protons produced during calcification are exported from coccolithophore cells[54], and in some culture experiments low pH reduces cellular Particulate Inorganic Carbon (PIC) to Particulate Organic Carbon (POC) ratio and increases the incidence of coccolith malformation (refs 6,8,55–57 but also see refs 7,58), the limited phenotypic plasticity of short (<20 generations) monoclonal experiments complicates their extrapolation to real-ocean responses.

Consistent with the long-term responses to high $[CO_{2aq}]$ and low pH since the Mocene shown here, a recent study in the Bay of Biscay found a dominance of heavily calcified *E. huxleyi* morphotypes during winter when $[CO_{2aq}]$ was highest and pH and $CaCO_3$ saturation were lowest[46] (Supplementary Fig. 8). In this study, the low winter calcite saturation state was driven primarily by an increase in dissolved inorganic carbon concentration and, to a minor extent, by reduced temperatures. However, preliminary results for the last 1 Myr suggest that at late Pleistocene $[CO_{2aq}]$ levels, increased alkalinity may have favoured a higher degree of cell calcification, consistent with culture experiments at constant $[CO_2]$, in which increased calcification in *E. huxleyi* accompanied increased alkalinity[6] (Supplementary Fig. 9). A better understanding of the evolution of ocean alkalinity and $[HCO_3^-]$ over the last 1.5 Myr as well as higher-resolution coccolith records may help disentangle the interplay of alkalinity and $[CO_{2aq}]$ on coccolithophore calcification. In addition, further studies with natural populations are required to establish whether $[CO_{2aq}]$ in the modern ocean is a significant driver of cellular calcification.

The long-term reduction in degree of cell-calcification between 14 and 1 Myr ago could potentially have influenced ocean biogeochemical cycles, if a reduction in coccolith ballast lowered the transfer efficiency of organic carbon to the deep ocean. The reduction in cell calcification we identify is coherent with a recent study documenting a global crash in coccolith $CaCO_3$ burial around 4 Myr ago[59]. One consequence of reduced transfer efficiency would be a shallower mean remineralization depth of organic matter. However, $\delta^{13}C$ gradients between surface and thermocline-dwelling foraminifers decrease in par with coccolith thickness from 6 to 4 Myr ago, a change driven primarily by the convergence of thermocline $\delta^{13}C$ values towards surface values (Fig. 6c,d), and suggestive of a deeper depth of organic matter remineralization in the upper water column[60,61]. This trend suggests that the effect of global cooling, which acts to slow remineralization rates, overrode any effect of reduced ballasting and led to a deepening of mean remineralization depth during the late Miocene[60]. If future global warming likewise leads to a shoaling of mean remineralization depth, it may act to counteract any shift towards enhanced ballasting by more heavily calcified coccolithophore cells.

In summary, our observations suggest that on long timescales, increased $[CO_{2aq}]$ and increased alkalinity may contribute selective pressures favouring thicker coccoliths of a given size and a higher degree of cell-calcification. As projected changes in surface ocean chemistry simulate increased $[CO_{2aq}]$ but diminished alkalinity, prediction of the sign of calcification will rely on better defining the thresholds of response to each parameter. In addition, it remains to be determined whether coccolithophore responses to rapid ocean chemistry changes in the future will be analogous to the geological-timescale adaptation studied here. The plasticity of coccolith thickness and potential selective pressures in the genetically diverse modern ocean thus warrant further investigation.

Methods

Cellular calcification and thickness in culture. Eight clonal strains of *E. huxleyi* and *Gephyrocapsa* (Supplementary Table 1) were maintained as dilute batch cultures in natural seawater from the Cantabrian Sea (Northern Spain). Prior to experiments, seawater was sterile filtered at 0.2 μm, heated to 80 °C for 3 h, cooled overnight in a sterile hood, and pH was adjusted to 8.3 by addition of NaOH. Media was enriched with major nutrients (P, N), trace metals and vitamins according to the K/2 recipe[62] modified by eliminating the Tris buffer and silicate. Media was then sterile filtered at 0.2 μm just prior to inoculation. Experiments were carried out under a light:dark cycle of 16:8 h at a constant temperature of 16 °C under saturated light growth conditions (80–150 μmol m^{-2} s^{-1} photon flux). A homogeneous distribution of cells was maintained by placing the cultures on a roller system providing gentle rotation during growth. Through serial dilution, cells were maintained in dilute cultures for 8–12 generations before sampling, to establish stable nutrient and carbon chemistry in the media. Each strain was grown in duplicate, and in some cases triplicate, culture bottles. One strain was cultured in two separate experiments and each experiment is reported separately because the two experiments had opposite trends of drift in pH. On collection, media pH and total alkalinity were measured with a Crison GLP-21pH metre calibrated with National Bureau of Standards (NBS) buffers and quadruplicate potentiometric titration of filtered, poisoned media samples on a Crison TitroMatic 1S, respectively[63,64]. Average media alkalinity was 2,572 μmol kg^{-1} (±115, 1 s.d.), pH 8.22 (±0.07, 1 s.d.) with drift in pH during experiments of <0.11 pH units. Cell density was maintained at biomass averaging 1.65 μg C ml^{-1} and in all cases <2.5 μg C ml^{-1}. Cell counts were determined at harvest with a Fuchs–Rosenthal haemocytometer.

For determination of cellular carbon quota, cells were harvested on pre-combusted GF/F or QFF filters. Following acidification to remove calcite, they were analysed for carbon content by flash combustion EA (Euro Vector EA-1108) at 1,020 °C coupled with a Nu Instruments mass-spectrometer (Nu-Horizon). For determination of cellular calcite, cells were harvested on polycarbonate filters. Filters were acidified in 2% HNO$_3$ and Ca concentration was measured by ICP-AES (Thermo ICAP DUO 6300).

Cell size (radius) and cell surface area (SA) were derived from measurements of cellular carbon quota using the regression of Popp *et al.*[65], which is similar to those derived from other studies[66,67]. We use PIC/SA as an optimal way to represent calcification across a range of cell sizes, because it is unaffected by the size scaling of surface area/volume (SA/V). The PIC/POC ratio is often used to describe the degree of calcification per cell, but scales with SA/V, which is dependent on cell size. In published culture studies, at the onset of the photoperiod and just after division, when cell size is smallest and SA/V is highest, the PIC/POC ratio is 30–36% higher whereas PIC normalized to cell surface area is only 10–12% higher[68]. For a given strain of coccolithophore, this inverse correlation between cell size and PIC/POC ratio is widespread in published culture studies (Supplementary Fig. 10) because of the large plasticity of cell size in response to changes in light regime, carbonate system parameters, and temperature. In our culture samples, because of the limited range in cell size, and larger range in coccolith morphology across different strains, both PIC/SA and PIC/POC yield similar results (Supplementary Fig. 1, Supplementary Table 2).

For determination of coccolith mass and thickness, cells were collected on polycarbonate filters. Coccoliths were extracted from the filters by addition of 0.5 ml ethanol with gentle ultrasonication, and evaporation of the ethanol suspension on a glass microscope slide. This ensured that coccoliths were in a single plane of focus for polarized microscopy. A total of 25–55 random fields of view (FOV) were imaged from microscope slides using a Nikon DS-Fi1 8-bit colour digital camera, Nis-Elements software and a Nikon Eclipse LV100 POL polarized light microscope equipped with a $\times$ 100 H/N2 objective set-up with circular polarization at Salamanca University (resultant area of one pixel = 0.035 μm^2). For full details of the circular polarization microscope set-up applied in this study, see ref. 69. The measurement of coccolith thickness from birefringence relies on the systematic relationship between the thickness of a calcite particle and the interference colour that it displays under polarized light[69–71]. In the thickness range 0–1.55 μm, calcite particles show first-order polarization colours ranging from black to white. Calcite particles thicker than 1.55 μm display first-order polarization colours ranging from yellow through to pink up to a thickness of 4.5 μm, beyond which second-order colours are observed[71]. For culture samples, we use a calibration based on a calcite wedge of known thickness to convert grey level to thickness. Images were processed with C-Calcita[69]. All coccoliths were analysed with a minimum of 100 per sample, and coccolith length, width and area were measured.

Sites and age models. Fifteen samples spanning the last 14 Myr ago were selected from two low-latitude sites to investigate the evolution of coccolith calcification: ODP Site 925 in the western tropical Atlantic Ocean (4°12.248′ N, 43°29.349′ W, water depth 2041 m) and Site NGHP-01-01A in the eastern Arabian Sea (Kerala–Konkan Basin, 15°18.366′ N, 070° 54.192′ E, water depth 2663 m; Fig. 1). Late Neogene sedimentary sections at both sites are primarily calcareous ooze, and sites were selected on the basis of their good coccolith preservation. The age model for Site NGHP-01-01A is based on calcareous nannofossil biostratigraphy[72], and ODP Site 925 ages are based on astronomical calibration of shipboard physical property data (ref. 73; S.J. Crowhurst and H. Pälike, personal communication, 2013).

Coccolith mass and size measurements in sediments. Microscope slides for image analysis were prepared using a quantitative decantation method producing a homogenous distribution of coccoliths[74]. A total of 20–50 random FOV were imaged as described above for cultures. Microscope light settings and camera parameters were kept constant during each imaging session, and calibration images of the same set of calcite particles (the same *Rhabdosphaera* coccoliths in our youngest ODP Site 925 Pleistocene sample) were taken at the start of each session to account for bulb ageing and the different light requirements for different sample groups. Because the method applied here uses grey scale images to estimate coccolith thickness, it is only applicable to coccoliths thinner than 1.55 μm. In the late Miocene, some large Noëlaerhabdaceae coccoliths display yellow or orange first-order interference colours. For this reason, although size data are presented for all Noëlaerhabdaceae coccoliths present in our samples (Fig. 3), we present only mass and thickness measurements for coccoliths in the length range 2–5 μm, which exhibit only grey scale colours throughout the time series (Fig. 4). All whole Noëlaerhabdaceae coccoliths were analysed with a minimum of 300 per sample. Coccolith length, width and area were also measured. Coccolith mass values in this study are comparable to published data using similar birefringence-based methods (Supplementary Fig. 11). To isolate the component of variation in coccolith thickness that represents a change in calcification and does not occur as a direct result of changes in coccolith size, we (1) examine changes in thickness within narrow size classes (2–3, 3–4 and 4–5 μm), and (2) calculate SN thickness of coccoliths within each size classes to further verify that changes in thickness are not a direct result of size changes (that is, the expectation that a larger coccolith is thicker as a result of isometric scaling). Following ref. 75, we use the equation:

$$SN\ thickness = [(ML - CL) \times S] + CT, \qquad (1)$$

where ML is the mean coccolith length within the size fraction in question over the whole time series at a given site (see Supplementary Data 1 for values used), CL is the length of coccolith X in Sample A, S is the slope of the regression between coccolith length and coccolith thickness for all coccoliths in Sample A and CT is the original thickness of coccolith X in Sample A.

Coccolith taxonomy and preservation. We identified coccoliths to genus level only because species-level classification of the smallest Noëlaerhabdaceae can be difficult under the light microscope, and because commonly used *Reticulofenestra* species assignations are primarily based on size[76]. Although most coccoliths were complete, some were found to be missing a piece of the outer rim cycle of one or both shields when observed under the SEM. To verify that such fragmentation did not result in underestimation of coccolith mass per unit area, we quantify this potential effect from SEM images of the 2–5 μm fraction of fossil samples and estimated the percentage of mass loss for individual Noëlaerhabdaceae coccoliths due to fragmentation (minimum 50 coccoliths per sample) using C-Calcita. In these calculations, we assumed that 50% of the total mass of each coccolith comes from the inner rim cycle and central area structure (bridge and/or grill), and that the outer rim cycle of each shield contributed 25% of the total mass. These assumptions were based on mass analyses of very well-preserved modern water column samples containing a mixture of *Gephyrocapsa* and *E. huxleyi* coccoliths. In these samples, mean contribution of inner rim plus central area to total mass was 63 ± 10% (1 s.d.), therefore our choice of 50% for fossil Noëlaerhabdaceae calculations is conservative.

Coccolith dimensions are used to infer cell size, based on relationships between coccolith length and cell size for Noelaerhabdaceae[13,52]. On this basis, we attribute geochemical results from large coccoliths (8–10 μm) as those characterizing larger cells, and those of smaller coccoliths (3–5 μm) as those characterizing small cells. k_s values (originally devised to estimate mass from coccolith shape), were calculated from fossil coccolith mass and length data using the equation of ref. 77:

$$k_s = mass / [2.71 \times length^3] \qquad (2)$$

We verify that temporal patterns in Noëlaerhabdaceae coccolith mass and thickness result only from primary biomineralization and not from abiogenic post-depositional overgrowth using qualitative preservation indices and SEM images (see Results, Supplementary Data 1; Supplementary Figs 4 and 5). Noëlaerhabdaceae with a closed central area occur in some samples older than 10 Myr ago at both sites, and because of the presence of some overgrowth near the central area in some specimens, we remain cautious in always identifying this as a primary morphological feature. Almost all *Reticulofenestra* in the 2–5 μm fraction of our 14 Myr ago sample at Site NGHP-01-01A have closed central areas and this appears to be a primary feature (Supplementary Fig. 5s,t), resulting in very high mean SN thickness and k_s values in this sample (Fig. 4). *Cyclicargolithus* coccoliths were seen sporadically in our oldest samples, although these were all > 5 μm.

Assumptions related to coccoliths per cell. Estimation of changing cellular calcite/SA from coccolith thickness requires that cellular calcite/SA be regulated more strongly by coccolith thickness than by the number of coccoliths per cell. Modern Noëlaerhabdaceae coccolithophore cells are typically covered with a monolayer comprised of interlocking coccoliths, with the exception of *E. huxleyi* that has a higher tendency to produce multi-layered coccospheres[78]. The number of coccoliths per cell has been shown to vary with the cell division cycle of a coccolithophore, with the accumulation of extra coccoliths immediately before cell division to ensure adequate coverage of the two daughter cells[68]. Recent work in culture[79] has confirmed that stationary phase cells, which are essentially paused in the stage just prior to cell division due to lack of nutrients, are more often covered with a higher number of coccoliths compared with exponentially growing cells. While natural populations in the ocean contain cells at a variety of life stages, that is, not synchronized in the same way as a culture population, populations of *Coccolithus pelagicus* recovered from the surface ocean during the North Atlantic spring bloom still have on average around 20–25% fewer coccolith per cell than populations growing in non-bloom conditions, analogous to culture[79]. These differences may be useful in characterizing major swings in productivity in fossil populations, in the rare cases where exceptional preservation conserves a large number of intact coccospheres. Quantification of changes in mean number of coccoliths per cell in fossil populations is not possible in this study because it requires significant numbers of whole coccospheres to be preserved, a phenomenon that is rare in deep, open-ocean sites such as ours (selected to be representative of global change) and on the long timescales at which we are working at a single site. However, in sediment populations such as those in our study, which integrate a time window of a few hundred to a thousand years (taking into account sedimentation rates and bioturbation), seasonal and inter-annual productivity changes that might cause higher representation of one growth phase versus another and concomitant changes in coccoliths per cell are likely averaged out. In addition, export of phytoplankton to the sediments is typically biased towards high production periods, so the significance of true stationary phase growth to export and the sediment population is likely diminished.

Stable isotopes in foraminifers. Isotope data for two planktic foraminiferal species at Sites NGHP-01-01A and ODP Site 925 were generated. Approximately 25 individuals of *G. sacculifer* (without sac-like final chamber) and *G. menardii* were picked from the 250–350-μm and the 300–400-μm size fractions, respectively. Foraminifers were broken open and ultrasonicated in methanol to remove fine fraction contamination, rinsed with MilliQ water, dried at 55 °C, and analysed on a Nu Instruments Perspective DI-IRMS connected to an automated carbonate preparation system (Nu Carb), with an analytical precision of 0.06‰ for $\delta^{18}O$ and 0.05‰ for $\delta^{13}C$ (1σ), at Oviedo University.

The variable depth habitats of planktic foraminifers allow us to reconstruct changes in upper water column properties via oxygen and carbon isotopic gradients. In tropical open-ocean settings such as those overlying our two sites, *G. sacculifer* is thought to live and calcify in the upper mixed layer of the ocean, whereas *G. menardii* favours the upper-middle thermocline[24,25,80,81]. To a first order, foraminiferal $\delta^{18}O$ gradients between the upper mixed layer and thermocline in well-stratified regions of the tropical ocean reflect the upper photic zone temperature and salinity gradients[24], although some additional physiological (vital effects) and environmental (for example, carbonate ion concentration) factors also affect isotopic fractionation[81]. Foraminiferal $\delta^{13}C$ gradients between the upper mixed layer, the thermocline and the deep ocean at both sites were used to evaluate the depth of organic matter remineralisation, following refs 60,61. Planktic foraminiferal $\delta^{13}C$ values were corrected to dissolved inorganic carbon (DIC) following ref. 81.

Stable isotopes in coccoliths. Samples were disaggregated and micro-filtered in 2% ammonia to separate coccolith size fractions (<3, 3–5, 5–8 and 8–10 μm), which were rinsed three times with MilliQ water and dried at 55 °C. All coccolith fractions were examined under the microscope, and fractions from Site NGHP-01-01A were found to be heavily contaminated with fragments of non-coccolith carbonate. For this reason, isotope data for these samples are not presented or considered further. At ODP Site 925, coccolith size fractions contain solely coccolith carbonate. In samples older than 3 Myr ago, coccolith fragments contribute significantly to the <3 μm fraction so data points were excluded. Coccolith samples were analysed as described above for foraminifers. Mean reproducibility, based on duplicate analyses of splits of 22 random coccolith samples from Site 925, is 0.08‰ for $\delta^{18}O$ and 0.07‰ for $\delta^{13}C$ (1σ). $\varepsilon_{coccolith}$ of small and large coccoliths was calculated from coccolith $\delta^{13}C$ relative to *G. menardii* $\delta^{13}C$ because this foraminiferal species calcifies in equilibrium with $\delta^{13}C$ DIC[81], and also has a similar depth habitat to the coccolithophores with maximum abundances in the chlorophyll maximum near the thermocline[82]; such that

$$\varepsilon_{coccolith} = \delta^{13}C_{coccolith size fraction} - \delta^{13}C_{G. menardii} \qquad (3)$$

Propagated analytical uncertainty of $\varepsilon_{coccolith} = \sqrt{((0.05^2) + (0.05^2))}. \qquad (4)$

For one sample around 2 Myr ago, there was an insufficient number of *G. menardii* individuals for analysis, therefore we were unable to calculate $\varepsilon_{coccolith}$.

See Supplementary Methods for details of carbon isotope determinations in alkenones, ε_p and $[CO_{2aq}]$ calculations, and details on ACTI-CO simulations to quantify the effect of changing active uptake on $[CO_{2aq}]$ estimates.

References

1. Klaas, C. & Archer, D. E. Association of sinking organic matter with various types of mineral ballast in the deep sea: implications for the rain ratio. *Global Biogeochem. Cycles* **16**, 1–14 (2002).
2. Hain, M., Sigman, D. & Haug, G. in *Treatise on Geochemistry* 2nd edn, Vol. 8(18) (eds Mottl, M. J. & Elderfield, Henry) 485–517 (2013).
3. Feely, R. A. *et al.* Impact of anthropogenic CO₂ on the CaCO₃ system in the oceans. *Science* **305**, 362–366 (2004).
4. Rost, B., Zondervan, I. & Wolf-Gladrow, D. Sensitivity of phytoplankton to future changes in ocean carbonate chemistry: current knowledge, contradictions and research directions. *Mar. Ecol. Prog. Ser.* **373**, 227–237 (2008).
5. Bach, L. T., Bauke, C., Meier, K. J. S., Riebesell, U. & Schulz, K. G. Influence of changing carbonate chemistry on morphology and weight of coccoliths formed by *Emiliania huxleyi*. *Biogeosciences* **9**, 3449–3463 (2012).
6. Bach, L. T. *et al.* Dissecting the impact of CO₂ and pH on the mechanisms of photosynthesis and calcification in the coccolithophore *Emiliania huxleyi*. *New Phytol.* **199**, 121–134 (2013).
7. Jones, B. M. *et al.* Responses of the *Emiliania huxleyi* proteome to ocean acidification. *PloS One* **8**, e61868 (2013).
8. Langer, G. *et al.* Species-specific responses of calcifying algae to changing seawater carbonate chemistry. *Geochem. Geophys. Geosyst.* **7**, Q09006 (2006).
9. Langer, G., Nehrke, G., Probert, I., Ly, J. & Ziveri, P. Strain-specific responses of *Emiliania huxleyi* to changing seawater carbonate chemistry. *Biogeosciences* **6**, 2637–2646 (2009).
10. Jin, P., Gao, K. & Beardall, J. Evolutionary responses of a coccolithophorid *Gephyrocapsa oceanica* to ocean acidification. *Evolution* **67**, 1869–1878 (2013).
11. Lohbeck, K. T., Riebesell, U. & Reusch, T. B. Adaptive evolution of a key phytoplankton species to ocean acidification. *Nat. Geosci.* **5**, 346–351 (2012).
12. Hannisdal, B., Henderiks, J. & Liow, L. H. Long-term evolutionary and ecological responses of calcifying phytoplankton to changes in atmospheric CO₂. *Global Change Biol.* **18**, 3504–3516 (2012).

13. Henderiks, J. Coccolithophore size rules - reconstructing ancient cell geometry and cellular calcite quota from fossil coccoliths. *Mar. Micropaleontol.* **67**, 143–154 (2008).
14. Bolton, C. T. & Stoll, H. M. Late Miocene threshold response of marine algae to carbon dioxide limitation. *Nature* **500**, 558–562 (2013).
15. Beaufort, L. *et al.* Sensitivity of coccolithophores to carbonate chemistry and ocean acidification. *Nature* **476**, 80–83 (2011).
16. Meier, K. J. S., Beaufort, L., Heussner, S. & Ziveri, P. The role of ocean acidification in *Emiliania huxleyi* coccolith thinning in the Mediterranean Sea. *Biogeosciences* **11**, 2857–2869 (2014).
17. Horigome, M. T. *et al.* Environmental controls on the *Emiliania huxleyi* calcite mass. *Biogeosciences* **11**, 2295–2308 (2014).
18. Meier, K. J. S., Berger, C. & Kinkel, H. Increasing coccolith calcification during CO₂ rise of the penultimate deglaciation (Termination II). *Mar. Micropaleontol.* **112**, 1–12 (2014).
19. Gibbs, S. J., Robinson, S. J., Bown, P. R., Dunkley Jones, T. & Henderiks, J. Comment on "Calcareous nannoplankton response to surface-water acidification around Oceanic Anoxic Event 1a". *Science* **332**, 175 (2011).
20. Riebesell, U. *et al.* Reduced calcification of marine plankton in response to increased atmospheric CO₂. *Nature* **407**, 364–367 (2000).
21. Read, B. A. *et al.* Pan genome of the phytoplankton *Emiliania* underpins its global distribution. *Nature* **499**, 209–213 (2013).
22. Kameo, K. & Bralower, T. J. in *Proceedings of the Ocean Drilling Program. Scientific Results* 165 (eds Leckie, R. M., Sigurdsson, H., Acton, G. D. & Draper, G.) 3–17 (Ocean Drilling Program, 2000).
23. Young, J. R. Size variation of Neogene *Reticulofenestra* coccoliths from Indian Ocean DSDP cores. *J. Micropalaeontol.* **9**, 71–85 (1990).
24. Steph, S., Regenberg, M., Tiedemann, R., Mulitza, S. & Nürnberg, D. Stable isotopes of planktonic foraminifera from tropical Atlantic/Caribbean core-tops: Implications for reconstructing upper ocean stratification. *Mar. Micropaleontol.* **71**, 1–19 (2009).
25. Chaisson, W. & Ravelo, A. In *Proceedings of the Ocean Drilling Program. Scientific Results* 255–268 (Ocean Drilling Program, 1997).
26. Rost, B., Zondervan, I. & Riebesell, U. Light-dependent carbon isotope fractionation in the coccolithophorid *Emiliania huxleyi*. *Limnol. Oceanogr.* **47**, 120–128 (2002).
27. Cassar, N., Laws, E. A. & Popp, B. N. Carbon isotopic fractionation by the marine diatom *Phaeodactylum tricornutum* under nutrient- and light-limited growth conditions. *Geochim. Cosmochim. Acta* **70**, 5323–5335 (2006).
28. Gupta, A. K., Singh, R. K., Joseph, S. & Thomas, E. Indian Ocean high-productivity event (10–8 Ma): linked to global cooling or to the initiation of the Indian monsoons? *Geology* **32**, 753–756 (2004).
29. Diester-Haass, L., Billups, K. & Emeis, K. C. In search of the late Miocene-early Pliocene "biogenic bloom" in the Atlantic Ocean (Ocean Drilling Program Sites 982, 925, and 1088). *Paleoceanography* **20** (2005).
30. Seki, O. *et al.* Alkenone and boron-based Pliocene pCO₂ records. *Earth. Planet. Sci. Lett.* **292**, 201–211 (2010).
31. Pagani, M. in *Treatise on Geochemistry 2nd edn* (2014).
32. Henderiks, J. & Pagani, M. Refining ancient carbon dioxide estimates: significance of coccolithophore cell size for alkenone-based pCO₂ records. *Paleoceanography* **22** (2007).
33. van de Wal, R. S., de Boer, B., Lourens, L. J., Köhler, P. & Bintanja, R. Reconstruction of a continuous high-resolution CO₂ record over the past 20 million years. *Clim. Past* **7**, 1459–1469 (2011).
34. Hopkinson, B. M., Dupont, C. L., Allen, A. E. & Morel, F. M. M. Efficiency of the CO₂-concentrating mechanism of diatoms. *Proc. Natl Acad. Sci. USA* **108**, 3830–3837 (2011).
35. Laws, E. A., Popp, B. N., Cassar, N. & Tanimoto, J. ¹³C discrimination patterns in oceanic phytoplankton: likely influence of CO₂ concentrating mechanisms, and implications for palaeoreconstructions. *Funct. Plant Biol.* **29**, 323–333 (2002).
36. Hönisch, B., Hemming, N. G., Archer, D., Siddall, M. & McManus, J. F. Atmospheric carbon dioxide concentration across the mid-Pleistocene transition. *Science* **324**, 1551–1554 (2009).
37. Zhang, Y. G., Pagani, M., Liu, Z., Bohaty, S. M. & DeConto, R. A 40 million-year history of atmospheric CO₂. *Phil. Trans. R. Soc. A* **371**, 20130096 (2013).
38. Lüthi, D. *et al.* High-resolution carbon dioxide concentration record 650,000–800,000 years before present. *Nature* **453**, 379–382 (2008).
39. Clark, P. U. *et al.* The middle Pleistocene transition: characteristics, mechanisms, and implications for long-term changes in atmospheric pCO₂. *Quat. Sci. Rev.* **25**, 3150–3184 (2006).
40. Markovic, S., Paytan, A. & Wortmann, U. G. Pleistocene sediment offloading and the global sulfur cycle. *Biogeosciences* **12**, 3043–3060 (2015).
41. Tyrrell, T. & Zeebe, R. E. History of carbonate ion concentration over the last 100 million years. *Geochim. Cosmochim. Acta* **68**, 3521–3530 (2004).

42. Beaufort, L., Couapel, M., Buchet, N., Claustre, H. & Goyet, C. Calcite production by coccolithophores in the south east Pacific Ocean. *Biogeosciences* **5**, 1101–1117 (2008).

43. Cubillos, J. et al. Calcification morphotypes of the coccolithophorid *Emiliania huxleyi* in the Southern Ocean: changes in 2001 to 2006 compared to historical data. *Mar. Ecol. Prog. Ser.* **348**, 47–54 (2007).

44. Henderiks, J. et al. Environmental controls on *Emiliania huxleyi* morphotypes in the Benguela coastal upwelling system (SE Atlantic). *Mar. Ecol. Prog. Ser.* **448**, 51–66 (2012).

45. Poulton, A. J., Young, J. R., Bates, N. R. & Balch, W. M. Biometry of detached *Emiliania huxleyi* coccoliths along the Patagonian Shelf. *Mar. Ecol. Prog. Ser.* **443**, 1–17 (2011).

46. Smith, H. E. K. et al. Predominance of heavily calcified coccolithophores at low $CaCO_3$ saturation during winter in the Bay of Biscay. *Proc. Natl Acad. Sci. USA* **109**, 8845–8849 (2012).

47. Poulton, A. J. et al. Coccolithophores on the north-west European shelf: calcification rates and environmental controls. *Biogeosciences* **11**, 3919–3940 (2014).

48. Ridgwell, A. et al. From laboratory manipulations to Earth system models: scaling calcification impacts of ocean acidification. *Biogeosciences* **6**, 2611–2623 (2009).

49. De Bodt, C., Van Oostende, N., Harlay, J., Sabbe, K. & Chou, L. Individual and interacting effects of pCO_2 and temperature on *Emiliania huxleyi* calcification: study of the calcite production, the coccolith morphology and the coccosphere size. *Biogeosciences* **7**, 1401–1412 (2010).

50. Iglesias-Rodriguez, M. D. et al. Phytoplankton calcification in a high-CO_2 world. *Science* **320**, 336–340 (2008).

51. Langer, G. & Bode, M. CO_2 mediation of adverse effects of seawater acidification in *Calcidiscus leptoporus*. *Geochem. Geophys. Geosyst.* **12**, Q05001 (2011).

52. Henderiks, J. & Pagani, M. Coccolithophore cell size and the Paleogene decline in atmospheric CO_2. *Earth. Planet. Sci. Lett.* **269**, 576–584 (2008).

53. Foster, G. L., Lear, C. H. & Rae, J. W. B. The evolution of pCO_2, ice volume and climate during the Middle Miocene. *Earth. Planet. Sci. Lett.* **341-344**, 243–254 (2012).

54. Taylor, A. R., Chrachri, A., Wheeler, G., Goddard, H. & Brownlee, C. A voltage-gated H^+ channel underlying pH homeostasis in calcifying coccolithophores. *PLoS Biol.* **9**, e1001085 (2011).

55. Langer, G., Probert, I., Nehrke, G. & Ziveri, P. The morphological response of *Emiliania huxleyi* to seawater carbonate chemistry changes: an inter-strain comparison. *J. Nannoplankton Res.* **32**, 29–34 (2011).

56. Rickaby, R. E. M., Henderiks, J. & Young, J. N. Perturbing phytoplankton: response and isotopic fractionation with changing carbonate chemistry in two coccolithophore species. *Clim. Past.* **6**, 771–785 (2010).

57. Riebesell, U. & Tortell, P. D. in *Ocean acidification*. (eds Gattuso, J. P. & Hanson, L.) 99–121 (Oxford University Press, 2011).

58. Young, J., Poulton, A. & Tyrrell, T. Morphology of *Emiliania huxleyi* coccoliths on the North West European shelf-is there an influence of carbonate chemistry? *Biogeosci. Discuss.* **11**, 4531–4561 (2014).

59. Suchéras-Marx, B. & Henderiks, J. Downsizing the pelagic carbonate factory: impacts of calcareous nannoplankton evolution on carbonate burial over the past 17 million years. *Glob. Planet. Change.* **123**, 97–109 (2014).

60. John, E. H. et al. Warm ocean processes and carbon cycling in the Eocene. *Phil. Trans. R. Soc. A* **371**, 20130099 (2013).

61. Kwon, E. Y., Primeau, F. & Sarmiento, J. L. The impact of remineralization depth on the air–sea carbon balance. *Nat. Geosci.* **2**, 630–635 (2009).

62. Keller, M., Selvin, R., Claus, W. & Guillard, R. Media for the culture of oceanic ultraphytoplankton. *J. Phycol.* **23**, 633–638 (1987).

63. Bradshaw, A., Brewer, P., Shafer, D. & Williams, R. Measurements of total carbon dioxide and alkalinity by potentiometric titration in the GEOSECS program. *Earth. Planet. Sci. Lett.* **55**, 99–115 (1981).

64. Brewer, P., Bradshaw, A. & Williams, R. in *The Changing Carbon Cycle: A Global Analysis* (eds Trabalka, J. R. & Reichle, D. E.) 348–370 (Springer, 1986).

65. Popp, B. N. et al. Effect of phytoplankton cell geometry on carbon isotopic fractionation. *Geochim. Cosmochim. Acta* **62**, 69–77 (1998).

66. Menden-Deuer, S. & Lessard, E. J. Carbon to volume relationships for dinoflagellates, diatoms, and other protist plankton. *Limnol. Oceanogr.* **45**, 569–579 (2000).

67. Montagnes, D. J., Berges, J. A., Harrison, P. J. & Taylor, F. Estimating carbon, nitrogen, protein, and chlorophyll a from volume in marine phytoplankton. *Limnol. Oceanogr.* **39**, 1044–1060 (1994).

68. Zondervan, I., Rost, B. & Riebesell, U. Effect of CO_2 concentration on the PIC/POC ratio in the coccolithophore *Emiliania huxleyi* grown under light-limiting conditions and different daylengths. *J. Exp. Mar. Bio. Ecol.* **272**, 55–70 (2002).

69. Fuertes, M. A., Flores, J. A. & Sierro, F. J. The use of circularly polarized light for biometry, identification and estimation of mass of coccoliths. *Mar. Micropaleontol.* **113**, 44–55 (2014).

70. Beaufort, L. Weight estimates of coccoliths using the optical properties (birefringence) of calcite. *Micropaleontology* **51**, 289–297(2005).

71. Beaufort, L., Barbarin, N. & Gally, Y. Optical measurements to determine the thickness of calcite crystals and the mass of thin carbonate particles such as coccoliths. *Nat. Protoc.* **9**, 633–642 (2014).

72. Flores, J. A. et al. Sedimentation rates from calcareous nannofossil and planktonic foraminifera biostratigraphy in the Andaman Sea, northern Bay of Bengal, and Eastern Arabian Sea. *Mar. Pet. Geol.* **58**, 425–437 (2014).

73. Shackleton, N., Crowhurst, S., Weedon, G. & Laskar, J. Astronomical calibration of Oligocene-Miocene time. *Phil. Trans. R. Soc. London Ser. A* **357**, 1907–1929 (1999).

74. Flores, J. & Sierro, F. Revised technique for calculation of calcareous nannofossil accumulation rates. *Micropaleontology* **43**, 321–324 (1997).

75. O'Dea, S. A. et al. Coccolithophore calcification response to past ocean acidification and climate change. *Nat. Commun.* **5**, 5363 (2014).

76. Young, J. in *Calcareous Nannofossil Biostratigraphy* 225–265 (1998).

77. Young, J. & Ziveri, P. Calculation of coccolith volume and its use in calibration of carbonate flux estimates. *Deep Sea Res. Part II* **47**, 1679–1700 (2000).

78. Hoffmann, R. et al. Insight into *Emiliania huxleyi* coccospheres by focused ion beam sectioning. *Biogeosciences* **12**, 825–834 (2015).

79. Gibbs, S. J. et al. Species-specific growth response of coccolithophores to Palaeocene-Eocene environmental change. *Nat. Geosci.* **6**, 218–222 (2013).

80. Farmer, E. C., Kaplan, A., de Menocal, P. B. & Lynch-Stieglitz, J. Corroborating ecological depth preferences of planktonic foraminifera in the tropical Atlantic with the stable oxygen isotope ratios of core top specimens. *Paleoceanography* **22** (2007).

81. Spero, H. J., Mielke, K. M., Kalve, E. M., Lea, D. W. & Pak, D. K. Multispecies approach to reconstructing eastern equatorial Pacific thermocline hydrography during the past 360 kyr. *Paleoceanography* **18** (2003).

82. Tedesco, K., Thunell, R., Astor, Y. & Muller-Karger, F. The oxygen isotope composition of planktonic foraminifera from the Cariaco Basin, Venezuela: Seasonal and interannual variations. *Mar. Micropaleontol.* **62**, 180–193 (2007).

83. Diester-Haass, L. et al. Mid-Miocene paleoproductivity in the Atlantic Ocean and implications for the global carbon cycle. *Paleoceanography* **24**, PA1209 (2009).

84. Pagani, M., Zachos, J. C., Freeman, K. H., Tipple, B. & Bohaty, S. Marked decline in atmospheric carbon dioxide concentrations during the Paleogene. *Science* **309**, 600–603 (2005).

85. Pagani, M., Arthur, M. A. & Freeman, K. H. Miocene evolution of atmospheric carbon dioxide. *Paleoceanography* **14**, 273–292 (1999).

86. Zachos, J. C., Dickens, G. R. & Zeebe, R. E. An early Cenozoic perspective on greenhouse warming and carbon-cycle dynamics. *Nature* **451**, 279–283 (2008).

Acknowledgements

We thank Torsten Bickert and Jeroen Groeneveld for access to unpublished data, and Heiko Pälike and Simon Crowhurst for advice regarding the ODP Site 925 age model. This paper benefitted from discussions with Luc Beaufort, Ian Bailey, Kaustubh Thirumalai, Samantha Gibbs and Paul Bown. We thank the Department of Biology of Organisms and Systems at the University of Oviedo for access to facilities for culturing coccolithophores. This is a contribution of the Asturias Marine Observatory. This work used samples provided by the (Integrated) Ocean Drilling Program (IODP) and the Indian National Gas Hydrate Program (NGHP). The IODP is sponsored by the US National Science Foundation and participating countries under management of the IODP Management International, Inc (IODP-MI). NGHP01 was planned and managed through collaboration between the Directorate General of Hydrocarbons (DGH) under the Ministry of Petroleum and Natural Gas (India), the US Geological Survey (USGS) and the Consortium for Scientific Methane Hydrate Investigations (CSMHI) led by Overseas Drilling Limited (ODL) and FUGRO McClelland Marine Geosciences (FUGRO). We thank those who contributed to the success of the NGHP Expedition 01 (NGHP01). C.T.B. acknowledges OSU-Institut Pythéas. Funding for this research was provided by the European Research Council under grant UE-09-ERC-2009-STG-240222-PACE (HMS), the Principado de Asturias under award FC-13-COF13-044 (HMS) and a French ANR infrastructure project EMBRC-France (IP).

Author contributions

H.M.S. designed the research. C.T.B., M.T.H.S., L.A. and A.M.V. prepared and analysed geochemical samples. I.P. provided coccolithophore strains and culturing expertise. S.G.L. cultured coccolithophores. S.G.L. and C.T.B. performed modern and fossil microscopy and image analysis, respectively. C.T.B., M.T.H.S. and H.M.S. analysed data and performed calculations. J.A.F. and M.A.F. contributed new analytical tools. L.G. and J.J. contributed expertise and samples from Site NGHP-01-01A. C.T.B. and H.M.S. wrote the paper with feedback from all authors.

Additional information

Competing financial interests: The authors declare no competing financial interests.

11

Biological and physical controls in the Southern Ocean on past millennial-scale atmospheric CO_2 changes

Julia Gottschalk[1], Luke C. Skinner[1], Jörg Lippold[2], Hendrik Vogel[2], Norbert Frank[3], Samuel L. Jaccard[2] & Claire Waelbroeck[4]

Millennial-scale climate changes during the last glacial period and deglaciation were accompanied by rapid changes in atmospheric CO_2 that remain unexplained. While the role of the Southern Ocean as a 'control valve' on ocean–atmosphere CO_2 exchange has been emphasized, the exact nature of this role, in particular the relative contributions of physical (for example, ocean dynamics and air–sea gas exchange) versus biological processes (for example, export productivity), remains poorly constrained. Here we combine reconstructions of bottom-water [O_2], export production and ^{14}C ventilation ages in the sub-Antarctic Atlantic, and show that atmospheric CO_2 pulses during the last glacial- and deglacial periods were consistently accompanied by decreases in the biological export of carbon and increases in deep-ocean ventilation via southern-sourced water masses. These findings demonstrate how the Southern Ocean's 'organic carbon pump' has exerted a tight control on atmospheric CO_2, and thus global climate, specifically via a synergy of both physical and biological processes.

[1] Godwin Laboratory for Palaeoclimate Research, Earth Sciences Department, University of Cambridge, Downing Street, Cambridge CB2 3EQ, UK. [2] Institute of Geological Sciences and Oeschger Center for Climate Change Research, University of Bern, Baltzerstr. 1-3, Bern 3012, Switzerland. [3] Institute of Environmental Physics, University of Heidelberg, Im Neuenheimer Feld 229, Heidelberg 69120, Germany. [4] Laboratoire des Sciences du Climat et de l'Environnement, LSCE/IPSL, CNRS-CEA-UVSQ, Université de Paris-Saclay, Domaine du CNRS, bât. 12, Gif-sur-Yvette 91198, France. Correspondence and requests for materials should be addressed to J.G. (email: jg619@cam.ac.uk).

The Southern Ocean is believed to play a key role in the global carbon cycle and millennial-scale variations in atmospheric CO_2 ($CO_{2,atm}$), which in turn may amplify the impacts of longer-term external climate forcing on global climate[1]. This role stems from the unique control the Southern Ocean is thought to exert on ocean–atmosphere CO_2 exchange[1-3] by both facilitating the upward transport of nutrient- and CO_2-rich water masses along outcropping density surfaces and their exposure to the atmosphere, and modulating the export of biologically fixed carbon into the ocean interior, where it is remineralized and may be effectively isolated from the atmosphere. It has been proposed that these two key aspects of the Southern Ocean's role in the marine carbon cycle may have exerted a dominant control on past $CO_{2,atm}$ change, for instance via variations of dust-driven biological carbon fixation in the sub-Antarctic[4], the extent of circum-Antarctic sea ice[5] impeding effective air–sea gas equilibration[6], and/or changes in the strength or position of southern hemisphere westerlies driving the residual overturning circulation in the Southern Ocean[7,8].

While all of these mechanisms for past $CO_{2,atm}$ change are compelling, observational evidence that might constrain the extent to which they have operated, in particular the balance of biological versus physical (that is, air–sea gas exchange or ocean dynamical) impacts, remains ambiguous. In the sub-Antarctic Atlantic north of the Polar Front (PF), decreased biological export production, along with a diminished aeolian supply of dust (and by inference iron) to the surface ocean, has been found to parallel millennial-scale increases in $CO_{2,atm}$. These observations suggest a significant impact of dust-driven variations of the strength of the 'organic carbon pump' on $CO_{2,atm}$ (refs 9–12). However, marked increases in $CO_{2,atm}$ are also accompanied by enhanced export productivity south of the PF (ref. 7). Polar- and sub-polar Southern Ocean export productivity changes thus appear to have opposed each other, raising questions concerning the overall magnitude and sign of the impact of Southern Ocean 'organic carbon pump' on $CO_{2,atm}$, when integrated across both regions[11,13]. On the other hand, while ^{14}C evidence has provided direct support for a link between Southern Ocean carbon sequestration (and millennial-scale $CO_{2,atm}$ variability) and physical/dynamical controls on air–sea CO_2 exchange[14], these data remain sparse and only extend across the last deglaciation.

Here we present sub-millennially resolved qualitative and quantitative proxy reconstructions of bottom-water [O_2] from sub-Antarctic Atlantic sediment core MD07-3076Q (14°13.7′W, 44°9.2′S, 3,770 m water depth; Fig. 1) to estimate the apparent oxygen utilization (AOU) in deep waters, which is closely (stoichiometrically) related to the amount of remineralized dissolved inorganic carbon (DIC) because of the consumption of oxygen during the degradation of organic carbon. We use two independent proxy approaches: first, we determined the redox-sensitive enrichment of uranium and manganese in authigenic foraminifer coatings[15], and second, we measured the difference in carbon isotopic composition between pore waters at the zero-oxygen boundary and overlying bottom waters, which is assumed to be reflected in $\delta^{13}C$ of the benthic foraminifer *Globobulimina affinis* and *Cibicides kullenbergi*, respectively ($\Delta\delta^{13}C_{C.\ kullenbergi-G.\ affinis}$; refs 16,17). Our deep sub-Antarctic Atlantic [O_2] reconstructions show a close correlation to $CO_{2,atm}$ variations during the last deglacial- and glacial periods. The combination of our [O_2] reconstructions with analyses of [230]Th-normalized opal fluxes, an indicator of biological export production[7,18], and deep water ^{14}C ventilation ages, along with a robust age model for our study core[14,19,20] (Methods), highlights that carbon sequestration changes in the southern high latitudes cannot be attributed solely to changes in local biological export production. Instead, they involve significant changes in Southern

Ocean vertical mixing and air–sea gas exchange, having direct implications for millennial-scale $CO_{2,atm}$ variations, since 65,000 years before present (BP).

Results

Redox-sensitive U and Mn enrichment in foraminifer coatings.

The uranium to calcium ratio in authigenic (that is, *in situ* generated) coatings (c), proposed to vary with changes in sedimentary redox-conditions, and therefore with bottom-water [O_2] (ref. 15), has been measured on weakly chemically cleaned ('host') calcium carbonate (cc) shells (hereafter referred to as U/Ca_{cc+c}) of the planktonic foraminifer *G. bulloides* and the benthic foraminifer *Uvigerina* spp. (Methods). The uranium concentration in the authigenic coatings of foraminifera strongly exceeds the concentration in the foraminiferal shell matrix[21,22]. Thus, the overall U/Ca_{cc+c} variability is marginally influenced by the uranium concentration in foraminifer shells, and has been proposed to primarily reflect coating-bound uranium variations instead that is inversely correlated with bottom-water oxygenation[15]. The co-variation of shell weights and U/Ca_{cc+c} levels of *G. bulloides*, however, indicates that shell size and/or wall thickness variations may bias U/Ca_{cc+c} ratios, via changes in the shell mass to surface-area ratio for example (Supplementary Fig. 1). The normalization of coating-bound uranium levels to manganese concentrations circumvents this bias for two reasons: manganese has generally an opposing redox-behaviour to that of uranium[23-25], and in particular manganese in weakly chemically cleaned foraminiferal tests mainly occurs in Fe-Mn-rich oxyhydroxides and/or Mn-rich carbonate overgrowths attached to the foraminiferal shell[22,26], which may be supported by the observed co-variation of Fe/Ca_{cc+c} and Mn/Ca_{cc+c} levels of *G. bulloides* (Supplementary Fig. 1). We propose that the U/Mn ratio of authigenic coatings in planktonic and benthic foraminifera, U/Mn_c, serves as reliable indicator of redox-conditions in marine sediments independent of shell matrix variations. The close agreement of planktonic and benthic foraminifer U/Mn_c suggests that it sensibly tracks early diagenetic redox-processes within the sediment consistent with previous findings[15] (Supplementary Fig. 2).

During the last glacial period, *G. bulloides* and *Uvigerina* spp. U/Mn_c are both found to vary with changes in $CO_{2,atm}$ (Fig. 2). During the last deglaciation, the large early deglacial decrease in U/Mn_c is clearly synchronous with the initial increase in $CO_{2,atm}$ before 15 kyr BP, while the second pulse in U/Mn_c in time with the $CO_{2,atm}$ increase during the following Antarctic warming period (that is, the northern-hemisphere Younger Dryas period) is more equivocal (Fig. 2). Our data are also in good agreement with changes in the authigenic enrichment of uranium in bulk sediments of Cape Basin core TN057-21 (ref. 27; location in Fig. 1), applying the most recent chronology of ref. 28 (Fig. 2). This suggests a basin-wide relevance of observed changes in sedimentary redox-conditions in the central sub-Antarctic Atlantic for variations in $CO_{2,atm}$.

Benthic foraminifer $\delta^{13}C$ gradients and bottom-water [O_2].

Redox-conditions in marine sediments generally reflect changes in organic carbon respiration within the sediment modulated by the downward diffusion of oxygen from bottom waters and/or the organic carbon supply to the sea floor[29]. Aerobic degradation of organic matter is the most efficient pathway of the respiration of organic carbon. Most of organic matter respiration therefore occurs above the sedimentary anoxic boundary. At the anoxic boundary, the diffusion of oxygen from the bottom water into the sediment is balanced by the rate of oxygen consumption during aerobic sedimentary organic carbon respiration in the sub-surface

a

b

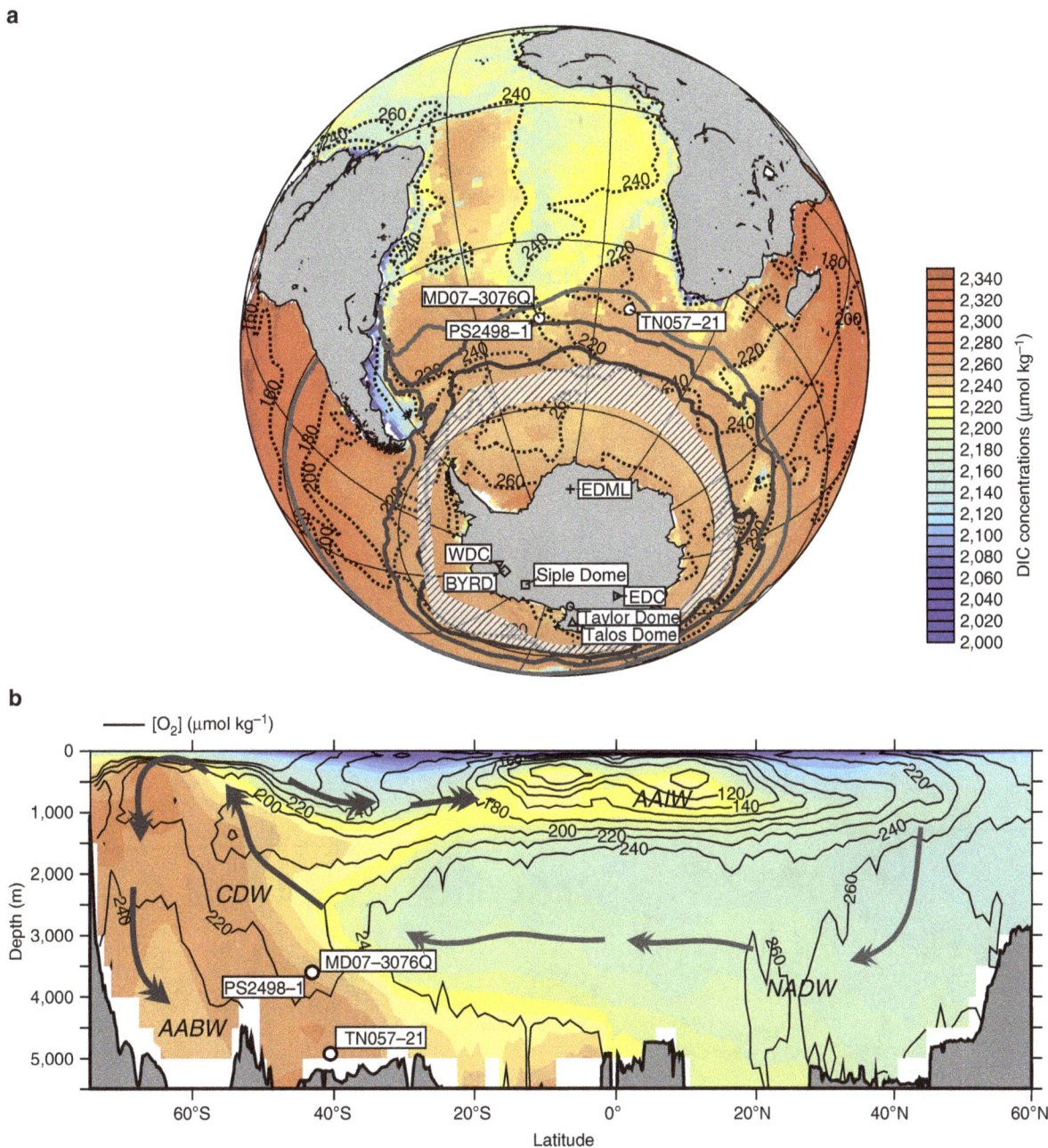

Figure 1 | Modern ocean DIC and oxygen concentrations. DIC levels (shaded) and $[O_2]$ (contours, in $\mu mol\,kg^{-1}$)[33,63] in (**a**) Southern Ocean- and Atlantic Ocean bottom waters and (**b**) in a meridional transect across the Atlantic (averaged between 70°W and 20°E). Hatched area broadly represents the region, where the deep DIC reservoir directly 'communicates' with the surface ocean and the atmosphere along steep density surfaces (equivalent to the area of strong positive CO_2 fluxes across the air–sea interface in austral winter in the Southern Ocean[64]), which is unique in the global ocean today. White circles show study cores and open symbols mark the location of ice cores that document past changes in atmospheric CO_2 ($CO_{2,atm}$; as in Figs 2 and 3). Thick lines show the modern positions of the PF, the sub-Antarctic Front (SAF) and the sub-Tropical Front (STF) (south to north)[65]. Arrows show general pathways of North Atlantic Deep Water (NADW), AABW (Antarctic Bottom Water), CDW (Circumpolar Deep Water) and Antarctic Intermediate Water (AAIW).

sediment column, such that $[O_2]$ becomes zero. As organic carbon has typical $\delta^{13}C$ values of about -22 ‰, the release of ^{13}C-depleted carbon during the degradation of organic matter substantially drives the $\delta^{13}C$ gradient in marine sub-surface pore waters[30]. The total amount of aerobic sedimentary organic carbon respiration is thus a function of bottom-water $[O_2]$ and is reflected in the $\delta^{13}C$ difference between bottom waters and pore waters at the zero-oxygen boundary[16,30].

The deep infaunal foraminifer *G. affinis* actively chooses the low-oxygen microhabitat near or at the anoxic boundary within

marine sub-surface sediments (in contrast to other benthic species)[31]. Assuming that *C. kullenbergi* $\delta^{13}C$ reflects bottom-water $\delta^{13}C$ (ref. 32), the offset of *G. affinis* $\delta^{13}C$ from bottom water (that is, *C. kullenbergi*) $\delta^{13}C$ thus sensitively records the relative depletion of pore-water $\delta^{13}C$ due to organic carbon respiraton[16,17,30] driven by the availability of oxygen in bottom waters. The occurrence of *G. affinis* in marine sediments may be in itself an indicator of an oxygen-limited sediment regime, where organic carbon is generally abundant and where the availability of oxygen is the main driver of organic matter respiration within the

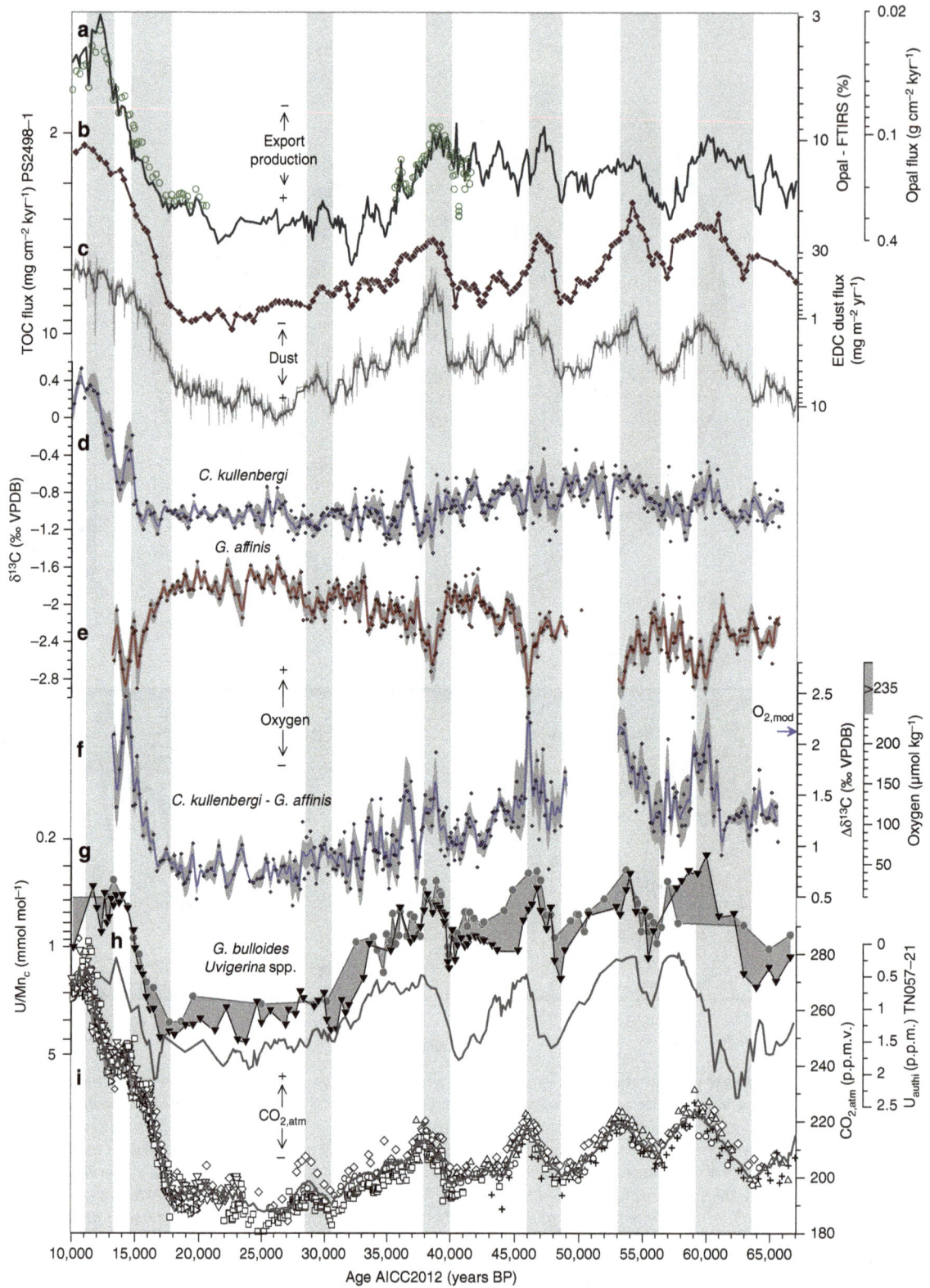

Figure 2 | Sub-Antarctic Atlantic bottom-water [O$_2$] and productivity changes during the last deglacial and glacial periods. (**a**) Sedimentary opal content (line) and ^{230}Thorium-normalized opal fluxes (circles), (**b**) flux of TOC in PS2498-1 (ref. 9; age scale adjusted as outlined in Methods), (**c**) Antarctic (EDC ice core) dust fluxes[66], (**d**) *C. kullenbergi* δ^{13}C (versus Vienna Pee Dee Belemnite (VPDB) standard), (**e**) *G. affinis* δ^{13}C (versus VPDB), (**f**) Δδ^{13}C$_{C.\ kullenbergi-G.\ affinis}$ and corresponding bottom-water [O$_2$] (ref. 16), arrow shows modern [O$_2$] at the core site[33], (**g**) *G. bulloides* (circles) and *Uvigerina* spp. (triangles) U/Mn$_c$, (**h**) authigenic uranium concentrations in TN057-21 (ref. 27), (**i**) CO$_{2,atm}$ variations recorded in the Antarctic ice cores BYRD (diamonds)[67,68], EDML (crosses)[47,69], EDC (right-pointed triangles)[70], Siple Dome (squares)[71], Talos Dome (triangles)[47], Taylor Dome (circles)[72] and WDC (inverted triangles)[73]. All data refer to the AICC2012 age scale[19,62]. Lines in **d-f** show 500 year-running averages with envelopes indicating the 500 year-window one-sigma standard deviation. Grey bars indicate periods of rising CO$_{2,atm}$.

sediment[29], because the characteristic zero-oxygen boundary in the shallow sub-surface of these sediments is the preferred habitat of *G. affinis*. The amount of pore water (that is, *G. affinis*) $\delta^{13}C$ depletion relative to bottom water (that is, *C. kullenbergi*) $\delta^{13}C$ is thus mostly insensitive to variations in organic carbon fluxes and scales instead with the amount of oxygen diffusing from the bottom water, allowing a quantification of bottom-water [O_2] (refs 16,17).

In sediment core MD07-3076Q, *G. affinis* $\delta^{13}C$ becomes markedly depleted by up to 1‰ relative to bottom-water (*C. kullenbergi*) $\delta^{13}C$ during decreases in U/Mn$_c$ (Fig. 2). The distinct negative offsets of *G. affinis* $\delta^{13}C$ from *C. kullenbergi* $\delta^{13}C$ mark millennial-scale increases in deep-water [O_2] (ref. 16) in the deep sub-Antarctic Atlantic that closely track rises in $CO_{2,atm}$ during the last deglacial and glacial periods (Fig. 2).

According to the modern $\Delta\delta^{13}C$-[O_2] calibration of ref. 16, bottom-water [O_2] in the deep sub-Antarctic Atlantic would have reached a minimum of about $40 \pm 20\,\mu mol\,kg^{-1}$ during the peak glacial, which translates into a bottom-water [O_2] reduction of $175 \pm 20\,\mu mol\,kg^{-1}$ from present-day levels of $\sim215\,\mu mol\,kg^{-1}$ at the core site[33] (Fig. 2). During the last glacial period, that is, Marine Isotope Stage (MIS) 3, deep sub-Antarctic Atlantic [O_2] would have varied between 90 ± 25 and $200 \pm 40\,\mu mol\,kg^{-1}$, in time with millennial-scale changes in $CO_{2,atm}$ (Fig. 2).

Our quantification of deep sub-Antarctic Atlantic [O_2] relies on the assumption that bottom-water $\delta^{13}C$ is reliably reflected in *C. kullenbergi* $\delta^{13}C$. This species has mostly been employed to reconstruct bottom-water $\delta^{13}C$ in the southern high latitudes (because of the low abundance of other benthic epifaunal species); yet a difference of up to ~0.6‰ has been observed between sparse glacial *C. kullenbergi* $\delta^{13}C$- and glacial *C. wuellerstorfi* $\delta^{13}C$ measurements at ODP site 1090 in the Cape Basin[34]. This may imply that *C. kullenbergi* $\delta^{13}C$ is anomalously depleted, for example, due to a slight infaunal habitat during glacial times[34], and/or that $\delta^{13}C$ measured on episodically occurring *C. wuellerstorfi* is anomalously enriched, for example, due to an affinity to anomalously well-ventilated water masses[35] and/or low carbon fluxes. If *C. kullenbergi* $\delta^{13}C$ in MD07-3076Q does not adequately represent bottom-water $\delta^{13}C$ at our core site, then absolute bottom-water [O_2] in the deep central sub-Antarctic Atlantic would be higher by up to $\sim40\,\mu mol\,kg^{-1}$ per 0.3‰-deviation of glacial bottom-water $\delta^{13}C$ from glacial *C. kullenbergi* $\delta^{13}C$ observed in MD07-3076Q (Supplementary Fig. 3). However, our *C. kullenbergi* $\delta^{13}C$ data are consistent with glacial benthic foraminifer (*C. kullenbergi* and *Cibicidoides* spp.) $\delta^{13}C$ measurements from different locations throughout the South Atlantic[34,36,37], suggesting that they are representative of deep-water $\delta^{13}C$. Regardless of these quantitative uncertainties, the co-variation of the U/Mn$_c$- and $\Delta\delta^{13}C$-based [O_2] reconstructions provides strong evidence for recurrent changes in deep sub-Antarctic oxygenation in parallel with $CO_{2,atm}$ over the last glacial and deglacial periods.

Changes in opal- and organic carbon fluxes. The flux of biogenic silica (opal) to marine sediments in the southern high latitudes is assumed to reflect changes in organic carbon flux to the sea floor and in the export of organic carbon from the euphotic zone (that is, export production)[9,38]. Variations in the weight percentages of opal observed in MD07-3076Q are tightly correlated with [230]Th-normalized opal fluxes ($R^2 = 0.94$, $P < 0.05$; Fig. 2; Supplementary Fig. 4), suggesting their accurate representation of past opal- (and therefore total organic carbon[9,38]; TOC) fluxes in the sub-Antarctic Atlantic. This is supported by synchronous variations in the TOC flux observed in the neighbouring core PS2498-1 (Fig. 2, location in Fig. 1), which

has been chronostratigraphically aligned to MD07-3076Q (Methods). As shown in Fig. 2, opal- and TOC fluxes in the sub-Antarctic Atlantic show a close link to dust flux variations in Antarctic ice cores and changes in dust supply to the sub-Antarctic region[9], which is consistent with earlier findings[9,10].

Estimates of radiocarbon ventilation ages. Two metrics for deep-water 'ventilation' (that is, deep ocean versus atmosphere gas/isotope equilibration) that provide a measure of the average time since carbon in the ocean interior last equilibrated with the atmosphere are considered here: [14]C age offsets between coexisting benthic (B) and planktonic (Pl) foraminifera (B-Pl ^{14}C ventilation ages), and benthic ^{14}C age offsets from contemporary atmospheric ^{14}C ages (B-Atm ^{14}C ventilation ages). While the first provide an estimate of deep-ocean ventilation relative to the local mixed layer, the latter provide a direct estimate of deep-ocean ventilation relative to the contemporary atmosphere. As shown in Fig. 3, B-Pl ^{14}C ventilation ages from sediment core MD07-3076Q broadly co-vary with changes in deep-ocean oxygenation (for example, with U/Mn$_c$: $R^2 = 0.31$, $P < 0.05$) and $CO_{2,atm}$ ($R^2 = 0.43$, $P < 0.05$), both statistically significant within the 95% significance interval (Supplementary Fig. 5). Parallel B-Atm ^{14}C ventilation age estimates agree with these observations, and confirm that B-Pl ^{14}C ventilation age fluctuations have not been significantly biased or masked by local surface-ocean radiocarbon disequilibrium effects (reservoir ages) (Fig. 3).

These findings are consistent with similar analyses in the central deep sub-Antarctic Atlantic for the last deglaciation[14]. Although B-Pl ^{14}C ventilation age variations are more strongly influenced by surface-ocean reservoir age variations during the last deglaciation, decreasing B-Atm ^{14}C age offsets are linked to deglacial increases in $CO_{2,atm}$, in particular during the early deglacial period[14].

Notably, absolute foraminifer ^{14}C ages appear to be slightly too young during the mid-glacial period, perhaps due to uncertainties associated with background corrections, which are especially important for old (>30 kyr BP) sample material. In practice, these background corrections are based on one radiocarbon-dead spar calcite sample measured in each sample batch (that is, an accelerator mass spectrometry (AMS) sample carousel), whose apparent radiocarbon content is subtracted from the measured radiocarbon content of all the foraminifer samples measured in that sample carousel. If the true background deviates from the measured background in this single sample, then B-Atm ^{14}C and Pl-Atm ^{14}C age offsets may deviate significantly from their true absolute values. Godwin Radiocarbon Laboratory-internal backgrounds compiled for the 4 years from April 2011 to January 2015 amount to $^{14}C/^{12}C_0 = 5.3 \pm 1.5 \times 10^{-15}$ (Supplementary Fig. 6). Considering a one-off estimate of the background that is slightly smaller (that is, $^{14}C/^{12}C_0 = 4 \times 10^{-15}$; within 1 s.d. of the mean) or larger (that is, $^{14}C/^{12}C_0 = 6 \times 10^{-15}$; within 1 s.d. of the mean), this would result in B-Atm ^{14}C and Pl-Atm ^{14}C age offsets that are shifted towards slightly lower and higher absolute values respectively, without affecting the overall variability in each time-series (Supplementary Fig. 7). As benthic and planktonic ^{14}C ages have been obtained from the same AMS sample carousels in this study, B-Pl ^{14}C ventilation ages are not affected by these uncertainties and are essentially the same irrespective of the applied background correction (Supplementary Fig. 7). Therefore, while our absolute B-Atm ^{14}C and Pl-Atm ^{14}C age offsets are dependent on the accuracy of our background corrections (which are arguably difficult to assess), relative changes in B-Atm ^{14}C ventilation ages and absolute variations in B-Pl ^{14}C ventilation ages remain robust. As shown in Fig. 3, these clearly co-vary with our estimates of bottom-water

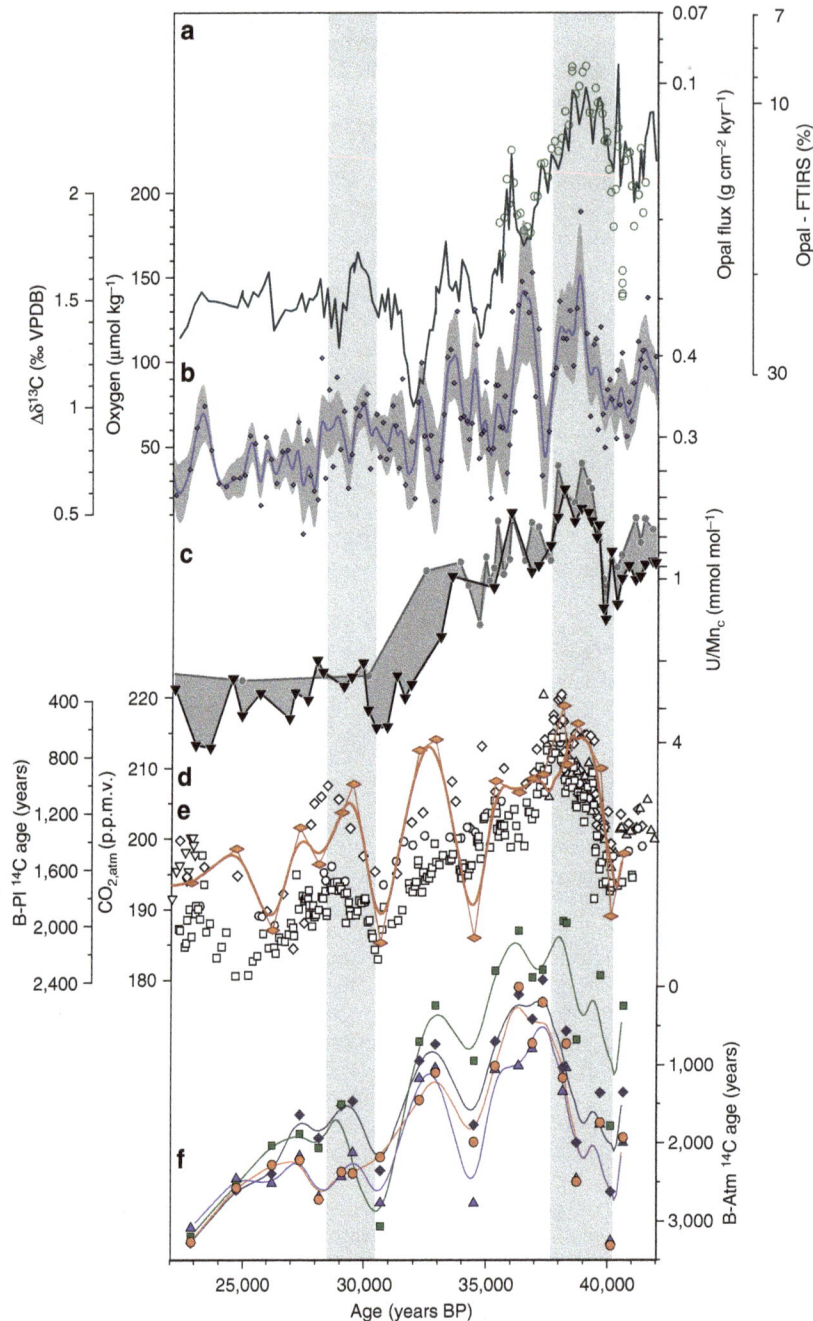

Figure 3 | Mid-glacial ventilation and carbon sequestration changes in the deep sub-Antarctic Atlantic. (a) Sedimentary opal content (line) and
^{230}Thorium-normalized opal fluxes (circles), (b) $\Delta\delta^{13}C_{C.\ kullenbergi-G.\ affinis}$ and corresponding bottom-water [O$_2$] (ref. 16), (c) *G. bulloides* (circles) and
Uvigerina spp. (triangles) U/Mn$_c$, (d) Benthic-Planktonic (B-Pl) ^{14}C ventilation ages and the corresponding 1,000 years-running mean (thick line) plotted on
top of (e) variations in CO$_{2,atm}$ recorded in the Antarctic ice cores (open symbols, refs as in Fig. 2), (f) benthic foraminifer ^{14}C age offset from atmospheric
^{14}C (Lake Suigetsu (green)[74], Cariaco Basin (orange)[75], Intcal09 (blue)[76] and Intcal13 (dark blue)[77]) shown as 1,000 years-running means (lines). Line
and grey envelope in **b** show a 500 year-running average and the 500 year-window one-sigma standard deviation, respectively. Grey bars indicate periods
of rising CO$_{2,atm}$.

oxygenation in the deep sub-Antarctic Atlantic (see also Supplementary Fig. 5).

Discussion

Changes in the elemental composition of foraminifer coatings and bottom-water versus pore-water δ^{13}C gradients, as described above, demonstrate that the amount of remineralized carbon sequestered in the deep sub-Antarctic Atlantic has varied

substantially and inversely with respect to millennial-scale CO$_{2,atm}$ changes (Fig. 2). These observations confirm a role for the Southern Ocean 'organic carbon pump' in regulating CO$_{2,atm}$ (refs 3,10,11). Below, we assess the quantitative impact of the inferred 'biological carbon pump' changes on CO$_{2,atm}$, as well as their governing biological and/or physical/dynamical controls.

Bottom-water [O$_2$] reconstructions at our core site via $\Delta\delta^{13}C_{C.\ kullenbergi-G.\ affinis}$ provide the basis for a quantification of the amount of respired carbon in the deep sub-Antarctic

Atlantic[16,30], provided the modern $\Delta\delta^{13}$C-[O_2] relationship holds for the past. In principle, seawater [O_2] consists of a saturated [O_2] component ([O_2]$_{sat}$) arising from the solubility-controlled O_2 exchange between the atmosphere and the surface ocean, a biological [O_2] component associated with the release and drawdown of [O_2] during photosynthesis and respiration ([O_2]$_{bio}$), and a preformed disequilibrium [O_2] component ([O_2]$_{diseq}$) due to inefficiencies in air–sea gas exchange ([O_2]$_{in\ situ}$ = [O_2]$_{sat}$ + [O_2]$_{bio}$ + [O_2]$_{diseq}$)[39]. Assuming that ocean [O_2] is in equilibrium with the atmosphere ([O_2]$_{diseq} \sim 0$) and that last glacial ocean [O_2]$_{sat}$ was slightly higher than today ([O_2]$_{sat,modern}$ = 345 µmol kg^{-1}) mostly due to a decrease in ocean temperature ([O_2]$_{sat,glacial}$ = 360 µmol kg^{-1}; Methods), the amount of [O_2] depletion at ocean depth (AOU = $-$[O_2]$_{bio}$ = [O_2]$_{sat}$ $-$ [O_2]$_{in\ situ}$) should scale with the formation of respired carbon according to a constant stoichiometric Redfield ratio of C:[O_2] = 117: $-$170 (ref. 40). Changes in bottom-water [O_2] (and AOU) in the sub-Antarctic Atlantic would therefore provide a direct quantitative measure of the amount of carbon sequestered in the southern high-latitude ocean, and thus the efficiency of the biological 'organic carbon pump'[16].

Converting our AOU estimates into respired carbon concentrations (AOU$_{Holocene}$ = (345 $-$ 215) µmol kg^{-1}, (AOU$_{MIS2}$ = (360 $-$ 40) $\pm$ 20 µmol kg^{-1}, ΔAOU$_{Deglaciation}$ = 190 $\pm$ 20 µmol kg^{-1}) based on the Redfield ratio of C:[O_2] = 117: $-$170 (ref. 40) gives a respired DIC contribution of 220 $\pm$ 14 µmol kg^{-1} to the total DIC pool at the core site during the last glacial maximum (LGM). This is higher by 130 $\pm$ 14 µmol kg^{-1} compared with the Holocene[33], indicating greater respired carbon accumulation during the LGM. During millennial-scale variations in CO$_{2,atm}$ during the last glacial, respired carbon levels varied by 75 $\pm$ 28 µmol kg^{-1} between 110 $\pm$ 28 µmol kg^{-1} (during peak CO$_{2,atm}$ levels) and 185 $\pm$ 14 µmol kg^{-1} (during minimum CO$_{2,atm}$ levels) assuming that [O_2]$_{sat}$ was not significantly different from LGM levels (that is, AOU$_{MIS3\ 'CO_2\ max'}$ = (360 $-$ 200) $\pm$ 40 µmol kg^{-1}, AOU$_{MIS3\ 'CO_2\ min'}$ = (360 $-$ 90) $\pm$ 25 µmol kg^{-1}, ΔAOU$_{MIS3}$ = 110 $\pm$ 40 µmol kg^{-1}).

If we assume that the respired carbon lost from the deep sub-Antarctic Atlantic, where it was sequestered away from the atmosphere was transferred to a non-respired marine carbon pool that in turn equilibrated with the atmosphere via a surface-ocean DIC 'buffer factor' (that is, Revelle factor) of $\sim$10 (ref. 41; Methods), our AOU and respired carbon estimates from the deep sub-Antarctic may only explain the full amplitude of observed CO$_{2,atm}$, if they are representative of a significant fraction of the global deep ocean, that is, at least $\sim$33% during the mid-glacial period and $\sim$45% during the early deglaciation (Methods). This would roughly correspond to the deep ocean below 2.9 and 2.3 km, respectively (Methods). These depths broadly agree with the depth of the putative glacial 'chemical divide' ($\sim$3 km water depth in the Atlantic)[42], and are supported by qualitative proxy data showing a decrease in oxygenation and radiocarbon ventilation in the global ocean below 2 km during the last peak glacial period[43,44]. The smaller the volume of the global deep ocean that experienced similar changes in AOU and respired carbon to our sub-Antarctic Atlantic site, the smaller the likely oceanic impact on CO$_{2,atm}$ concentrations.

Our calculations have two major caveats. First, we ignore possible open-system effects due to the interaction of deep waters with sediments, and second, we may have underestimated glacial deep sub-Antarctic Atlantic [O_2], in the case that deep sub-Antarctic Atlantic C. kullenbergi δ^{13}C values are strongly negatively biased versus bottom-water δ^{13}C. Any open-system effects involving a degree of 'carbonate compensation' (on multi-millennial timescales) would tend to enhance the impact of

marine respired carbon inventory changes on CO$_{2,atm}$ (Methods). If true glacial bottom-water [O_2] were higher than estimated in MD07-3076Q for instance via an anomalous depletion of C. kullenbergi δ^{13}C from bottom-water δ^{13}C by 0.3‰, as mentioned above, deglacial changes in respired carbon and AOU (and thus the oceanic impact on CO$_{2,atm}$) would be reduced by $\sim$20%, as LGM AOU values would be lower (Supplementary Fig. 3). In contrast, our estimates of respired carbon changes during the last mid-glacial period remain to a large extent similar as they are based on relative [O_2] changes (Supplementary Fig. 3). Our calculations are rough estimates that are intended only to provide a first indication of the potential impact of our observed marine carbon sequestration changes on CO$_{2,atm}$. To determine the full impact of changes in deep-ocean respired carbon levels on CO$_{2,atm}$ concentrations, our estimates would need to be corroborated by further reconstructions of past bottom-water oxygen- and DIC concentrations from throughout the global ocean, in particular in the volumetrically most significant Pacific Ocean.

The analysis above demonstrates the potential quantitative significance of the oxygenation changes that we observe, and more specifically of the role of the Southern Ocean 'organic carbon pump' in regulating CO$_{2,atm}$ (refs 3,10,11). However, it remains to be shown whether the observed decreases in 'organic carbon pump' efficiency resulted primarily from decreases in export productivity (allowing oxygen to increase due to reduced organic carbon remineralization in the ocean interior) or primarily from increases in ocean 'ventilation' (causing carbon loss to the atmosphere with direct oxygen gain of the ocean interior). Below, we address this question by reference to our export productivity- and ^{14}C ventilation age estimates.

The observed correlation between changes in the dust supply to the southern high-latitude regions and in export production in the central sub-Antarctic Atlantic (as recorded by variations in opal- and TOC fluxes[9]; Fig. 2) supports earlier findings of a dust-driven biological organic carbon pump in the sub-Antarctic Atlantic[9,10]. The close relationship between variations in sub-Antarctic export production and CO$_{2,atm}$ changes (Fig. 2) would be consistent with a significant impact of the efficiency of the sub-Antarctic biological organic carbon pump on surface-ocean DIC levels, and thus on CO$_{2,atm}$ (refs 9–11).

The correlation between opal- and TOC fluxes and bottom-water [O_2] in the sub-Antarctic Atlantic (Fig. 2) may point to a role of organic carbon respiration at depth driving deep sub-Antarctic Atlantic bottom-water oxygenation. To test whether export production was the major driver of our observed deep-ocean [O_2] changes (and therefore of the associated changes in deep-ocean respired carbon sequestration), we make use of the unique microhabitat of G. affinis near the anoxic boundary in marine sediments and the associated mechanisms that drive its δ^{13}C signature. Notably, negative excursions of G. affinis δ^{13}C are observed during each of the marked CO$_{2,atm}$ rises during MIS 3 and the last deglaciation. These excursions indicate that total organic carbon respiration within deep sub-Antarctic Atlantic sediments increased at times of reduced opal- and TOC fluxes, that is, reduced export production (Fig. 2). An increase in sedimentary organic carbon respiration (that is, pore water/ G. affinis δ^{13}C depletions) would be driven by an increase in organic carbon flux, an increase in bottom-water [O_2], or both of these together. As the first is evidently not the case (Fig. 2), we conclude that sedimentary carbon respiration must have instead been driven by enhanced deep-ocean 'ventilation' (that is, circulation/ convection rates and/or air–sea gas exchange) supplying oxygen to the deep sub-Antarctic Atlantic.

Alternatively, a decreased oxygen demand in bottom waters due to diminished organic carbon fluxes and less respiration of

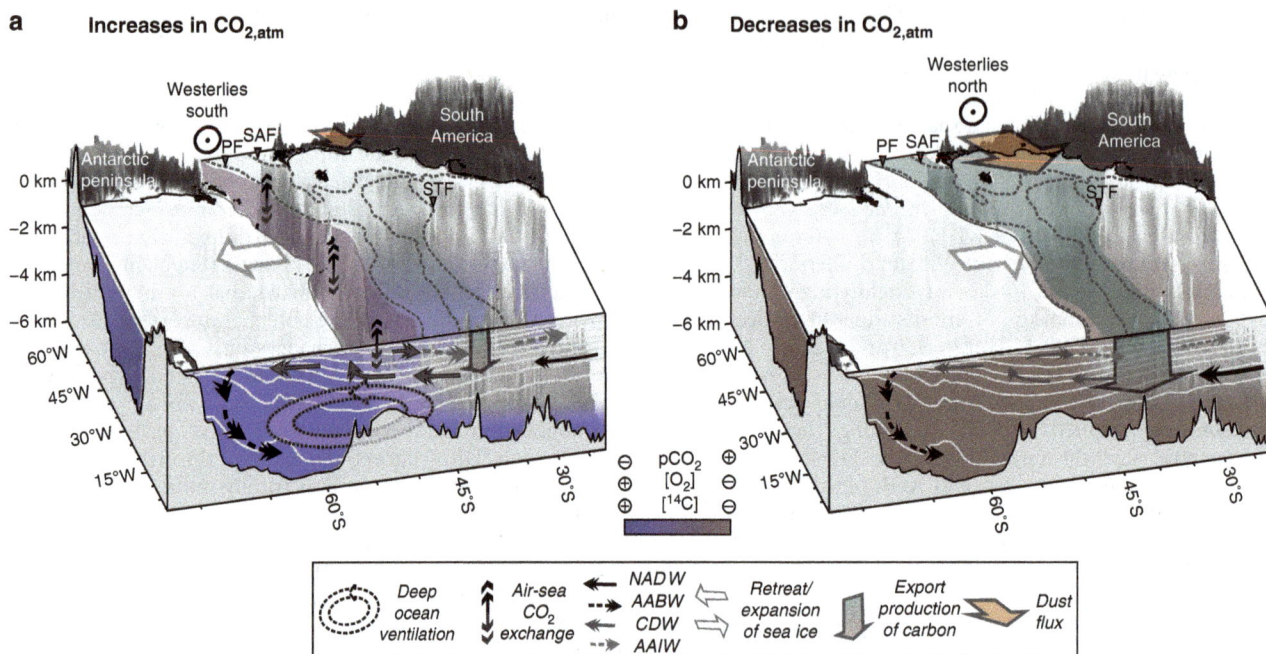

a   **Increases in CO$_{2,atm}$**

b   **Decreases in CO$_{2,atm}$**

Figure 4 | Schematic view on the southern high-latitude Atlantic during millennial-scale CO$_{2,atm}$ variations based on new and existing proxy evidence.
(**a**) Dust-driven decreases of export production in the sub-Antarctic Atlantic[9,10] during the last glacial and deglacial periods were accompanied by decreases in deep carbon storage in the Southern Ocean (this study and ref. 48). The latter was further promoted by increases in the air-sea CO$_2$ exchange south of the PF and in the ventilation of the deep carbon pool (this study and ref. 48), causing millennial-scale increases in CO$_{2,atm}$, as postulated earlier[3,7]. (**b**) Enhanced dust-driven, biological export of carbon to the deep sub-Antarctic Atlantic[9,10] paralleled increases in deep Southern Ocean respired carbon levels during the last glacial period and the last deglaciation (this study and ref. 48). The enhanced Southern Ocean carbon pool was effectively isolated from the atmosphere by decreases in air-sea CO$_2$ equilibration in the Antarctic region and a poor 'ventilation' of the deep-ocean during these times (this study and ref. 48), leading to decreases in CO$_{2,atm}$ during the last 70,000 years, as proposed previously[3,7]. Accompanying changes in sea ice[5,6] and the westerly position/strength[7,8] are debated and remain speculative. The modern positions of ocean fronts (as in Fig. 1) and ocean density surfaces (white lines) are shown as reference.

organic matter in a benthic 'fluff' layer could facilitate the diffusion of oxygen into the sediment, and drive the *G. affinis* δ^{13}C signal more negative. However, a poor inverse correlation between epibenthic and deep infaunal benthic foraminifer δ^{13}C over past millennial timescales (Fig. 2; $R^2 = 0.0001$, $N = 258$, Supplementary Fig. 8) would appear to rule out this scenario. We therefore conclude that the observed changes in 'organic carbon pump' efficiency and deep sub-Antarctic carbon storage were not only controlled by changes in export productivity but must also have involved biology-independent processes that contributed to past CO$_{2,atm}$ changes specifically by enhancing ocean–atmosphere CO$_2$ exchange in the Antarctic region (Fig. 2).

Our interpretation is confirmed by parallel estimates of deep-water ^{14}C 'ventilation ages' (Fig. 3). We observe that the marked CO$_{2,atm}$ rise around 38 kyr BP is paralleled by a decrease in B-Atm ^{14}C ventilation ages of ~2,000 ^{14}C years. A consistent link between deep-ocean (B-Atm and B-Pl) ^{14}C ventilation and CO$_{2,atm}$ variability is further supported by a high and statistically significant correlation coefficient between them (up to $R^2 = 0.6$, $P < 0.05$; Supplementary Fig. 5). The good correlation between (B-Atm and B-Pl) ^{14}C ventilation ages, deep-water [O$_2$] and CO$_{2,atm}$ provides strong independent support for changes in the air-sea equilibration of deep waters in the Southern Ocean and their link to changes in respired carbon storage.

It has previously been shown that the incursion of well-ventilated northern-sourced waters into the sub-Antarctic Atlantic was reduced during intervals of rising CO$_{2,atm}$ (refs 20,28). On this basis, the periods of increased ^{14}C ventilation that we observe would therefore specifically reflect periods of increased local dominance of southern-sourced deep

waters and an 'improvement' of their ventilation state. Numerous processes have been suggested to have caused changes in vertical mixing in the southern high latitudes, including for instance the intensity and/or the position of the southern hemisphere westerlies[7,8], a retreat of circum-Antarctic sea ice[6], a decline in the formation and advection of northern component waters[45] and/or changes in surface buoyancy fluxes[46]. It remains currently impossible to evaluate the relative importance of these specific processes and their controls on CO$_{2,atm}$. Nevertheless, the strong co-variations of our abyssal oxygenation and ventilation proxies with CO$_{2,atm}$ confirm that some combination of dynamical (that is, residual circulation and shallow mixing) and/or physical (gas exchange efficiency) processes in the southern high-latitude region indeed had a significant impact on deep-ocean carbon sequestration[3,7,19,45,47] (Fig. 4).

Furthermore, our findings are entirely consistent with recently published sedimentary redox-sensitive trace element data from the Antarctic Zone of the Atlantic Ocean[48]. These data show that the accumulation of authigenic uranium (and therefore oxygenation) in the Antarctic Atlantic is generally inversely correlated with opal fluxes (that is, organic carbon fluxes) over the past 80,000 years, ruling out a dominant control of local surface-ocean productivity on deep Antarctic Atlantic [O$_2$] and deep-ocean respired carbon levels south of the PF (ref. 48). The combination of our sub-Antarctic study with the Antarctic study of ref. 48 provides strong evidence for millennial-scale changes in the respired carbon concentrations across the entire deep high-latitude South Atlantic, varying in parallel with CO$_{2,atm}$ during the last glacial period and deglaciation, and for a significant impact of physical 'ventilation' processes (that is, overturning circulation, mixing and/or air–sea gas exchange) on changes in deep-ocean respired carbon sequestration and millennial-scale $CO_{2,atm}$ in the past.

In conclusion, our results show that pulses of $CO_{2,atm}$ during the last glacial- and deglacial periods coincided with increases in the ventilation of the southern high-latitude deep ocean (specifically via regions of deep-water formation in the Southern Ocean[7,48]), in addition to reductions in sub-Antarctic export productivity. By ruling in a role for variations in both the strength and the efficiency of the biological carbon pump via changes in the biological carbon export as well as the air–sea CO_2 exchange and Southern Ocean vertical mixing, the findings reconcile two opposing theories for the Southern Ocean's role in past millennial-scale $CO_{2,atm}$ variability[3,7,10–12,47]. Further work, for example using numerical model simulations will be required to quantify more precisely the contributions of (sub-polar zone) biological export productivity changes and (polar zone) physical/dynamical changes to deep-ocean carbon sequestration, as well as their down-stream effects on low-latitude export production[49]. Nevertheless, our data emphasize that while biological carbon export to the deep ocean is ultimately what permits ocean dynamics and air–sea exchange to impact on $CO_{2,atm}$ by continually tending to 'recharge' the abyssal carbon pool, the rate of equilibration of the deep ocean with the atmosphere will ultimately determine whether or not the biological 'organic carbon pump' is efficient or not at sequestering CO_2 (Fig. 4). Thus, ocean physics and marine biology acted together, synergistically, to repeatedly nudge the Southern Ocean from carbon sink to carbon source, with a direct impact on global climate over the last ~65,000 years.

Methods

Regional setting and chronology. Sediment core MD07-3076Q (14°13.7'W, 44°9.2'S, 3,770 m water depth) is bathed in Lower Circumpolar Deep Water, which is formed by the entrainment of northward spreading DIC- and preformed nutrient-rich Circumpolar Deep Water into southward flowing DIC-low and regenerated nutrient-rich North Atlantic Deep Water[50]. Chronological control of sediment core MD07-3076Q is based on ^{14}C measurements of mono-specific planktonic foraminifer samples, which have been adjusted for variations in surface-ocean reservoir ages[14]. The ^{14}C-based age constraints are complemented by the stratigraphic alignment of abundance variations of the cold-water species *Neogloboquadrina pachyderma* (sinistral-coiling) with rate changes in Antarctic temperature over time[19]. Age model uncertainties, mainly a function of age marker density, amount to 1,600 ± 500 years during the last glacial period and to 1,200 ± 400 years after 27 kyr BP (ref. 19). Resulting sedimentation rates range between 5 cm kyr^{-1} during the last deglaciation and 15 cm kyr^{-1} during MIS 3.

Element composition of authigenic foraminifer coatings. Down-core measurements of U/Ca$_{cc+c}$ and U/Mn$_c$ have been made on 18–25 specimens of the planktonic foraminifer *G. bulloides* (250-300 μm size fraction) and the 5–13 specimens of the benthic infaunal foraminifer *Uvigerina* spp. (250-300 μm size fraction). Foraminifera have been weakly chemically cleaned (clay removal and silicate picking) to maintain foraminiferal coatings but to remove extraneous detritus[15]. Cleaned foraminifera have been dissolved in 0.1 M nitric acid for inductively coupled plasma (ICP)-atomic emission spectroscopy analyses. The samples were subsequently re-diluted to 10 p.p.m. Ca^{2+} concentration and elemental concentrations have been determined by ICP-mass spectrometry[15]. Mean s.d. of U/Mn$_c$ of six duplicate samples is 0.08 ± 0.06 mmol mol^{-1}. Given the high sedimentation rates of 15 cm kyr^{-1}, the impact of potential sedimentary re-oxidation processes ('burn-down' effects) of already precipitated uranium complexes is negligible for the interpretation of U/Ca$_{cc+c}$ and U/Mn$_c$ ratios.

Reconstruction of bottom- to pore-water δ^{13}C gradients. Stable isotopic analyses on *G. affinis* and *C. kullenbergi* have been performed on 1–4 specimens (>150 μm size fraction) on Finnigan Δ+ and Elementar Isoprime mass spectrometers. The results are reported with reference to the international Vienna Pee Dee Belemnite (VPDB) standard. VPDB is defined with respect to the NBS-19 calcite standard. The mean external reproducibility of carbonate standards is σ ± 0.03 ‰.

In MD07-3076Q, $\Delta\delta^{13}C_{C.\ kullenbergi-G.\ affinis}$ has been determined from δ^{13}C measurements of benthic foraminifera from the same sediment sample, and has been converted into bottom-water [O$_2$] after ref. 16. The calibration error

associated with bottom-water [O$_2$] reconstructions using this method is ± 17 μmol kg^{-1} (ref. 16). Analytical uncertainties of benthic δ^{13}C analyses (two-sigma) translate into a bottom-water [O$_2$] uncertainty of ± 8 μmol kg^{-1}. We have smoothed our high-resolution record by a running 500 year-window (solid line in Fig. 2) to reduce such biases and those from intra-specific δ^{13}C variations. Mean bottom-water [O$_2$] have been determined for the LGM (23-18 kyr BP) as well as CO$_2$ minima (40.2-39.9 kyr BP, 48.4-47.6 kyr BP, 56.7-55.7 kyr BP, 63.6-63.0 kyr BP) and -maxima (38.8-38.0 kyr BP, 46.3-45.8 kyr BP, 53.6-53.3 kyr BP and 59.3-58.8 kyr BP) during MIS 3. Errors reported in our study are one-sigma standard deviations of our bottom-water [O$_2$] estimates during these periods.

Calculation of deep-ocean and atmospheric carbon budgets. [O$_2$] saturation levels are calculated according to ref. 51 assuming a glacial increase in salinity from present-day (~35 p.s.u.) by ~2 p.s.u. and a decrease in deep-ocean temperatures from modern-day values (~1 °C) by 2 °C in the deep Southern Ocean[52]. [O$_2$] saturation in the glacial deep Southern Ocean increased by ~15 μmol kg^{-1} from modern-day levels (~345 μmol kg^{-1}) (ref. 33).

To estimate the amount of carbon that is transferred to the atmosphere from the ocean's remineralized carbon pool (sequestered in the deep ocean), via the ocean's non-remineralized carbon pool (in equilibrium with the atmosphere), we adopt the conceptual framework of ref. 41, whereby:

$$\frac{dpCO_2}{pCO_2} = -0.0053\,\Delta c_{soft} + 0.0034\,\Delta c_{carb} \tag{1}$$

Here, Δc_{soft} and Δc_{carb} are DIC changes for the ocean's total remineralized carbon pool (that is not in equilibrium with the atmosphere), due to changes in the soft-tissue pump and the carbonate pump (for instance via changes in the export of organic carbon and carbonate to the ocean interior), respectively. Our estimate of Δc_{DIC} during the last deglacial increase in $CO_{2,atm}$ ($\Delta c_{DIC} = 130 \pm 14$ μmol kg^{-1}) and during mid-glacial $CO_{2,atm}$ changes ($\Delta c_{DIC} = 75 \pm 28$ μmol kg^{-1}) determined above from oxygenation estimates provides an estimate of Δc_{soft}, during these time intervals, and we assume that the associated Δc_{soft} is approximately three times smaller (for example, as observed spatially in the modern ocean)[41], yielding:

$$\frac{dpCO_2}{pCO_2} = -0.004167\,\overline{\Delta c_{DIC}} \tag{2}$$

where $\overline{\Delta c_{DIC}}$ is the whole-ocean average change in remineralized carbon during the investigated time intervals. It is given by the product of the change observed at our core location and the fraction (f) of the total ocean volume that also experienced this magnitude of change:

$$\overline{\Delta c_{DIC}} = f\Delta c_{DIC} \tag{3}$$

Assuming that the rest of the ocean volume experienced no significant change in respired DIC, remaining well-equilibrated with the atmosphere, the fraction of the ocean f, and therefore the deep-ocean volume V_d and the upper 'boundary' of the deep-ocean z', may be calculated that would account for the last early deglacial and mid-glacial atmospheric pCO_2 changes of ~50 and ~20 p.p.m. (for glacial background pCO_2 levels of 190–200 p.p.m.), if affected by similar changes in AOU and respired DIC levels as our sub-Antarctic Atlantic core site.

We have calculated the deep-ocean volume V_d and z' based on the GEBCO bathymetric data set (excluding the Arctic Ocean) archived by the British Oceanographic Data Centre (http://www.gebco.net/), according to:

$$V_d = \sum \left(\left|\left(\frac{\pi\cos(\phi)r\Delta\phi}{180}\right)\right| * \left(\frac{\pi r\Delta\lambda}{180}\right) * (z - z')\right) \tag{4}$$

that is the sum of all volumes of grid boxes (distance in west-east direction (km) times distance north-south direction (km) times depth), where ϕ is latitude, λ is longitude, $\Delta\phi$ and $\Delta\lambda$ represent the grid spacing of the bathymetric data set, r is the Earth's radius, z the water depth and z' the upper limit of the deep ocean.

Opal measurements. Opal concentrations were measured on ~400 samples by means of Fourier transform infrared (FTIR) spectroscopy[53] using a Vertex 70 FTIR-spectrometer (Bruker Optics Inc.) at the Institute of Geological Sciences at the University of Bern (CH). The FTIR spectra have been independently calibrated based on FTIRS analyses of artificial sand/opal mixtures[54]. Opal concentrations determined by means of FTIR spectroscopy show excellent agreement with conventional photometric-based[55] opal concentration determinations ($R^2 = 0.91$; Supplementary Fig. 4) that have been performed on one quarter of the total number of samples ($N = 101$). However, an increasing offset between photometric and FTIR-based opal measurements towards increasing opal values (Supplementary Fig. 4) might point at incomplete alkaline opal dilution during photometric measurements[55], potentially caused by a significant fraction of radiolarian skeletons in MD07-3076Q sediments[56].

Opal fluxes have been determined by normalizing the opal data with measured [230]Th concentrations[57]. For these analyses, U- and Th- isotopes were analysed by means of ICP–quadrupole mass spectrometry (iCAP-Q ICP-MS, ThermoFisher) at the Institute for Environmental Physics in Heidelberg, Germany. The contribution of detrital [230]Th has been estimated by assuming a [238]U/[232]Th ratio of 0.6 and a correction[58] for the detrital [234]U/[238]U not in secular equilibrium of 0.96. The quality of the analyses and the sample digestion and purification process has been monitored by blanks, certified UREM-11 standard material and replicate measurements of samples. Full replicates ($N = 5$) yielded an average uncertainty of 2.8 % (two-sigma) of the excess ^{230}Th concentrations (Supplementary Table 1). The chosen parameter set for the measurements of marine sediments applied here for the first time using an iCAP-Q ICP-MS (Supplementary Tables 1 and 2) puts emphasis on time efficiency for high-matrix sample analyses.

Radiocarbon measurements. The previously published set of foraminiferal ^{14}C dates in sediment core MD07-3076Q (ref. 14) has been extended by additional paired ^{14}C measurements of mixed benthic and mono-specific planktonic foraminifera (*N. pachyderma* s.). The conventional ^{14}C ages are reported in Supplementary Tables 3 and 4. The mean ^{14}C age uncertainty of the new ^{14}C data set amounts to 650 ± 270 ^{14}C years (Supplementary Table 3).

Foraminifer samples had a mean weight of 5.1 ± 1.0 mg, and weighed always more than 3.4 mg. They have been gently cleaned in methanol, and were subsequently transferred to sealed septum vials after they were completely dry. After evacuation 0.5 ml dry phosphoric acid has been injected into the vials. The acid-carbonate reaction has been sustained for at least 0.5 h at 60 °C. The CO_2 samples were graphitized in the Godwin Radiocarbon Laboratory at the University of Cambridge (UK), along with standards and radiocarbon-dead spar calcite (backgrounds), following a standard hydrogen/iron catalyst protocol[59]. Pressed graphite targets were subsequently analysed by AMS at the ^{14}Chrono Centre, University of Belfast (UK). Measured ^{14}C ages have been corrected for mass-dependent fractionation (normalization to $\delta^{13}C = -25$‰) and the background radiocarbon content by analysing radiocarbon-dead spar calcite with each sample batch. Paired planktonic and benthic samples have been measured in the same AMS sample carousel.

Four paired measurements have resulted in younger benthic than planktonic foraminifera (Supplementary Fig. 5). We have omitted these data from the initial analyses, but including these samples does not alter the general trend of the data (Supplementary Fig. 5).

Correlation of marine proxy records with $CO_{2,atm}$ variations. To calculate correlation coefficients R^2 between $CO_{2,atm}$ variations and ^{14}C-based deep sub-Antarctic ventilation ages during the last glacial period, that is, 41-22 kyr BP (Supplementary Fig. 5e), we interpolated the mean $CO_{2,atm}$ record[19] at the sampling resolution of the ^{14}C proxy data. Similarly, the mean $CO_{2,atm}$ has been interpolated at the resolution of the mean U/Mn$_c$- and the $\Delta\delta^{13}$C-based [O_2] records in order to estimate the correlation (R^2) between changes in bottom-water oxygenation and ^{14}C ventilation in the deep sub-Antarctic Atlantic (Supplementary Fig. 5f,g). For these calculations, the mean U/Mn$_c$ has been obtained by averaging *G. bulloides* and *Uvigerina* spp. U/Mn$_c$ (stippled line in Supplementary Fig. 2a) and the $\Delta\delta^{13}$C-derived [O_2] record is based on a 500 year-running average (solid line in Fig. 2f).

Chronostratigraphy of other sub-Antarctic Atlantic cores. The most recent age model of sediment core PS2498-1 has been established based on an alignment of variations in lithogenic fluxes with the EPICA Dome C dust record[9]. Because sediment cores MD07-3076Q and PS2498-1 are in close proximity (Fig. 1), we have compared the magnetic susceptibility records and noticed stratigraphic offsets of ± 900 years. To allow a faithful inter-core comparison, we have adjusted the chronology of PS2498-1 by aligning the magnetic susceptibility record of PS2498-1 (ref. 60) to the magnetic susceptibility record of MD07-3076Q, which has been measured with the GEOTEK Multi-Sensor-Core-Logger aboard *R/V Marion Dufresne* using a low field susceptibility (Bartington) sensor. For TN057-21, we rely on the most recently established chronology of ref. 28, which is based on the GICC05 age scale[61] that is equivalent to the AICC2012 age scale used in this study within decades to few hundred years[62].

References

1. Sigman, D. M. & Boyle, E. A. Glacial/interglacial variations in atmospheric carbon dioxide. *Nature* **407**, 859–869 (2000).
2. Ito, T. & Follows, M. J. Preformed phosphate, soft tissue pump and atmospheric CO_2. *J. Mar. Res.* **63**, 813–839 (2005).
3. Schmittner, A. & Galbraith, E. D. Glacial greenhouse-gas fluctuations controlled by ocean circulation changes. *Nature* **456**, 373–376 (2008).
4. Martin, J. H. Glacial-interglacial CO_2 change: the iron hypothesis. *Paleoceanography* **5**, 1–13 (1990).
5. Ferrari, R. *et al.* Antarctic sea ice control on ocean circulation in present and glacial climates. *Proc. Natl. Acad. Sci. USA* **111**, 8753–8758 (2014).
6. Stephens, B. B. & Keeling, R. F. The influence of Antarctic sea ice on glacial-interglacial CO_2 variations. *Nature* **404**, 171–174 (2000).
7. Anderson, R. F. *et al.* Wind-driven upwelling in the Southern Ocean and the deglacial rise in atmospheric CO_2. *Science* **323**, 1443–1448 (2009).
8. Toggweiler, J. R., Russell, J. L. & Carson, S. R. Midlatitude westerlies, atmospheric CO_2, and climate change during the ice ages. *Paleoceanography* **21**, 2005 (2006).

9. Anderson, R. F. *et al.* Biological response to millennial variability of dust and nutrient supply in the Subantarctic South Atlantic Ocean. *Philos. Trans. R. A Math. Phys. Eng. Sci.* **372**, 20130054 (2014).
10. Martínez-García, A. *et al.* Iron fertilization of the Subantarctic Ocean during the last ice age. *Science* **343**, 1347–1350 (2014).
11. Jaccard, S. L. *et al.* Two modes of change in Southern Ocean productivity over the past million years. *Science* **339**, 1419–1423 (2013).
12. Ziegler, M., Diz, P., Hall, I. R. & Zahn, R. Millennial-scale changes in atmospheric CO_2 levels linked to the Southern Ocean carbon isotope gradient and dust flux. *Nat. Geosci.* **6**, 457–461 (2013).
13. Frank, M. *et al.* Similar glacial and interglacial export bioproductivity in the Atlantic sector of the Southern Ocean: multiproxy evidence and implications for glacial atmospheric CO_2. *Paleoceanography* **15**, 642–658 (2000).
14. Skinner, L. C., Fallon, S., Waelbroeck, C., Michel, E. & Barker, S. Ventilation of the deep Southern Ocean and deglacial CO_2 rise. *Science* **328**, 1147–1151 (2010).
15. Boiteau, R., Greaves, M. & Elderfield, H. Authigenic uranium in foraminiferal coatings: a proxy for ocean redox chemistry. *Paleoceanography* **27**, PA3227 (2012).
16. Hoogakker, B. A. A., Elderfield, H., Schmiedl, G., McCave, I. N. & Rickaby, R. E. M. Glacial – interglacial changes in bottom-water oxygen content on the Portuguese margin. *Nat. Geosci.* **8**, 40–43 (2015).
17. McCorkle, D. C., Keigwin, L. D., Corliss, B. H. & Emerson, S. R. The influence of microhabitats on the carbon isotopic composition of deep-sea benthic foraminifera. *Paleoceanography* **5**, 161–185 (1990).
18. Anderson, R. F. *et al.* Biological response to millennial variability of dust supply in the Subantarctic South Atlantic Ocean. *Philos. Trans. R. A Math. Phys. Eng. Sci.* **372**, 20130054 (2014).
19. Gottschalk, J., Skinner, L. C. & Waelbroeck, C. Contribution of seasonal sub-Antarctic surface water variability to millennial-scale changes in atmospheric CO_2 over the last deglaciation and Marine Isotope Stage 3. *Earth Planet. Sci. Lett.* **411**, 87–99 (2015).
20. Gottschalk, J. *et al.* Abrupt changes in the southern extent of North Atlantic Deep Water during Dansgaard-Oeschger events. *Nat. Geosci.* **8**, 950–955 (2015).
21. Russell, A. D., Hönisch, B., Spero, H. J. & Lea, D. W. Effects of seawater carbonate ion concentration and temperature on shell U, Mg, and Sr in cultured planktonic foraminifera. *Geochim. Cosmochim. Acta* **68**, 4347–4361 (2004).
22. Yu, J., Elderfield, H., Greaves, M. & Day, J. Preferential dissolution of benthic foraminiferal calcite during laboratory reductive cleaning. *Geochem. Geophys. Geosyst.* **8**, Q06016 (2007).
23. Klinkhammer, G. P. & Palmer, M. R. Uranium in the oceans: where it goes and why. *Geochim. Cosmochim. Acta* **55**, 1799–1806 (1991).
24. Froelich, P. N. *et al.* Early oxidation of organic matter in pelagic sediments of the eastern equatorial Atlantic: suboxic diagenesis. *Geochim. Cosmochim. Acta* **43**, 1075–1090 (1979).
25. Barnes, C. E. & Cochran, J. K. Uranium removal in oceanic sediments and the oceanic U balance. *Earth Planet. Sci. Lett.* **97**, 94–101 (1990).
26. Boyle, E. A. Manganese carbonate overgrowths on foraminifera tests. *Geochim. Cosmochim. Acta* **47**, 1815–1819 (1983).
27. Sachs, J. P. & Anderson, R. F. Fidelity of alkenone paleotemperatures in southern Cape Basin sediment drifts. *Paleoceanography* **18**, 1082 (2003).
28. Barker, S. & Diz, P. Timing of the descent into the last ice age determined by the bipolar seesaw. *Paleoceanography* **29**, 489–507 (2014).
29. Emerson, S., Fischer, K., Reimers, C. & Heggie, D. Organic carbon dynamics and preservation in deep-sea sediments. *Deep Sea Res.* **32**, 1–21 (1985).
30. McCorkle, D. C. & Emerson, S. R. The relationship between pore water carbon isotopic composition and bottom water oxygen concentration. *Geochim. Cosmochim. Acta* **52**, 1169–1178 (1988).
31. Geslin, E., Heinz, P., Jorissen, F. & Hemleben, C. Migratory responses of deep-sea benthic foraminifera to variable oxygen conditions: laboratory investigations. *Mar. Micropaleontol.* **53**, 227–243 (2004).
32. Duplessy, J.-C. *et al.* ^{13}C Record of benthic foraminifera in the last interglacial ocean: Implications for the carbon cycle and the global deep water circulation. *Quat. Res.* **21**, 225–243 (1984).
33. Garcia, H. E. *et al. World Ocean Atlas 2009* Vol. 3: Dissolved Oxygen, Apparent Oxygen Utilization, and Oxygen Saturation (Ed. Levitus, S.) 344 pp NOAA Atlas NESDIS 70, U.S. Government Printing Office, Washington, D.C. (2010).
34. Hodell, D. A., Venz, K. A., Charles, C. D. & Ninnemann, U. S. Pleistocene vertical carbon isotope and carbonate gradients in the South Atlantic sector of the Southern Ocean. *Geochem. Geophys. Geosyst.* **4**, 1–19 (2003).
35. Schmiedl, G. & Mackensen, A. Late quaternary paleoproductivity and deep water circulation in the eastern South Atlantic Ocean: evidence from benthic foraminifera. *Palaeogeogr. Palaeoclimatol. Palaeoecol.* **130**, 43–80 (1997).

36. Ninnemann, U. S. & Charles, C. D. Changes in the mode of Southern Ocean circulation over the last glacial cycle revealed by foraminiferal stable isotopic variability. *Earth Planet. Sci. Lett.* **201**, 383–396 (2002).

37. Mackensen, A., Rudolph, M. & Kuhn, G. Late Pleistocene deep-water circulation in the subantarctic eastern Atlantic. *Glob. Planet. Change* **30**, 197–229 (2001).

38. Ragueneau, O. *et al.* A review of the Si cycle in the modern ocean: recent progress and missing gaps in the application of biogenic opal as a paleoproductivity proxy. *Glob. Planet. Change* **26**, 317–365 (2000).

39. Jaccard, S. L., Galbraith, E. D., Frölicher, T. L. & Gruber, N. Ocean (de)oxygenation across the last deglaciation: insights for the future. *Oceanography* **27**, 26–35 (2014).

40. Anderson, L. A. & Sarmiento, J. L. Redfield ratios of remineralization determined by nutrient data analysis. *Global Biogeochem. Cycles* **8**, 65–80 (1994).

41. Kwon, E. Y., Sarmiento, J. L., Toggweiler, J. R. & DeVries, T. The control of atmospheric pCO$_2$ by ocean ventilation change: the effect of the oceanic storage of biogenic carbon. *Global Biogeochem. Cycles* **25**, GB3026 (2011).

42. Curry, W. B. & Oppo, D. W. Glacial water mass geometry and the distribution of δ^{13}C of ΣCO$_2$ in the western Atlantic Ocean. *Paleoceanography* **20**, PA1017 (2005).

43. Jaccard, S. L. & Galbraith, E. D. Large climate-driven changes of oceanic oxygen concentrations during the last deglaciation. *Nat. Geosci.* **5**, 151–156 (2012).

44. Sarnthein, M., Schneider, B. & Grootes, P. M. Peak glacial ^{14}C ventilation ages suggest major draw-down of carbon into the abyssal ocean. *Clim. Past* **9**, 2595–2614 (2013).

45. Schmittner, A., Brook, E. J. & Ahn, J. in *Ocean Circulation: Mechanisms and Impacts* (eds. Schmittner, A., Chiang, J. C. H. & Hemming, S. R.) **173**, 209–246 (American Geophysical Union, Geophysical Monograph Series, 2007).

46. Watson, A. J. & Naveira Garabato, A. C. The role of Southern Ocean mixing and upwelling in glacial-interglacial atmospheric CO$_2$ change. *Tellus B* **58**, 73–87 (2006).

47. Bereiter, B. *et al.* Mode change of millennial CO$_2$ variability during the last glacial cycle associated with a bipolar marine carbon seesaw. *Proc. Natl. Acad. Sci. USA* **109**, 9755–9760 (2012).

48. Jaccard, S. L., Galbraith, E. D., Martínez-Garcia, A. & Anderson, R. F. Covariation of abyssal Southern Ocean oxygenation and pCO$_2$ throughout the last ice age. *Nature* **530**, 207–210 (2016).

49. Sarmiento, J. L., Gruber, N., Brzezinski, M. A. & Dunne, J. P. High-latitude controls of thermocline nutrients and low latitude biological productivity. *Nature* **427**, 56–60 (2004).

50. Carter, L., McCave, I. N. & Williams, M. J. M. Circulation and water masses of the Southern Ocean: a review. *Dev. Earth Environ. Sci.* **8**, 85–114 (2009).

51. Weiss, R. F. The solubility of nitrogen, oxygen and argon in water and seawater. *Deep Sea Res.* **17**, 721–735 (1970).

52. Adkins, J. F., McIntyre, K. & Schrag, D. P. The salinity, temperature, and δ^{18}O of the glacial deep ocean. *Science* **298**, 1769–1773 (2002).

53. Vogel, H., Rosén, P., Wagner, B., Melles, M. & Persson, P. Fourier transform infrared spectroscopy, a new cost-effective tool for quantitative analysis of biogeochemical properties in long sediment records. *J. Paleolimnol.* **40**, 689–702 (2008).

54. Meyer-Jacob, C. *et al.* Independent measurement of biogenic silica in sediments by FTIR spectroscopy and PLS regression. *J. Paleolimnol.* **52**, 245–255 (2014).

55. DeMaster, D. J. The supply and accumulation of silica in the marine environment. *Geochim. Cosmochim. Acta* **45**, 1715–1732 (1981).

56. Mortlock, R. A. & Froelich, P. N. A simple method for the rapid determination of biogenic opal in pelagic marine sediments. *Deep Sea Res.* **36**, 1415–1426 (1989).

57. François, R., Frank, M., van der Loeff, M. M. R. & Bacon, M. P. ^{230}Th normalization: an essential tool for interpreting sedimentary fluxes during the late Quaternary. *Paleoceanography* **19**, 16 (2004).

58. Bourne, M. D., Thomas, A. L., Mac Niocaill, C. & Henderson, G. M. Improved determination of marine sedimentation rates using ^{230}Th$_{xs}$. *Geochemistry Geophys. Geosystems* **13**, Q09017 (2012).

59. Vogel, J. S., Southon, J. R., Nelson, D. E. & Brown, T. A. Performance of catalytically condensed carbon for use in accelerator mass spectrometry. *Nucl. Instrum. Methods Phys. Res.* **5**, 289–293 (1984).

60. Kuhn, G. Susceptibility raw data of sediment core PS2498-1. http://dx.doi.org/10.1594/PANGAEA.87282 (2002).

61. Svensson, A. *et al.* A 60000 year Greenland stratigraphic ice core chronology. *Clim. Past* **4**, 47–57 (2008).

62. Veres, D. *et al.* The Antarctic ice core chronology (AICC2012): an optimized multi-parameter and multi-site dating approach for the last 120 thousand years. *Clim. Past* **9**, 1733–1748 (2013).

63. Key, R. M. *et al.* A global ocean carbon climatology: Results from Global Data Analysis Project (GLODAP). *Global Biogeochem. Cycles* **18**, GB4031 (2004).

64. Takahashi, T. *et al.* Global sea-air CO$_2$ flux based on climatological surface ocean pCO$_2$, and seasonal biological and temperature effects. *Deep Sea Res.* **49**, 1601–1622 (2002).

65. Orsi, A. H., Whitworth, T. & Nowlin, W. D. On the meridional extent and fronts of the Antarctic Circumpolar Current. *Deep Sea Res.* **42**, 641–673 (1995).

66. Lambert, F. *et al.* Dust-climate couplings over the past 800,000 years from the EPICA Dome C ice core. *Nature* **452**, 616–619 (2008).

67. Ahn, J. & Brook, E. J. Atmospheric CO$_2$ and climate on millennial time scales during the last glacial period. *Science* **322**, 83–85 (2008).

68. Blunier, T. & Brook, E. J. Timing of millennial-scale climate change in Antarctica and Greenland during the last glacial period. *Science* **291**, 109 (2001).

69. Lüthi, D. *et al.* CO$_2$ and O$_2$/N$_2$ variations in and just below the bubble-clathrate transformation zone of Antarctic ice cores. *Earth Planet. Sci. Lett.* **297**, 226–233 (2010).

70. Monnin, E. *et al.* Atmospheric CO$_2$ concentrations over the last glacial termination. *Science* **291**, 112 (2001).

71. Ahn, J. & Brook, E. J. Siple Dome ice reveals two modes of millennial CO$_2$ change during the last ice age. *Nat. Commun.* **5**, 3723 (2014).

72. Indermühle, A., Monnin, E., Stauffer, B., Stocker, T. F. & Wahlen, M. Atmospheric CO$_2$ concentration from 60 to 20 kyr BP from the Taylor Dome ice core, Antarctica. *Geophys. Res. Lett.* **27**, 735–738 (2000).

73. Marcott, S. A. *et al.* Centennial-scale changes in the global carbon cycle during the last deglaciation. *Nature* **514**, 616–619 (2014).

74. Ramsey, C. B. *et al.* A complete terrestrial radiocarbon record for 11.2 to 52.8 kyr BP. *Science* **338**, 370–374 (2012).

75. Hughen, K., Southon, J., Lehman, S., Bertrand, C. & Turnbull, J. Marine-derived ^{14}C calibration and activity record for the past 50,000 years updated from the Cariaco Basin. *Quat. Sci. Rev.* **25**, 3216–3227 (2006).

76. Reimer, P. J. *et al.* IntCal09 and Marine09 radiocarbon age calibration curves, 0-50,000 years cal BP. *Radiocarbon* **51**, 1111–1150 (2009).

77. Reimer, P. J. *et al.* IntCal13 and Marine13 radiocarbon age calibration curves 0-50,000 years cal BP. *Radiocarbon* **55**, 1869–1887 (2013).

Acknowledgements

We are very indebted to Sambuddha Misra, Stephen Barker and Emma Freeman for fruitful discussions. Fabien Dewilde, Gülay Isguder, Margret Bayer, Verena Lanny, Lena Thöle, Emma Freeman, Ron Reimer, María de la Fuente and Benny Antz are thanked for the technical support. We also acknowledge Andreas Mackensen and Rainer Gersonde for sharing their expertise in South Atlantic coring locations. J.G. and L.C.S. acknowledge support from the Gates Cambridge Trust, the Royal Society, the Cambridge Newton Trust and NERC grant NE/J010545/1. J.L. was supported by Marie Curie Fellowship FP7-PEOPLE-2013-IEF (Marie Curie proposal 622483). S.L.J. was funded through the Swiss National Science Foundation (grant PP00P2-144811). C.W. acknowledges support from the European Research Council grant ACCLIMATE/no 339108. This is LSCE contribution no. 4488.

Author contributions

J.G. and L.C.S. designed the study. C.W. collected the core material. J.G., J.L. and H.V. performed the analyses with support from S.L.J. and N.F. J.G. and L.C.S. analysed the proxy data and wrote this manuscript with contributions from all authors.

Additional information

12

Tornado outbreak variability follows Taylor's power law of fluctuation scaling and increases dramatically with severity

Michael K. Tippett[1,2] & Joel E. Cohen[3,4]

Tornadoes cause loss of life and damage to property each year in the United States and around the world. The largest impacts come from 'outbreaks' consisting of multiple tornadoes closely spaced in time. Here we find an upward trend in the annual mean number of tornadoes per US tornado outbreak for the period 1954–2014. Moreover, the variance of this quantity is increasing more than four times as fast as the mean. The mean and variance of the number of tornadoes per outbreak vary according to Taylor's power law of fluctuation scaling (TL), with parameters that are consistent with multiplicative growth. Tornado-related atmospheric proxies show similar power-law scaling and multiplicative growth. Path-length-integrated tornado outbreak intensity also follows TL, but with parameters consistent with sampling variability. The observed TL power-law scaling of outbreak severity means that extreme outbreaks are more frequent than would be expected if mean and variance were independent or linearly related.

[1] Department of Applied Physics and Applied Mathematics, Columbia University, New York, New York 10027, USA. [2] Center of Excellence for Climate Change Research, Department of Meteorology, King Abdulaziz University, Jeddah 21589, Saudi Arabia. [3] Laboratory of Populations, Rockefeller University, New York, New York 10065, USA. [4] The Earth Institute, Columbia University, New York, New York 10027, USA. Correspondence and requests for materials should be addressed to M.K.T. (email: mkt14@columbia.edu) or to J.E.C. (email: cohen@rockefeller.edu).

Hazardous convective weather (tornadoes, hail and damaging wind) associated with severe thunderstorms affects large portions of the United States. Tornadoes cause particularly intense damage. Over a recent 10-year period (2005–2014), tornadoes in the United States resulted in an average of 110 deaths per year and annual losses ranging from $500 million to $9.6 billion[1]. The largest societal impacts from tornadoes are from 'outbreaks' in which multiple tornadoes occur in a single weather event. Tornado outbreaks across the eastern two-thirds of the United States were associated with 79% of all tornado fatalities over the period 1972–2010 (ref. 2) and are routinely responsible for billion-dollar loss events[3].

Whether US tornado activity will change in the future remains uncertain. Climate change projections indicate that environments favourable to severe thunderstorms will be more frequent in a warmer climate[4], and high-resolution convection-permitting numerical modelling indicates increased activity and year-to-year variability of March–May US tornado occurrence as measured by severe weather proxies derived from explicitly depicted storms[5]. However, to date, there is no upward trend in the number of reliably reported US tornadoes per year[6]. Interpretation of the US tornado report data requires some caution. For instance, the total number of US tornadoes reported each year has increased dramatically over the last half century, but most of that increase is due to more reports of weak tornadoes and is believed to reflect changing reporting practices and other non-meteorological factors rather than increased tornado occurrence[7]. The variability of reported tornado occurrence has increased over the last few decades with more tornadoes being reported on days when tornadoes are observed[8,9]. In addition, greater year-to-year variability in the number of tornadoes reported per year has been associated with consistent changes in the monthly averaged atmospheric environments favourable to tornado occurrence[10]. Likewise, US large-event severe thunderstorm losses and the frequency of the most extreme environments have increased[11]. Regional changes in the seasonality of tornado occurrence have also been reported[12,13].

Despite their importance, relatively little is known about the current or projected statistics of tornado outbreak severity beyond their climatology[2]. Most studies have considered only the statistics of tornadoes occurring during a single day[14] and have not considered outbreaks over multiple dates. Here we show that the annual mean number of tornadoes per outbreak increased during 1954–2014, and the annual variance increased more than four times faster than the mean. We show that the mean and variance of the number of tornadoes per outbreak are related by Taylor's power law of fluctuation scaling (TL)[15,16] with parameters that are consistent with multiplicative growth[17]. TL scaling in tornado outbreak statistics was not previously known. Although power-law scaling is present in the probability distributions of various tornado characteristics[18,19], the presence of power-law scaling in such probability distributions is neither necessary nor sufficient for TL scaling of the variance in relation to the mean (see also Supplementary Discussion for examples of TL scaling without power-law probability distributions). We also find that a tornado-related atmospheric proxy shows a similar power-law scaling and multiplicative growth. Path-length-integrated tornado outbreak intensity follows TL as well, but with parameters predicted by sampling variability[20]. The findings are similar when we restrict the analysis to the more recent period 1977–2014 and to more intense tornadoes. The observed TL scaling of outbreak severity means that extreme outbreaks are more frequent than would otherwise be expected if mean and variance were independent or linearly related.

Results

Data and sensitivity analyses. We use data from 1954 to 2014, which are generally considered reliable. Because of concerns regarding the data before 1977 ('Methods' section, Supplementary Fig. 1), we repeat some analysis using the more recent period 1977–2014 to test the robustness of the results (Supplementary Figs 2–5). We exclude the weakest tornadoes from our analysis and denote the remaining tornadoes as F1+ tornadoes ('Methods' section). We repeat some of the analysis restricted to more intense tornadoes (F2+; Supplementary Figs 7–10).

Number of tornadoes per outbreak. The annual number of F1+ tornadoes shows no significant trend over the period 1954–2014 (Fig. 1a). Generally, trends have not been found in the number of severe tornadoes when severity is defined using the Fujita scale, but upward trends have been found when severity is defined using path length[18]. The percentage of F1+ tornadoes that occur in outbreaks ('Methods' section) is increasing by 0.34 percentage points ± 0.13 percentage points per year (Fig. 1b), consistent with upward trends in the proportion of tornadoes occurring on days with many tornadoes[8,9]. Here and in all results, ± intervals are 95% confidence intervals. The fraction of F1+ tornadoes that occur in outbreaks is less than one because not all F1+ tornadoes occur in outbreaks. During the period 1977–2014, the number of F1+ tornadoes also shows no significant trend, and the percentage of F1+ tornadoes occurring in outbreaks is also increasing, at a larger estimated rate (Supplementary Fig. 2). The US tornado reports show no statistically significant trend in the frequency of tornado outbreaks (Fig. 2a). Since the number of

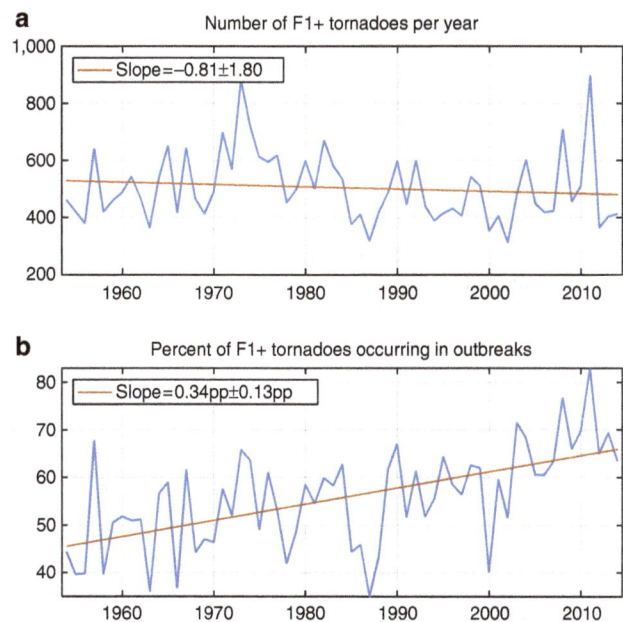

a

Number of F1+ tornadoes per year

Slope=−0.81±1.80

b

Percent of F1+ tornadoes occurring in outbreaks

Slope=0.34pp±0.13pp

Figure 1 | Time series of counts and clustering of F1+ tornadoes 1954–2014 in the contiguous US. (a) Number of F1+ tornadoes per year. The slope of the least-squares regression indicates that the number of F1+ tornadoes per year declined by 0.81 per year on average from 1954 to 2014 inclusive. This rate of decline is not statistically significantly different from 0 (no change). **(b)** Annual percentage of F1+ tornadoes occurring in outbreaks. The slope of the least-squares regression indicates that the percentage of F1+ tornadoes per year that occurred as part of outbreaks increased by 0.34 percentage points (pp) per year on average from 1954 to 2014 inclusive. This increase is statistically significantly greater than 0. In both **a** and **b**, ± intervals are 95% confidence intervals.

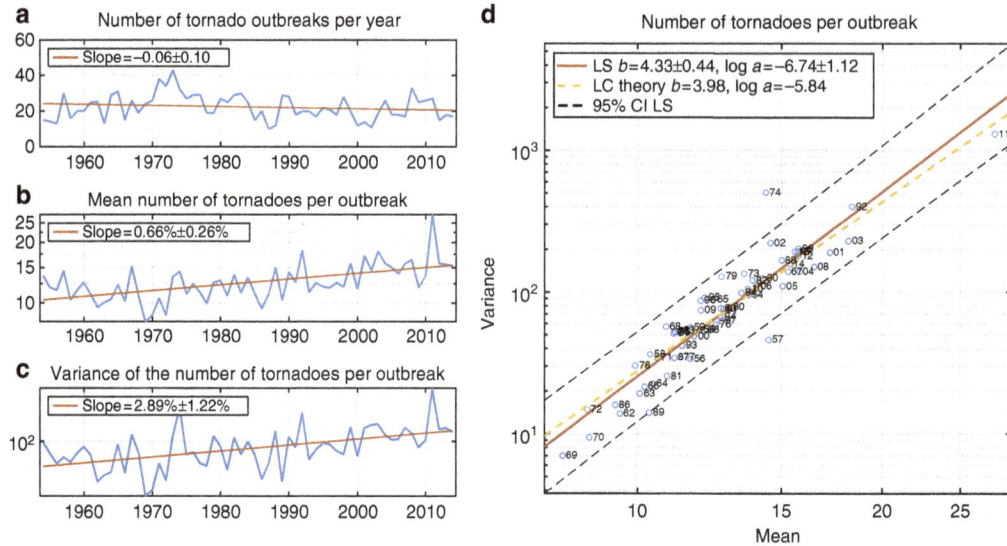

Figure 2 | Numbers of F1+ tornadoes per outbreak 1954-2014. (**a**) Number of tornado outbreaks per year. The rate of decline is not statistically significantly different from 0 (no change). (**b**) Annual mean number of tornadoes per outbreak. Vertical axis is on logarithmic scale, so the rate of increase in the annual mean is expressed as a percentage per year. This rate of increase is statistically significantly greater than 0. (**c**) Annual variance of the number of tornadoes per outbreak. Vertical axis is on logarithmic scale, so the rate of increase in the annual mean is expressed as a percentage per year. This rate of increase is statistically significantly greater than 0. (**d**) Scatter plot of the annual mean number of tornadoes per outbreak versus the annual variance of the number of tornadoes per outbreak. Both axes are on logarithmic scale. The solid red line is the least-squares (LS) regression line (Taylor's power law of fluctuation scaling) and the dashed yellow line has the slope and intercept predicted by LC theory[17]. The two-digit number following the plotting symbol 'o' gives the calendar year in the second half of the twentieth century or first half of the twenty-first century. In all the panels, ± intervals are 95% confidence intervals.

F1+ tornadoes and the number of outbreaks are not changing (on average, over time), the increasing percentage of F1+ tornadoes occurring in outbreaks means that the number of F1+ tornadoes per outbreak must be increasing, and indeed, the annual mean number of F1+ tornadoes per outbreak shows a significant upward trend (Fig. 2b). The annual mean number of tornadoes per outbreak is increasing by 0.66% ± 0.26% per year, and the variance is increasing more than four times as fast, 2.89% ± 1.22% per year (Fig. 2b,c) over the period 1954–2014. The growth rates are greater over the recent period 1977–2014, with similar ratio between the growth rates of mean and variance (Supplementary Fig. 3b,c).

The fact that the variance is increasing several times faster than the mean is especially noteworthy: it indicates a changing distribution in which the likelihood of extreme outbreaks is increasing faster than what the trend in mean alone would suggest. The coefficient of dispersion of a probability distribution with a positive mean is the ratio of its variance to its mean. Values greater than one (over-dispersion) indicate more clustering than a Poisson variable. For instance, European windstorms exhibit over-dispersion and serial clustering that increases with intensity[21] with implications for the return intervals of rare events[22]. Taylor's law (TL) relates the mean and variance of a probability distribution by

$$\text{variance} = a(\text{mean})^b, a > 0, \qquad (1)$$

where a and b are constants[15,16]. A value of $b > 1$ indicates that the coefficient of dispersion increases with the mean. The annual mean and annual variance of the number of tornadoes per outbreak approximately satisfy TL with $b = 4.3 \pm 0.44$ and log $a = -6.74 \pm 1.12$ (Fig. 2d); consistent values are seen over the period 1977–2014 (Supplementary Fig. 3d). (Throughout log is the natural logarithm.) The value of b here is remarkable since in most ecological applications, the TL exponent seldom exceeds 2. The TL exponent can be greater than 2 for lognormal

distributions with changing parameters (Supplementary Discussion and Supplementary Fig. 11). The TL scaling of tornado outbreak severity reveals a remarkably regular relation between annual mean and annual variance that extends over the full range of the data, even for years like 2011 which are extreme in mean and variance. The data from 1974 deviate most from TL scaling, with the excessive variance reflecting the 3–4 April 'Super Outbreak.'

The upward trend in the number of tornadoes per outbreak provides an interpretation for the observed TL scaling since TL scaling arises in models of stochastic multiplicative growth[17]. In such models, the quantity $N(t+1)$ at time $t+1$ is related to its previous value $N(t)$ by

$$N(t+1) = A(t)N(t), \qquad (2)$$

where $A(t)$ is the random multiplicative factor by which $N(t)$ grows or declines from one time to the next. Here $N(t)$ is the annual average number of tornadoes per outbreak, and each integer value of t represents one calendar year. The Lewontin–Cohen (LC) model for stochastic multiplicative growth assumes that the $A(t)$ are independently and identically distributed for all $t \geq 0$ with finite mean $M > 0$ and finite variance V. If $M \neq 1$, $N(t)$ follows TL asymptotically with[17]

$$a = \frac{E[(N(0))^2]}{[E(N(0))]^b} \text{ and } b = \frac{\log[V + M^2]}{\log M}. \qquad (3)$$

Here we estimate ('Methods' section) $M = 1.03$ and $V = 0.068$, which leads to TL parameters $b = 3.98$ and $\log(a) = -5.84$. Both values are consistent with the least-squares (LS) estimates of the corresponding parameters of TL (Fig. 2d). The LS estimates are also consistent with the values from LC theory during 1977–2014 (Supplementary Fig. 3d). 95% confidence intervals for M and V show that the hypothesis of no growth ($M = 1$) under which equation (3) is not valid cannot be rejected (Supplementary Table 1). The Supplementary Discussion provides additional

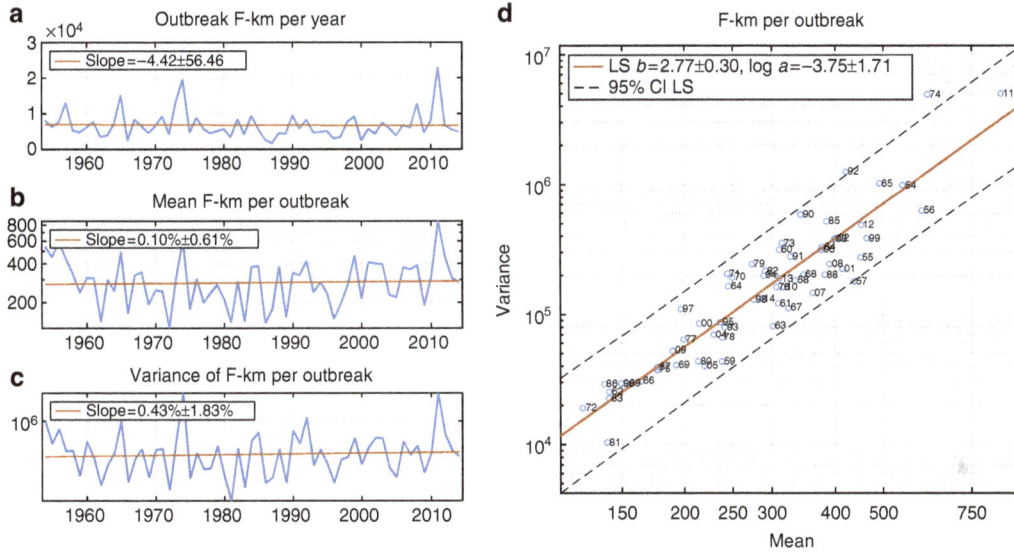

Figure 3 | F-km per outbreak 1954–2014. (**a**) Total outbreak F-km per year. The rate of decline is not statistically significantly different from 0 (no change) among F1 + tornadoes. (**b**) Annual mean F-km per outbreak. Vertical axis is on logarithmic scale, so the rate of increase in the annual mean is expressed as a percentage per year. This rate of increase is not statistically significantly greater than 0. (**c**) Annual variance of F-km per outbreak. Vertical axis is on logarithmic scale, so the rate of increase in the annual mean is expressed as a percentage per year. This rate of increase is not statistically significantly greater than 0. (**d**) Scatter plot of the annual mean of F-km per outbreak versus the annual variance of F-km per outbreak. Both axes are on logarithmic scale. The solid red line is the least-squares (LS) regression line (Taylor's power law of fluctuation scaling). The two-digit number following the plotting symbol 'o' gives the calendar year in the second half of the twentieth century or first half of the twenty-first century. In all the panels, ± intervals are 95% confidence intervals.

description of how the LC model leads to TL scaling with exponent approximately 4.

Fujita-kilometers per outbreak. Another measure of outbreak severity is Fujita-kilometers (ref. 2; F-km) which is the sum (over all tornadoes in an outbreak) of each tornado's path length in kilometers multiplied by its Fujita or Enhanced Fujita rating ('Methods' section). Annual totals of outbreak F-km, mean number of F-km per outbreak and the variance of F-km per outbreak do not show significant trends over the period 1954–2014 (Fig. 3a–c). The mean number of F-km per outbreak and the variance of F-km per outbreak show marginally significant trends over the recent period 1977–2014 (Supplementary Fig. 4a–c). The TL parameters relating the mean and variance of F-km per outbreak are $b = 2.77 \pm 0.30$ and log $a = -3.75 \pm 1.71$ (Fig. 3d). The lack of robust trends means that LC theory is not appropriate to explain the TL scaling of F-km. However, TL scaling also arises from the sampling of stationary skewed distributions[20]. For a distribution with mean m, variance v, skewness γ_1 and coefficient of variation CV, theory[20] predicts

$$a = \log v - \frac{\gamma_1}{CV} \log m \text{ and } b = \frac{\gamma_1}{CV}. \quad (4)$$

Here, excluding two outlier outbreaks from the calculation of the distribution parameters ('Methods' section, Supplementary Fig. 5), equation (4) gives $b = 2.71$ and log $a = -3.05$, both of which are consistent with the LS estimates of the TL parameters for F-km per outbreak. (We use 'outlier' to indicate values far from other observations, not to suggest that the unusual values are the result of measurement error.) Therefore TL scaling of F-km per outbreak could be explained by sampling variability.

A tornado environment proxy. A reasonable concern is that the findings here represent properties of the tornado report database

that are not meteorological in origin, especially since other prominent features of the tornado report database are not meteorological in origin[7]. Environmental proxies for tornado occurrence and number of tornadoes per occurrence provide an independent, albeit imperfect, measure of tornado activity for the period 1979–2013 ('Methods' section). At a minimum, the environmental proxies provide information about the frequency and severity of environments favourable to tornado occurrence. The correlation between the annual average number of tornadoes per outbreak and the proxy for number of tornadoes per occurrence is 0.56 (Supplementary Fig. 6a). This correlation falls to 0.34, still significant at the 95% level, when the data from 2011 are excluded. Applying a 5-year moving average to the data highlights their common trends and increases the correlation to 0.88 (Supplementary Fig. 6b). The annual mean of the occurrence proxy, a surrogate for number of tornadoes per year, shows a marginally significant upward trend (Fig. 4a). The annual mean and annual variance of the proxy for number of tornadoes per occurrence show upward trends of $0.63 \pm 0.30\%$ and $2.43 \pm 1.12\%$, respectively (Fig. 4b,c), values strikingly similar to those for number of tornadoes per outbreak (Fig. 2b,c). Moreover, the TL parameters of the proxy for number of tornadoes per occurrence are 3.54 ± 0.42 and log $a = -5.87 \pm 1.43$, which are consistent with the LC multiplicative growth theory estimates for the proxy (Fig. 4d) and quite similar to those for the number of tornadoes per outbreak. Extreme environments associated with tornado occurrence display the TL scaling and multiplicative growth similar to those of the number of tornadoes per outbreak. This similarity plausibly suggests that the changes in the number of tornadoes per outbreak reflect changes in the physical environment.

Sensitivity to outbreak definition. Another concern is that the results are sensitive to the details of the outbreak definition. We

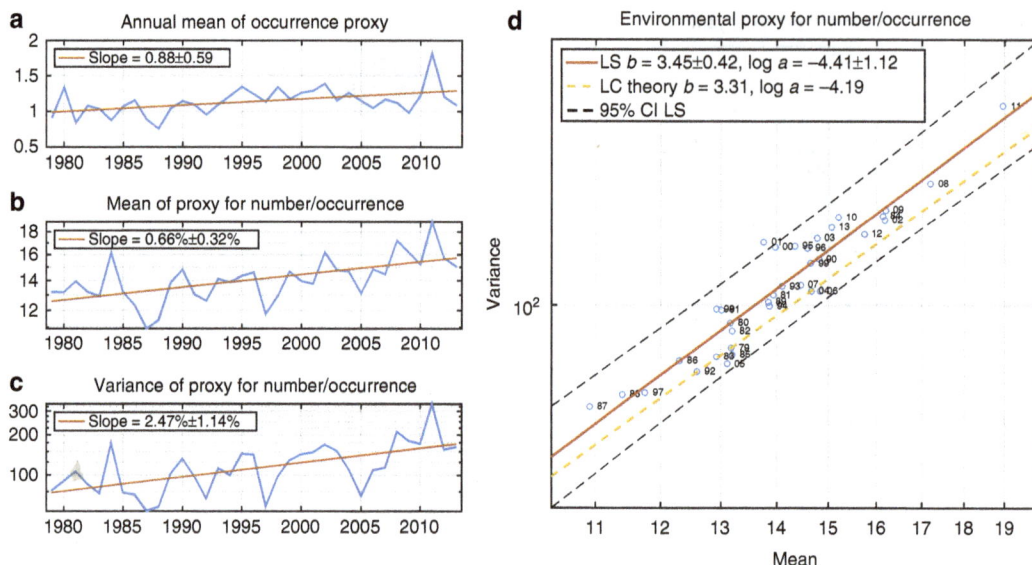

Figure 4 | Environmental proxies 1979-2013. (a) Annual mean of occurrence proxy in per cent. The rate of increase is statistically significantly greater than 0. **(b)** Annual mean of the environmental proxy for number of tornadoes per occurrence. Vertical axis is on logarithmic scale, so the rate of increase in the annual mean is expressed as a percentage per year. This rate of increase is statistically significantly greater than 0. **(c)** Annual variance of the environmental proxy for number of tornadoes per occurrence. Vertical axis is on logarithmic scale, so the rate of increase in the annual mean is expressed as a percentage per year. This rate of increase is statistically significantly greater than 0. **(d)** Scatter plot of the annual mean of proxy for number of tornadoes per occurrence versus the annual variance of proxy for number of tornadoes per occurrence. Both axes are on logarithmic scale. The solid red line is the least-squares (LS) regression line and the dashed yellow line has slope and intercept predicted by LC theory[17]. The two-digit number following the plotting symbol 'o' gives the calendar year in the second half of the twentieth century or first half of the twenty-first century. In all the panels, ± intervals are 95% confidence intervals.

assess the robustness of the results to the E/F1 threshold by repeating the analysis with tornadoes rated E/F2 and higher, denoted F2+ (Supplementary Figs 7–10). We use the period 1977–2014 because the annual number of F2+ tornadoes display a substantial decrease (not shown) around the 1970s that is likely related to the introduction of the F-scale. Overall the F2+ results are remarkably similar to the F1+ ones. The annual number of F2+ tornadoes has an insignificantly negative trend during 1977–2014 (Supplementary Fig. 7a), and the percentage of F2+ tornadoes occurring in F2+ outbreaks has a significant positive trend (Supplementary Fig. 7b). Although the number of F2+ outbreaks shows no significant trend, the mean number of tornadoes per F2+ outbreak and its variance both have significant upward trends (Supplementary Fig. 8a–c). The TL exponent for number of tornadoes per F2+ outbreak is 3.65 and is consistent with LC theory (Supplementary Fig. 8d). Annual totals of F2+ outbreak F-km have no significant trend (Supplementary Fig. 9a). Mean F-km per F2+ outbreak does have a significant upward trend, but variance does not (Supplementary Fig. 9b,c). The TL scaling of F2+ outbreak F-km (Supplementary Fig. 9d) is consistent with sampling variability when 2011 is excluded (Supplementary Fig. 10).

Discussion

These findings have important implications for tornado risk in the United States and perhaps elsewhere, though we have examined only the US data. First, the number of tornadoes per outbreak is increasing. However, there is less evidence that F-km per outbreak are increasing. Both the number of tornadoes per outbreak and F-km per outbreak follow TL, which relates mean and variance. We find that TL scaling for the number of tornadoes per outbreak is compatible with multiplicative growth, and that TL scaling for F-km per outbreak could be due to

sampling variability. Finally, the key implication of TL scaling is that both number of tornadoes per outbreak and F-km per outbreak exhibit extreme over-dispersion which increases with mean. When the average tornado outbreak severity gets worse, the high extreme of severity rises even faster and the low extreme falls even faster, by either measure of severity.

Methods

Outbreak data. Tornado reports come from the NOAA Storm Prediction Center (http://www.spc.noaa.gov/wcm/#data). There are no reports from either Alaska or Hawaii; one report from Puerto Rico is excluded. Tornadoes are rated by estimated or reported damage using the Fujita (F) scale, introduced in the mid-1970s, and since March 2007, the enhanced Fujita (EF) scale, with 0 being the weakest and 5 the strongest. Only reports of tornadoes rated F/EF1 or greater are used and are denoted as F1+. To compute outbreak statistics, tornado reports are first sorted in chronological order taking into account the time zone. Outbreaks are sequences of 6 or more F1+ tornadoes (regardless of location in the contiguous United States) whose successive start times are separated by no more than 6 h (ref. 2). A total 1,361 outbreaks are found over the period 1954–2014. Outbreaks spanning more than 1 day are possible with this definition. The median outbreak duration is about 8 h, and 95% of the outbreaks last less than 24 h. Additional climatological features can be found in ref. 2 The number of tornadoes and F-km are computed for each outbreak. Then the annual mean and variance of two outbreak severity measures, number of tornadoes per outbreak and F-km per outbreak, are calculated. The distribution of outbreak tornadoes by F/EF-scale shows considerable variation before 1977 (Supplementary Fig. 1) and may well reflect lower reliability in the earlier period[23] before damage surveys were a routine part of the rating procedure. Therefore, we repeat some of our analysis using the recent period 1977–2014 (Supplementary Figs 2,3).

The same outbreak calculation procedure is repeated but considering only reports of tornadoes rated F/EF2 or greater, denoted as F2+. These outbreak events are referred to as F2+ outbreaks (Supplementary Figs 7–10). Only F2+ tornadoes are used to calculate tornado numbers and F-km of F2+ outbreaks.

Trends. All trends and 95% confidence intervals are assessed using linear regression and ordinary least squares, assuming approximately normal distributions of residuals. The growth rates of the annual mean and variance of the outbreak severity measures in Fig. 2b,c, Fig. 3b,c and Supplementary Figs 3b,c; 4b,c; 8b,c and 9b,c are computed by assuming exponential growth and fitting a linear trend to the logarithms of the data. All the other trends are fitted using untransformed data.

TL parameters. The TL parameters in Figs 2d, 3d and 4d and Supplementary Figs 3d, 4d, 8d and 9d and 95% confidence intervals are estimated using ordinary least-squares regression with the logarithms of the mean and variance.

TL parameters implied by LC theory. The growth factor in the LC model is computed from $A(t) = N(t+1)/N(t)$, where $N(t)$ is the average number of tornadoes per outbreak in calendar year t. The 95% confidence intervals for the mean and variance of $A(t)$ are computed from 10,000 bootstrap samples and reported in Supplementary Table 1. The mean and variance of $A(t)$ are used to compute a prediction of the slope b using equation (3). The TL parameter a is estimated from equation (1) evaluated at the initial year, either 1954 or 1977.

TL parameters implied by sampling variability. There are 1,361 cases of outbreak F-km, and their distribution is highly right-skewed (Supplementary Figs 5a and 10a). The F-km values for the 1974 Super Outbreak and the 25–28 April 2011 tornado outbreak are more than 26 standard deviations above the mean of the data on an arithmetic scale and more than six standard deviations above the mean of the log-transformed data, when means and standard deviations are calculated after withholding the two extreme values. These outliers (values that are far from other observations) have a substantial impact on the estimates of the mean, variance and skewness of the F-km distribution. Despite their rarity, about 86% $(1 - (1 - 2/1,361)^{1361})$ of the bootstrap samples will contain one or both of these two events. The presence of the outliers results in bimodal distributions of the TL slope and intercept estimates (Supplementary Figs 5b,c and 10b,c) computed from equation (4), depending on whether or not the outlier values are in the particular bootstrap sample. Removal of the outliers results in unimodal distributions, whose ranges are consistent with the least-squares estimates of the TL slope and intercept for F-km (Supplementary Figs 5d,e and 10d,e).

Environmental proxies. We use two environmental proxies: one for tornado occurrence and one for the number of tornadoes. Tornadoes are often associated with elevated values of convective available potential energy (CAPE; $J kg^{-1}$) and a measure of vertical wind shear, called storm relative helicity (SRH; $m^2 s^{-2}$). Here 0–180 hPa CAPE and 0–3,000 m SRH data are taken from the North American Regional Reanalysis[24]. The data are interpolated to a 1×1 degree latitude-longitude grid over the continental United States and daily averages are computed. The environmental proxy for tornado occurrence is defined using the energy-helicity index[25] (EHI), which is the product of CAPE and SRH divided by $160,000 J kg^{-1} m^2 s^{-2}$. Values of EHI greater than one indicate the potential for supercell thunderstorms and tornadoes[26], and accordingly we take our proxy for tornado occurrence to be the condition that EHI is greater than 1. Selection of an environmental proxy for the number of tornadoes is more challenging. The only example of a proxy calibrated to predict the number of tornadoes from the surrounding environment uses monthly averaged environment and monthly number of tornadoes[27,28], and therefore is not suited for outbreaks, which last much less than a month. However, previous research does provide some indications for the functional form of such a proxy. For instance, on subdaily timescales the likelihood of significant severe weather is nearly twice as sensitive to vertical wind shear as to CAPE[29,30]. This sensitivity would argue for a proxy based on the product of CAPE and the square of vertical wind shear or equivalently the product of the square root of CAPE and vertical wind shear[31]. Likewise, proxies for the monthly number of tornadoes contain SRH with an exponent ranging from 1.89 to 4.36 depending on region[27,28]. The supercell composite parameter[32] is used in weather forecasting and is the product of CAPE, SRH and vertical shear, again indicating a scaling of severe weather with the square of vertical wind shear measures. On the basis of this evidence, we take as the proxy for number of tornadoes

$$\frac{CAPE \times SRH^2}{3,600,000 \, m^4 \, s^{-4}}, \tag{5}$$

which differs from the EHI in that the square of SRH appears. The normalizing factor is chosen to match overall annual outbreak numbers. The proxy for the number of tornadoes per occurrence is therefore the mean of equation (5) conditional on the occurrence proxy of EHI being greater than one.

References

1. Data from Summary of U.S. Natural Hazard Statistics 2005-2014 compiled by the National Weather Service Office of Climate, Water and Weather Services and the National Climatic Data Center. http://www.nws.noaa.gov/om/hazstats.shtml.
2. Fuhrmann, C. M. *et al.* Ranking of tornado outbreaks across the United States and their climatological characteristics. *Wea. Forecasting* **29**, 684–701 (2014).
3. Smith, A. & Matthews, J. Quantifying uncertainty and variable sensitivity within the US billion-dollar weather and climate disaster cost estimates. *Nat. Hazards* **77**, 1829–1851 (2015).
4. Diffenbaugh, N. S., Scherer, M. & Trapp, R. J. Robust increases in severe thunderstorm environments in response to greenhouse forcing. *Proc. Natl Acad. Sci. USA* **110**, 16361–16366 (2013).
5. Gensini, V. & Mote, T. Downscaled estimates of late 21st century severe weather from CCSM3. *Clim. Change* **129**, 307–321 (2015).
6. Tippett, M. K., Allen, J. T., Gensini, V. A. & Brooks, H. E. Climate and hazardous convective weather. *Curr. Clim. Change Rep.* **1**, 60–73 (2015).
7. Verbout, S. M., Brooks, H. E., Leslie, L. M. & Schultz, D. M. Evolution of the U.S. tornado database: 1954-2003. *Wea. Forecasting* **21**, 86–93 (2006).
8. Brooks, H. E., Carbin, G. W. & Marsh, P. T. Increased variability of tornado occurrence in the United States. *Science* **346**, 349–352 (2014).
9. Elsner, J. B., Elsner, S. C. & Jagger, T. H. The increasing efficiency of tornado days in the United States. *Clim. Dyn.* **45**, 651–659 (2015).
10. Tippett, M. K. Changing volatility of U.S. annual tornado reports. *Geophys. Res. Lett.* **41**, 6956–6961 (2014).
11. Sander, J., Eichner, J. F., Faust, E. & Steuer, M. Rising variability in thunderstorm-related U.S. losses as a reflection of changes in large-scale thunderstorm forcing. *Wea. Climate Soc.* **5**, 317–331 (2013).
12. Long, J. A. & Stoy, P. C. Peak tornado activity is occurring earlier in the heart of 'Tornado Alley'. *Geophys. Res. Lett.* **41**, 6259–6264 (2014).
13. Lu, M., Tippett, M. & Lall, U. Changes in the seasonality of tornado and favorable genesis conditions in the central United States. *Geophys. Res. Lett.* **42**, 4224–4231 (2015).
14. Doswell, III C. A., Edwards, R., Thompson, R. L., Hart, J. A. & Crosbie, K. C. A simple and flexible method for ranking severe weather events. *Wea. Forecasting* **21**, 939–951 (2006).
15. Taylor, L. R. Aggregation, variance and the mean. *Nature* **189**, 732–735 (1961).
16. Eisler, Z., Bartos, I. & Kertész, J. Fluctuation scaling in complex systems: Taylor's law and beyond. *Adv. Phys.* **57**, 89–142 (2008).
17. Cohen, J. E., Xu, M. & Schuster, W. S. F. Stochastic multiplicative population growth predicts and interprets Taylor's power law of fluctuation scaling. *Proc. R. Soc. B* **280**, 20122955 (2013).
18. Malamud, B. D. & Turcotte, D. L. Statistics of severe tornadoes and severe tornado outbreaks. *Atmos. Chem. Phys.* **12**, 8459–8473 (2012).
19. Elsner, J. B., Jagger, T. H., Widen, H. M. & Chavas, D. R. Daily tornado frequency distributions in the United States. *Environ. Res. Lett.* **9**, 024018 (2014).
20. Cohen, J. E. & Xu, M. Random sampling of skewed distributions implies Taylor's power law of fluctuation scaling. *Proc. Natl Acad. Sci. USA* **112**, 7749–7754 (2015).
21. Vitolo, R., Stephenson, D. B., Cook, I. M. & Mitchell-Wallace, K. Serial clustering of intense European storms. *Meteor. Z.* **18**, 411–424 (2009).
22. Karremann, M. K., Pinto, J. G., Reyers, M. & Klawa, M. Return periods of losses associated with European windstorm series in a changing climate. *Environ. Res. Lett.* **9**, 124016 (2014).
23. Agee, E. & Childs, S. Adjustments in tornado counts, F-scale intensity, and path width for assessing significant tornado destruction. *J. Appl. Meteor. Climatol.* **53**, 1494–1505 (2014).
24. Mesinger, F. *et al.* North American regional reanalysis. *Bull. Am. Meteorol. Soc.* **87**, 343–360 (2006).
25. Davies, J. M. in *17th Conference on Severe Local Storms*, 107–111 (Amer. Meteor. Soc., St. Louis, MO, 1993).
26. Rasmussen, E. N. & Blanchard, D. O. A baseline climatology of sounding-derived supercell and tornado forecast parameters. *Wea. Forecasting* **13**, 1148–1164 (1998).
27. Tippett, M. K., Sobel, A. H. & Camargo, S. J. Association of U.S. tornado occurrence with monthly environmental parameters. *Geophys. Res. Lett.* **39**, L02801 (2012).
28. Tippett, M. K., Sobel, A. S., Camargo, S. J. & Allen, J. T. An empirical relation between U.S. tornado activity and monthly environmental parameters. *J. Clim.* **27**, 2983–2999 (2014).
29. Brooks, H. E., Lee, J. W. & Craven, J. P. The spatial distribution of severe thunderstorm and tornado environments from global reanalysis data. *Atmos. Res.* **67–68**, 73–94 (2003).
30. Brooks, H. E. Severe thunderstorms and climate change. *Atmos. Res.* **123**, 129–138 (2013).
31. Gilleland, E., Brown, B. G. & Ammann, C. M. Spatial extreme value analysis to project extremes of large-scale indicators for severe weather. *Environmetrics* **24**, 418–432 (2013).
32. Thompson, R. L., Mead, C. M. & Edwards, R. Effective storm-relative helicity and bulk shear in supercell thunderstorm environments. *Wea. Forecasting* **22**, 102–115 (2007).

Acknowledgements

M.K.T. was partially supported by a Columbia University Research Initiatives for Science and Engineering (RISE) award, the Office of Naval Research award N00014-12-1-091, NOAA's Climate Program Office's Modeling, Analysis, Predictions and Projections program award NA14OAR4310185, and the Willis Research Network. J.E.C. was partially supported by U.S. National Science Foundation grant DMS-1225529. The views expressed herein are those of the authors and do not necessarily reflect the views of any of the sponsoring agencies.

Author contributions

All the authors were involved in conducting the research and the scientific discussion equally throughout the study. The data were prepared by M.K.T. The manuscript was written, edited and reviewed by both the authors.

Additional information

Competing financial interests: The authors declare no competing financial interests.

13

European land CO_2 sink influenced by NAO and East-Atlantic Pattern coupling

Ana Bastos[1,2], Ivan A. Janssens[3], Célia M. Gouveia[2], Ricardo M. Trigo[2], Philippe Ciais[1], Frédéric Chevallier[1], Josep Peñuelas[4,5], Christian Rödenbeck[6], Shilong Piao[7], Pierre Friedlingstein[8] & Steven W. Running[9]

Large-scale climate patterns control variability in the global carbon sink. In Europe, the North-Atlantic Oscillation (NAO) influences vegetation activity, however the East-Atlantic (EA) pattern is known to modulate NAO strength and location. Using observation-driven and modelled data sets, we show that multi-annual variability patterns of European Net Biome Productivity (NBP) are linked to anomalies in heat and water transport controlled by the NAO–EA interplay. Enhanced NBP occurs when NAO and EA are both in negative phase, associated with cool summers with wet soils which enhance photosynthesis. During anti-phase periods, NBP is reduced through distinct impacts of climate anomalies in photosynthesis and respiration. The predominance of anti-phase years in the early 2000s may explain the European-wide reduction of carbon uptake during this period, reported in previous studies. Results show that improving the capability of simulating atmospheric circulation patterns may better constrain regional carbon sink variability in coupled carbon-climate models.

[1] Laboratoire des Sciences du Climat et de l'Environnement, LSCE/IPSL, CEA-CNRS-UVSQ, Université Paris-Saclay, F-91191 Gif-sur-Yvette, France. [2] Instituto Dom Luiz, IDL, Faculdade de Ciências, Universidade de Lisboa, Lisboa 1749-016, Portugal. [3] Department of Biology, University of Antwerp, Universiteitsplein 1, 2610 Wilrijk, Belgium. [4] CREAF, Cerdanyola del Vallès, Catalonia, 08193 Barcelona, Spain. [5] CSIC, Global Ecology Unit CREAF-CSIC-UAB, Cerdanyola del Vallès, Catalonia, 08193 Barcelona, Spain. [6] Max Planck Institute for Biogeochemistry, Jena 07701, Germany. [7] Department of Ecology, College of Urban and Environmental Sciences, Peking University 5 Yiheyuan Road, Haidian District, Beijing 100871, China. [8] College of Engineering, Mathematics and Physical Sciences, University of Exeter, Exeter EX4 4QF, UK. [9] Numerical Terradynamic Simulation Group, University of Montana, Missoula, Montana 59812, USA. Correspondence and requests for materials should be addressed to A.B. (email: ana.bastos@lsce.ipsl.fr).

Fundamental patterns of atmosphere-ocean variability are explained by teleconnections at the global[1] and regional[2-4] scales. At the global scale, the El-Niño/Southern-Oscillation has been shown to influence inter-annual variability in the global land carbon sink[5,6] with an evident fingerprint in atmospheric CO_2 growth rate fluctuations[7], highlighting their impact on ecosystem functioning. Hallet et al.[8] proposed that teleconnections explain ecological processes even better than single climate variables, because they influence simultaneously the range of weather variables that elicit interacting, and sometimes opposing, responses by ecosystems[9-12].

European and North-American climate is particularly influenced by the North-Atlantic Oscillation (NAO), which corresponds to a meridional dipole of sea-level pressure (SLP) variability between Iceland and the Azores that controls the location and strength of storm tracks in the North-Atlantic[3] (Fig. 1a). NAO drives patterns and extremes in temperature, precipitation, snow cover and wind in Europe, especially during winter[2,4], although its effects may propagate through the following seasons[13]. NAO is usually associated with warmer and wetter (colder and drier) winters in northern (southern) Europe during positive phase and approximately the reverse pattern in negative phase[2]. Some studies have analysed impacts of NAO on European ecosystems[14,15]. In the Iberian Peninsula, Carnicer et al.[16] found an increase in tree defoliation in response to a period of drought (1990–1995) that coincided with the persistence of positive NAO. However, other teleconnections are known to also influence the European climate, particularly the East-Atlantic (EA) pattern that has a similar configuration as NAO, albeit displaced southwards (Fig. 1a). Although EA and NAO indices are computed as independent variables (Fig. 1b), the EA pattern has been proposed to modulate the multi-decadal variability of the location and strength of the NAO dipole[17]. Comas-Bru and McDermott[4] showed that the combined analysis of NAO and EA more appropriately describes winter climate variability in Europe.

Northern Hemisphere ecosystems are important carbon sinks[18,19] that partially offset anthropogenic CO_2 emissions[20]. The inter-annual variability in the terrestrial carbon sink is poorly understood[21], hindering better future land-sink strength projections[7,22] and making regional trends in sinks difficult to detect and hence controversial[23,24]. Various estimates of European Net Biome Productivity (NBP) have shown that European ecosystems have been a CO_2 sink during the past decades (1980–2005)[25,26], although signs of saturation in temperate forest carbon uptake in the early 2000s have been reported[27]. Moreover, Piao et al.[28] have found a weakening relationship between temperature and growing season vegetation greenness in Northern ecosystems, especially in Europe, during 1982–2011, attributable to an increase in drought conditions. Their results highlight the importance of evaluating inter-annual variability of NBP within a framework that takes into account the co-variations of the climate variables that drive ecosystem dynamics.

Here we analyse the combined impact of NAO and EA on NBP over the European area extending to the Ural Mountains (Supplementary Fig. 1), using three state-of-the-art atmospheric CO_2 inversions: two versions of the Monitoring Atmospheric Composition and Climate—Interim Implementation (MACCII) inversion[29,30] and the Jena s81 v3.6 inversion[31] (Supplementary Table 1). Atmospheric inversions are known to provide consistent estimates of inter-annual variability of the land sink on large aggregated regions[19]. The uncertainty of the annual anomalies is expected to be lower than that of the long-term mean, since these errors on annual fluxes are positively correlated from 1 year to the next, and Europe is one of the regions with most robust inter-annual variability estimates[32]. As inversions present increasing uncertainty for smaller scale regions[33] we use a complementary bottom-up estimation of NBP from eleven Dynamic Global Vegetation Models (DGVMs)[20] (Supplementary Table 2). The DGVMs were forced with climate observations and include land-use changes (LUCs), but differ on a number of characteristics, which makes them independent estimations of NBP. These data sets now report more than 30 years of CO_2 flux variability, a timely opportunity to revisit the relationship between carbon fluxes and climate signals. Our results show that the combined NAO–EA variability explains variations in the European CO_2 sink from 1982 to 2012, through their control on heat and water-vapour transport towards the continent and consequent climate anomalies. We find that the European sink is enhanced only when the two modes are in negative phase due to enhanced photosynthesis, while for both anti-phase combinations NBP is kept below average through different responses of photosynthesis and respiration.

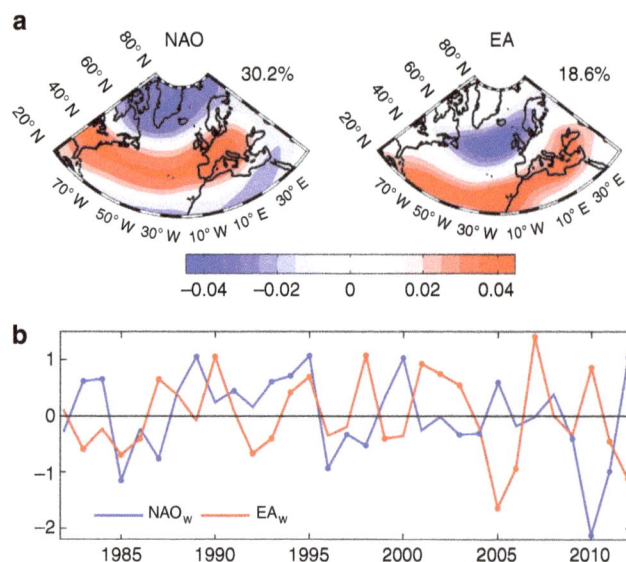

Figure 1 | Main large-scale atmospheric circulation patterns in Europe. (a) Spatial patterns of the two first components of 500 mb geopotential height (from NCEP/DOEII Reanalysis) variability over the North-Atlantic: NAO and EA teleconnections. (b) Time series of the winter composites of NAO (blue line) and EA (red line) from NOAA; circle markers indicate the positive and negative phases of each mode, defined by the upper and lower tercile thresholds (Supplementary Table 3).

Results

Net biome production variability. European NBP from inversions confirms that European ecosystems were on average a net CO_2 sink between 1982 and 2012 (Supplementary Fig. 2), in line with previous estimates[19,25], with strong inter-annual and multi-year variability (Fig. 2a). The late 1980s and early 1990s were characterized by generally positive NBP anomalies, that is, CO_2 uptake above average. A strong sink was observed during the years 1996–1997, followed by a decade of mostly below-average NBP, consistent with the satellite observations of Normalized Difference Vegetation Index (NDVI) data, a proxy for Gross Primary Productivity (GPP)[27]. From 2009 onwards, results from inversions suggest that European ecosystems became a stronger C sink again. The set of DGVMs suggests a weaker and less-variable European sink than the inversions (Supplementary Fig. 2), although the average NBP anomalies from these models

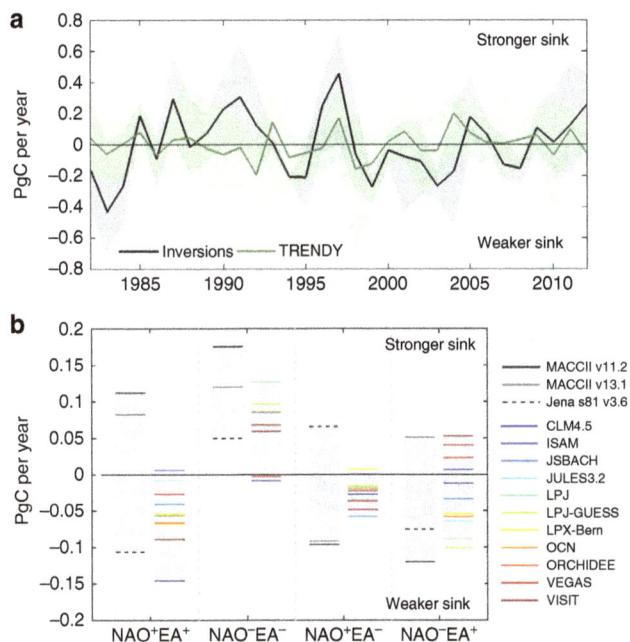

Figure 2 | Impact of NAO-EA variability on NBP. (**a**) Annual NBP anomalies integrated over Europe (bold lines) and corresponding spread, that is, the minimum and maximum NBP estimated by each set (shaded area), for the atmospheric transport model inversions (black) and the DGVMs from TRENDY project (green). (**b**) Average European NBP anomalies for the four NAO-EA composites (Supplementary Table 3), assessed by inversions (greyscale lines) and DGVMs (colour lines).

this large spread of the NBP response to current year's climate, we here analyse the multi-year average NBP values within the four NAO–EA phase composites (NAO^+EA^+, NAO^-EA^-, NAO^+EA^- and NAO^-EA^+), rather than individual years (Fig. 2b; Supplementary Table 3).

Our analysis reveals that the sole use of NAO does not suffice to explain NBP variability, because its impacts are strongly modulated by the sign of the EA. Inversions and DGVMs agree on significant differences in NBP composite anomalies during NAO^- years (Supplementary Tables 4 and 5), depending on the state of the EA. NBP is enhanced when EA is also negative, and is weaker than average when EA is positive (Fig. 2a; compare NAO^-EA^- with NAO^-EA^+). In fact, both sets of NBP estimates identify NAO^-EA^- as the single combination with consistent estimates of NBP enhancement (except two DGVMs), even for an extended record from 1950 to 2012 (Supplementary Fig. 5). During the other phase combinations, inversions and DGVMs report NBP anomalies of similar magnitude for the two anti-phase states (NAO^+EA^- and NAO^-EA^+), which are significantly lower than for NAO^-EA^-. For NAO^+EA^+, MACCII inversions disagree with DGVMs and Jena inversion on the sign of the anomalies. However, the limited number of years (only four) in this composite highlights the need for caution in the interpretation of NBP anomalies for this combination. It is worth pointing that inversions present different sensitivities to NAO–EA variability: while the sign of NBP anomaly from the two MACCII versions tends to depend on the phase/anti-phase dichotomy, NBP anomalies from the Jena inversion appear to respond more strongly to EA phases.

Mechanism. Having shown the need to consider both NAO and EA when assessing European CO_2 uptake variability, we now propose the main components of the physical mechanisms associated, that is, linking variations in atmospheric circulation and climate variables to the anomalies observed in NBP for each NAO–EA phase combination.

Variability in the land sink is linked to that in the teleconnections through the control of the latter on one or more of the meteorological variables influencing NBP. The four NAO–EA phases are characterized by distinct winter SLP anomaly patterns (December–April (Dec–Apr); Fig. 3a, black contours) and 500 mb geopotential height anomalies (Fig. 3b, black contours) during late winter/early spring. Although NAO clearly controls the sign of the North/South SLP anomaly dipole, with NAO^+ (NAO^-) imposing negative (positive) anomalies in the North and positive (negative) in the south, the EA modulates their strength and configuration, consistent with (ref. 4). During in-phase combinations, a sharp North/South pressure gradient is observed at the surface and in altitude, while during anti-phase combinations anomalies tend to be less intense and present a meandering configuration.

By influencing the strength and shape of the North-Atlantic jet, the interplay between NAO and EA controls the atmospheric transport of heat and humidity over the region (Fig. 3a,b; colours). The stronger pressure gradients during in-phase combinations create a channel of enhanced (weaker) eastward transport of heat and water-vapour spanning from the UK to eastern Europe and including Scandinavia, and reduced (higher) heat and moisture advection over the Mediterranean during NAO^+ (NAO^-). In contrast, during anti-phase periods, the weaker pressure gradients at the surface and in altitude leads to a northward displacement and attenuation of eastward heat flux, and water-vapour transport is close to average values in most of the continental Europe. The impact of these fluxes on cloud cover was evaluated as a means to estimate the combined variability of

remain in most years within the uncertainty range of the inversions and also present no significant trend (Fig. 2a).

Despite the lack of long-term NBP trend at continental scale, Supplementary Fig. 3 shows distinct temporal patterns in annual NBP anomalies in the four large regions encompassing most of continental Europe. The continental integral is mainly influenced by the dynamics in central Europe (33% and 34% of variance explained on average for inversions and DGVMs, respectively) and western Russia (29%, inversions). A principal component analysis (PCA) applied on NBP fields from inversions and DGVMs confirms the importance of central Europe and western Russia dynamics, by identifying a predominant variability pattern (explaining about 30% of spatio-temporal variance) with two centres located in these regions (Supplementary Fig. 4). Particularly, anomalies in central Europe dominate the multi-year pattern observed at the continental scale, while in western Russia NBP presents an increasing trend in the first decade followed by a peak in 1996–1997 and a stalling onwards (Supplementary Fig. 3). However, it must be kept in mind that atmospheric inversions do not perform as well at the regional scale as at the continental scale due to the limited, albeit still greater than in most regions, number of observation sites in Europe and to limitations of transport models.

Impact of NAO and EA on NBP. A large spread in NBP response to climate is expected from the natural variability of the climate conditions (for example, NAO^+ years differ in NAO strength and therefore in its impacts), as well as from carryover effects from previous year's climate conditions. For example, warm and moist conditions in 1 year promote organic matter decomposition, thereby increasing nutrient availability with impact on plant growth during subsequent years[12]. Because of

Figure 3 | Impact of NAO–EA variability on atmospheric circulation. Anomaly fields during extended winter (Dec–Apr) for each of the four NAO–EA phases of (**a**) sea-level pressure (black contours, in mb) and eastward water-vapour transport (colours); (**b**) 500-mb geopotential height (black contours, in g.p.m.) and heat transport (colours); (**c**) Cloud cover fraction.

precipitation and sunlight, both of which are relevant (with opposing effects) for vegetation growth during these months[9]. The patterns of winter cloud cover anomalies roughly match the ones of heat and vapour transport, with increased cloudiness in the regions affected by stronger heat and water flux anomalies. It must be noted that although precipitation anomalies are expected to be related to cloud cover, the dependency is not direct, since precipitation also depends on vertical instability of the atmosphere.

The differences in regional advection of heat and humidity transport associated with a given phase impose distinct and spatially heterogeneous climate conditions during late winter/early spring, which may propagate until summer (May–September) due to the memory effects of snow cover[13] and soil moisture[34] (Fig. 4). During NAO⁻EA⁻, the inhibition of heat and water-vapour transport in winter establishes very cold conditions over most of Europe, which lead to more precipitation falling as snow rather than rain, explaining the negative soil moisture anomalies and the generally higher-than-average snow depth registered in most areas. In the Mediterranean region, soil moisture is enhanced during December–April due to the stronger advection of water vapour and prevalence of cloudiness (Fig. 3). In contrast to the lower soil moisture during winter/spring, soil water was enhanced during spring/summer in the NAO⁻EA⁻ composite in most of central Europe (Fig. 4; Supplementary Table 6). This shift in water availability may be related to later snow melt and lower evaporative losses during the colder winter/spring periods. Higher soil moisture levels support stronger latent heat exchange and thus evaporative cooling in most regions, except the Iberian Peninsula, Scandinavia and the southern section of western Russia, in which dry conditions promote higher summer temperatures. For NAO⁺EA⁺ these patterns are approximately the opposite.

The meandering SLP pattern and the corresponding energy transport during anti-phase periods impose a north-east/south-west gradient (rather than north/south as during in-phase combinations) in temperature anomalies with warmer-than-average (colder) winter/spring conditions over Scandinavia and western Russia and cooler (warmer) temperatures in the western-Mediterranean sector during NAO⁺EA⁻ (NAO⁻EA⁺). The patterns of soil moisture for NAO⁺EA⁻ (NAO⁻EA⁺)

reflect the northward displacement and attenuation of water-vapour transport anomalies: most regions register close to average winter soil moisture, with drier (wetter) winter conditions being registered only in western Russia and Iberian Peninsula. Most of central Europe registers lower (higher) levels of snow cover, followed by dry (wet) soil conditions in summer. As for NAO⁻EA⁻, the snow-depth winter patterns generally match the ones observed in summer soil moisture and temperature anomalies, although the moisture–temperature coupling is most evident for NAO⁻EA⁺. The Iberian peninsula is also affected by very low winter cloud cover during NAO⁺EA⁻, which promotes higher evaporation rates, explaining the strong water deficits in winter.

The anomalies in NBP are related to two largely offsetting fluxes, photosynthesis and respiration, which respond differently to climate anomalies[12]. It is worth comparing the continental NBP response to NAO–EA phases with observations of vegetation greenness anomalies assessed by annually integrated NDVI fields (Fig. 5a) from the Global Inventory Modeling and Mapping Studies (GIMMS)[35], as well as the corresponding seasonal evolution (Fig. 5b, black lines). Since DGVMs allow partitioning the NBP anomalies into GPP and respiration responses, the seasonal dynamics of these two components of NBP is also evaluated (Fig. 5b, coloured lines). It must be noted that since the climate anomalies are spatially heterogeneous, and Europe is composed of diverse biomes, these responses must also be assessed at the regional scale (Supplementary Figs 6 and 7, Supplementary Table 4). Regional NBP anomalies from inversions and DGVMs are compared, although inversions present higher uncertainty for smaller regions and in most cases do not present significant results. At the regional scale, DGVMs are expected to better capture ecological variability patterns in response to climate anomalies (see Methods).

Inversions and most DGVMs consistently estimate enhanced NBP during NAO⁻EA⁻ years in all regions, except in the Iberian Peninsula, where inversions and DGVMs diverge. The clear positive NBP anomaly during NAO⁻EA⁻ cannot be explained solely by positive NDVI anomalies, since large areas in Europe present NDVI anomalies of small magnitude, and in many regions even slight browning. On the seasonal scale, the cold and dry winter and spring conditions affecting most of

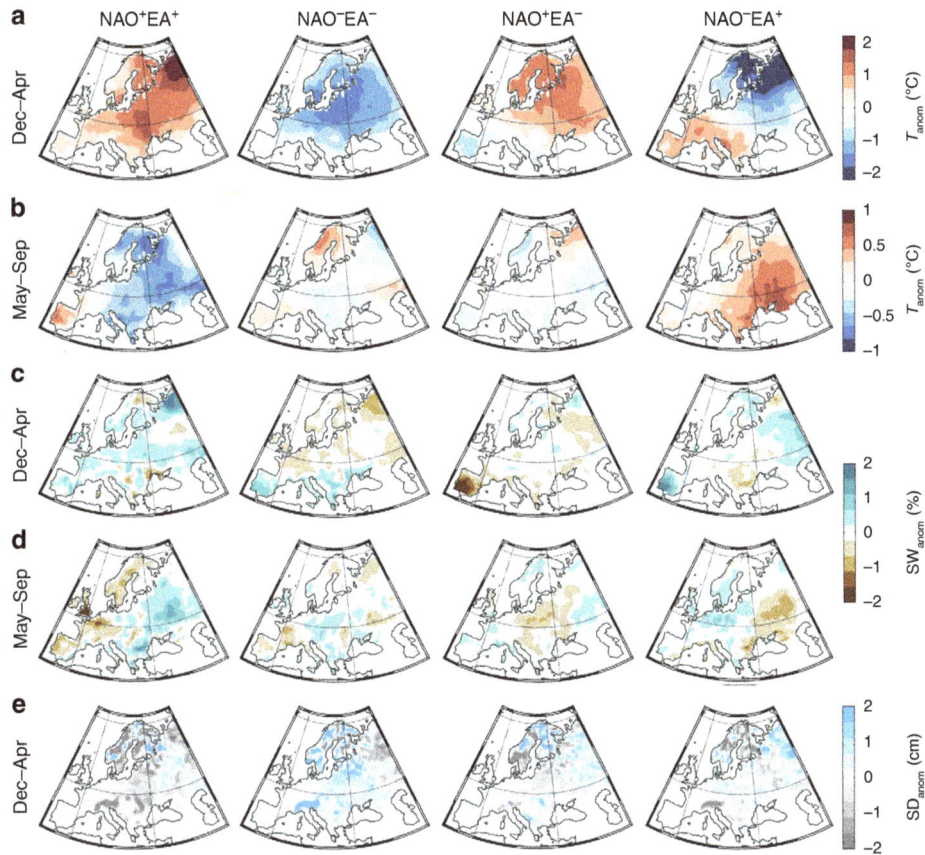

Figure 4 | Impact of NAO-EA variability on climate variables. Average anomalies in late winter/early spring (Dec–Apr) and summer (May–Sep) of temperature (T, (**a,b**)), soil water content (SW, (**c,d**)) and snow depth (SD, only Dec–Apr, (**e**)) for the four NAO–EA composites.

Europe impose negative NDVI anomalies and lower-than-average GPP (especially in western Russia; Fig. 5b; Supplementary Fig. 6). High water availability due to retarded snow melt in spring and reduced evaporative soil-water loss during the colder spring sustains increased GPP from late spring to early autumn, while respiration is kept close to average values due to the mild summer temperatures, especially in parts of central Europe and western Russia. Indeed, most DGVMs estimate a strong positive dependence of summer GPP in soil water in all regions except Scandinavia (Supplementary Table 7; Supplementary Fig. 8), while only four are able to capture the known dependence of GPP in winter water availability in the Iberian Peninsula[36]. The only two DGVMs that estimate negative NBP anomalies at continental scale for NAO^-EA^- (ISAM and VEGAS) present very weak control of water availability on GPP. The enhanced CO_2 uptake during the NAO^-EA^- phase combination is thus mostly due to enhanced GPP during the growing season, not to reduced respiration.

The highest positive integrated NDVI anomalies are observed for NAO^+EA^+ in most of Europe (except Iberia). However, the higher annual NDVI does not coincide with increased GPP, mainly because positive NDVI anomalies occur when radiation is sub-optimal (that is, late winter and early spring) and hence photosynthesis is weakly stimulated. DVGMs report higher GPP during spring, which offsets the increase in respiration during these warmer conditions. In summer, NDVI is about average, however, DGVMs estimate strong negative GPP and respiration anomalies, the former being stronger than the latter, and leading the negative NBP response (Fig. 2b). This is mainly due to the response observed in central Europe, where very dry

conditions lead to very low GPP, despite the low temperatures inducing lower-than-average respiration rates (Supplementary Figs 6 and 7).

Despite both anti-phase composites resulting in negative NBP anomalies, they diverge in the corresponding climate patterns (Fig. 4). For NAO^+EA^-, DGVMs (inversions are consistent, but not significant) report a sink reduction in all regions except western Russia, which matches the marked vegetation 'browning' and decrease in GPP during the year (Fig. 5) imposed by the cold and dry winter/spring in central Europe and the Mediterranean, especially in the Iberian Peninsula, and the drier summers in central Europe. This decrease in photosynthesis is accompanied by a smaller reduction in respiration, explaining the reduced NBP during this phase combination (Fig. 2; Supplementary Fig. 7). On the contrary, NAO^-EA^+ is characterized by overall high annual NDVI fields, associated with increased photosynthesis during spring (Fig. 5), promoted by warm and wet conditions over most of Europe, except in western Russia, where very cold late winter and spring inhibits vegetation activity (Supplementary Fig. 7; Supplementary Table 6). However, the GPP enhancement during spring is offset by greater increase in respiration during summer due to widespread positive temperature anomalies that reinforce, and are amplified by, soil dryness. This is especially evident in western Russia, where below average summer GPP is accompanied by enhanced respiration.

Discussion

Changes in NBP are the result of variation in land cover, in land management and/or in ecosystem activity. Changes in land cover

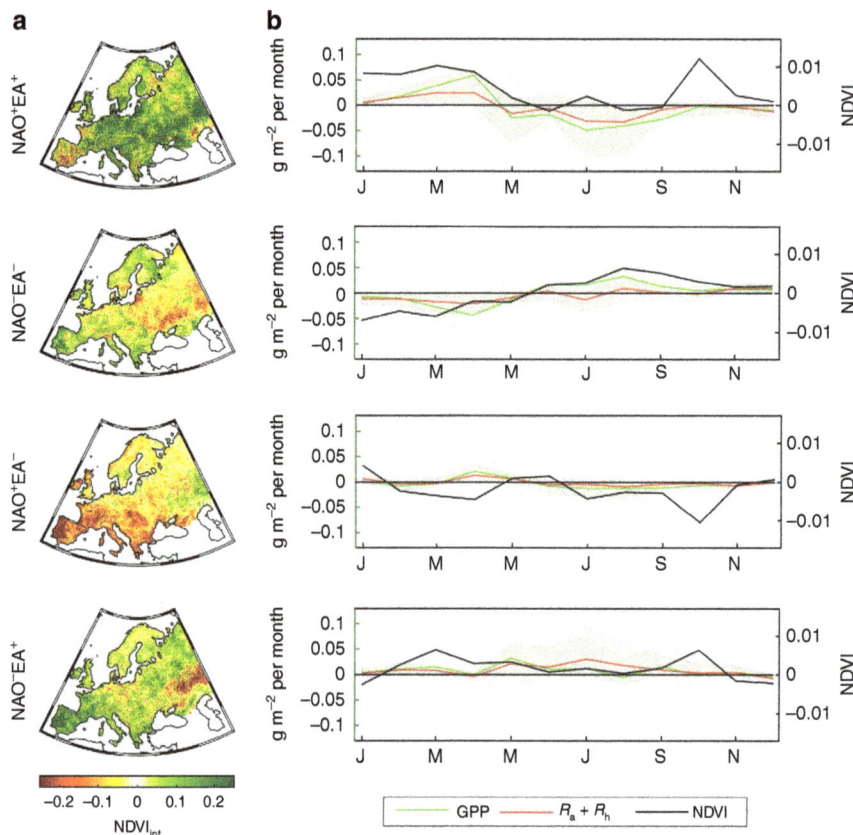

Figure 5 | Ecological response to NAO–EA phases. (a) Anomaly fields of annually integrated NDVI from GIMMS for each of the four NAO–EA composites; **(b)** seasonal anomalies of NDVI (black line, right yy-axis), GPP (green, left yy-axis) and aggregated autotrophic and heterotrophic respiration (red, left yy-axis). For GPP and respiration, bold line corresponds to the average of the eleven DGVMs and the shaded area indicates the model spread.

and land management are slow and not very intense in Europe, being more likely to induce long-term trends rather than short-term variation in NBP. On the inter-annual scale, NBP responds to a number of physical drivers which present co-variability patterns that are ultimately controlled by large-scale atmospheric circulation. Here, a mechanism driving part of the inter-annual and multi-year variability in the European sink is proposed, based on the interplay between the two main patterns of atmospheric circulation in the North-Atlantic sector (NAO and EA) and their control on heat and water-vapour transport across Europe.

The NBP enhancement during NAO^-EA^- conditions suggests that the highest NBP (largest uptake) should occur during periods with predominantly negative phases of both teleconnections. In fact, the largest NBP peak observed in 1996–1997 in the inversions corresponds to two consecutive years of NAO^-EA^-. The strong increase in NBP during 1985 also occurred during this combination of NAO and EA. In contrast, the years with lowest NBP are associated with years of opposing phases of NAO and EA: the first 3 years of the period (1982–1984) and 1998–1999, when a strong decrease in NBP is observed. Anti-phase combinations dominated the late 1990s and the 2000s, with some exceptions, with a period of 6 consecutive years of anti-phase lasting from 1998 to 2003, coinciding with a weaker sink over the European continent[11] (Fig. 2).

When computing future trends in NBP, the presence of multi-annual variability and trends in the most prominent modes of atmospheric circulation should be taken into account. In this context, anthropogenic forcing has been shown to lead to more frequent positive phases of NAO^{37}, associated with lower NBP,

especially when EA is negative. However CMIP3 and CMIP5 models predict opposite trends in the NAO sign over the 21st century under similar scenarios[38]. Shepherd[39] pointed out that atmospheric dynamics still constitutes an important source of uncertainty in the state-of-the-art CMIP5 models, and showed that the climate-change response of the wintertime North-Atlantic jet largely differs among models, both in magnitude and sign. Other works have shown the influence of other teleconnections such as the Scandinavian Pattern[17] or the Pacific Decadal Oscillation[40] on North-Atlantic sea-level pressure, adding further complexity to the study of long-term NBP variability.

This study shows that the non-stationary relationship between teleconnections may strongly affect the relationships found between climate and ecosystem activity. This highlights the importance of correctly reproducing the mid-latitude atmospheric circulation patterns in Earth System Models to account for inter-annual and decadal variability in the land carbon sink. More long-term high-quality observation-driven data sets are needed to further improve our understanding of the multi-year variability of the land sink and its relation to climate variability.

Methods

Inversions. Atmospheric CO_2 inversions estimate spatial distribution of surface CO_2 fluxes using a top-down approach, in which surface fluxes are estimated from atmospheric CO_2 concentration measurements across the globe. An atmospheric transport model that represents the global atmospheric circulation is required, as well as prior-information about the fluxes. Flux estimates over land include the carbon exchange in ecosystems, LUC related fluxes and forest fire CO_2 emissions as well as fossil-fuel emissions. To estimate NBP, the latter are subtracted from the flux estimates.

We use monthly CO_2 fluxes of two versions of the atmospheric inversion of the MACCII from the LSCE[29,30]. Version 11.2 covers the period from 1979 until 2011 and is used here from January 1982 onwards. It has a 2.5×3.75 (latitude, longitude) spatial resolution. Version 13.1 has a slightly higher latitudinal resolution (1.9) and covers year 2012 as well (Supplementary Table 1). Apart from the archive length, the main difference between v11.2 and v13.1 is the number of vertical layers in the underlying atmospheric transport model, the LMD General Circulation Model (LMDZ)[29], that was doubled in the latter (39 versus 19), thus allowing a better representation of the variability of CO_2 in the planetary boundary layer, where the assimilated measurements are located.

Since MACC inversions are based on a large but variable number of stations during the study period (134 in total), with fewer observations in the beginning of the time series, part of the variability observed on the estimated fluxes may be affected by the changes in the number of sites[41]. Thus, we use the Jena s81 (v3.6) inversion that is based on fewer stations and is provided at 4×5 spatial resolution (Supplementary Table 1), but uses a consistent set of sites for the 1982–2012 period. Jena inversion computes NBP on each grid cell using the atmospheric transport model TM3, as described in ref. 31.

To subtract fossil fuel emissions, MACCII inversions use the EDGAR3.2 FastTrack 2000 emission database[42] with scaling factor to account for trends: MACC uses the annual global totals of the Carbon Dioxide Information Analysis Center and prescribes an increase of 1.4% per year after 2001. Jena inversion uses EDGAR 4.2, with FastTrack 2010 for 2009 and 2010, extrapolation based on British Petrol global totals for 2011 and 2012, and 2% year increase afterwards.

Uncertainties of annual NBP at the continental scale are given, for each model, in Supplementary Table 1. Uncertainties of the annual fluxes account for errors in the prior fluxes, network configuration (number of sites, Supplementary Fig. 1) and in the transport model. The uncertainty of the annual anomalies is expected to be lower than that of the absolute fluxes, since these errors are positively correlated from 1 year to the next[32]. We use two inversions computed with the same model but variable number of sites, and a second with a fixed number of observations, and use the spread between inversions as an indicator of the uncertainty in the anomalies.

In order to test the available signals from the smaller number of sites in Europe used by the Jena s81 inversion, we performed three synthetic-data runs (1982–2010) by transporting net ecosystem exchange (NEE) fields as simulated by the Biome-BGC DGVM in the atmosphere using the same transport model as in the Jena inversion (v3.7), though on coarser 10×8 spatial resolution. To create pseudo data, modelled atmospheric CO_2 is sampled at the same locations and times as the real data for three different station sets: as in Jena s81 (14 sites) and s90 (25 sites) as well as mostly the same sites used for MACCII inversion (Supplementary Table 2). The CO_2 concentrations sampled in this way were inverted back using the Jena scheme (setting ocean and fossil fuel priors to zero). As the same transport model is used for synthetic data and inversion, results are not contaminated by model errors. By comparing with the original fluxes (Supplementary Fig. 9), the sensitivity of the inversion skill to the site network can then be evaluated. Despite the presence of discrepancies, the run with fewer stations (Jena s81) is able to capture the variability patterns of the CO_2 fluxes and differences in the magnitude of the fluxes are on average 0.04–0.05 PgC per year for each of the three runs.

Dynamic global vegetation models. DGVMs simulate the main processes of vegetation dynamics and decomposition associated with energy, water and carbon balances at the ecosystem scale and provide a bottom-up approach to evaluate NBP variability, as well as the corresponding GPP and respiration components.

The TRENDY project[43] compiles outputs from a group of DGVMs to evaluate trends and drivers in land-atmosphere carbon exchange[20]. We use 11 DGVMs (Supplementary Table 2) from simulation S3, in which all models are forced by the same values of changing atmospheric CO_2 concentrations from ice core data and observations, historical climate observations from the CRU-NCEP v4 data set and LUC from the History Database of the Global Environment (HYDE) database of human-induced land-cover changes[44].

Supplementary Table 2 provides a summary of the characteristics of the 11 DGVMs used in simulation S3, which is performed over the period 1860–2012. All models include deforestation, afforestation and regrowth, but differ in the way they represent disturbances (fire), nutrient limitation and do not realistically simulate land-management practices or crop seasonality, which strongly interplays with climate in determining NBP of European ecosystems. Nevertheless, DGVMs provide an independent data set to evaluate the variability of the land carbon sink. NBP from DGVMs was selected for two periods: the common period with the inversions and satellite data (1982–2012) for the main analysis, and the common period with the NAO and EA indices (1950–2012), except for JSBACH whose record ends in 2005. To partition NBP, continental and regional GPP and respiration (computed as the sum of autotrophic and heterotrophic respiration components) anomalies from the DGVMs were also evaluated between 1982 and 2012.

NDVI. Biweekly NDVI fields over 1982–2012 from the GIMMS NDVI data set, provided at 8 km spatial resolution[35]. NDVI values were integrated over each year on a pixel basis from the biweekly fields. NDVI anomalies were then computed as

the departure from the average annual integral values, and used as a proxy for GPP anomalies[28].

Monthly NBP values from inversions and DGVMs were first deseasonalized (mean seasonal-cycle removed) on a pixel basis to compute monthly anomaly and integrated over the year for annual NBP anomaly fields, for the European continent and for the regional boxes (Supplementary Fig. 3). A positive (negative) sign of NBP anomalies correspond to either an enhanced (weaker) sink or a smaller (increased) source. To evaluate the main pattern of NBP variability, a PCA was performed on NBP fields for inversions and DGVMs. Since results from PCA depend on the resolution of the data set, all fields were resampled to a common 1° spatial resolution. The first principal component of NBP variability (PC_1), the corresponding empirical orthogonal function (EOF_1) and explained variance were selected for each inversion and model. EOF_1 presented in all cases dipolar pattern. The consistency of these patterns was compared by locating the corresponding centres of action for each data set (Supplementary Fig. 4).

All climate variables used for this analysis were extracted from ERA-Interim reanalysis[45]: vertical integral of eastward transport water vapour (VT, in $g\,km^{-1}\,s^{-1}$) and heat (HT, in $MW\,km^{-1}$), fraction of cloud cover (%), SLP (in mb), 500-mb geopotential height (z500, in geopotential metres, g.p.m.), average air temperature at 2 m (T, in °C), volumetric soil-water content (SW, in % of volume) in the top layer[46] and average snow depth (SD, in cm). The data comprise monthly average fields from 1982 to 2012, at 0.75 spatial resolution.

As complementary information about soil-water conditions, monthly values of the self-calibrated Palmer Drought Severity Index (PDSI)[47], provided by the NOAA/OAR/ESRL PSD, Boulder, CO, USA at 2.5 spatial resolution was used. Since PDSI presented regional dependence (negative for Iberian Peninsula and central Europe and positive for western Russia and Scandinavia), we estimated drought conditions as the PDSI departure from the regional average (Supplementary Table 6).

Atmospheric circulation. We focus our analysis on the impacts of winter state Dec-Feb (DJF) of two teleconnections, using the monthly indices from NOAA's Climate Prediction Centre, which are calculated by rotated PCA of monthly means of 500-mb geopotential height anomalies from NCEP/DOE II (ref. 48) as described in (ref. 49). This procedure allows the calculation of orthogonal (that is, non-correlated) indices for each month. The first mode corresponds to the NAO pattern, the leading mode of SLP variability in the North Atlantic and the main climate teleconnection affecting European weather. The second mode corresponds to EA circulation pattern. These modes affect more significantly European weather during winter[2,50]. For visualization of the spatial patterns associated with the two teleconnections used in this work, we performed a PCA on standardized monthly means of 500-mb geopotential height from NCEP/DOEII Reanalysis over the North-Atlantic region between 20 to 80 N and 90 W to 50 E for the winter months (DJF) as in ref. 49.

The phases of an index are defined as those exceeding the lower (negative) or upper (positive) terciles, and identified in superscript (for example, NAO^+ for a positive phase of winter NAO). To evaluate the combined impacts of both teleconnections, we identify four subsets of years corresponding to the four possible NAO-EA phase combinations, together with a neutral composite (Supplementary Table 3).

Regional fluxes. Since data sets are provided at very different spatial resolutions (Supplementary Tables 1 and 2), and because inversions perform better on larger scales, we consider regionally integrated NBP from inversions and DGVMs for the European continent defined in the TransCom inter-comparison project[51] (Supplementary Fig. 1). We additionally define four large regions encompassing most of Europe: Iberian Peninsula extending from $-11°$ to 3.5° E and 34° to 44° N; continental central Europe encompassing the region between $-5°$ to 25° E and 44° to 53° N, but excluding Great Britain; western Russia and eastern Europe extending from 29° to 60° E and 46° to 62° N; Scandinavia covering the region between 4.5° to 29° E and 56° to 71° N (Supplementary Fig. 1).

To evaluate the skill of inversions in capturing regional fluxes, we used the same synthetic runs from Jena inversion to evaluate: (i) the ability of the inversion to reproduce fluxes in large regions over Europe; (ii) the influence of the observational network. Despite the very coarse resolution, results (Supplementary Fig. 10) indicate that inversions do have moderate skill in distinguishing sub-regions within Europe. As expected from the better observational constrain, the skill increases with increasing number of stations. Nevertheless, even the s81 run (with few sites) is able to capture part of the regional differences, especially in eastern, central and western Europe. In south-western Europe (corresponding to the Iberian Peninsula), the inversion is not able to reproduce the fluxes, as expected from the very low observational constraint (only two sites in the complete MACCII set, located in the Pyrenees).

NBP response to NAO-EA. All data fields were selected for the study region (Supplementary Fig. 1). Continental and regional NBP anomalies were calculated for each inversion and DGVM and evaluated for each of the four NAO-EA combinations (Fig. 2). In the case of MACC v11.2 inversion and JSBACH, the data sets do not encompass the whole period, therefore, the composites were calculated with fewer years. To partition annual NBP anomalies, continental and regional GPP and total respiration (autotrophic plus heterotrophic respiration components) were calculated at monthly scale, and compared with the corresponding NDVI anomalies (Supplementary Fig. 7).

For each composite, a one-sided analysis of variance (ANOVA) was performed to test the significance of the average anomaly value on the continental and regional scales (Supplementary Table 4), separately for inversions and DGVMs. Absolute NBP anomalies for each composite are easier to associate with NBP variations from year to year. However, it is worth assessing whether the inversions (or DGVMs) agree on the relative response, that is, the difference in NBP anomalies between two NAO-EA combinations. Therefore, a two-sided ANOVA was carried out to test the difference between NBP anomaly estimates for pairs of the four combinations (Supplementary Table 5), on the continental scale. Supplementary Tables 4 and 5 also include results of the ANOVA analyses for the neutral composite. To assess whether relationships hold for longer periods, the same analysis was extended for the period 1950–2012 using DGVMs (Supplementary Fig. 5).

Since the synthetic runs of Jena inversion present some dependence of absolute fluxes in the observational network, it is worth assessing how the number of sites may influence the results of NBP anomalies in response to NAO–EA variations. As the observational network has increased with time, Jena inversion is provided for different periods using an increasing number (always constant for each data set) of sites. Here we compare the other two inversions from Jena v3.6 that still encompass at least 20 years and use the same model but more sites: s85 (1985–2012, 19 sites), s90 (1990–2012, 25 sites). We also performed a run with the newer version of Jena inversion (v3.7) using the MACCII sites, at the same resolution as Jena s81 v3.6. The corresponding anomalies for each NAO–EA combination are presented in Supplementary Fig. 11 and show that, despite results depending on the size of the observational network, the response of continental NBP to the phases of NAO and EA for the data sets with more observations is consistent with the anomalies estimated by the two MACCII inversions and Jena s81 v3.6. In most cases, the inversion run using the MACCII sites is closer to the anomalies estimated by MACCII, highlighting the dependence of the anomalies on the observational constrains, but also the importance of other sources of uncertainty, for example, differences in the transport models.

References

1. McPhaden, M. J., Zebiak, S. E. & Glantz, M. H. ENSO as an integrating concept in earth science. *Science* **314**, 1740–1745 (2006).
2. Trigo, R. M., Osborn, T. J. & Corte-Real, J. M. The North Atlantic Oscillation influence on Europe: climate impacts and associated physical mechanisms. *Clim. Res* **20**, 9–17 (2002).
3. Scaife, A. a., Folland, C. K., Alexander, L. V., Moberg, A. & Knight, J. R. European climate extremes and the North Atlantic Oscillation. *J. Clim.* **21**, 72–83 (2008).
4. Comas-Bru, L. & McDermott, F. Impacts of the EA and SCA patterns on the European twentieth century NAO-winter climate relationship. *Quart. J. R. Meteorol. Soc.* **140**, 354–363 (2013).
5. Behrenfeld, M. J. *et al.* Biospheric primary production during an ENSO transition. *Science* **291**, 2594–2597 (2001).
6. Bastos, A., Running, S. W., Gouveia, C. & Trigo, R. M. The global NPP dependence on ENSO: La Niña and the extraordinary year of 2011. *J. Geophys. Res. Biogeosci.* **118**, 1247–1255 (2013).
7. Friedlingstein, P. & Prentice, I. Carbon-climate feedbacks: a review of model and observation based estimates. *Curr. Opin. Environ. Sustainability* **2**, 251–257 (2010).
8. Hallett, T. B. *et al.* Why large-scale climate indices seem to predict ecological processes better than local weather. *Nature* **430**, 71–75 (2004).
9. Nemani, R. R. *et al.* Climate-driven increases in global terrestrial net primary production from 1982 to 1999. *Science* **300**, 1560–1563 (2003).
10. Angert, A. *et al.* Drier summers cancel out the CO_2 uptake enhancement induced by warmer springs. *Proc. Natl Acad. Sci. USA* **102**, 10823–10827 (2005).
11. Piao, S. *et al.* Net carbon dioxide losses of northern ecosystems in response to autumn warming. *Nature* **451**, 49–52 (2008).
12. Xia, J. *et al.* Terrestrial carbon cycle affected by non-uniform climate warming. *Nat. Geosci.* **7**, 173–180 (2014).
13. Barriopedro, D., García-Herrera, R. & Hernández, E. The role of snow cover in the Northern Hemisphere winter to summer transition. *Geophys. Res. Lett.* **33**, L14708 (2006).
14. Gouveia, C., Trigo, R. M., DaCamara, C. C., Libonati, R. & Pereira, J. M. C. The North Atlantic Oscillation and European vegetation dynamics. *Int. J. Clim.* **28**, 1835–1847 (2008).
15. Peters, W. *et al.* Seven years of recent European net terrestrial carbon dioxide exchange constrained by atmospheric observations. *Global Change Biol.* **16**, 1317–1337 (2010).
16. Carnicer, J. *et al.* Widespread crown condition decline, food web disruption, and amplified tree mortality with increased climate change-type drought. *Proc. Natl Acad. Sci. USA* **108**, 1474–1478 (2011).
17. Moore, G. W. K., Renfrew, I. A. & Pickart, R. S. Multidecadal mobility of the North Atlantic Oscillation. *J. Clim.* **26**, 2453–2466 (2013).
18. Pan, Y. *et al.* A large and persistent carbon sink in the world's forests. *Science* **333**, 988–993 (2011).
19. Peylin, P. *et al.* Global atmospheric carbon budget: results from an ensemble of atmospheric CO_2 inversions. *Biogeosciences* **10**, 6699–6720 (2013).
20. Le Quéré, C. *et al.* The global carbon budget 1959-2011. *Earth Syst. Sci. Data* **5**, 165–185 (2013).
21. Heimann, M. & Reichstein, M. Terrestrial ecosystem carbon dynamics and climate feedbacks. *Nature* **451**, 289–292 (2008).
22. Cox, P. M. *et al.* Sensitivity of tropical carbon to climate change constrained by carbon dioxide variability. *Nature* **494**, 341–344 (2013).
23. Zhao, M. & Running, S. W. Drought-induced reduction in global terrestrial net primary production from 2000 through 2009. *Science* **329**, 940–943 (2010).
24. Ahlström, A., Miller, P. A. & Smith, B. Too early to infer a global NPP decline since 2000. *Geophys. Res. Lett.* **39**, L15403 (2012).
25. Janssens, I. A. *et al.* Europe's terrestrial biosphere absorbs 7 to 12% of European anthropogenic CO_2 emissions. *Science* **300**, 1538–1542 (2003).
26. Schulze, E. D. *et al.* Importance of methane and nitrous oxide for Europe's terrestrial greenhouse-gas balance. *Nat. Geosci.* **2**, 842–850 (2009).
27. Piao, S. *et al.* Changes in satellite-derived vegetation growth trend in temperate and boreal Eurasia from 1982 to 2006. *Global Change Biol.* **17**, 3228–3239 (2011).
28. Piao, J. *et al.* Evidence for a weakening relationship between interannual temperature variability and northern vegetation activity. *Nat. Commun.* **5**, 5018 (2014).
29. Chevallier, F. *et al.* CO_2 surface fluxes at grid point scale estimated from a global 21 year reanalysis of atmospheric measurements. *J. Geophys. Res.* **115**, D21307 (2010).
30. Chevallier, F. *et al.* Toward robust and consistent regional CO_2 flux estimates from in situ and spaceborne measurements of atmospheric CO_2. *Geophys. Res. Lett.* **41**, L058772 (2014).
31. Rödenbeck, C. *Estimating CO_2 sources and sinks from atmospheric mixing ratio measurements using a global inversion of atmospheric transport.* Technical Report No. 6 (Max Planck Institute for Biogeochemistry, Jena, 2005) Available at http://www.bgc-jena.mpg.de/mpg/websiteBiogeochemie/Publikationen/Technical_Reports/tech_report6.pdf.
32. Baker, D. F. *et al.* TransCom 3 inversion intercomparison: impact of transport model errors on the interannual variability of regional CO_2 fluxes, 1988-2003. *Global Biogeochem. Cycles* **20**, GB1002 (2006).
33. Kaminski, T. & Heimann, M. Inverse modeling of atmospheric carbon dioxide fluxes. *Science* **294**, 259–259 (2001).
34. Quesada, B., Vautard, R., Yiou, P., Hirschi, M. & Seneviratne, S. I. Asymmetric European summer heat predictability from wet and dry southern winters and springs. *Nat. Clim. Change* **2**, 736–741 (2012).
35. Tucker, C. *et al.* An extended AVHRR 8km NDVI dataset compatible with MODIS and SPOT vegetation NDVI data. *Int. J. Remote Sens.* **26**, 4485–4498 (2005).
36. Peñuelas, J. *et al.* Complex spatiotemporal phenological shifts as a response to rainfall changes. *New Phytol.* **161**, 837–846 (2004).
37. Gillett, N. P., Zwiers, F. W., Weaver, A. J. & Stott, P. A. Detection of human influence on sea-level pressure. *Nature* **422**, 292–294 (2003).
38. Cattiaux, J., Douville, H. & Peings, Y. European temperatures in CMIP5: origins of present-day biases and future uncertainties. *Clim. Dyn.* **41**, 2889–2907 (2013).
39. Shepherd, T. G. Atmospheric circulation as a source of uncertainty in climate change projections. *Nat. Geosci* **7**, 703–708 (2014).
40. Trenberth, K. E., Fasullo, J. T., Branstator, G. & Phillips, A. S. Seasonal aspects of the recent pause in surface warming. *Nat. Clim. Change* **4**, 911–916 (2014).
41. Rödenbeck, C., Houweling, S., Gloor, M. & Heimann, M. CO_2 flux history 1982-2001 inferred from atmospheric data using a global inversion of atmospheric transport. *Atmos. Chem. Phys.* **3**, 1919–1964 (2003).
42. Olivier, J. G. J. & Berdowski, J. J. M. *The Climate System* 33–78 (Swets & Zeitlinger Publishers, 2001).
43. Sitch, S. *et al.* Trends and drivers of regional sources and sinks of carbon dioxide over the past two decades. *Biogeosci. Discuss.* **10**, 20113–20177 (2013).
44. Klein Goldewijk, K., Beusen, A., van Drecht, G. & de Vos, M. The HYDE 3.1 spatially explicit database of human-induced global land-use change over the past 12.000 years. *Global Ecol. Biogeogr.* **20**, 73–86 (2011).
45. Dee, D. P. *et al.* The ERA-Interim reanalysis: configuration and performance of the data assimilation system. *Quart. J. R. Meteorol. Soc.* **137**, 553–597 (2009).
46. Balsamo, G. *et al.* A revised hydrology for the ECMWF model: verification from field site to terrestrial water storage and impact in the integrated forecast system. *J. Hydrometeorol.* **10**, 623–643 (2009).
47. Dai, A. Characteristics and trends in various forms of the Palmer Drought Severity Index during 1900-2008. *J. Geophys. Res. Atmospheres* **116**, D12115 (2011).
48. Kanamitsu, M. *et al.* NCEP/DOE AMIP-II Reanalysis (R-2). *Bull. Am. Meteorol. Soc.* **83**, 1631–1643 (2002).
49. Barnston, A. G. & Livezey, R. E. Classification, seasonality and persistence of low-frequency atmospheric circulation patterns. *Mon. Weather Rev.* **115**, 1083–1126 (1987).

50. Hurrell, J. W. Decadal Trends in the North Atlantic Oscillation: regional temperatures and precipitation. *Science* **269,** 676–679 (1995).
51. Gurney, K. R. *et al.* Towards robust regional estimates of CO_2 sources and sinks using atmospheric transport models. *Nature* **415,** 626–630 (2002).

Acknowledgements

Ana Bastos was partially funded by the Portuguese Foundation for Science and Technology (SFRH/BD/78068/2011). Célia M. Gouveia and Ricardo M. Trigo were supported by QSECA (PTDC/AAGGLO/4155/2012). This research was also supported by the European Research Council Synergy grant ERC-2013-SyG 610028 IMBALANCE-P.

Author contributions

A.B. conceived the study, conducted the analysis and wrote the paper; I.A.J., R.M.T. and C.M.G. conceived the study and supervised the project; S.P., F.C. and C.R. developed the data sets and provided technical and conceptual advice; C.R. also performed the sensitivity tests for Jena inversion. P.C, J.P., P.F. and S.W.R. provided expert advice on the different topics. All authors contributed to the revision of the paper.

Additional information

14

Contrasting scaling properties of interglacial and glacial climates

Zhi-Gang Shao[1] & Peter D. Ditlevsen[2]

Understanding natural climate variability is essential for assessments of climate change. This is reflected in the scaling properties of climate records. The scaling exponents of the interglacial and the glacial climates are fundamentally different. The Holocene record is mono-fractal, with a scaling exponent $H \sim 0.7$. On the contrary, the glacial record is multifractal, with a significantly higher scaling exponent $H \sim 1.2$, indicating a longer persistence time and stronger nonlinearities in the glacial climate. The glacial climate is dominated by the strong multi-millennial Dansgaard–Oeschger (DO) events influencing the long-time correlation. However, by separately analysing the last glacial maximum lacking DO events, here we find the same scaling for that period as for the full glacial period. The unbroken scaling thus indicates that the DO events are part of the natural variability and not externally triggered. At glacial time scales, there is a scale break to a trivial scaling, contrasting the DO events from the similarly saw-tooth-shaped glacial cycles.

[1] Guangdong Provincial Key Laboratory of Quantum Engineering and Quantum Materials, SPTE, South China Normal University, Guangzhou 510006, China.
[2] Centre for Ice and Climate, Niels Bohr Institute, University of Copenhagen, Juliane Maries Vej 30, Copenhagen 2100, Denmark. Correspondence and requests for materials should be addressed to P.D.D. (email: pditlev@nbi.ku.dk).

The climate system is characterized by complex interactions between atmosphere, oceans, ice, landmasses and the biosphere over a large range of temporal and spatial scales. For understanding natural climate variability and the character of climate change, assessing correlation and persistence times is important. These are reflected in the scaling properties of the climatic records. Scaling was first noted in the seminal work by Hurst[1] on reservoir capacity and runoff in the Nile. Hereafter, it was realized that time series of complex systems such as the climate system are characterized by fractal[2] or even multifractal[3] scaling properties. It has been a long-standing discussion to which extent the fractal nature of the climate dynamics is universal or if it is more specific to the processes and range of scales observed[3–6].

To identify the underlying dynamics reflected in scale breaks or robust scaling relations, records covering a large range of temporal scales are necessary. On the atmospheric variability, a range of instrumental records of temperatures and other meteorological parameters have been investigated[3,4,7]. On short time scales the turbulent nature of the atmospheric fields is measured in airplane and drop-sonde campaigns. From these measurements, multifractal scaling of the fields has been reported[8].

On longer time scales, instrumental records exist for ~ 150 years, typically with daily or twice daily resolution. By filtering out the annual cycle, the scaling properties of temperature variations, covering four to five decades, have been investigated. These indicate universal persistence laws for atmospheric variability[4], where station data from around the globe shows monofractal (Hurst)-scaling exponents around 0.7. This is significantly different from the value 0.5 characteristic for a trivial white noise process. On even longer time scales we rely on proxy records, where the ice core records are especially suited, as they can be understood as high-resolution sedimentation records from the atmosphere. It was in ice core records that it was first realized that the glacial climate was dominated by millennial scale instabilities, the Dansgaard–Oeschger (DO) events[9]. These events occur all the way from the last inception through to the termination of the last glacial period. One striking feature although is that in the period around the last glacial maximum (LGM, 27–15 kyr) the record only contains a single small event (DO2).

Here we find, by analysing climate periods separately, that the Holocene and the glacial climates have distinctly different scaling properties. The Holocene is monofractal with a scaling exponent $H \sim 0.7$, whereas the glacial climate is multifractal with $H_2 \sim 1.2$. The longer persistence time in the glacial period is expected, owing to the presence of the pronounced millennium scale DO events. However, the same scaling is found for the LGM period, indicating that the DO events are not the cause and should be seen as an intrinsic part of the glacial climate. The DO events have a characteristic saw-tooth shape in the records, with rapid warming and slow cooling, similar to the shape of the externally forced glacial cycles with fast terminations and slow inceptions. By analysing the much longer 5 Myr ocean sediment climate record[10] and the 800-kyr Antarctic EPICA (European Project for Ice Coring in Antarctica) ice core record[11], we find a scale break around 20 kyr, such that on glacial time scales (> 20 kyr) we have a trivial scaling $H = 0.5$. We might speculate that this reflects that DO events are internally generated, whereas glacial cycles are externally forced. It is noteworthy that for a trivial red noise process, the so-called Ornstein–Uhlenbeck process, the scaling spectrum shows a cross-over from $H = 3/2$ for time scales shorter than the correlation time to $H = 1/2$ for time scales longer than the correlation time. The cross-over time scale indicates the internal time scale of built-up of the large ice sheets.

Results

The climate record. To determine to which extent the scaling behaviour extends to scales beyond the length of the instrumental temperature records, we thus rely on paleoclimatic proxies. The stable isotope ($\delta^{18}O$ and δD) records obtained from the Greenland and Antarctic ice sheets constitute such temperature proxies, with a sufficiently linear relationship to the average atmospheric temperature mixed with an independent noise, which the scaling properties of the atmospheric temperature can be assumed to be reflected in the record[12].

The issue of nonlinearity and multifractality on multi-glacial cycle time scales has been addressed before in the analysis of the Antarctic Vostok record[13] and comparison with Greenland Icecore Project (GRIP), Greenland Ice Sheet Project (GISP) (Greenland) and Taylor dome (Antarctica) ice core records. The analysis show slightly different scaling exponents between the different ice core records, indicating that dating accuracy and period analysed are important for the results (GRIP and GISP cores should, due to close proximity, give the same result). The influence of chronology on the scaling properties is confirmed by an analysis of the North Greenland Icecore Project (NGRIP) record[14,15] (Fig. 1c), where the scaling properties of the recent 2,000-year $\delta^{18}O$ record is different from the properties of the record older than 2,000 years[16]. This should be expected, as the climate is influenced by different processes operating at different time scales. The NGRIP ice core[14] has been dated with unprecedented accuracy over the past 122 kyr (ref. 15) This enables us to accurately calculate the scaling properties for the different climate states separately.

On multi-millennial time scales, it is known that glacial cycles are linked, in a nonlinear manner[13], to periodic and quasi-periodic changes in the insolation (incoming solar radiation) from variations in Earth's orbit around the Sun. The climate response is ~ 100 kyr glacial cycles since the middle Pleistocene transition around 1 Myr ago, where the climate has shifted regularly between the glacial and interglacial climate states.

The temperature and proxy records ranging from the instrumental record to five million years, obtained from the stacked deep ocean sediment record[10], over a huge range of scales are shown in Fig. 1. As there is a connection between spatial and temporal scales, the records shown will be more local in the top panels and more global in the bottom panels. The Greenland ice core record represent a composite of a local and a Northern Atlantic climate signal. To eliminate the influence of long-term trends on the scaling properties, we employ the multifractal detrended fluctuation analysis (MF-DFA) method[17] for the analysis of the records. We have employed the DFA analysis of both first and second order (DFA1 and DFA2), and found that our results are robust in the sense that there is very little difference between the two for the analysed data (see Supplementary Figs 1 and 2). Here we report for DFA1 (see Methods section for description of the MF-DFA). The MF-DFA is complementary to a power spectrum analysis, where the discrete components such as the diurnal and seasonal cycle, as well as the orbital periods on Milankovitch timescales, are mixed with the continuous part of the spectrum[18,19]. If the continuous spectrum scales with frequency $P(f) \sim f^{-\beta}$, there is a direct linear relation between the scaling exponents for the spectrum and for the fluctuations[20]; $H = (\beta + 1)/2$. Thus, the trivial white noise power spectrum $\beta = 0$ corresponds to $H = 0.5$, whereas the trivial red noise spectrum $\beta = 2$ corresponds to $H = 1.5$. This simply follows from the power spectrum being the Fourier transform of the autocorrelation function. The multifractality is not captured in the powerspectrum.

Multifractal detrended fluctuation analysis. We first analyse the Holocene climate (0–11.7 kyr B2K). The scaling of the Holocene

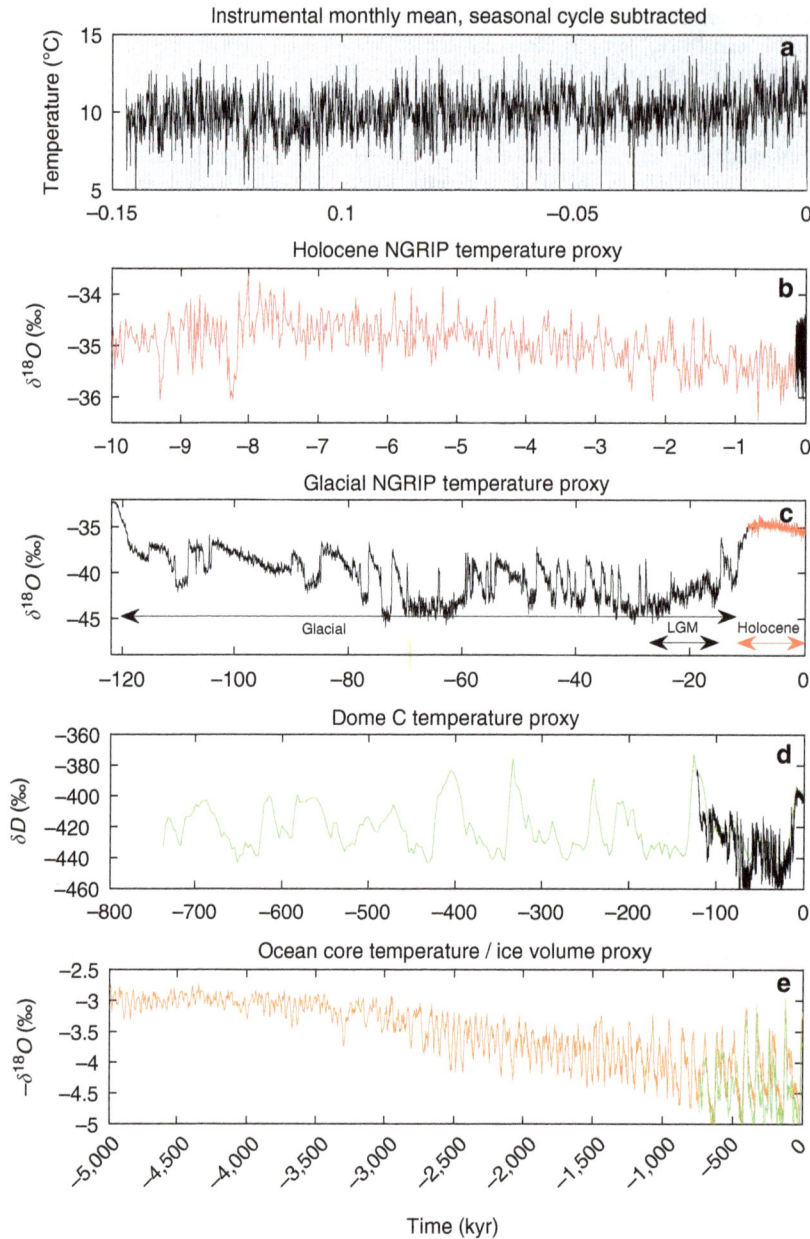

Figure 1 | Temperature variations from monthly to geological time scales. (**a**) A European meteorological station record (http://www.metoffice.gov.uk/ pub/data/weather/uk/climate/stationdata/oxforddata.txt (accessed 2015). URL http://www.metoffice.gov.uk/pub/data/weather/uk/climate/stationdata/ oxforddata.txt). ranging over 150 years (light blue, full range not shown). The record with the seasonal cycle subtracted is shown in blue. (**b**) The Holocene part of the NGRIP isotope record[6]. The instrumental record normalized to the ice core record (arbitrary) is shown in blue. (**c**) The full NGRIP record, dated using the 'GICC05modelext' chronology. The $\delta^{18}O$ is a linear proxy for temperature. The warm Holocene period 11.7 kyr to present (red, corresponding to plot (**b**)) is remarkably stable in comparison with the previous glacial period 12-120 kyr B2K (long black arrow). Before that the end of the previous warm period (Eem) is seen. LGM (15-27 kyr B2K) experienced only one small DO event. (**d**) The Antarctic EPICA δD record[21] spanning almost eight glacial cycles at 3 kyr resolution. The black curve covering the last glacial period is the (normalized) record shown in **c**. (**e**) The stacked marine benthic foraminiferal isotope record[16] (minus). This is a proxy for global ice volume and global deep ocean temperature. In green is the (normalized) curve in **d**.

represented by the ice core compares well with the instrumental record (Fig. 2); this shows a remarkable range of scaling over more than five decades, from a day to a few thousand years. In the insert, the corresponding power spectra are shown. Here the pronounced spectral peaks, the year and the Milankovitch periods, are mixed with the continuous part of the spectra. The Holocene climate shows monofractal scaling, with a scaling exponent $H \sim 0.7$, significantly different from the trivial value $H = 0.5$. Figure 3a shows the spectra $F_q(s)$ for $q = \pm 2$, for both Holocene and the glacial records. For Holocene, the black line

corresponds to $H = 0.7$, which fits for both values of q. This is in contrast to the glacial record that shows multifractal scaling with $H_{-2} = 1.4$ and $H_2 = 1.2$. The Holocene record is tested against a Monte Carlo reshuffling, which preserved the probability density of the data, see Fig. 3b and blue histogram in Fig. 3c. We have also tested the record against an autoregressive process with identical lack-one autocorrelation, with similar results (not shown). To further investigate the reliability, we have generated time series of same length as the record from a fractional Brownian motion, with Hurst exponent 0.7 (Supplementary

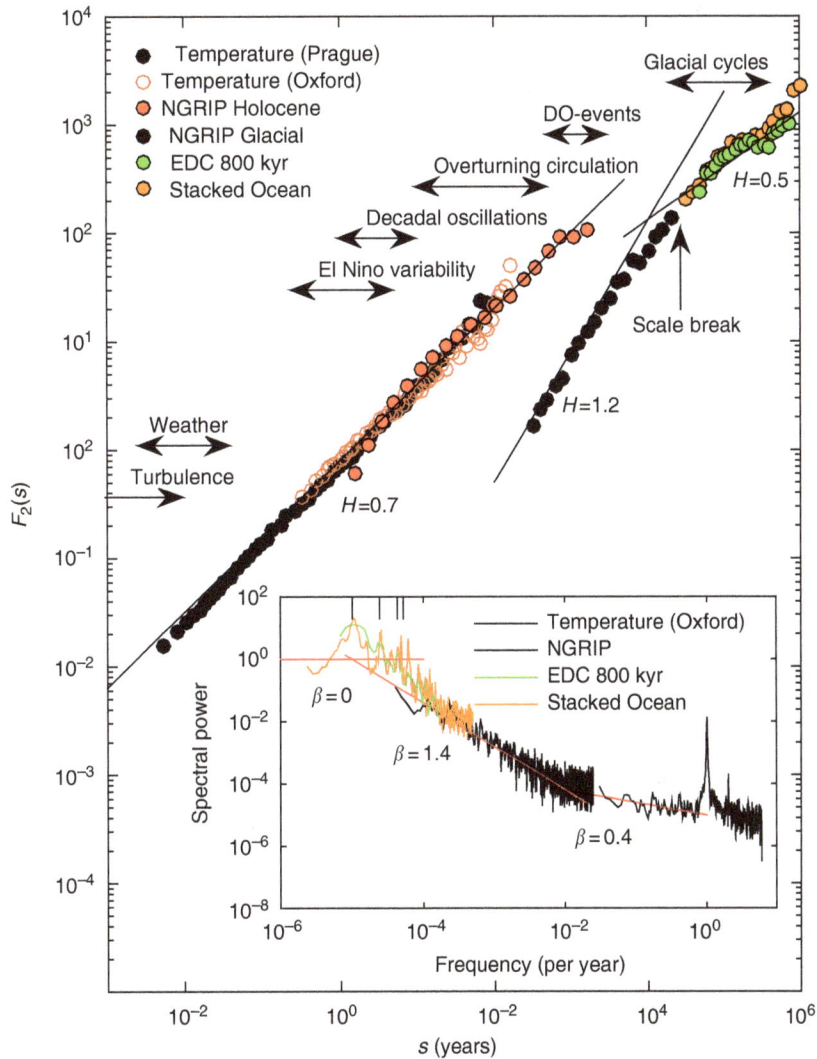

Figure 2 | The mean variance in a time window scale with the length of the window. Main panel shows the fractal spectrum $F_2(s)$ of the climate records. The scaling for the Holocene range from days to millennia (blue and red dots). The scaling exponent $H = H_2$ is the slope of the line. The typical time scales of different processes in the climate system are indicated. The scale break around 20 kyr is noteworthy. The insert shows the power spectra of the climatic time series (same colours). The slopes of the continuous part of the spectra corresponds to the (monofractal) Hurst exponent through $H = (\beta + 1)/2$. The power spectra mix the continuous (scaling) part and the discrete peaks corresponding to periodic and quasi-periodic components. The small black bars in the top left corner of the insert indicates the Milankovitch periods at 19, 23, 41 and 100 kyrs.

Figs 3 and 4). The probability density of measured exponents is shown in Fig. 3c (orange histogram). It is noteworthy that the consistency with the record (red bar) is a much weaker result than rejection of the null hypothesis above. It indicates the uncertainty in the estimated scaling exponent due to the limited length of the record. As we do not have a full theory of the underlying climate processes generating the fractal structure, we cannot rule out that the Holocene is weakly multifractal but the time series is to short to detect the multifractality. To obtain some indications on this possibility, we have simulated a known weakly multifractal process of similar length to the record (Supplementary Figs 5–7), where the multifractallity can indeed be detected. This we interpret as further supporting the observation that the Holocene record is monofractal.

The findings for the Holocene is in agreement with findings from climate model millennial simulations[21]. In Supplementary Fig. 8 we present an analysis of two of the CESM1-CAM5 Last Millennium Ensemble runs.

For longer time scales we enter the glacial climate state and on even longer time scales we observe a scale break at the

Milankovitch time scales (> 20 kyr). Analysis of the 800-kyr Antarctic Epica Dome C (EDC) core[11] with 3 kyr resolution and the 5.3-Myr stacked ocean benthic foraminiferal isotope record[10] (green and orange curves in Fig. 1d,e) does indeed show that these records have a trivial scaling with a Hurst exponent close to 0.5 (green and orange dots in Fig. 2). As a further consolidation of the robustness we also analysed GRIP ice core on the GICC05 time scale, in agreement with the results for NGRIP (Supplementary figs 9 and 10). Furthermore, we have analysed the 420-kyr Vostok core[22], which agreed with the results for EDC (not shown).

The ice core record (Fig. 1c) shows that the warmer climate of the Holocene period (0–11.7 kyr B2K) is more stable than that of the last glaciation (12–120 kyr B2K)[23]. The difference in climate states is reflected in the scaling properties of the proxy temperature signal; thus, we split the NGRIP $\delta^{18}O$ signal into two parts covering the Holocene warm period and the last glacial period. The glacial period was characterized by millennium scale sudden climate shifts, the DO events[9], the cause of which is still unknown. The predominant assumption for the cause of the DO

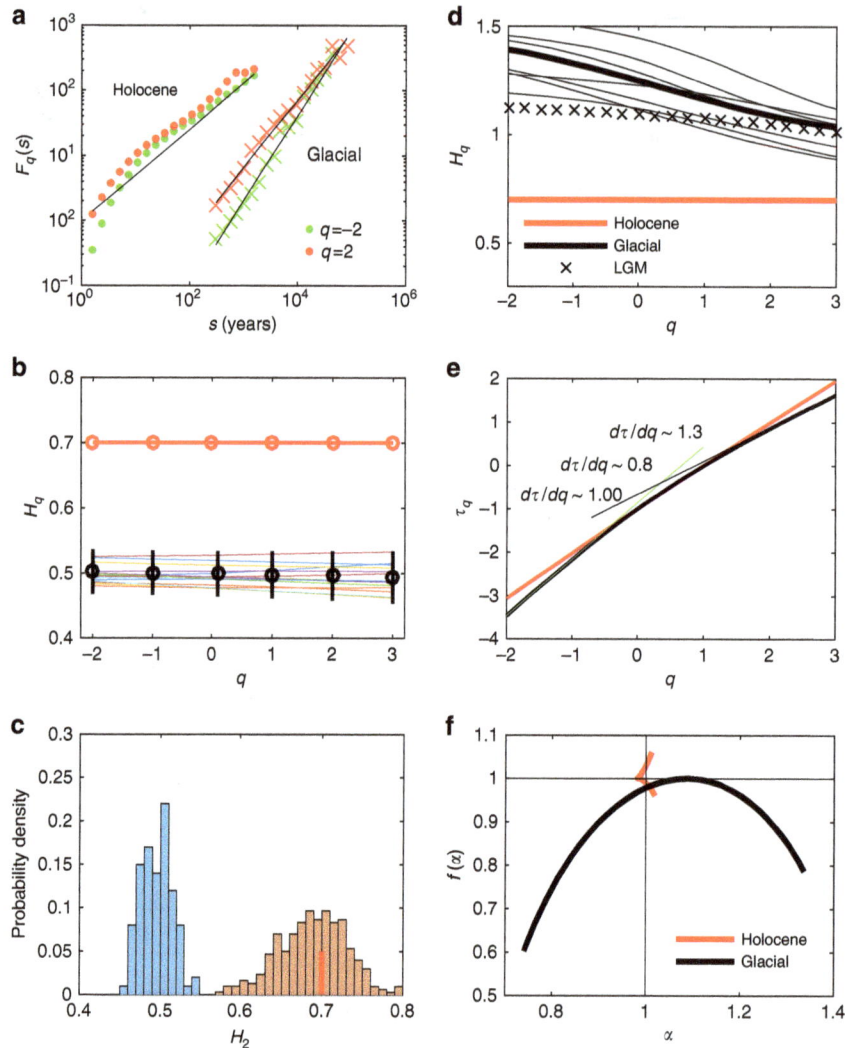

Figure 3 | Contrasting monofractal scaling for Holocene and multifractal scaling for glacial periods. The scaling spectra $F_q(s)$ for the Holocene and glacial periods are shown in (**a**). Holocene shows monofractal scaling, whereas the glacial is multifractal, as there is a significant difference in the slopes for $q = -2$ and $q = 2$. (**b**) The scaling in a set of 100 reshuffled versions of the record (thin lines, not all shown). The black dots are the means for the reshuffled data (true value is 0.5) and the bars are the 2σ spreads. (**c**) The distribution of Hurst exponent H_2 for the 100 reshuffled series, centred around $H_2 = 0.5$ (blue histogram). The Holocene record is marked by the red bar. The orange histogram is the distribution for a set of 100 simulations of fractional Brownian motion with $H_2 = 0.7$. (**d**) The generalized Hurst exponent H_q is shown for Holocene (red, same as in **b**) and the glacial period (thick black). The thin black curves are for non-overlapping 12 kyr sections of the glacial periods, indicating the range of uncertainty. The black crosses is the LGM, where only one short DO event occurs. (**e**) The multifractal scaling exponent τ_q for Holocene (red) and the glacial period (black), where a change in slope from negative (green line) to positive (blue line) values of q is observed. The Holocene curve is indistinguishable from a straight line with slope 1.00. (**f**) The multifractal spectrum $f(\alpha)$ versus $\alpha = d\tau/dq$. The small red wedge indicates that the Holocene record is almost monofractal (for which the spectrum $(\alpha, f(\alpha))$ would collapse to the point (1, 1)). This is also seen in the perfect linear fit in **b**. The black curve is for the glacial record, indicating strong multifractality.

events is abrupt changes in the Atlantic meridional overturning circulation[24,25] perhaps triggered by (negative) changes in freshwater forcing. Many mechanisms have been proposed as a trigger, from oscillations in the ice sheets[26], ice shelf breakup[27] or sea ice switching[28], or changes in solar output[29]. The occurrence of DO events influences the correlation time and thus the scaling properties of the record. One could speculate that the larger glacial Hurst exponent is a consequence of the presence of the DO events alone. This is not the case, as the LGM period, 15–27 kyr B2K, with only one short DO event (DO2) does also show a scaling, which is significantly different from the Holocene climate (black crosses in Fig. 3d). To assess the robustness, we furthermore analyse non-overlapping 12 kyr glacial periods (same duration as the Holocene). For all these periods and for the full glacial period, we get multifractal scaling with $H = H_2 \sim 1.2$ (thin

black lines in Fig. 3d). As the DO events are a part of the glacial record scaling, this indicates that they are part of the internal variability and not externally caused, in contrast to the glacial cycles, which are forced by the Milankovitch cycles and show trivial scaling. Confirming the robustness of the results is rather delicate, as there is only a limited set of truly independent paleoclimatic records, each influenced by independent noise processes, which might mask the scaling properties inferred for the climate. One possible test of the observed difference between the interglacial and the glacial climate is to investigate the previous interglacial and glacial periods separately. This we have done by splitting the EDC record into interglacial and glacial periods, and analysing them separately. The result is shown in Fig. 4 and are consistent with the results found for the Greenland ice cores. Furthermore, we have confirmed the results by

a

b

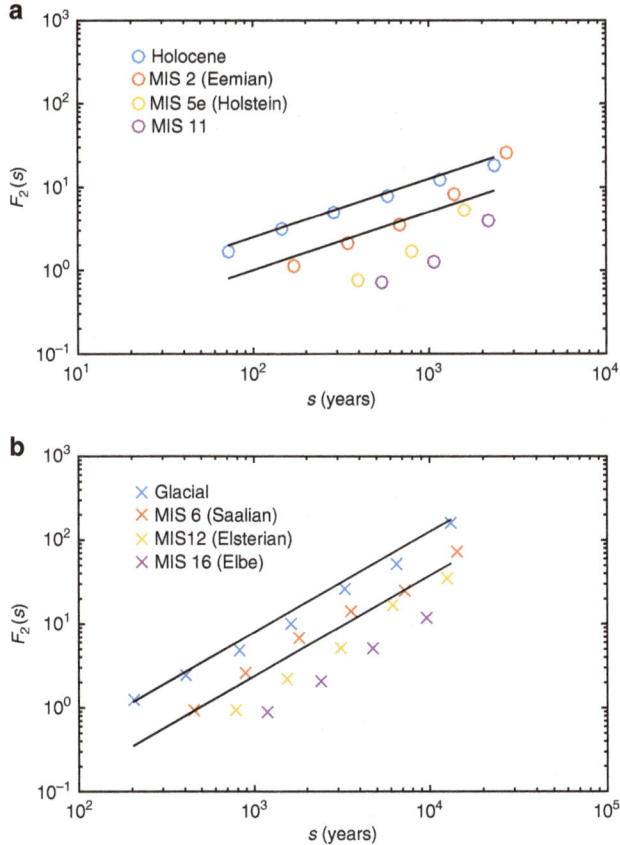

Figure 4 | Comparing scaling for interglacial and glacial periods. The scaling spectra $F_2(s)$ for the EDC core split into (**a**) interglacial periods: Holocene (0–12 kyr BP), MIS 2 (113–130 kyr BP), MIS 5e (235–243 kyr BP) and MIS 11 (321–335 kyr BP); and (**b**) glacial periods: last glacial (16–113 kyr BP), MIS 6 (136–235 kyr BP), MIS 12 (246–321 kyr BP) and MIS 16 (340–396 kys BP). The straight lines have slopes $H_2 = 0.7$ for interglacials and $H_2 = 1.2$ for glacials.

repeating the same analysis for the EDML and Vostok cores (Supplementary Figs 11 and 12).

The Greenland stable isotope ice core records are proxies for some large-scale North Atlantic mean temperature. Nearby North Atlantic sea surface temperatures has been reconstructed for the Holocene from ocean sediment cores[30,31] Despite proximity to the Greenland ice core sites, these show much more variability in the Holocene climate. The scaling exponents are $H \sim 1.1$ for LO09-14 from the North Atlantic current at the Reykjanes Ridge south of Iceland and $H \sim 1.4$ for MD95-2011 in the Norwegian Atlantic current (Supplementary Fig. 13). The two records are slightly anticorrelated (corr $= -0.25$), indicating that they locally monitor fractions of the northward Atlantic heat transport, which can be much more variable than the mean climate of the North Atlantic.

To investigate further their multifractality and difference, Fig. 3e shows the multifractal scaling exponent τ_q (see Methods section) for Holocene and the glacial period, where a change in slope from negative to positive values of q is observed. This means that there is an asymmetry in the scaling properties of large and small fluctuations. This has been quoted a multifractal phase transition[32]. In multifractal phase transition, two critical q orders defined as q_{s+} and q_{s-} exist[16]. When $q > q_{s+}$ or $q < q_{s-}$, τ_q will be linear as a function of q, as the largest fluctuations dominate the empirical moments[6]. We observe that $q_{s+} \approx 3$ and $q_{s-} \approx -2$ for the glacial record; thus, we can limit the plotting interval for q from -2 to 3 in Fig. 3e. Figure 3f shows the multifractal spectrum $f(\alpha)$ versus $\alpha = d\tau/dq$. For monofractals the $(\alpha, f(\alpha))$

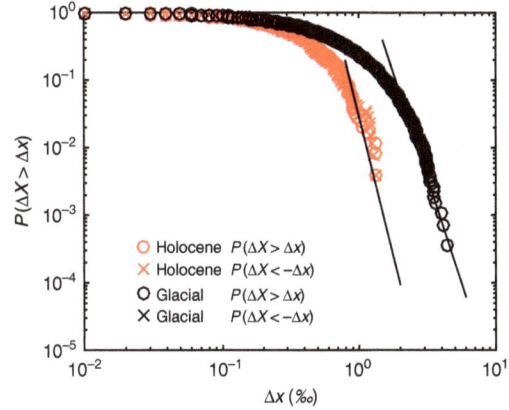

Figure 5 | Comparing the extreme tails of the distributions for Holocene and glacial records. Comparison of probability distribution $Pr(\Delta X > \Delta x)$ versus Δx between the Holocene (red) and last glacial (black) records. $\Delta x_t = x(t + 20\text{yrs}) - x(t)$, where $x(t)$ is the $\delta^{18}O$ record of NGRIP ice core. For both parts of the record, the extreme tail is a power law, $P(\Delta X' > \Delta) \sim \Delta^\gamma$, represented by the straight lines with slopes $\gamma \approx -8.3$ (Holocene) and $\gamma \approx -6.3$ (glacial).

spectrum collapses to the point (1, 1). The small red wedge indicates that the Holocene record is monofractal, whereas the black curve shows strong multifractality in the glacial climate.

The tail of the probability distribution. The difference of multifractality in the records of the two periods reflects that the climate of the interglacial is quite different from the glacial climate: The DO events introduce long-range correlations related to the waiting times of several thousand years for jumping between the stadial and the interstadial states. This jumping between states is absent in the Holocene climate. As the DO events do not lead to a scale break in the scaling of the glacial climate signal, we speculate that this is an indication that they do not have a trigger, which is disconnected (such as changing solar radiation) from the climate dynamics giving rise to the scaling properties. The occurrence of the DO events seems random in nature[33], which agrees well with internal fluctuations in the Atlantic meridional overturning circulation as the cause of these events. The reason why DO variability is absent in the Holocene climate could be attributed to the absence of the large ice sheets. However, this does not necessarily imply ice sheet instability to be the trigger. It could well be that the larger short-term variability in the glacial climate strongly enhance the triggering. The short-term fluctuations can be represented by $\Delta x_t = x(t + \Delta t) - x(t)$, where $x(t)$ is the given evenly spaced time series. A generic character of multifractal processes is that they have fat-tailed probability distributions, that is, $P(\Delta X > \Delta x) \sim (\Delta x)^{-\gamma}$, for large Δx (and similar for the negative tail: $P(\Delta X < -\Delta x)$), where γ is the corresponding probability exponent[6]. Figure 5 shows the comparison of probability distributions $P(|\Delta X| > \Delta)$ versus Δx for the the Holocene (red) and last glacial (black) (positive and negative tails have identical distributions). The sizes of the fluctuations are larger in the glacial period but interestingly the probability distributions for both the Holocene and the glacial show fat-tail scaling (straight black line) for large Δx with $\gamma \approx -8.3$ (Holocene) and $\gamma \approx -6.3$ (glacial). One could speculate that the less extreme climate of the Holocene prevents triggering of DO events.

Discussion

In summary, the interglacial climate shows scaling over a remarkable range of scales from daily to millennial. The generalized Hurst exponent of $H \sim 0.7$ is significantly different from the trivial value $H = 0.5$. The glacial climate state has a distinctly more fractal characteristics, with a much larger generalized Hurst exponent $H \sim 1.2$. Although neither the Antarctic- nor the Greenland ice core records represent a global climate signal, the differences in scaling exponents reported for different records[6] in the range 1.2–1.4 are in our judgement identical within the uncertainty of measurement.

The glacial record also shows a clear multifractal scaling, with an asymmetry between small and large fluctuations. The DO events are a part of the scaling process, indicating that they are part of the internal variability, and not externally caused, in contrast to the glacial cycles, which are forced by the Milankovitch cycles and show trivial scaling.

Methods

Multifractal detrended fluctuation analysis. The MF-DFA[17] is a robust and easily implemented analysis of the scaling properties in strongly fluctuating or non-stationary time series. It is performed in five steps as follows: (i) determine the cumulated data series

$$y(k) = \sum_{i=1}^{k} [x_i - \langle x \rangle], \tag{1}$$

where $\langle x \rangle$ is the mean value of the time series x_k ($k = 1, ..., N$). (ii) The profile is divided into $N_s = \text{int}(N/s)$ disjoint segments with same size s. As the congruence between N and s is often not zero, a part will remain after division. To preserve this part, the same dividing procedure is repeated from the opposite end. As a consequence, $2N_s$ segments are generated. (iii) The variance is calculated as

$$F^2(s, v) = \frac{1}{s} \sum_{i=1}^{s} \{y[(v-1)s+i] - y_v(i)\}^2, \tag{2}$$

where $v = 1, ..., N_s$ and

$$F^2(s, v) = \frac{1}{s} \sum_{i=1}^{s} \{y[N - (v - N_s)s + i] - y_v(i)\}^2, \tag{3}$$

where $v = N_s + 1, ..., 2N_s$. $y_v(i)$ is the least square-fitting line in segment v. Next, the q-th order fluctuation function is

$$\begin{cases} F_q(s) = \left\{ \frac{1}{2N_s} \sum_{v=1}^{2N_s} [F^2(s, v)]^{q/2} \right\}^{1/q}, \text{ for } q \neq 0; \\ F_0(s) = \exp\left\{ \frac{1}{4N_s} \sum_{v=1}^{2N_s} \ln[F^2(s, v)] \right\}, \text{ for } q = 0. \end{cases} \tag{4}$$

(iv) The above steps (ii) and (iii) are repeated as the segment size s increases. (v) The scaling exponent is determined by fitting log-log plots of $F_q(s)$ versus s as

$$F_q(s) \sim s^{H_q}, \tag{5}$$

where H_q is the generalized Hurst exponent. Next, the multifractal spectrum ($f(\alpha)$ versus α) can be derived using the following relationship[16,17]:

$$\begin{cases} \tau_q = q(H_q - \xi) - 1, \\ \alpha = \frac{d\tau}{dq}, \\ f(\alpha) = q\alpha - \tau_q, \end{cases} \tag{6}$$

where α is the singularity strength, τ_q is the multifractal scaling exponent and $f(\alpha)$ is the dimension as a function of the α. $\xi = H_1 - 1$ is the scaling exponent of the mean fluctuations. For monofractal time series, the $(\alpha, f(\alpha))$ spectrum collapses to the point $(1, 1)$. In practice, for finite time series $f(\alpha)$ versus α will be a tiny arc solely due to random fluctuation. For multifractal time series, the multifractal spectrum will typically have a single-humped parabolic shape[17].

References

1. Hurst, H. E. Long term storage capacity of reservoirs. *Trans. Am. Soc. Civil Eng.* **116**, 770–808 (1951).
2. Mandelbrot, B. & Wallis, J. R. Computer experiments with fractional gaussian noises, averages and variances. *Wat. Resour. Res.* **5**, 228–241 (1969).
3. Kantelhardt, J. *et al.* Long-term persistence and multifractality of precipitation and river runoff records. *J. Geophys. Res.* **111**, D01106 (2006).
4. Koscielny-Bunde, E. *et al.* Indication of a universal persistence law governing atmospheric variability. *Phys. Rev. Lett.* **81**, 729–732 (1998).
5. Pelletier, J. D. & Turcotte, D. L. Long-range persistence in climatological and hydrological time series: analysis, modeling and application to drought hazard assessment. *J. Hydrol.* **203**, 198–208 (1997).
6. Lovejoy, S. & Schertzer, D. *The Weather and Climate: Emergent Laws and Multifractal Cascades* (Cambridge Univ. Press, 2013).
7. Bodri, L. Fractal analysis of climatic data: mean annual temperature records in hungary. *Theor. Appl. Climatol.* **49**, 53–57 (1994).
8. Lovejoy, S., Tuck, A. F., Hovde, S. J. & Schertzer, D. Vertical cascade structure of the atmosphere and multifractal dropsonde outages. *J. Geophys. Res. Atmos.* **114**, D07111 (2009).
9. Dansgaard, W. *et al.* Evidence for general instability of past climate from a 250-kyr ice-core record. *Nature* **364**, 218–220 (1993).
10. Lisiecki, L. E. & Raymo, M. E. A pliocene-pleistocene stack of 57 globally distributed benthic D18O records. *Paleoceanography* **20**, 1–17 (2005).
11. EPICA Community Members. Eight glacial cycles from an Antarctic ice core. *Nature* **429**, 623–628 (2004).
12. Schmitt, F., Lovejoy, S. & Schertzer, D. Multifractal analysis of the greenland ice-core project climate data. *Geophys. Res. Lett.* **22**, 1689–1692 (1995).
13. Ashkenazy, Y., Baker, D. R., Gildor, H. & Havlin, S. Nonlinearity and multifractality of climate change in the past 420,000 years. *Geophys. Res. Lett.* **30**, 2146–2149 (2003).
14. North GRIP Members. High resolution climate record of the northern hemisphere reaching into the last glacial interglacial period. *Nature* **431**, 147–151 (2004).
15. Wolff, E. W., Chappellaz, J., Blunier, T., Rasmussen, S. O. & Svensson, A. Millennial-scale variability during the last glaciation: the ice core record. *Quat. Sci. Rev.* **29**, 2828–2838 (2010).
16. Shao, Z.-G. & Wang, H.-H. Multifractal detrended fluctuation analysis of the δ18o record of ngrip ice core. *Clim. Dynam.* **43**, 2105–2109 (2014).
17. Kantelhardt, J. *et al.* Multifractal detrended fluctuation analysis of nonstationary time series. *Phys. A* **316**, 87–114 (2002).
18. Huybers, P. & Curry, W. Links between annual, milankovitch and continuum temperature variability. *Nature* **441**, 329–332 (2006).
19. Marsh, N. D. & Ditlevsen, P. D. Observation of atmospheric and climate dynamics from a high resolution ice core record of a passive tracer over the last glaciation. *J. Geophys. Res.* **102**, 11219–11224 (1997).
20. Hansen, A., Schmittbuhl, J. & Batrouni, G. G. Distinguishing fractional and white noise in one and two dimensions. *Phys. Rev. E* **63**, 062102 (2001).
21. Rybski, D., Bunde, A. & von Storch, H. Long-term memory in 1000-year simulated temperature records. *J. Geophys. Res. Atmos.* **113**, D02106 (2008).
22. Basile, I. *et al.* Climate and atmospheric history of the past 420,000 years from the Vostok ice core, Antarctica. *Nature* **399**, 429–436 (1999).
23. Ditlevsen, P. D., Svensmark, H. & Johnsen, S. Contrasting atmospheric and climate dynamics of the last-glacial and holocene periods. *Nature* **379**, 810–812 (1996).
24. Marotzke, J. Abrupt climate change and thermohaline circulation: mechanisms and predictability. *Proc. Natl Acad. Sci. USA* **97**, 1347–1350 (2000).
25. Ganopolski, A. & Rahmstorf, S. Rapid changes of glacial climate simulated in a coupled climate model. *Nature* **409**, 153–158 (2001).
26. MacAyeal, D. R. Binge/purge oscillations of the laurentide ice sheet as a cause of the north atlantic's heinrich events. *Paleoceanography* **8**, 775–783 (1993).
27. Petersen, S. V., Schrag, D. P. & Clark, P. U. A new mechanism for dansgaard-oeschger cycles. *Paleoceanography* **28**, 24–30 (2013).
28. Gildor, H. & Tziperman, E. Sea-ice switches and abrupt climate change. *Phil. Trans. R. Soc. Lond. A* **361**, 1935–1944 (2003).
29. Braun, H. *et al.* Possible solar origin of the 1,470-year glacial climate cycle demonstrated in a coupled model. *Nature* **438**, 208–211 (2005).
30. Berner, K. S., Ko, N., Divine, D., Godtliebsen, F. & Moros, M. A decadal-scale holocene sea surface temperature record from the subpolar north atlantic constructed using diatoms and statistics and its relation to other climate parameters. *Paleoceanography* **23**, PA2210 (2008).
31. Berner, K. S., Ko, N., Godtliebsen, F. & Divine, D. Holocene climate variability of the norwegian atlantic current during high and low solar insolation forcing. *Paleoceanography* **26**, PA2220 (2011).
32. Lee, J. & Stanley, H. E. Phase transition in the multifractal spectrum of diffusion-limited aggregation. *Phys. Rev. Lett.* **61**, 2945–2948 (1988).
33. Ditlevsen, P. D., Andersen, K. K. & Svensson, A. The do-climate events are probably noise induced: statistical investigation of the claimed 1470 years cycle. *Clim. Past* **3**, 129–134 (2007).

Acknowledgements

Z.-G.S. is supported by the National Natural Science Foundation of China (grant numbers 11105054 and 11274124) and PCSIRT (grant number IRT1243). P.D.D. is supported by the RiskChange project funded by the Danish Strategic Research Council.

Author contributions

Z.-G.S. and P.D.D. contributed to all stages of the work. Z.-G.S. wrote the first draft of the paper and P.D.D. wrote the final paper.

Additional information

Competing financial interests: The authors declare no competing financial interests.

15

Recent increases in Arctic freshwater flux affects Labrador Sea convection and Atlantic overturning circulation

Qian Yang[1], Timothy H. Dixon[1], Paul G. Myers[2], Jennifer Bonin[3], Don Chambers[3] & M.R. van den Broeke[4]

The Atlantic Meridional Overturning Circulation (AMOC) is an important component of ocean thermohaline circulation. Melting of Greenland's ice sheet is freshening the North Atlantic; however, whether the augmented freshwater flux is disrupting the AMOC is unclear. Dense Labrador Sea Water (LSW), formed by winter cooling of saline North Atlantic water and subsequent convection, is a key component of the deep southward return flow of the AMOC. Although LSW formation recently decreased, it also reached historically high values in the mid-1990s, making the connection to the freshwater flux unclear. Here we derive a new estimate of the recent freshwater flux from Greenland using updated GRACE satellite data, present new flux estimates for heat and salt from the North Atlantic into the Labrador Sea and explain recent variations in LSW formation. We suggest that changes in LSW can be directly linked to recent freshening, and suggest a possible link to AMOC weakening.

[1] School of Geosciences, University of South Florida, 4202 E Fowler Avenue, Tampa, Florida 33620, USA. [2] Department of Earth and Atmospheric Sciences, University of Alberta, 1-26 ESB, Edmonton, Alta, Canada T6G 2E3. [3] College of Marine Science, University of South Florida, St. Petersburg, Florida 33701, USA. [4] Institute for Marine and Atmospheric Research Utrecht, Utrecht University, P.O. Box 80.005, 3508 TA, Utrecht, The Netherlands. Correspondence and requests for materials should be addressed to Q.Y. (email: qianyang@mail.usf.edu).

It has long been accepted that the Atlantic Meridional Overturning Circulation (AMOC) has two stable modes[1-3], and that anthropogenic warming could weaken or shut down the AMOC[4,5]. Recent accelerated melting of the Greenland ice sheet is freshening the North Atlantic[6-10]. So-called 'hosing experiments', where freshwater may be distributed over broad or narrow regions of the North Atlantic in numerical models, have been used to study the sensitivity of the AMOC to freshwater flux[11-17]. Some of these studies suggest that AMOC strength is sensitive to Greenland melting[11,17], while others do not[12,14,16]. A few studies suggest that freshwater additions of 0.1 Sv (100 mSv)[18-20] or possibly less[11,17] could affect the AMOC.

Changes in the AMOC are difficult to measure directly: currents that comprise the deeper, southward flowing portions can be diffuse and/or spatially and temporally variable, and instrumental drift can mask subtle, long-term changes in oceanic properties. It is also challenging to separate changes forced by anthropogenic warming from natural variability. The AMOC is difficult to model numerically: model grids may be too coarse to reflect realistic oceanic processes and geographic constraints, and feedbacks among atmosphere, ocean and cryosphere (land and sea ice) are poorly known.

The Labrador Sea is a key location for the formation of one of the dense, deep-water components of the AMOC via winter convection; however, the process is sensitive to surface conditions[21]. Wood et al.[5] suggest the possibility of a shutdown in Labrador Sea convection in response to global warming. Kuhlbrodt et al.[22] provide a theoretical stability analysis, and suggest that winter convection in the Labrador Sea can be turned off by increased freshwater input. Unfortunately, winter convection here is difficult to observe directly because of extreme conditions and its small spatial scale.

Here we consider recent Labrador Sea changes associated with an increased freshwater flux. We derive a new estimate for recent increased freshwater flux into the sub-polar North Atlantic, and suggest that because of the clockwise nature of ocean circulation around Greenland[23], most of this increase is being focused towards the Labrador Sea (Fig. 1), magnifying its impact and increasing the likelihood of significant effects on the AMOC.

Results

Recent accelerated melting of the Greenland ice sheet. Numerous studies have described recent acceleration of Greenland's ice mass loss[6-10]. We use GRACE data updated to October 2014 to derive a new acceleration estimate and its onset time (Methods). GRACE data and uncertainty estimates follow Bonin and Chambers[24]. We fit a constant acceleration model to the data, and extrapolate the best-fit model back to the time of zero mass loss rate, obtaining 20-Gt per square year acceleration with a start time of 1996 ± 1.4 years (Fig. 2). Several lines of evidence suggest that the ice sheet was relatively stable from 1980 to the early 1990s (refs 25,26), and we use that assumption in our modelling of GRACE data and freshwater flux calculations (below and Methods section).

Irminger Water heat and salt fluxes. Warming of sub-polar mode waters including Irminger Water in the mid- to late-1990s (refs 27,28) is thought to influence coastal mass loss in Greenland by increasing submarine melting of outlet glaciers and related dynamic effects[29-31]. Here we examine the variability of heat and salt fluxes of Irminger Water along three sections (Fig. 1) offshore southwest coastal Greenland for the period 1949–2013 (Methods). Currents associated with the sub-polar gyre here are quite compact as they round the southern tip of Greenland, limiting spatial variability and facilitating accurate flux

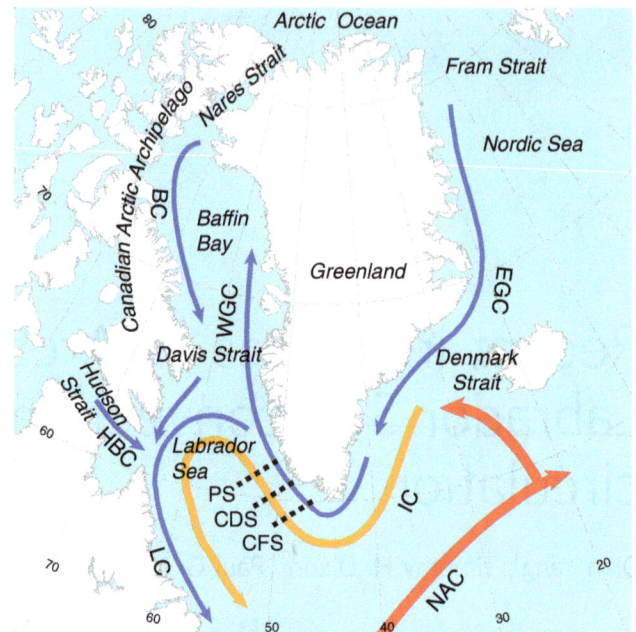

Figure 1 | Study region showing oceanographic sections and major currents around Greenland. Red and orange arrows indicate Atlantic-origin water and blue arrows indicate Arctic-origin water. BC, Baffin Current; CDS, Cape Desolation Section; CFS, Cape Farewell Section; EGC, East Greenland Current; HBC is Hudson's Bay Current; IC, Irminger Current; LC, Labrador Current; NAC, North Atlantic Current; PS, Paamiut Section; WGC is West Greenland Current. Three-dimensional structure of major water masses in the Labrador Sea is shown in Supplementary Fig. 1.

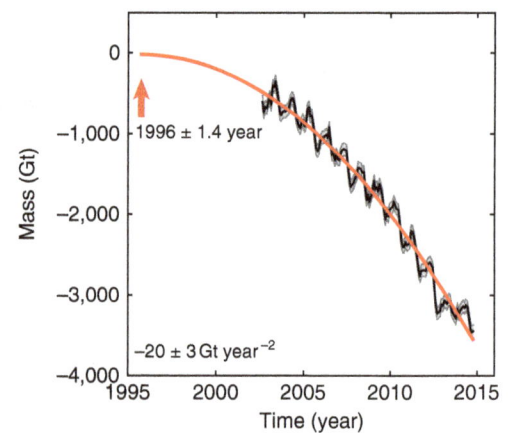

Figure 2 | Mass change of Greenland estimated from GRACE for the period 2002-2014. Black curve shows data, grey shading indicates monthly uncertainty and red curve shows the best fitting constant acceleration model. Onset time of acceleration defined when the rate of mass change is zero, in 1996 (red arrow), with mass arbitrarily set to zero.

measurements because the cross-section area of current is well defined. Note that, while the flux (sensu stricto) is flow rate per unit area and transport (or total flux) represents the flux integrated over the larger area of interest, the terms 'flux' and 'transport' are often used interchangeably in the oceanographic literature. We follow the broader (sensu lato) usage here.

We carry out our analysis on the upper 700 m, the greatest depth common to all years, binned on a 2-m vertical grid. Time series of heat and salt fluxes at the three sections are shown in Fig. 3. At the southernmost Cape Farewell section, both heat and salt fluxes experienced a large multi-year anomaly around 1995,

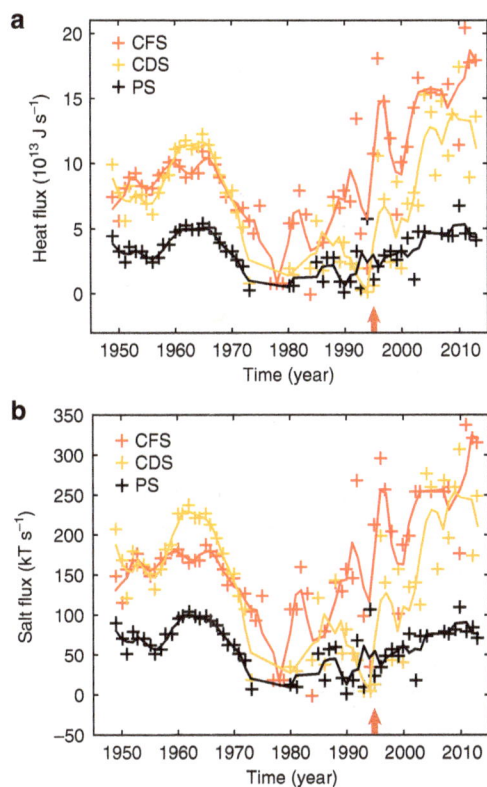

Figure 3 | Heat and salt fluxes of Irminger Water for the period 1949-2013. (**a**) Heat and (**b**) salt fluxes of Irminger Water are measured at three sections in southwest Greenland. Locations of three sections are shown in Fig. 1. CDS, Cape Desolation Section; CFS, Cape Farewell Section; PS, Paamiut Section. Solid line represents a 3-year running average, yearly data shown by plus signs. Red arrow marks the onset time of accelerated mass loss for Greenland estimated from GRACE (Fig. 2).

followed by another in the late 1990s. The heat flux was 80% higher than a previous multi-year anomaly in the 1960s. Similar variability is seen at the more northerly Cape Desolation section, although salinities and heat are generally lower, and only exceed previous levels after 2000. No significant anomalies were observed at the northernmost Paamiut section during these times; however, the heat and salt fluxes are still roughly 50% higher after 2000 than they were in the 1980s, and approach levels that are not seen since the 1960s. Thus, we conclude that Irminger Water became warmer and saltier in the mid-late 1990s, which agrees well with the onset time of recent accelerated Greenland mass loss (Fig. 2). This is consistent with the idea that accelerating ice mass loss in the mid-late 1990s reflects, at least in part, the appearance of warmer Irminger Water on the peripheral continental shelf at that time[29]. The anomalous heat flux we observe off southern Greenland in the mid-1990s can be directly tied to warming of the North Atlantic (Supplementary Fig. 2; see also ref. 31).

Northward reduction in heat and salt transport between the Cape Desolation and Paamiut sections likely reflects strong offshore eddy transport[32], advecting Irminger Water into the interior of the Labrador Sea. However, since the sections are only occupied once a year in summer, some seasonal aliasing is possible. The eddies also transport fresh shelf water into the Labrador Sea[33].

Estimates of the freshwater flux into the Labrador Sea. Major sources of freshwater entering the Labrador Sea include precipitation, oceanic transport and melt from the Greenland ice

sheet, glaciers in the Canadian Arctic Archipelago (CAA) and Arctic sea ice. Precipitation in the Labrador Sea region is about 20–30 mSv (ref. 34), and there has been a general increase over the North Atlantic region in the last few decades as the hydrologic cycle accelerates[35]. Oceanic transport from the Arctic Ocean is the largest source of Labrador Sea freshwater and is derived from several sources, including the difference between precipitation and evaporation, river discharge and fractionation associated with annual sea ice formation. Peterson et al.[36] show that the average annual river discharge from six rivers in Eurasia into the Arctic Ocean has increased by 7% (~ 4 mSv) from 1936 to 1999. The Arctic Ocean exports low-salinity water to the North Atlantic through two main pathways: Fram Strait east of Greenland and the CAA west of Greenland. The CAA pathway has three main routes: Barrow Strait, Nares Strait and Cardigan Strait-Hell Gate. Roughly, 100 mSv of freshwater is exported through each of the east and west pathways, relative to a reference salinity of 34.80 (ref. 37). Within broad error bars, oceanic transport from the Arctic Ocean is relatively stable on the decadal timescale, although there has been some reduction through the CAA and then Davis Strait, and shorter-term fluctuations are common[37–39].

Here we focus on three Arctic freshwater sources that are undergoing rapid increases, which likely contribute freshwater to the Labrador Sea, and which can be estimated from remote observations: the Greenland ice sheet, CAA glaciers and Arctic sea ice. We also consider snowmelt runoff from tundra in Greenland and the CAA as they follow directly from the same models used to quantify Greenland ice sheet and CAA glacier melt[40,41]. As we are not considering the large Arctic oceanic transport term and several other sources, our estimate is a minimum estimate.

The freshwater flux from Greenland is composed of ice and tundra runoff plus ice discharge; this quantity is equal to accumulation minus mass balance (Methods). We derive mass balance for Greenland from GRACE, while accumulation is obtained from the RACMO2.3 model[42,43]. Our GRACE data suggest that mass loss of the Greenland ice sheet accelerates from 1996 onwards (Fig. 2; Methods). Our mass balance estimate agrees with the estimate of Box and Colgan[26], with the Greenland ice sheet in near balance from 1980 to about 1996, after which it starts to lose mass (Supplementary Fig. 3). Therefore, we assume that between 1980 and 1996 the freshwater flux from Greenland is approximately equal to accumulation; after 1996, the freshwater flux from Greenland equals the sum of mass loss and accumulation (Supplementary Fig. 3). Since the accumulation is highly variable from year to year, we smooth it with a 5-year running mean. Figure 4 shows the resulting freshwater flux estimates from Greenland. This approach yields freshwater flux estimates that agree with those of Bamber et al.[40] during the period of data overlap, once a correction for solid ice discharge is applied[8] (Supplementary Fig. 4). Freshwater from the CAA is approximated by ice and tundra runoff predicted by RACMO2.3 since ice discharge (0.16 mSv) is negligible[44].

Large amounts of Arctic sea ice and freshwater are exported from the Arctic Ocean to the North Atlantic through several pathways. Of these, Fram Strait and the CAA are the major ones; nearly all ($\sim 98\%$) Arctic Ocean exports drain through them[37]. However, there are large uncertainties in these fluxes[37]. We focus on changes in the freshwater flux as inferred from recent accelerated melting of Arctic sea ice, assuming that the change is partitioned the same way as the total export, that is, 98% of the change is advected through Fram Strait and the CAA. Changes in the annual minimum of Arctic sea ice volume are a relevant indicator (see Methods and Supplementary Methods). We first use the annual minimum volume predicted by the Pan-Arctic Ice

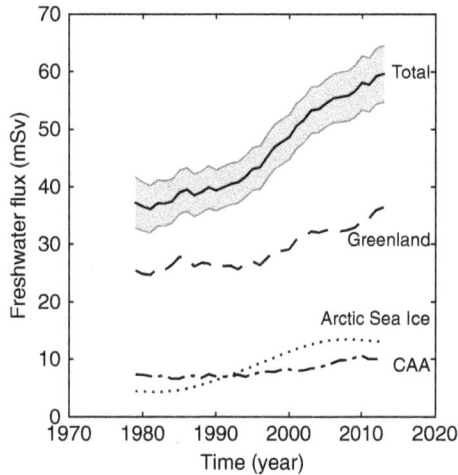

Figure 4 | Freshwater flux from Greenland and CAA and Arctic sea ice for the period 1979-2013. For Arctic sea ice, we plot only changes in flux (see text). The sum of these sources (Total) is also plotted. Grey shading indicates propagated uncertainty (see Supplementary Note 1).

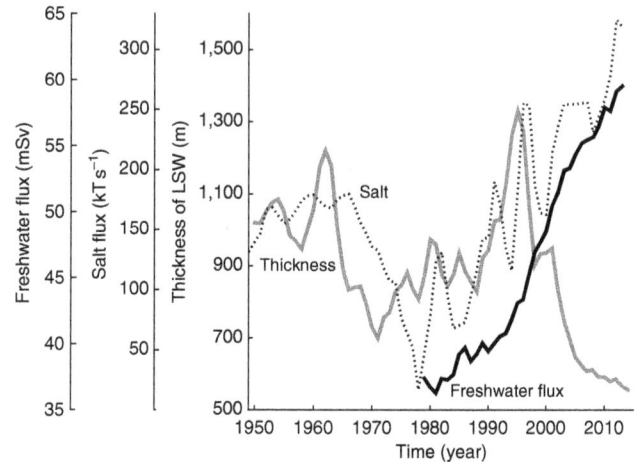

Figure 5 | Thickness of LSW and total freshwater flux and salt flux of Irminger Water. Grey solid line indicates the thickness of LSW, black solid line indicates total freshwater flux and dotted line indicates salt flux of Irminger Water. Thickness and salt flux are smoothed with a 3-year running mean. Thickness is obtained from the objective analysis of EN4.0.2 data set from the UK Met Office Hadley Center[52]. Thickness is averaged over 50° N-65° N and 38° W-65° W. Expression of salt flux in terms of freshwater flux is shown in Supplementary Fig. 7.

Ocean Modeling and Assimilation System (PIOMAS) model[45]. We also apply the same method to the Arctic sea ice extent and sea ice area data sets[46], where 'extent' defines a region as either 'ice-covered' or 'not ice-covered' using a threshold of 15%; 'area' is a more conservative estimate, defined as the percentage of actual sea ice within a given data cell. We assume a standard ice thickness of 2 m (ref. 47) to convert ice extent and ice area to volume, obtaining results that are somewhat smaller than the PIOMAS volume model. Figure 4 shows results from the PIOMAS volume model. Results from the other two approaches are shown in Supplementary Figs 5 and 6.

Figure 4 also shows the summed result from these various freshwater sources (recall that this summed value does not include several major sources and is therefore a minimum estimate), which is our estimate of the freshwater flux into the sub-polar North Atlantic. The freshwater flux from Greenland is relatively stable from 1979 to the mid-late 1990s and then increases. The freshwater flux from the CAA is relatively stable until the early 2000s and then increases. Freshwater flux from Arctic sea ice increases mainly during the period 1990-2000. The total freshwater flux for the sub-polar North Atlantic from these sources is about 60 mSv by 2013, with an increase of 20 mSv during the last two decades. Of this, ~12 mSv comes from the Greenland ice sheet and CAA glaciers, whereas ~8 mSv represents excess melting of Arctic sea ice.

Focused freshwater flux into the Labrador Sea has the potential to disrupt the AMOC by increasing the buoyancy of surface waters and reducing the formation of dense, deep water[13]. How much of the enhanced freshwater flux that we calculate actually winds up in the Labrador Sea?

Myers et al.[33,48] showed that a significant fraction of freshwater originating in and around Greenland is transported to the Labrador Sea: melt water from eastern Greenland is entrained in the East Greenland Current, where it moves south and merges with the Irminger Current as it rounds Cape Farewell; melt water from southwestern Greenland joins the West Greenland Current, similarly merging with the Irminger Current (Fig. 1). Melt water from the CAA enters the Labrador Sea through Davis and Hudson straits, either directly or indirectly[49]. The pattern of boundary currents and eddy activity around Greenland and Labrador insures that at least 75 per cent of the freshwater flux from the Greenland ice sheet and CAA eventually winds up in the Labrador Sea (Supplementary Methods). Freshwater and sea ice

drained from the Arctic Ocean moves south through Fram Strait and the CAA[37], also contributing to freshening of the Labrador Sea both remotely and locally[50,51]. We estimate that at least 65 per cent of freshwater and sea ice exported from the Arctic Ocean through Fram Strait and the CAA ultimately makes it to the Labrador Sea (Supplementary Methods). Assuming that these estimates are correct, of the 20-mSv freshwater flux increase that we estimate, at least 14 mSv (70%) winds up in the Labrador Sea (Supplementary Methods). Given typical coastal current velocities, most of this freshwater is transported to the Labrador Sea within 3-12 months. Some freshwater from the CAA may take 2-3 years to reach the Labrador Sea due to recirculation and storage in Baffin Bay and/or recirculation in the sub-polar gyre.

Impact of increased the freshwater flux on deep water formation. To investigate effects of increased freshwater flux on deep water formation in the Labrador Sea, we can either look at the mean density of Labrador Sea Water (LSW) within a given depth range or look at the thickness of LSW as defined by a given density range. We used both approaches, obtaining similar results. Figure 5 shows results from the second approach, where we calculate the thickness of LSW, defined by $\sigma_\theta = 27.74 - 27.80 \, \mathrm{kg \, m}^{-3}$, from 1950 to 2013, using the objective analyses of the EN4.0.2 data set from the UK Met Office Hadley Center[52]. The data set includes monthly temperature and salinity, with a spatial resolution of $1° \times 1°$ and 42 depth intervals (5-5,350 m) from 1900 to present. Results for density over a fixed depth range (1,000-2,500 m) are shown in Supplementary Fig. 8.

Figure 5 shows the time series of LSW thickness, compared with our estimate of freshwater flux and with the Irminger salt flux time series. From 1950 to the mid-1990s, Irminger salt flux and LSW thickness are weakly correlated ($R = 0.3$, $P = 0.03$), and both show multidecadal oscillations, with highs in the 1960s, lows in the 1980s and highs in the 1990s. In particular, LSW thickness increased significantly (by 65%) between 1990 and 1995 when the salt flux increased, consistent with the idea that dense deep water in the Labrador Sea originates from warm, saline North Atlantic water that subsequently experiences winter cooling. However, this relationship begins to break down in the mid- to late-1990s, when the freshwater flux from Greenland and other sources increased rapidly. Since then, LSW thickness decreased continuously, reaching lows not observed since the early 1970s, despite continued high salt flux. One interpretation of this is that the increased freshwater flux has now overwhelmed increased salt flux from the Atlantic, and has begun to influence LSW formation. Recall that the increased salt flux from the Atlantic is accompanied by an increased heat flux (Fig. 3), which promotes melting of marine-terminating outlet glaciers in southern Greenland[29,53] and an increased freshwater flux.

Our data are consistent with recent studies, showing a decline in the thickness of the dense mode of LSW since 1994/95, with a switch to a less dense and presumably fresher and warmer upper mode[54,55]. Yashayaev et al.[56] show declining upper salinity since the mid-2000s and suggest that salinity is the controlling factor for ocean stratification in this region. Declining upper layer salinity would weaken or even prevent Labrador Sea convection. However, cold winter air also plays a role in LSW formation. Severe winter conditions and strong air–sea heat exchange for the period 1990–1995 may have contributed to the increased LSW thickness[57], while milder winter conditions and weaker cooling since 1995 may have contributed to LSW decline[58]. The Labrador Sea is also sensitive to multidecadal climate variations. Hydrographic properties in the Labrador Sea exhibit multidecadal variability that resemble the Atlantic Multidecadal Oscillation and the North Atlantic Oscillation[56], and these variations are obvious in the flux (Fig. 3) and LSW thickness (Fig. 5) time series. Bidecadal variability in the Labrador Sea forced by volcanic activity has also been proposed[59]. Despite these complications, our data clearly show a steep, recent increase in the freshwater flux into the Labrador Sea and a steep decline in LSW thickness (and density) at the same time (Fig. 5), which is inconsistent with the estimated salt flux into the region. This suggests a potential impact on the formation of North Atlantic Deep Water.

Discussion

Our reconstructed annual freshwater flux for the sub-polar North Atlantic reaches 60 mSv in 2013, with an increase of 20 mSv in the last two decades (Fig. 4). At least 70 per cent (14 mSv) of this increased freshwater is focused towards the Labrador Sea (Supplementary Methods). This is a minimum estimate since we do not consider other major sources. LSW formation may reflect a delicate balance between this cold freshwater and warm, salty North Atlantic water from the Irminger Current. The flux of freshwater from Greenland may in turn be influenced by warm Atlantic water and its influence on the regional ocean and atmosphere, a potentially important feedback in the system.

Since LSW is an important component of the dense southward return flow of the AMOC, factors influencing LSW formation may in turn have an impact the AMOC. Hosing experiments show different sensitivities of the AMOC to freshwater fluxes at high latitudes[11–17]. Hu et al.[14] suggest that freshwater inputs much larger than we observe are required. On the other hand, Fichefet et al.[11] suggest that freshwater flux anomalies as small as 15 mSv will affect the AMOC. Brunnabend et al.[17] suggest that freshwater flux anomalies as small as 7 mSv applied over 30 years could have an impact on the AMOC. Different model outcomes partly reflect their spatial resolution, the degree to which freshwater fluxes are focused towards the Labrador Sea, and the timescale over which the anomalous flux is applied. Swingedouw et al.[15] compared different model responses to freshwater release around Greenland, assuming freshwater focusing into the Labrador Sea. They show significant AMOC weakening after several decades with a flux anomaly of 100 mSv.

If our inference that the sub-polar gyre's coastal currents focus melt water from Greenland, CAA glaciers and Arctic sea ice into the Labrador Sea is correct, then present rates of increased freshwater flux may be sufficient to influence convection in the Labrador Sea and, by implication, the AMOC. Northward decreases in heat and salt fluxes across our three sections in southwest Greenland indicate a strong mixing of coastal water and westward advection into the Labrador Sea. Eddy kinetic energy reaches a local maximum offshore Cape Desolation and Paamiut, where a front develops between Irminger Water and fresh shelf water, promoting baroclinic instability and eddy formation; these eddies propagate westwards into the Labrador Sea. Local bathymetric structures, especially the sill at Davis Strait, also promote westward propagation of coastal water from southwestern Greenland. Recent high-resolution eddy-permitting or eddy-resolving numerical models support this type of spatial focusing, and indicate decline or even shutdown of Labrador Sea convection with an enhanced freshwater flux from Greenland[60] or from the Arctic Ocean through the CAA[61]. Since freshwater lenses can retain their integrity for some time, 'temporal focusing' may also be important. Summer (June, July and August) freshwater fluxes from Greenland and CAA's ice and snow runoff greatly exceed the annual mean. Summer freshwater flux from Greenland and the CAA increased by ∼50 mSv from mid-late 1990s to 2013, reaching a height of 150 mSv in 2012 (Supplementary Fig. 9), potentially limiting convection during the subsequent winter.

We suggest that recent freshening in the vicinity of Greenland is reducing the formation of dense LSW, potentially weakening the AMOC. Recent observations are beginning to document declines in the AMOC[62–64], consistent with our hypothesis. Longer time series will be required to confirm this link, but our preliminary results suggest that detailed studies of Labrador Sea hydrography and proximal sources of freshwater, including Greenland, have the potential to improve our understanding of AMOC variability and sensitivity to anthropogenic warming.

Methods

GRACE data. The GRACE time series were created via the least squares inversion method described in ref. 24. Release-05 GRACE data from the Center for Space Research were used, with the standard post-processing applied as described in that paper: C_{20} is replaced by Satellite Laser Ranging estimates, a geocentre model is added, GIA is corrected for and the monthly averages of the Atmosphere and Ocean Dealiasing product are restored.

The inversion technique is designed to localize the mass change signal, such that coastal mass loss from Greenland does not leak into the ocean or into interior Greenland because of GRACE's inherently low spatial resolution. Briefly, the method involves breaking Greenland and the surrounding area into pre-defined regions (Greenland drainage basins; Supplementary Fig. 10). Each region is assumed to have a uniform mass distribution when gridded as $1° \times 1°$-binned kernels. The transformation to degree/order 60 spherical harmonics is then made on each individual regional kernel, resulting in a smoothed version of each region that mimics what GRACE would see from its limited resolution, if a uniform mass of 1 was placed over the kernel, with zeroes elsewhere.

The goal is to find a set of multipliers for each region that most closely describes mass distribution over Greenland, given the assumption of uniform weights across each pre-defined shape. A least squares method is used to fit an optimal multiplier to each basin simultaneously, such that the combination of the multiplier times the smoothed basin kernels best fits the actual (smoothed) GRACE data for that month. An optimal amount of process noise is added to stabilize the solution[24].

The GRACE mass balance in this paper is the sum of the individual signals from the 16 Greenland regions (Supplementary Fig. 10).

Irminger Water heat and salt flux analysis. Details of the data collection and analysis are discussed in Myers et al.[27] and summarized here. Before 1984, the estimates are based on a climatological analysis of the Labrador Sea. The 1984–2013 observations are collected on a set of standard sections by the Danish Meteorological Institute. Each section (Fig. 1) involves the same five stations; however, in some years only three or four stations could be occupied. The sections are occupied annually in most years, in late June or early July. Direct sampling using bottles was performed in 1984–1987, while conductivity–temperature–depth data were collected in later years. We carry out our analysis on the upper 700 m, the deepest depth common to all years, binned on a 2-m vertical grid. For current motions, we determine the geostrophic velocity, relative to 700 dbar (~ 700 m depth) or the bottom in shallower water, for each pair of stations at each depth, and add an estimate of the barotropic velocity[33]. If data are missing, we do not include that point in the calculation. We calculate heat flux (Q_t) and salt flux (Q_s) at each depth and then sum those whose temperature and salinity are consistent with Irminger Water to obtain the total transport:

$$Q_t = \rho \cdot C_p \cdot \int_{s=1}^{s=5} \int_{z=-700}^{z=0} v(s,z) \cdot (T(s,z) - T_{ref}) dz ds \tag{1}$$

$$Q_s = \int_{s=1}^{s=5} \int_{z=-700}^{z=0} v(s,z) \cdot (S(s,z) - S_{ref}) dz ds \tag{2}$$

where ρ and C_p are ocean water density and heat capacity, respectively; $v(s,z)$, $T(s,z)$ and $S(s,z)$ are velocity, temperature and salinity in station s at depth z, respectively; T_{ref} is the reference temperature (0 °C) and S_{ref} is the reference salinity (34.80). Here we choose a broad definition including both pure and modified Irminger Water, with temperatures warmer than 3.5 °C and salinity higher than 34.88 (ref. 27).

Freshwater flux. To estimate the freshwater flux from Greenland, we first use a simple constant acceleration model to fit the monthly GRACE mass balance data:

$$M(t_i) = a + bt_i + \frac{1}{2}ct_i^2 \tag{3}$$

where $M(t_i)$ ($i = 1,2,3\ldots\ldots n$) are GRACE monthly solutions, a is the initial mass of Greenland, b is the initial mass balance and c is the acceleration term. Given the estimated parameters, the mass balance (MB) of Greenland can be represented by:

$$MB(t_i) = b + ct_i \tag{4}$$

The start time of recent accelerated mass loss is the time t_i when $MB(t_i)$ is zero. The mass balance of Greenland is:

$$MB = SMB - D \tag{5}$$

where SMB is surface mass balance and D is discharge, related to freshwater flux (FWF) by:

$$SMB = A - R \tag{6}$$

$$FWF = R + D \tag{7}$$

where A is the accumulation and R is runoff.

We then calculate freshwater flux from Greenland using the above relations, rewriting them as:

$$FWF = A - MB \tag{8}$$

where accumulation (A) is predicted by RACMO2.3 and MB is estimated from the GRACE data. Note that accumulation is defined over ice and tundra, and mass balance is the total mass balance of Greenland, including ice and tundra. Therefore, freshwater flux from Greenland is composed of ice mass loss and tundra runoff (Supplementary Fig. 11). Mass balance is considered equal to zero before the recent acceleration phase, beginning in 1996. Since mass balance is the long-term average, accumulation is smoothed with 5-year running average.

For the CAA, we assume FWF = R when estimating freshwater flux since ice discharge from the CAA is negligible compared with runoff[44]. Thus, freshwater flux from the CAA is derived from runoff predicted by RACMO2.3. Note that both ice runoff and tundra runoff are considered in the freshwater flux calculation (Supplementary Fig. 12).

For Arctic sea ice, we focus just on recent accelerated melting of multi-year ice, which results in the loss of ice area and extent, rather than the much larger contribution from the annual freeze-thaw cycle, which forms significant freshwater through fractionation (Supplementary Methods), but is more difficult to quantify with remote methods. We use three data sets (area, extent and volume; see Supplementary Methods and Supplementary Fig. 5) to estimate freshwater flux from accelerated melting of Arctic sea ice. All three approaches give similar results (Supplementary Fig. 6). The one based on volume is shown in Fig. 4. To convert area and extent to mass, we assume that sea ice thickness is 2 m (ref. 47) and sea ice density is 900 kg m^{-3}. Annual melting of Arctic sea ice is estimated by fitting annual minimum Arctic sea ice mass estimates with a linear state space model (Supplementary Methods).

References

1. Stommel, H. Thermohaline convection with two stable regimes of flow. *Tellus* **13**, 224–230 (1961).
2. Rooth, C. Hydrology and ocean circulation. *Prog. Oceanogr.* **11**, 131–149 (1982).
3. Broecker, W. S., Peteet, D. M. & Rind, D. Does the ocean-atmosphere system have more than one stable mode of operation. *Nature* **315**, 21–26 (1985).
4. Broecker, W. S. Unpleasant surprises in the greenhouse? *Nature* **328**, 123–126 (1987).
5. Wood, R. A., Keen, A. B., Mitchell, J. F. B. & Gregory, J. M. Changing spatial structure of the thermohaline circulation in response to atmospheric CO2 forcing in a climate model. *Nature* **401**, 508–508 (1999).
6. Jiang, Y., Dixon, T. H. & Wdowinski, S. Accelerating uplift in the North Atlantic region as an indicator of ice loss. *Nat. Geosci.* **3**, 404–407 (2010).
7. Rignot, E., Velicogna, I., van den Broeke, M. R., Monaghan, A. & Lenaerts, J. T. M. Acceleration of the contribution of the Greenland and Antarctic ice sheets to sea level rise. *Geophys. Res. Lett.* **38**, L05503 (2011).
8. Enderlin, E. M. *et al.* An improved mass budget for the Greenland ice sheet. *Geophys. Res. Lett.* **41**, 866–872 (2014).
9. Velicogna, I., Sutterley, T. C. & van den Broeke, M. R. Regional acceleration in ice mass loss from Greenland and Antarctica using GRACE time-variable gravity data. *Geophys. Res. Lett.* **41**, 8130–8137 (2014).
10. Yang, Q., Dixon, T. H. & Wdowinski, S. Annual variation of coastal uplift in Greenland as an indicator of variable and accelerating ice mass loss. *Geochem. Geophys. Geosyst.* **14**, 1569–1589 (2013).
11. Fichefet, T. *et al.* Implications of changes in freshwater flux from the Greenland ice sheet for the climate of the 21st century. *Geophys. Res. Lett.* **30**, 1911 (2003).
12. Jungclaus, J. H., Haak, H., Esch, M., Roeckner, E. & Marotzke, J. Will Greenland melting halt the thermohaline circulation? *Geophys. Res. Lett.* **33**, L17708 (2006).
13. Stouffer, R. J. *et al.* Investigating the causes of the response of the thermohaline circulation to past and future climate changes. *J. Clim.* **19**, 1365–1387 (2006).
14. Hu, A. X., Meehl, G. A., Han, W. Q. & Yin, J. J. Effect of the potential melting of the Greenland Ice Sheet on the Meridional Overturning Circulation and global climate in the future. *Deep Sea Res. II* **58**, 1914–1926 (2011).
15. Swingedouw, D. *et al.* Decadal fingerprints of freshwater discharge around Greenland in a multi-model ensemble. *Clim. Dyn.* **41**, 695–720 (2013).
16. Ridley, J. K., Huybrechts, P., Gregory, J. M. & Lowe, J. A. Elimination of the Greenland ice sheet in a high CO2 climate. *J. Clim.* **18**, 3409–3427 (2005).
17. Brunnabend, S. E., Schröter, J., Rietbroek, R. & Kusche, J. Regional sea level change in response to ice mass loss in Greenland, the West Antarctic and Alaska. *J. Geophys. Res. Oceans* **120**, 7316–7328 (2015).
18. Rahmstorf, S. Bifurcations of the Atlantic thermohaline circulation in response to changes in the hydrological cycle. *Nature* **378**, 145–149 (1995).
19. Rahmstorf, S. *et al.* Thermohaline circulation hysteresis: a model intercomparison. *Geophys. Res. Lett.* **32**, L23605 (2005).
20. Hawkins, E. *et al.* Bistability of the Atlantic overturning circulation in a global climate model and links to ocean freshwater transport. *Geophys. Res. Lett.* **38**, L10605 (2011).
21. Yashayaev, I. & Loder, J. W. Enhanced production of Labrador Sea Water in 2008. *Geophys. Res. Lett.* **36**, L01606 (2009).
22. Kuhlbrodt, T., Titz, S., Feudel, U. & Rahmstorf, S. A simple model of seasonal open ocean convection. *Ocean Dyn.* **52**, 36–49 (2001).
23. Joyce, T. M. & Proshutinsky, A. Greenland's Island Rule and the Arctic Ocean circulation. *J. Mar. Res.* **65**, 639–653 (2007).
24. Bonin, J. & Chambers, D. Uncertainty estimates of a GRACE inversion modelling technique over Greenland using a simulation. *Geophys. J. Int.* **194**, 212–229 (2013).
25. Howat, I. M. & Eddy, A. Multi-decadal retreat of Greenland's marine-terminating glaciers. *J. Glaciol* **57**, 389–396 (2011).
26. Box, J. E. & Colgan, W. Greenland ice sheet mass balance reconstruction. part III: marine ice loss and total mass balance (1840–2010). *J. Clim.* **26**, 6990–7002 (2013).
27. Myers, P. G., Kulan, N. & Ribergaard, M. H. Irminger water variability in the west greenland current. *Geophys. Res. Lett.* **34**, L17601 (2007).
28. Thierry, V., de Boisséson, E. & Mercier, H. Interannual variability of the subpolar mode water properties over the Reykjanes Ridge during 1990–2006. *J. Geophys. Res.* **113**, C04016 (2008).
29. Holland, D. M., Thomas, R. H., de Young, B., Ribergaard, M. H. & Lyberth, B. Acceleration of Jakobshavn Isbrae triggered by warm subsurface ocean waters. *Nat. Geosci.* **1**, 659–664 (2008).
30. Joughin, I., Alley, R. B. & Holland, D. M. Ice-sheet response to oceanic forcing. *Science* **338**, 1172–1176 (2012).
31. Straneo, F. & Heimbach, P. North Atlantic warming and the retreat of Greenland's outlet glaciers. *Nature* **504**, 36–43 (2013).
32. Jakobsen, P. K., Ribergaard, M. H., Quadfasel, D., Schmith, T. & Hughes, C. W. Near-surface circulation in the northern North Atlantic as inferred from Lagrangian drifters: variability from the mesoscale to interannual. *J. Geophys. Res. Oceans* **108**, 3251 (2003).
33. Myers, P. G., Donnelly, C. & Ribergaard, M. H. Structure and variability of the West Greenland Current in Summer derived from 6 repeat standard sections. *Prog. Oceanogr.* **80**, 93–112 (2009).

34. Myers, P. G., Josey, S. A., Wheler, B. & Kulan, N. Interdecadal variability in Labrador Sea precipitation minus evaporation and salinity. *Prog. Oceanogr.* **73**, 341–357 (2007).

35. Josey, S. A. & Marsh, R. Surface freshwater flux variability and recent freshening of the North Atlantic in the eastern subpolar gyre. *J. Geophys. Res. Oceans* **110**, C05008 (2005).

36. Peterson, B. J. et al. Increasing river discharge to the Arctic Ocean. *Science* **298**, 2171–2173 (2002).

37. Haine, T. W. N. et al. Arctic freshwater export: status, mechanisms, and prospects. *Global Planet. Change* **125**, 13–35 (2015).

38. Castro de la Guardia, L., Hu, X. M. & Myers, P. G. Potential positive feedback between Greenland Ice Sheet melt and Baffin Bay heat content on the west Greenland shelf. *Geophys. Res. Lett.* **42**, 4922–4930 (2015).

39. Curry, B., Lee, C. M., Petrie, B., Moritz, R. E. & Kwok, R. Multiyear volume, liquid freshwater, and sea ice transports through Davis Strait, 2004–10*. *J. Phys. Oceanogr.* **44**, 1244–1266 (2014).

40. Bamber, J., van den Broeke, M., Ettema, J., Lenaerts, J. & Rignot, E. Recent large increases in freshwater fluxes from Greenland into the North Atlantic. *Geophys. Res. Lett.* **39**, L19501 (2012).

41. Lenaerts, J. T. M. et al. Irreversible mass loss of Canadian Arctic Archipelago glaciers. *Geophys. Res. Lett.* **40**, 870–874 (2013).

42. Ettema, J. et al. Higher surface mass balance of the Greenland ice sheet revealed by high-resolution climate modeling. *Geophys. Res. Lett.* **36**, L12501 (2009).

43. Noël, B. et al. Summer snowfall on the Greenland Ice Sheet: a study with the updated regional climate model RACMO2.3. *Cryosphere Disc.* **9**, 1177–1208 (2015).

44. Gardner, A. S. et al. Sharply increased mass loss from glaciers and ice caps in the Canadian Arctic Archipelago. *Nature* **473**, 357–360 (2011).

45. Zhang, J. L. & Rothrock, D. A. Modeling global sea ice with a thickness and enthalpy distribution model in generalized curvilinear coordinates. *Mon. Weather Rev.* **131**, 845–861 (2003).

46. Fetterer, F., Knowles, K., Meier, W. & Savoie, M. *Sea Ice Index* (National Snow and Ice Data Center, 2002).

47. Laxon, S., Peacock, N. & Smith, D. High interannual variability of sea ice thickness in the Arctic region. *Nature* **425**, 947–950 (2003).

48. Myers, P. G. Impact of freshwater from the Canadian Arctic Archipelago on Labrador Sea Water formation. *Geophys. Res. Lett.* **32**, L06605 (2005).

49. McGeehan, T. & Maslowski, W. Evaluation and control mechanisms of volume and freshwater export through the Canadian Arctic Archipelago in a high-resolution pan-Arctic ice-ocean model. *J. Geophys. Res.* **117**, C00D14 (2012).

50. Koenigk, T., Mikolajewicz, U., Haak, H. & Jungclaus, J. Variability of Fram Strait sea ice export: causes, impacts and feedbacks in a coupled climate model. *Clim. Dyn.* **26**, 17–34 (2006).

51. Peterson, B. J. et al. Trajectory shifts in the Arctic and subarctic freshwater cycle. *Science* **313**, 1061–1066 (2006).

52. Good, S. A., Martin, M. J. & Rayner, N. A. EN4: Quality controlled ocean temperature and salinity profiles and monthly objective analyses with uncertainty estimates. *J. Geophys. Res. Oceans* **118**, 6704–6716 (2013).

53. Rignot, E. & Kanagaratnam, P. Changes in the velocity structure of the Greenland ice sheet. *Science* **311**, 986–990 (2006).

54. Rhein, M. et al. Deep water formation, the subpolar gyre, and the meridional overturning circulation in the subpolar North Atlantic. *Deep Sea Res. II* **58**, 1819–1832 (2011).

55. Kieke, D. & Yashayaev, I. Studies of Labrador Sea Water formation and variability in the subpolar North Atlantic in the light of

56. Yashayaev, I., Seidov, D. & Demirov, E. A new collective view of oceanography of the Arctic and North Atlantic basins. *Prog. Oceanogr.* **132**, 1–21 (2015).

57. Lazier, J., Hendry, R., Clarke, A., Yashayaev, I. & Rhines, P. Convection and restratification in the Labrador Sea, 1990–2000. *Deep Sea Res. I* **49**, 1819–1835 (2002).

58. Vage, K. et al. Surprising return of deep convection to the subpolar North Atlantic Ocean in winter 2007–2008. *Nat. Geosci.* **2**, 67–72 (2009).

59. Swingedouw, D. et al. Bidecadal North Atlantic ocean circulation variability controlled by timing of volcanic eruptions. *Nat. Commun.* **6**, 6545 (2015).

60. Weijer, W., Maltrud, M. E., Hecht, M. W., Dijkstra, H. A. & Kliphuis, M. A. Response of the Atlantic Ocean circulation to Greenland Ice Sheet melting in a strongly-eddying ocean model. *Geophys. Res. Lett.* **39**, L09606 (2012).

61. McGeehan, T. & Maslowski, W. Impact of shelf basin freshwater transport on deep convection in the western Labrador sea. *J. Phys. Oceanogr.* **41**, 2187–2210 (2011).

62. Robson, J., Hodson, D., Hawkins, E. & Sutton, R. Atlantic overturning in decline? *Nat. Geosci.* **7**, 2–3 (2013).

63. Smeed, D. A. et al. Observed decline of the Atlantic meridional overturning circulation 2004-2012. *Ocean Sci.* **10**, 29–38 (2014).

64. Rahmstorf, S. et al. Exceptional twentieth-century slowdown in Atlantic Ocean overturning circulation. *Nature Clim. Change* **5**, 475–480 (2015).

Acknowledgements

West Greenland Current data were provided by the Danish Meteorological Institute (DMI) as well as the Greenland Institute of Natural Resource (GINR) for more recent years. We thank three anonymous reviewers for their insightful comments. This research was funded by NASA grants to T.H.D. and D.C.

Author contributions

Q.Y. conducted freshwater flux calculation and wrote the manuscript. T.H.D. designed the study and edited the manuscript. P.G.M. calculated the heat and salt fluxes of Irminger Water. J.B. and D.C. processed the GRACE data. M.R.v.d.B. provided the RACMO2.3 data. All authors discussed and commented on the manuscript.

Additional information

16

Modelled drift patterns of fish larvae link coastal morphology to seabird colony distribution

Hanno Sandvik[1], Robert T. Barrett[2], Kjell Einar Erikstad[1,3], Mari S. Myksvoll[4], Frode Vikebø[4], Nigel G. Yoccoz[3,5], Tycho Anker-Nilssen[6], Svein-Håkon Lorentsen[6], Tone K. Reiertsen[3], Jofrid Skarðhamar[4], Mette Skern-Mauritzen[4] & Geir Helge Systad[3]

Colonial breeding is an evolutionary puzzle, as the benefits of breeding in high densities are still not fully explained. Although the dynamics of existing colonies are increasingly understood, few studies have addressed the initial formation of colonies, and empirical tests are rare. Using a high-resolution larval drift model, we here document that the distribution of seabird colonies along the Norwegian coast can be explained by variations in the availability and predictability of fish larvae. The modelled variability in concentration of fish larvae is, in turn, predicted by the topography of the continental shelf and coastline. The advection of fish larvae along the coast translates small-scale topographic characteristics into a macroecological pattern, *viz.* the spatial distribution of top-predator breeding sites. Our findings provide empirical corroboration of the hypothesis that seabird colonies are founded in locations that minimize travel distances between breeding and foraging locations, thereby enabling optimal foraging by central-place foragers.

[1] Centre for Biodiversity Dynamics, Department of Biology, Norwegian University of Science and Technology, 7491 Trondheim, Norway. [2] Department of Natural Sciences, Tromsø University Museum, PO Box 6050 Langnes, 9037 Tromsø, Norway. [3] Norwegian Institute for Nature Research, FRAM—High North Research Centre for Climate and the Environment, 9296 Tromsø, Norway. [4] Institute of Marine Research and Hjort Centre for Marine Ecosystem Dynamics, PO Box 1870 Nordnes, 5817 Bergen, Norway. [5] Department of Arctic and Marine Biology, University of Tromsø, PO Box 6050 Langnes, 9037 Tromsø, Norway. [6] Norwegian Institute for Nature Research, PO Box 5685 Sluppen, 7485 Trondheim, Norway. Correspondence and requests for materials should be addressed to H.S. (email: hanno@evol.no).

Advantages and disadvantages of colonial breeding are associated with social interactions, predator avoidance and food acquisition (exchange of information), for which selection pressures vary across species[1-3]. Whereas 13% of all birds are colonial breeders, ~98% of the some 330 seabird species breed in colonies comprised of often very dense aggregations of breeding territories[4,5]. Possible explanations include the minimization of travel distances between the nest and foraging locations (geometrical model[1,6]), enhanced food finding efficiency through information transfer (information centre hypothesis[7]), limited nest-site availability[1], predator avoidance[8] and mate-choice mechanisms[2].

The distance between neighbouring colonies is often negatively related to their size[9], indicating a regulating role of intraspecific competition and emphasising the importance of food availability[10-13]. The mechanisms behind dispersal of individuals from and their immigration into already existing colonies are increasingly understood[14,15]. However, the question remains what determined the placement of seabird colonies along a coast in the first place. Whereas recent colonization of empty spaces may have been determined by human-induced habitat changes[16], the initial colonization of an area can be expected to be linked to the predictability of habitat quality[17]. Habitat quality may be rather straightforward to quantify in terrestrial ecosystems[18-20], but this is not as easily accomplished for species feeding on oceanic prey[21,22]. Of the main factors contributing to nesting habitat suitability, features such as substrate and the absence of terrestrial predators are predictably determined by a site's physical properties, whereas food availability in the local marine environment is much less predictable in time and space. The formation and maintenance of large seabird colonies in relation to semi-stable physical oceanic phenomena, such as upwellings, fronts, gyres and polynyas, nonetheless indicate that there may, under some conditions, be a certain degree of predictability of food supply within the foraging range of seabirds[1,23,24]. Until now, however, no studies have documented more than theoretical associations between food availability and the distribution of seabird colonies, and little data exist to evaluate the predictions empirically[21,22,25,26].

We carried out a quantitative test of the association between modelled food availability/predictability and colony locations, using empirical data from the coast of Northern Norway (66–71°N). More than 90% of the two million pairs of Norwegian cliff-nesting seabirds (mainly Atlantic puffin *Fratercula arctica*, black-legged kittiwake *Rissa tridactyla* and common guillemot *Uria aalge*) occur north of 66°N (Fig. 1) (ref. 27). This distribution has been attributed to favourable physical oceanographic conditions linked to the northward-flowing Norwegian Coastal Current (NCC) and to bathymetry, because the exceptional marine productivity at the continental shelf break is close to the coastline (Fig. 1) (refs 27,28). Such productivity is essential to sustain the large numbers of breeding seabirds, whose foraging range is constrained by flight costs and the energetic demands of raising offspring[29].

Along the long stretch of Northern Norway's coast—roughly 1,200 km along its shortest line—there are 20 large seabird colonies (here defined as 10,000 breeding pairs or more). As suitable habitat (high cliffs, turf-covered islands and promontories that are all but inaccessible to terrestrial predators) is available along almost all the coast, and human disturbance is minimal for most of its parts, nest-site availability is not a factor limiting the distribution of seabird colonies in Northern Norway. We hypothesize that large seabird colonies are located in those areas where large numbers of planktonic organisms, for example, small crustaceans (especially *Calanus finmarchicus*), fish and their spawning products (eggs and larvae), occur predictably within the

foraging range of breeding birds during their breeding period. Such organisms are important food items for both seabirds and some of their fish prey (for example, capelin *Mallotus villosus*, 0- and I-group herring *Clupea harengus*, sandeels *Ammodytes* spp.), and an independent quantification of their abundance is thus a potential proxy of food availability for seabirds.

Coupled ocean circulation and biophysical models allow the quantification of seasonally predictable biological production[30,31]. Here, we use an ocean circulation model hindcast to quantify the drift of particles, such as eggs and larvae, from fish spawning grounds along the coast and their interaction with the physical environment[32], providing a proxy of the abundance of ichthyoplankton throughout the study area with a spatial resolution of 4 km[2]. The output of this model has been validated against empirical observations in the Barents Sea[32,33], and the proxy has already proven to be useful by giving novel insights into several aspects of seabird ecology, such as the day-to-day and interannual variation in stress hormones[34], fledging weight[35] and population dynamics[36] of common guillemots.

Using this approach, we relate the positions of seabird colonies directly to the modelled small-scale temporal and spatial variation of plankton along a coastline. Our findings provide an empirical corroboration of the hypothesis that the initial formation of colonies minimizes travel distances between breeding and foraging locations in a marine habitat.

Results

Physical correlates of high larval density. Despite the fact that the NCC transports spawning products along the entire

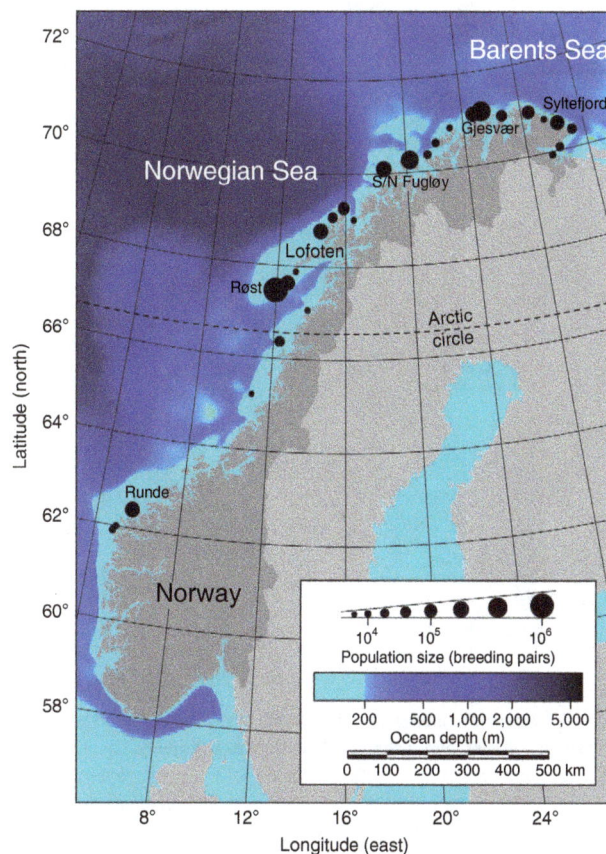

Figure 1 | Map of Norway displaying the largest seabird colonies and sea bottom topography. All colonies with more than 5,000 breeding pairs of cliff-nesting species are shown. The entire continental shelf (turquoise) north of 61°N was used as initiation area of the generic larval drift model.

Norwegian coast into the Barents Sea (Supplementary Fig. 1a), the larval drift models show clearly that these particles are not uniformly distributed along the coast. Instead, their distribution is rather patchy with widely varying concentrations (Fig. 2a). Physical features explain 39% of the variation in particle density (Supplementary Table 1). This is mainly due to a gradient on a large spatial scale, *viz* a decrease in particle concentration towards the east. Since the particle trajectories spread out as they approach the Barents Sea, the maximum is reached at 68° 15′ N, where the currents are funnelled round the Lofoten Islands by the narrowing continental shelf.

The smaller-scale patchiness is caused by the interaction between the NCC and coastal morphology, two aspects of which have been quantified: in areas where the continental shelf is narrow, the NCC is constricted near the coast, so that bathymetry explains 7% of the variation in particle density. A further 10% are explained by the structure of the coastline. This is likely because the NCC interacts with bank structures, islets and promontories

along the coast, generating stationary eddies that increase the residence time of passive particles such as crustaceans, fish eggs and larvae.

Relation of seabird colonies to larval density. The existing seabird colonies are located in or close to areas where particle concentrations are higher than the average for the respective coastal segments (Fig. 2a), with the association between seabird colonies and grid cells with high particle abundance being stronger than expected by chance ($P = 0.014$). The association between seabird colonies and grid cells with low interannual variability in prey abundance is even stronger ($P < 0.001$; Figs 2b and 3a). This means that seabird colonies along the North Norwegian coast are systematically located to give a much better and more stable, that is, predictable, access to suitable food than randomly chosen locations. The importance of low temporal variability in prey abundance is further highlighted by the fact that the colony locations are strongly associated with grid cells that have high minimum food availability across years ($P = 0.013$), but less so with grid cells that have high maximum food availability ($P = 0.059$; Supplementary Table 2).

Simulations based on fish species rather than generic particles, that is, simulations that take into account the spawning sites, growth patterns and vertical migration behaviour of particular species, showed divergent results. Herring larvae alone did not yield a significant association with colony sites ($P = 0.36$), whereas simulations based on cod (*Gadus morhua*) larvae ($P = 0.003$) or combinations of both species did ($P < 0.025$; Supplementary Table 2). Cod larvae thus produced a model with a much stronger association between larval abundance and colony placement than the generic particle model. A *post-hoc* analysis of the different spawning grounds of cod revealed that this very significant pattern was due mainly to one spawning ground, the Lofoten area ($P = 0.005$; Supplementary Table 3). This indicates that any perturbations to cod spawning in this area may have serious consequences for seabirds breeding in and to the north of Lofoten, an area that holds 19 out of the 21 large seabird colonies in Norway. The poorer performance of the herring simulation is explained by the spawning grounds of herring being further south, resulting in considerably lower concentrations of herring than cod larvae along the northern and eastern parts of the coast.

Significance of foraging distance. The strength of the association between seabird colonies and grid cells with high particle abundance varies depending on the foraging range of the seabirds (Fig. 3b,c). In the initial model, a foraging radius of 10 km is assumed. However, significant associations are obtained for foraging radii of up to 25 km, and associations are strongest for radii between 6 and 8 km ($P < 0.01$, Fig. 3b,c). This can be taken to indicate that seabird colonies are usually situated within 5–25 km of productive food patches.

Although recent tracking studies have revealed that several seabirds are capable of foraging at much larger distances from their breeding colony (> 400 km)[37], average trip length under normal conditions is much shorter (< 50 km)[38,39]. Long trips are costly and are most likely forced upon birds only under unusual food conditions[37]. Our results indicate that seabirds are rather successful in choosing breeding sites that minimize travel distances between food patches and the nest.

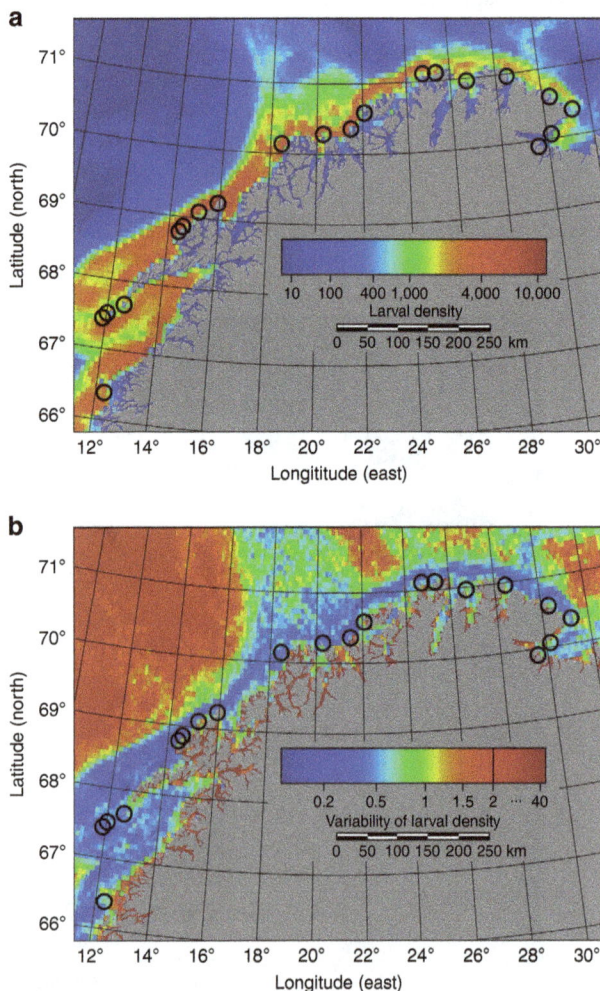

Figure 2 | Association between abundance of fish larvae and the location of seabird colonies along the coast of Northern Norway. Circles indicate foraging ranges with radii of 10 km around the 20 largest seabird colonies (at least 10,000 breeding pairs). (**a**) Map of seabird colonies and sites of high modelled larval abundance. There are more high-abundance grid cells within the circles than expected by chance ($P = 0.014$). (**b**) Map of seabird colonies and sites of high predictability (low variability) of modelled larval density. Variability is measured as quartile coefficients of dispersion. There are more low-variability grid cells within the circles than expected by chance ($P = 0.0003$).

Robustness of results. The association between seabird colonies and particle abundance, which is estimated using randomization tests, holds for a broad range of model parameterizations and is further corroborated by *post-hoc* tests (Fig. 3 and Supplementary

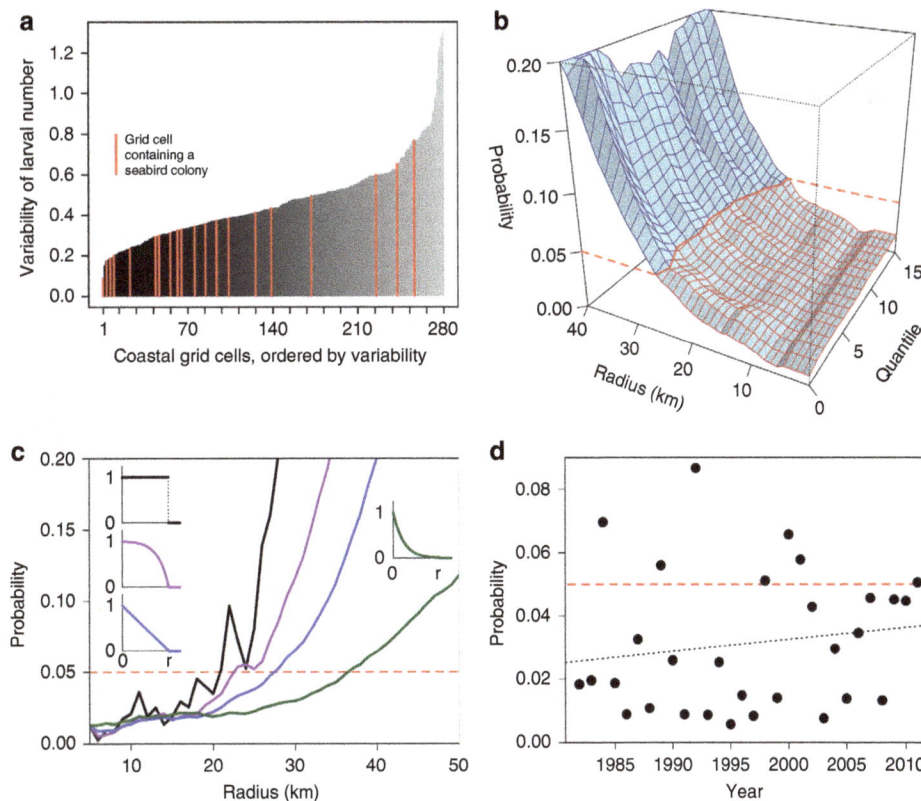

Figure 3 | Results of simulations and relaxation of the assumptions on which the initial model is based. (a) Histogram of the 280 grid cells along the coast, sorted by increasing variability of larval abundance (as measured by the quartile coefficient of dispersion). Grid cells containing the 20 largest seabird colonies, highlighted as red bars, are concentrated in the left part of the histogram, that is, they have significantly lower variability in larval abundance than randomly chosen grid cells. (If uniformly distributed, the 20 red bars would lie exactly between the 21 ticks of the x-axis.) (b-d) The panels report the probability that a sample of 20 randomized colony positions obtains a better availability of fish larvae than the actual positions of the 20 largest seabird colonies in Northern Norway. (Each point in the parameter space is based upon simulations with 100,000 replicates. The portion of the parameter space below the red line is statistically significant at the 5% level). (b) Probabilities given different radii of feeding ranges around seabird colonies (relaxation of assumption (vii)) and given different 30-quantiles of particle abundance for each grid cell (relaxation of assumption (iv)). The 0th 30-quantile corresponds to the minimum particle density within the 30-year period, the 15th 30-quantile to the median. The initial model uses a radius of 10 km and the 3rd 30-quantile. (c) Probabilities given different feeding radii (r) and different weighting schemes of particles within the feeding radius specified (relaxation of assumption (viii)): none-or-all threshold (black), convex weighting function (purple), linear weighting function (blue), concave weighting function (green). Small panels visualize the weighting functions. The initial model uses a linear weighting scheme. (d) Probabilities for single years. There is no significant trend in the probabilities (dotted line; slope 0.00038 ± 0.00047, $R^2 = 0.02$, $P = 0.42$).

Tables 2 and 3), indicating that it is very robust to changes in the background assumptions that had to be made. By modifying some of the initial assumptions, model fit could be substantially improved (all details in Supplementary Table 2), for example, by considering conditions in July alone rather than the average of the breeding season ($P = 0.002$), by using a wider definition of suitable coastline ($P < 0.001$ for the widest definition) or by using smaller or larger size thresholds for the inclusion of colonies than 10,000 breeding pairs ($P = 0.006$ for colonies with at least 5,000 breeding pairs). On the other hand, results were rather insensitive to transformation of larval abundance data (Supplementary Table 2), to the quantile used to represent larval counts (Fig. 3b) or to the weighting function of different foraging distances within the maximum foraging radius (Fig. 3c).

The *post-hoc* tests showed that the association was not merely due to one influential colony location; and that data for a single year of an even more highly resolved larval model (800 m) are compatible with the 4 km model (Supplementary Table 3). When decreasing the grid resolution of the simulations, the association becomes gradually weaker as sample size decreases and the grid

cell size exceeds foraging radius; however, even for a fourfold decrease in resolution, the association remains significant for cod ($P = 0.025$) and marginally so for the generic particles ($P = 0.051$). When considering probabilities for single years, the association between particles and seabird colonies is significant at the 5% level in 23 out of 30 years in the period from 1982 to 2011, the non-significant years do not cluster, and there is no temporal trend in the probabilities (Fig. 3d).

Discussion

During their breeding season, seabirds are central-place foragers depending on spatially patchy and temporally variable food. According to the geometrical model and optimal foraging theory, such conditions favour the evolution of colonial breeding[6]. Our findings accord with the hypothesis that seabird colonies are established in locations that minimize distances between breeding and foraging locations: colonies are considerably closer to the nearest high-density prey patch than suggested by the potential flight range of seabirds and than expected by chance. Although documented for colonial breeders in terrestrial ecosystems[18-20], this has been difficult to show in marine ecosystems, presumably because the small-scale distribution of prey items is harder to quantify (but see refs 21,22).

This analysis has been made possible by a coupled ocean circulation and biophysical model of ichthyoplankton abundance. At the same time, the use of a modelling approach introduces some additional uncertainty, because the empirical validation[32] is necessarily restricted to a fraction of the area and time period covered by the models. It is also obvious that the findings are open to alternative interpretations, and that the model does not identify all factors relevant for the establishment and maintenance of seabird colonies. There is, for instance, a suspicious absence of seabird colonies in two regions where Fig. 2a suggests high levels of larvae (viz, 67.7°–68.7°N and 69.3°–70.1°N). However, the interannual variability of larval abundance is somewhat higher in these regions (0.39 ± 0.12 (quartile coefficient of dispersion (QCD) ± s.d.; see the Methods: Simulation: iv), $N = 52$ grid cells) than in the surrounding colony locations (0.27 ± 0.02, $N = 4$; cf. Fig. 2b). Of course, a number of other causes can be involved in the absence of colonies, including habitat quality, predation pressure, human disturbance and stochastic factors.

One major cost of coloniality is increased competition among foragers at food patches[10-13]. To sustain coloniality, the food patches must thus be sufficiently large that sharing by foragers is not prohibitively costly, yet they must also be sufficiently scarce and ephemeral that information exchange about their location will result in a higher food intake relative to the alternative strategy of breeding alone[24]. In our case, the areas of temporary food retention are continually resupplied with fish larvae during their drift northwards in the NCC, thus reducing prey depletion and enhancing the food source for seabirds. Areas of particle accumulation may also exhibit high primary production caused by turbulence, which might synergistically improve the foraging conditions for seabird prey.

Although the data on larval abundance used here spanned a 30-year period, the spatial larvae aggregation patterns we have identified are almost certainly more permanent in nature, especially to the degree they are linked to coastal topography and bathymetry. According to historical[40], archaeological[41,42] and sedimentological evidence[43-45], seabird colonies are known to have been occupied for hundreds, and even thousands, of years. No such datings are available for Norway. However, the long-term importance of the Lofoten area as a cod spawning area is documented by the fact that this area has exported stockfish at least since the Viking age (that is, for 900 years)[46]. It is therefore possible that the distribution of large colonies along the Norwegian coast constitutes a macroecological pattern that has been stable for a long time, potentially since the end of the last glaciation, c. 9,000 years ago.

Methods

Oceanography of the study system. Variability in oceanographic features of the Northeastern North Atlantic causes pronounced variations in the distribution and abundance of larvae and juvenile fish in the marine pelagic ecosystem[32]. The physical conditions in the model domain are dominated by the influx of relatively warm Atlantic water through the Faeroe–Shetland Channel[47], continuing as a two-branch flow, with the outer branch known as the Norwegian Atlantic Slope Current (NASC) along the shelf break (Supplementary Fig. 1a)[48]. On the shelf, the less saline NCC, originating from south of Norway but fed by rivers all along the coast, is trapped between the NASC and the coast[49]. Interannual variations in its physical features affect the dispersal of ichthyoplankton[50]. The frontal zone area near the northern Norwegian shelf break separates the northward-flowing NASC and NCC, and is associated with meso-scale meanders, filaments and eddies[51]. Short-term variability of the flow is forced by variable winds and tide, whereas long-term features are controlled by topography such as banks and troughs.

Larval drift model. By using a daily hydrodynamic model archive with 4 km × 4 km horizontal resolution (vertical resolution into 32 sigma-levels; baroclinic time step 160 s; barotropic time step 5 s), we simulate the horizontal dispersion of passive particles along the Norwegian coast. The ocean circulation model provides a 30-year time series of hydrography from 1982 to 2011 stored on a daily basis[33]. This spatial resolution is sufficient in order to resolve the eddy field adequately[52]. An individual-based larval drift model was used to advect passive particles from potential spawning areas along the entire Norwegian coast. A full description of the features of this model has been provided by Myksvoll et al.[35] The larval drift model, as well as the ocean circulation model underlying it, have been validated empirically using observations from the Norwegian and Barents Sea[32,33].

In a generic model, the entire continental shelf (that is, all grid points with bottom depth less than 200 m; Fig. 1) from 61°N and northwards were used as initial release locations for passive particles (6,156 grid points in total). One particle was released per grid point per day from 1 March to 30 April, adding up to 369,360 passive particles in total. After release, the particles drifted freely with the ocean currents for 3 weeks, representing the fish egg stage, followed by a phase with vertical migration depending on swimming capability and light availability, representing the fish larval stage[53]. Although designed to represent the physiology and behaviour of fish larvae, the model also captures the drift of other, more or less passively drifting organisms, such as crustaceans or fish at the small fry stage.

Particle abundance was interpolated to a gridded data set with a latitudinal resolution of 0.075° (c. 8.3 km) and a longitudinal resolution of 0.150° (c. 5.4 to 6.8 km). The same grid was used as the basis for aggregation of seabird numbers and for simulation (cf. Supplementary Fig. 1b).

In addition to the generic model, we run two models for specific target species, viz Northeast Arctic cod and Norwegian spring-spawning herring. The cod and herring spawning areas overlap with the original release areas in the generic model, but they consist of several separated areas of limited geographical extent (Supplementary Fig. 1c)[54,55]. The cod larvae model includes a three weeks long free drifting stage before the eggs hatch into larvae with vertical migration[53]. The herring larvae model only includes the larval stage, as herring have adhesive benthic eggs. The models released a total of 94,500 cod eggs and 230,400 herring eggs from 1 March to 30 April.

Coastal morphology. The near-shore accumulation of particles is partly governed by physical properties of the coast. We quantified two aspects of coastal morphology for each coastal grid cell, viz the width of the continental shelf (bathymetry) and the degree of protrusion of the coastline (topography). Width of the continental shelf was defined as the shortest distance between the points of the coastline within the grid cell and the 200-m isobath (depth-contour), rounded to the nearest kilometre (cf. Fig. 1).

Protrusion was quantified by fitting a straight line through portions of coastline as follows: separately for each grid cell, other grid cells within a 40-km radius were considered, sampling from each grid cell the point protruding most towards the ocean (Supplementary Fig. 1d). The straight line was fitted to minimize squared distances for this sample of points (major axis regression), excluding the focus cell. The shortest distance (km) of the most protruding point in the focus cell to this line is referred to as the degree of protrusion of this grid cell. Headlands and outlying islands are characterized by positive protrusions, bays and fjords by negative protrusions. This algorithm could not be unambiguously applied to the southernmost grid cell of the Lofoten archipelago and the islands southwest of Lofoten (Røst, Værøy and others; in total, 9 grid cells, hereafter referred to as 'ambiguous'; cf. Supplementary Fig. 1d). In these cases, protrusion was defined as the distance from the nearest coastal point.

The effect of coastal morphology on particle density was investigated using linear models. Model selection was based on Akaike's Information Criterion. To account for ambiguities regarding protrusion, models were rerun excluding ambiguous grid cells. To correct for spatial autocorrelation, P-values were obtained using randomization tests (see below, 'Spatial autocorrelation'). The proportion of randomized linear models obtaining an equal or larger R^2 than the original linear model was used as a corrected P-value.

Seabird colony data. The position and size of all seabird colonies north of 66°N that at any one time comprised 1,000 or more pairs were extracted from the Norwegian seabird database (http://www.seapop.no; cf. Supplementary Table 4). Entries in this database date back to the mid-twentieth century. Species included in the counts were cliff-nesting seabirds known to feed at least partly on fish larvae or the forage fish predators of larvae, that is, razorbill (Alca torda), Atlantic puffin (Fratercula arctica), northern fulmar (Fulmarus glacialis), black-legged kittiwake (Rissa tridactyla), common guillemot (Uria aalge) and Brünnich's guillemot (Uria lomvia). Apart from the fulmar that may forage far out at sea[56] (but whose breeding numbers are small in Norway), these are species with, under normal conditions, foraging ranges of c. 20–50 km from the nest[38,39], that is, considerably larger than the dimension of grid cells.

A grid cell of the larval drift model was considered to contain a seabird colony if the maximum sum of breeding pairs of all species within the grid cell at any point in time was at least 10,000 (but other population thresholds were also considered;

see 'Simulation'). For each grid cell, we considered the maximum overall population count available across years, rather than the count from a specific year, because we were interested in the position of colonies in a long-term perspective, independent of short-term variation in the size of breeding populations. This procedure yielded a sample of 20 seabird colonies of at least 10,000 breeding pairs.

Spatial autocorrelation. The spatial autocorrelation of observations introduces a statistical dependence that precludes the use of standard tests. Probabilities were therefore estimated using randomization tests, that is, a Monte Carlo method involving the repeated random reordering of the data. Based on at least 100,000 replications, probabilities were defined as the proportion of reordered data sets that obtained a test statistic as extreme as, or more extreme than, the one observed.

As the former method may not remove all covariation between variables, it was supplemented by the toroidal-shift method as a *post-hoc* test. This test entails shifting or rotating one point pattern against the other, while leaving both point patterns in their original order[57,58]. This was accomplished by shifting grid cells along the coastline, treating the latter as a closed ring (torus), that is, by connecting the two ends of the coastline. To avoid edge effects in models involving colony locations, both ends of the coastline were defined as the grid cell placed midway between the ultimate and penultimate colonies, reducing the number of colonies included by 2. The sample size of this test equals the number of coastal grid cells.

Simulation. Simulations were used in order to estimate the probability that randomly distributed seabird colonies would have equally good or better access to food particles than actual seabird colonies. Each simulation consisted of a randomization test that randomly defined n different coastal grid cells as seabird colonies (where n is the number of actual colonies of a certain size in Northern Norway), aggregated larval abundance within a certain feeding radius around each colony, summed the n values to obtain the total aggregation, and compared this value to the total aggregation around the actual colonies. This procedure was repeated 1,000,000 times for the initial model.

Before simulation, certain assumptions had to be made, involving (i) the prey species modelled; (ii) the months considered; (iii) how to quantify larval abundance; (iv) how to take the inter-annual variability of larval abundance into account; (v) the selection of grid cells deemed suitable for seabird colonies; (vi) the size threshold for actual colonies included; (vii) the radius of the foraging range and (viii) the weighting of larval numbers within the feeding range. The optimal parameterization and its importance for the results were unknown in advance. Therefore, the initial model was built upon a reasonable set of *a priori* assumptions, all of which were relaxed in separate analyses (*cf.* Supplementary Table 2), each based on simulations with 100,000 replicates:

(i) The initial model was based on generic particles rather than specific fish species. This means that the model captures features common to all planktonic prey items. It was contrasted with separate simulations based on cod larvae, herring larvae and combinations of these two. In one combined model, fish larvae abundance in each grid cell was defined as the sum of cod and herring larvae in each grid cell, scaled by the overall geometric mean for each species. In the other combinations, herring was considered up to a latitude of 68° 30′ N or 69° 30′ N, respectively, and cod north of these thresholds.

(ii) Output from larval drift models was extracted for May, June and July of each year, which constitute the main breeding season of the seabirds considered. The initial model is based on the mean larval abundance for this 3-month period for each year. Single months were also tested.

(iii) The larval abundances across grid cells were heavily skewed, with many grid cells having low abundances, and a few cells with very high abundances. A selection of grid cells containing one of the latter cells could theoretically bias the simulations, if the larval abundance in this cell overshadows the abundances of the remaining grid cells. To avoid this potential bias, the initial model was based on log-transformed larval abundances. This model was contrasted with (a) simulations using untransformed larval abundances and (b) simulations that merely considered the ordering of grid cells (that is, grid cells were ranked in the order of their larval abundances, and ranks, rather than larval abundances, were compared).

(iv) We hypothesize that optimal colony sites are those with high mean abundance and low inter-annual variability in abundance of larvae. As a measure that captures both aspects of central tendency and variability, we used the 1st decile (3rd 30-quantile) of larval abundances for each grid cell. This means that, for each grid cell and in 9 out of 10 years, larval abundances were equal to or higher than the specific value identified. This simulation was contrasted with others in which larval abundance was represented by 16 quantiles from the minimum (0th 30-quantile) to the median (15th 30-quantile) of the 30 annual estimates for each of the grid cells (*cf.* Fig. 3b). In addition, the arithmetic and geometric mean, the maximum value and three measures of variability were tested explicitly, *viz*, the variance, the coefficient of variation (CV = standard deviation/mean) and the QCD (QCD = interquartile range/median). The latter is more robust than the CV when distributions are skewed, and results reported for variability are therefore based on the QCD.

(v) The result of simulation depends on the selection of coast cells deemed suitable. We therefore produced four sets of grid cells, progressively omitting cells that did not have direct contact with oceanic cells or were placed inside fjords or sounds (Supplementary Fig. 1b). Our main results are reported for the most conservative selection of grid cells (that is, the smallest set). Grid cells outside the outermost colonies (that is, grid cells south/west of the southernmost actual colony on the Norwegian Sea coast and south/east of the southeasternmost actual colony on the Barents Sea coast) were disregarded.

(vi) In the initial model, the threshold for actual colonies included was arbitrarily set at $\geq 10,000$ breeding pairs (20 colonies). In addition, we used other thresholds for the inclusion of actual seabird colonies, *viz*, $\geq 5,000$ pairs, $\geq 20,000$ pairs, $\geq 50,000$ pairs and $\geq 100,000$ pairs, resulting in samples of 9 to 27 colonies (Fig. 1 and Supplementary Fig. 1b).

(vii) Different seabird species have different foraging radii, and foraging ranges reported in the literature are likely to be greater than the distances preferred by the birds (assuming that seabirds are selected to minimize the energetic costs of flight). We therefore had no *a priori* estimate of what the relevant radius might be, and assumed 10 km as a reasonable first guess. In addition, radii of foraging range were varied by 1 km between 5 and 50 km (*cf.* Fig. 3b).

(viii) Different assumptions can be made about how seabirds choose between prey patches in the vicinity of their breeding colony. It is likely that prey close to the colony is preferred over distant prey, but the shape of this preference function has not been quantified. In the initial model, we assume that the degree of preference decreases linearly from 100% at a distance of 0 to 1% at the maximum allowed feeding distance, by weighting larval numbers accordingly. Simulations based on this assumption were contrasted with models with alternative weighting schemes, *viz*, a threshold shape (all prey patches within the feeding radius are weighted equally), a concave and a convex shape (both based on exponential functions reaching 1% at the predefined maximum radius; see Fig. 3c for details).

Post-hoc tests. Based on the results obtained for the initial model and the set of models relaxing its assumptions, a number of further tests were devised (*cf.* Supplementary Table 3). Each of these *post-hoc* tests was based on simulations with 100,000 replicates.

Exclusion of the most influential colony. To ensure that the patterns found are not merely driven by one colony with extremely high larval numbers in its vicinity, the most influential colony was excluded. This was done for the initial model as well as the model of particle variability (QCD) and the model for cod larvae. In each case, the colony that obtained the highest (in case of variability, lowest) number of particles was identified, and the respective grid cell removed from the simulation.

Higher-resolution larval drift model. A sensitivity test was performed using an ocean circulation model with higher resolution (horizontal resolution 800 m × 800 m; vertical resolution into 35 sigma-levels; baroclinic time step 60 s; barotropic time step 1 s; output fields of model variables stored on hourly basis)[59,60]. As it was not feasible to process data for 30 seasons at this resolution, the model was run for cod in a single year (2010), the results were interpolated to the same grid as the original data set, and contrasted with the results of the 4-km model for the same fish species and year.

Lower-resolution coastal grid. To ensure that the associations found do not critically depend on the grid chosen, we varied the grid scale by collapsing neighbouring grid cells into larger cells and base the simulations on the new grid obtained. This was done for the generic model and the cod model. We collapsed pairs of grid cells (thus obtaining a latitudinal resolution of 0.075° and a longitudinal resolution of 0.300°) and quadruples of grid cells (thus obtaining a latitudinal resolution of 0.15° and a longitudinal resolution of 0.30°). As there are two different ways to collapse pairs and four different ways to collapse quadruples, all versions were simulated (each 100,000 times), and the average probability is reported. Averaging probabilities is adequate in this case, because each probability expresses a proportion (*viz*, the fraction of simulations that obtained equal or higher larval counts around randomized colonies than around actual colonies).

Toroidal-shift method. The toroidal-shift method (see above, 'Spatial autocorrelation') was used as an alternative test that takes the spatial autocorrelation pattern of the variables into account. It was applied to the initial model, the model of particle variability (QCD) and the model for cod larvae. As the sample size of this test is the number of grid cells rotated ($N = 260$), the P-value cannot be lower than 0.004.

Separate analyses of cod spawning grounds. As the distribution of cod larvae gave an even better fit to seabird colonies than the distribution of generic particles, and because the spawning grounds of cod are well known, we quantified the relative importance of the latter. Spawning grounds were aggregated in seven areas, and tested separately.

Coastal characteristics. As both fine-scale measures of coastal topography (protrusion, breadth of continental shelf; *cf.* 'Coastal topography' above) turned out to be important for the distribution of particles along the coast, we quantified their direct importance for the location of seabird colonies. Simulations were carried out excluding the grid cells with ambiguous definitions of protrusion (see the section 'Coastal morphology').

Single years. Finally, each year of the 30-year-period was tested separately. This was done in order to ensure that the results were not driven by a few anomalous years (*cf.* Fig. 3d).

References

1. Wittenberger, J. F. & Hunt, G. L., Jr in *Avian Biology* Vol VIII (eds Farner, D. S., King, J. R. & Parkes, K. C.) 1–78 (Academic, 1985).
2. Danchin, É. & Wagner, R. H. The evolution of coloniality: the emergence of new perspectives. *Trends Ecol. Evol.* **12**, 342–347 (1997).
3. Rolland, C., Danchin, É. & de Fraipont, M. The evolution of coloniality in birds in relation to food, habitat, predation, and life-history traits: a comparative analysis. *Am. Nat.* **151**, 514–529 (1998).
4. Hamer, K. C., Schreiber, E. A. & Burger, J. in *Biology of Marine Birds* (eds Schreiber, E. A. & Burger, J.) 217–261 (CRC, 2002).
5. Coulson, J. C. in *Biology of Marine Birds* (eds Schreiber, E. A. & Burger, J.) 87–113 (CRC, 2002).
6. Horn, H. S. The adaptive significance of colonial nesting in the Brewer's blackbird (*Euphages cyanocephalus*). *Ecology* **49**, 682–694 (1968).
7. Ward, P. & Zahavi, A. The importance of certain assemblages of birds as "information centres" for food-finding. *Ibis* **115**, 517–534 (1973).
8. Barrett, R. T. Recent establishments and extinctions of Northern Gannet *Morus bassanus* colonies in North Norway, 1995–2008. *Ornis Nor.* **31**, 172–182 (2008).
9. Furness, R. W. & Birkhead, T. R. Seabird colony distributions suggest competition for food supplies during the breeding season. *Nature* **311**, 655–656 (1984).
10. Ashmole, N. P. The regulation of numbers of tropical oceanic birds. *Ibis* **103b**, 458–473 (1963).
11. Cairns, D. K. The regulation of seabird colony size: a hinterland model. *Am. Nat.* **134**, 141–146 (1989).
12. Wakefield, E. D., Phillips, R. A. & Matthiopoulos, J. Habitat-mediated population limitation in a colonial central-place forager: the sky is not the limit for the black-browed albatross. *Proc. R. Soc. B* **281**, 20132883 (2014).
13. Oppel, S. *et al.* Foraging distribution of a tropical seabird supports Ashmole's hypothesis of population regulation. *Behav. Ecol. Sociobiol.* **69**, 915–926 (2015).
14. Boulinier, T., McCoy, K. D., Yoccoz, N. G., Gasparini, J. & Tveraa, T. Public information affects breeding dispersal in a colonial bird: kittiwakes cue on neighbours. *Biol. Lett.* **4**, 538–540 (2008).
15. Cristofari, R. *et al.* Spatial heterogeneity as a genetic mixing mechanism in highly philopatric colonial seabirds. *PLoS ONE* **10**, e0117981 (2015).
16. Oro, D. & Ruxton, G. D. The formation and growth of seabird colonies: Audouin's gull as a case study. *J. Anim. Ecol.* **70**, 527–535 (2001).
17. Kildaw, S. D., Irons, D. B., Nysewander, D. R. & Buck, C. L. Formation and growth of new seabird colonies: the significance of habitat quality. *Mar. Ornithol.* **33**, 49–58 (2005).
18. Gibbs, J. P., Woodward, S., Hunter, M. L. & Hutchinson, A. E. Determinants of Great Blue Heron colony distribution in coastal Maine. *Auk* **104**, 38–47 (1987).
19. Gibbs, J. P. Spatial relationships between nesting colonies and foraging areas of Great Blue Herons. *Auk* **108**, 764–770 (1991).
20. Gibbs, J. P. & Kinkel, L. K. Determinants of the size and location of Great Blue Heron colonies. *Colon. Waterbirds* **20**, 1–7 (1997).
21. Ainley, D. G., Ford, R. G., Brown, E. D., Suryan, R. M. & Irons, D. B. Prey resources, competition, and geographic structure of Kittiwake colonies in Prince William Sound. *Ecology* **84**, 709–723 (2003).
22. Ford, R. G., Ainley, D. G., Brown, E. D., Suryan, R. M. & Irons, D. B. A spatially explicit optimal foraging model of Black-legged Kittiwake behavior based on prey density, travel distance, and colony size. *Ecol. Model.* **204**, 335–348 (2007).
23. Brown, R. G. B. & Nettleship, D. N. The biological significance of polynyas to Arctic colonial seabirds. *Can. Wildl. Serv. Occas. Pap.* **45**, 59–65 (1981).
24. Buckley, N. J. Spatial-concentration effects and the importance of local enhancement in the evolution of colonial breeding in seabirds. *Am. Nat.* **149**, 1091–1112 (1997).
25. Buckley, F. G. & Buckley, P. A. in *Behavior of Marine Animals: Current Perspectives in Research. Volume 4: Marine Birds* (eds Burger, J., Olla, B. L. & Winn, H. E.) 69–112 (Plenum, 1980).
26. Brown, C. R., Brown, M. B. & Ives, A. R. Nest placement relative to food and its influence on the evolution of avian coloniality. *Am. Nat.* **139**, 205–217 (1992).
27. Barrett, R. T., Lorentsen, S.-H. & Anker-Nilssen, T. The status of breeding seabirds in mainland Norway. *Atl. Seab.* **8**, 97–126 (2006).
28. Skjoldal, H. R., Dalpadado, P. & Dommasnes, A. in *The Norwegian Sea Ecosystem* (ed. Skjoldal, H. R.) 447–506 (Tapir, 2004).
29. Ellis, H. I. & Gabrielsen, G. W. in *Biology of Marine Birds* (eds Schreiber, E. A. & Burger, J.) 359–408 (CRC, 2002).
30. Svendsen, E. *et al.* An ecosystem modeling approach to predicting cod recruitment. *Deep Sea Res. 2* **54**, 2810–2821 (2007).
31. Daewel, U. *et al.* Coupling ecosystem and individual-based models to simulate the influence of environmental variability on potential growth and survival of larval sprat (*Sprattus sprattus* L.) in the North Sea. *Fish. Oceanogr.* **17**, 333–351 (2008).
32. Vikebø, F. B. *et al.* Real-time ichthyoplankton drift in Northeast Arctic cod and Norwegian spring-spawning herring. *PLoS ONE* **6**, e27367 (2011).
33. Lien, V. S., Gusdal, Y. & Vikebø, F. B. Along-shelf hydrographic anomalies in the Nordic Seas (1960–2011): locally generated or advective signals? *Ocean Dyn.* **64**, 1047–1059 (2014).
34. Barrett, R. T. *et al.* The stress hormone corticosterone in a marine top-predator reflects short-term changes in food availability. *Ecol. Evol.* **5**, 1306–1317 (2015).
35. Myksvoll, M. S., Erikstad, K. E., Barrett, R. T., Sandvik, H. & Vikebø, F. Climate-driven ichthyoplankton drift model predicts growth of top predator young. *PLoS ONE* **8**, e79225 (2013).
36. Erikstad, K. E., Reiertsen, T. K., Barrett, R. T., Vikebø, F. & Sandvik, H. Seabird–fish interactions: the fall and rise of a common guillemot *Uria aalge* population. *Mar. Ecol. Prog. Ser.* **475**, 267–276; **481**, 305 (2013).
37. Ponchon, A. *et al.* When things go wrong: intra-season dynamics of breeding failure in a seabird. *Ecosphere* **5**, art4 (2014).
38. Coulson, J. C. *The Kittiwake* (Poyser, 2011).
39. Harris, M. P. & Wanless, S. *The Puffin* 2nd ed (Poyser, 2011).
40. Petersen, Æ. Rita í Breiðafjarðareyjum: varpdreifing, stofnbreytingar, landnám og talningaraðferðir. *Náttúrufræðingurinn* **79**, 45–56 (2010).
41. Falk, K. *et al.* Seabirds utilizing the Northeast Water polynya. *J. Mar. Syst.* **10**, 47–65 (1997).
42. Simeone, A. & Navarro, X. Human exploitation of seabirds in coastal southern Chile during the mid-Holocene. *Rev. Chil. Hist. Nat.* **75**, 423–431 (2002).
43. Hiller, A., Wand, U., Kämpf, H. & Stackebrandt, W. Occupation of the Antarctic continent by petrels during the past 35000 years: inferences from a ^{14}C study of stomach oil deposits. *Polar Biol.* **9**, 69–77 (1988).
44. Gaston, A. J. & Donaldson, G. Peat deposits and thick-billed murre colonies in Hudson Strait and Northern Hudson Bay: clues to post-glacial colonization of the area by seabirds. *Arctic* **48**, 354–358 (1995).
45. Sun, L., Xie, Z. & Zhao, J. A 3,000-year record of penguin populations. *Nature* **407**, 858 (2000).
46. Barrett, J. *et al.* Detecting the medieval cod trade: a new method and first results. *J. Archaeol. Sci.* **35**, 850–861 (2008).
47. Eldevik, T. *et al.* Observed sources and variability of Nordic Seas overflow. *Nat. Geosci.* **2**, 406–410 (2009).
48. Orvik, K. A. & Skagseth, Ø. Heat flux variations in the eastern Norwegian Atlantic Current toward the Arctic from moored instruments, 1995–2005. *Geophys. Res. Lett.* **32**, L14610 (2005).
49. Skagseth, Ø., Drinkwater, K. & Terrile, E. Wind and buoyancy induced transport of the Norwegian Coastal Current in the Barents Sea. *J. Geophys. Res.* **116**, C08007 (2011).
50. Opdal, A. F., Vikebø, F. B. & Fiksen, Ø. Parental migration, climate, and thermal exposure of larvae: spawning in southern regions gives Northeast Arctic cod a warm start. *Mar. Ecol. Prog. Ser.* **439**, 255–262 (2011).
51. Orvik, K. A. A case-study of Doppler-shifted inertial oscillations in the Norwegian Coastal Current. *Cont. Shelf Res.* **15**, 1369–1381 (1995).
52. Nurser, A. J. G. & Bacon, S. The Rossby radius in the Arctic Ocean. *Ocean Sci.* **10**, 967–975 (2014).
53. Vikebø, F., Jørgensen, C., Kristiansen, T. & Fiksen, Ø. Drift, growth, and survival of larval Northeast Arctic cod with simple rules of behaviour. *Mar. Ecol. Prog. Ser.* **347**, 207–219 (2007).
54. Sætre, R., Toresen, R., Søiland, H. & Fossum, P. The Norwegian spring-spawning herring – spawning, larval drift and larval retention. *Sarsia* **87**, 167–178 (2002).
55. Sundby, S. & Nakken, O. Spatial shifts in spawning habitats for Arcto-Norwegian cod related to multidecadal climate oscillations and climate change. *ICES J. Mar. Sci.* **65**, 953–962 (2008).
56. Weimerskirch, H., Chastel, O., Cherel, Y., Henden, J.-A. & Tveraa, T. Nest attendance and foraging movements of northern fulmars rearing chicks at Bjørnøya Barents Sea. *Polar Biol.* **24**, 83–88 (2001).
57. Lotwick, H. W. & Silverman, B. W. Methods for analysing spatial processes of several types of points. *J. R. Stat. Soc. B* **44**, 406–413 (1982).
58. Fortin, M.-J., Drapeau, P. & Jacquez, G. M. Quantification of the spatial co-occurrences of ecological boundaries. *Oikos* **77**, 51–60 (1996).
59. Albretsen, J. *et al.* NorKyst-800 report no. 1: user manual and technical descriptions. *Fisken Havet* **2**, 1–46 (2011).
60. Skarðhamar, J., Skagseth, Ø. & Albretsen, J. Diurnal tides on the Barents Sea continental slope. *Deep Sea Res. 1* **197**, 40–51 (2015).

Acknowledgements

The study was funded by grants from the Norwegian Research Council (project number 216547 to K.E.E. and Centres of Excellence funding scheme, project number 223257, to H.S.) and the FRAM Centre (Flagship 'Fjord and Coast' to K.E.E.). We thank the Norwegian Environment Agency, the SEAPOP programme (http://www.seapop.no)

and all who have collected, collated and organized seabird distribution data on mainland Norway.

Author contributions
R.T.B., K.E.E. and H.S. designed the study; M.S.M., J.S. and F.V. carried out the larval drift models; H.S. performed simulations and statistical tests; H.S., R.T.B., M.S.M. and F.V. wrote the paper. All authors discussed the results and commented on the manuscript.

Additional Information

17

Oxygen depletion recorded in upper waters of the glacial Southern Ocean

Zunli Lu[1], Babette A.A. Hoogakker[2], Claus-Dieter Hillenbrand[3], Xiaoli Zhou[1], Ellen Thomas[4], Kristina M. Gutchess[1], Wanyi Lu[1], Luke Jones[2] & Rosalind E.M. Rickaby[2]

Oxygen depletion in the upper ocean is commonly associated with poor ventilation and storage of respired carbon, potentially linked to atmospheric CO_2 levels. Iodine to calcium ratios (I/Ca) in recent planktonic foraminifera suggest that values less than $\sim 2.5\,\mu mol\,mol^{-1}$ indicate the presence of O_2-depleted water. Here we apply this proxy to estimate past dissolved oxygen concentrations in the near surface waters of the currently well-oxygenated Southern Ocean, which played a critical role in carbon sequestration during glacial times. A down-core planktonic I/Ca record from south of the Antarctic Polar Front (APF) suggests that minimum O_2 concentrations in the upper ocean fell below $70\,\mu mol\,kg^{-1}$ during the last two glacial periods, indicating persistent glacial O_2 depletion at the heart of the carbon engine of the Earth's climate system. These new estimates of past ocean oxygenation variability may assist in resolving mechanisms responsible for the much-debated ice-age atmospheric CO_2 decline.

[1] Department of Earth Sciences, Syracuse University, Syracuse, New York 13244, USA. [2] Department of Earth Sciences, University of Oxford, Oxford OX1 3AN, UK. [3] British Antarctic Survey, Cambridge CB3 0ET, UK. [4] Department of Geology and Geophysics, Yale University, New Haven, Connecticut, USA. Correspondence and requests for materials should be addressed to Z.L. (email: zunlilu@syr.edu) or to R.E.M.R. (email: rosr@earth.ox.ac.uk).

The Southern Ocean is widely considered to be critical to global nutrient and carbon cycling, including over glacial–interglacial time scales[1]. As an area of incomplete nutrient utilization, it is a major source of CO_2 to the atmosphere today[2]. At present, old (CO_2 — and nutrient-rich and relatively O_2 — depleted) deep waters upwell along most of the Antarctic continental margin[3,4] (Fig. 1), release CO_2 into, and recharge O_2 from surface waters before they down-well in distinct areas, such as the Weddell and Ross seas, to form Antarctic Bottom Water (AABW). In the glacial Southern Ocean, strengthening of the biological pump due to enhanced iron supply[5,6], increased stratification[7], and expanded sea–ice cover[8], were among the dominant players in reducing atmospheric CO_2 by ~ 90 p.p.m.V. Each of these mechanisms could counterbalance the increased O_2 solubility due to lower glacial temperatures, leading to a reduction in the O_2 concentration of the seawater. Since the Southern Ocean is thought to have reduced its CO_2 leakage during glacial periods[1], it provides an ideal location to search for evidence of deoxygenation linked to CO_2 sequestration in the upper ocean.

During the last glacial period, deep waters surrounding Antarctica were less ventilated, and older than today (relative to the atmosphere)[9]. A recent quantitative O_2 proxy study based on benthic foraminiferal $\delta^{13}C$ indicates that decreased ventilation linked to a reorganization of glacial ocean circulation and a strengthened global biological pump significantly enhanced the ocean storage of respired carbon in the deep North Atlantic[10]. Early box-models hypothesized very low-oxygen levels in the high latitude Southern Ocean[11,12]. Proxies did not paint a clear picture for bottom-water O_2 concentrations in the glacial Southern Ocean[13]. Only a few studies on marine sediment cores south of the APF have found evidence for substantially lowered bottom water O_2 concentrations. There, authigenic uranium concentrations were elevated in sediments deposited during glacial Marine Isotope Stages (MIS) 2 and 6 (refs 14,15). By contrast, another study highlighted a transient stagnation event during the early stage of the last interglacial (MIS 5e)[16].

Bottom water or porewater redox proxies cannot capture upper ocean O_2 levels far from the continental shelf, so there is scant constraint on upper ocean oxygenation conditions in vast tracts of the open ocean[13]. A novel proxy, the I/Ca composition of marine carbonates, especially planktonic and benthic foraminiferal tests, has demonstrated its potential to reconstruct paleo-oxygenation levels in both the upper ocean[17-20] and bottom waters[21], respectively. The thermodynamically stable forms of iodine in seawater are iodate (IO_3^-) and iodide (I^-)[22].

The total concentrations of IO_3^- and I^- are relatively uniform in the world ocean at around $0.45\,\mu mol\,l^{-1}$ due to the residence time of $\sim 300\,kyr$ (ref. 23), supported by a more recent compilation of iodine concentrations in global rivers[24]. Therefore, the total iodine concentration in the global ocean likely remained invariant over the duration of a glacial termination ($\sim 6\,kyr$).

Iodate is taken up by marine organisms as a micronutrient in surface waters[25], but its concentration does not increase during the aging of deep waters[26,27], in contrast to those of the major nutrients nitrate and phosphate, probably due to the low I/C_{org} ratio of plankton[25]. Iodine speciation is strongly redox sensitive. IO_3^- is completely converted to I^- when oxygen is depleted[28]. Because IO_3^- is the only chemical form of iodine that is incorporated into the structure of carbonate[17], calcareous tests precipitated closer to an oxygen minimum zone (OMZ) will record lower I/Ca and vice versa. An OMZ is defined by $O_2 < 20\,\mu mol\,kg^{-1}$ in the Pacific Ocean and $O_2 < 50\,\mu mol\,kg^{-1}$ in the Atlantic Ocean[29].

In this paper, we use recent planktonic foraminifera and modern water column data to establish typical I/Ca values for the presence of an OMZ or O_2-depleted water. On the basis of this proxy development, the down-core record of planktonic foraminifera I/Ca obtained at site TC493/PS2547 indicates the persistent presence of oxygen-depletion in the upper waters of high latitude Southern Ocean during the last two glacial periods.

Results

Site selection. We measured I/Ca values on eleven planktonic foraminiferal species in modern to Holocene samples, and in one sample from a previous interglacial (Supplementary Table 1 and Supplementary Figs 1 and 2). We chose sites from well-oxygenated areas (for example, the North and sub-Antarctic South Atlantic), and sites located beneath OMZs, including Ocean Drilling Program (ODP) Sites 658, 709, 720 (Site 720: last interglacial samples), 849 and 1242. First, we use these data to further establish the foundations of the I/Ca proxy. Subsequently, we focus on an I/Ca down-core record on *Neogloboquadrina pachyderma* sinistral deposited during the last two glacial cycles in two sediment cores (PS2547 and TC493) recovered from the same location (71°09′ S, 119°55′ W, water depth 2,096 m) on a seamount in the Amundsen Sea (Fig. 1)[30]. The excellent carbonate preservation at this site[30] provides a unique window to reconstruct past upper ocean conditions south of the APF. Site

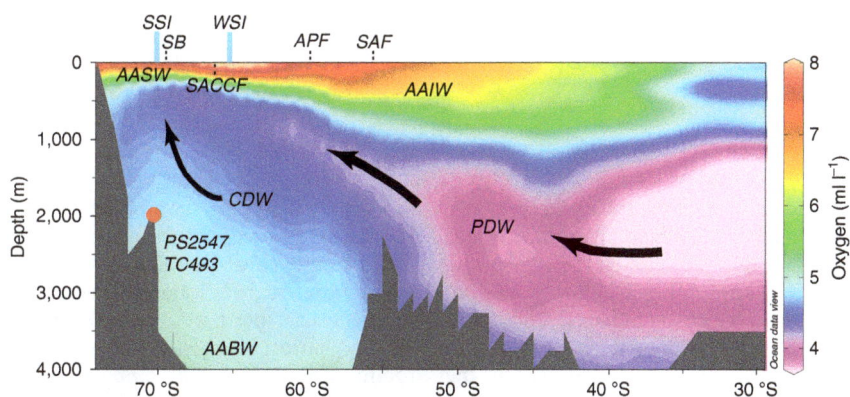

Figure 1 | Hydrographic section of Southern Ocean in the Pacific sector. Dissolved oxygen concentrations showing major water masses[50] and boundaries, average modern summer (SSI) and winter (WSI) sea-ice extent[62], and core site PS2547/TC493. The locations of the Antarctic Circumpolar Current (ACC) fronts are marked as SB, Southern Boundary of the ACC; SACCF, Southern ACC Front; APF, Antarctic Polar Front; SAF, Sub-Antarctic Front. AABW, Antarctic Bottom Water; AAIW, Antarctic Intermediate Water; AASW, Antarctic Surface Water; CDW, Circumpolar Deep Water; PDW, Pacific Deep Water. This graph is generated in Ocean Data View, using the Southern Ocean Atlas data set[63].

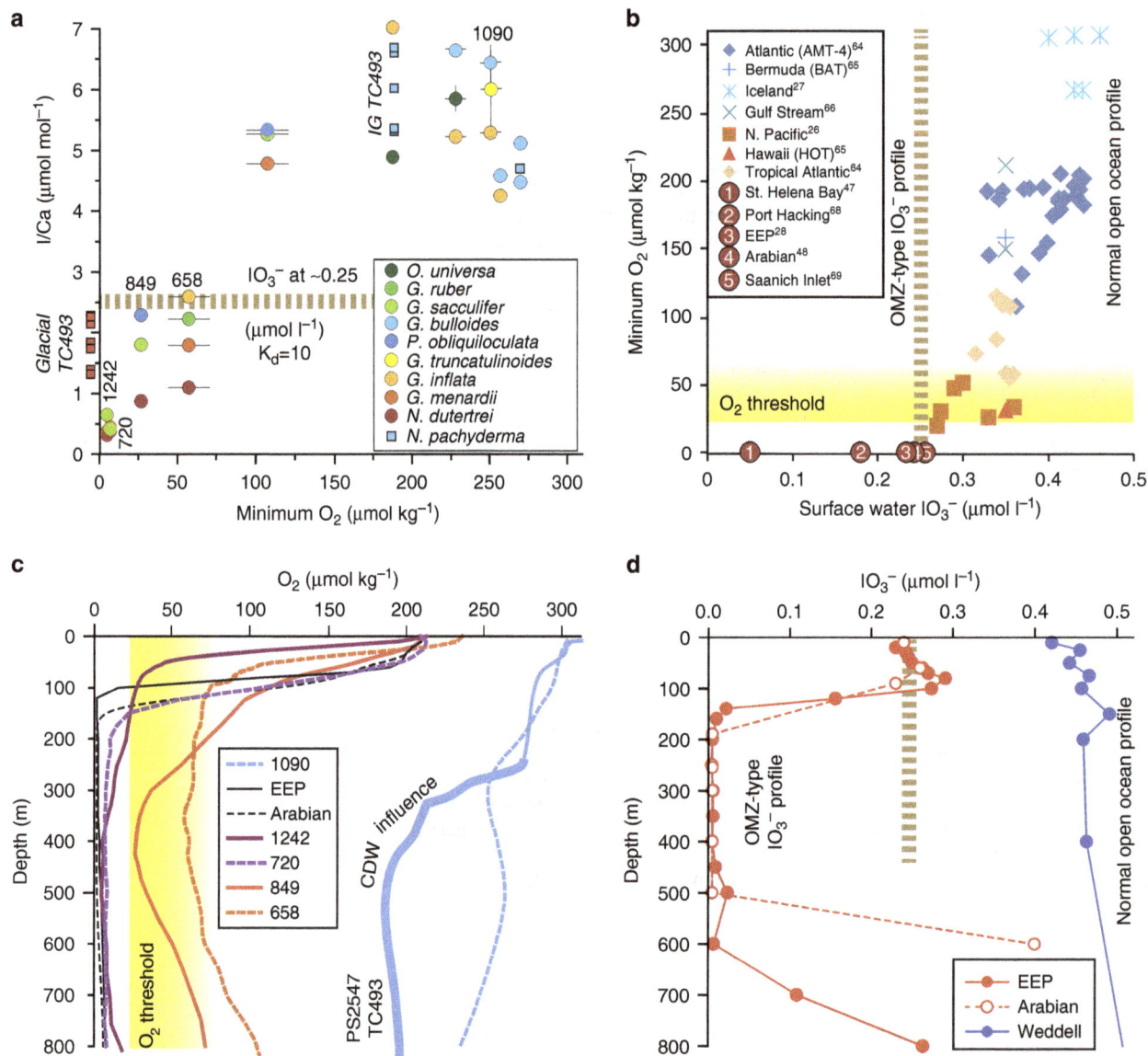

Figure 2 | I/Ca and modern OMZs. (**a**) Modern and late Holocene I/Ca in planktonic foraminiferal tests versus minimum O_2 concentrations in the upper 500 m of the water column (Note: I/Ca at Site 720 is from a MIS 5 sample). Error bars for y axis indicate the s.d. (1 s.d.) of triplicate measurements. Blue squares show down-core interglacial (IG) I/Ca data on *N. pachyderma* (s) from site TC493/PS2547 plotted against minimum O_2 concentrations in the modern water column, indicating well-oxygenated conditions. I/Ca for glacial *N. pachyderma* (s) tests are marked as red squares, indicating O_2 depletion. (**b**) Compilation of modern ocean surface water IO_3^- concentrations compared with minimum O_2 concentrations[26-28,47-48,64-69]. Brown dashed line indicates the surface water IO_3^- concentration of $\sim 0.25 \, \mu mol \, l^{-1}$ as a threshold value for differentiating OMZ-type and normal open ocean type of IO_3^- depth profiles. (**c**) O_2 depth profiles. Yellow shading marks 20-70 $\mu mol \, kg^{-1}$ O_2 concentration as the threshold for complete iodate reduction. (**d**) IO_3^- depth profiles at OMZ sites from the Eastern Equatorial Pacific (EEP)[28] and the Arabian Sea (station N8)[48] and at a well-oxygenated high-latitude site near the Weddell Sea (station PS71/179-1)[54].

TC493/PS2547 is currently bathed by Circumpolar Deep Water (CDW), which is overlain by a layer of Antarctic Surface Water (AASW), or Winter Water[31-33], and is located on the edge of the average modern summer sea–ice limit[34] (Fig. 1). During the Last Glacial Maximum (LGM), the sea–ice boundaries within the Southern Ocean shifted significantly northwards[35,36]. Thus, it is highly likely that site TC493/PS2547 was located within the permanent sea–ice zone during past glacial periods[34].

Age model and glacial polynyas. The sediments of core TC493/PS2547 consist mainly of foraminiferal ooze and sandy mud, with *N. pachyderma* (s) tests forming the primary carbonate component[30]. The age model of the record is based on

magnetostratigraphy combined with benthic foraminiferal (*Cibicides* cf. *wuellerstorfi*) oxygen isotope ($\delta^{18}O$) stratigraphy[30], tuned to the global benthic $\delta^{18}O$ stack[37]. Continuous deposition of foraminifera[30] indicates at least episodic opening of polynyas during glacial periods[34], because of its seamount location[38,39]. This scenario is consistent with the occurrence of the benthic foraminifera species *Epistominella exigua*, which is adapted to highly episodic phytodetritus supply[40].

I/Ca in foraminifera. I/Ca values in the modern and late Holocene samples are lower than $\sim 2.5 \, \mu mol \, mol^{-1}$ at sites with O_2 minima $< 70 \, \mu mol \, kg^{-1}$ in the upper ocean (0–500 m) (Fig. 2a).

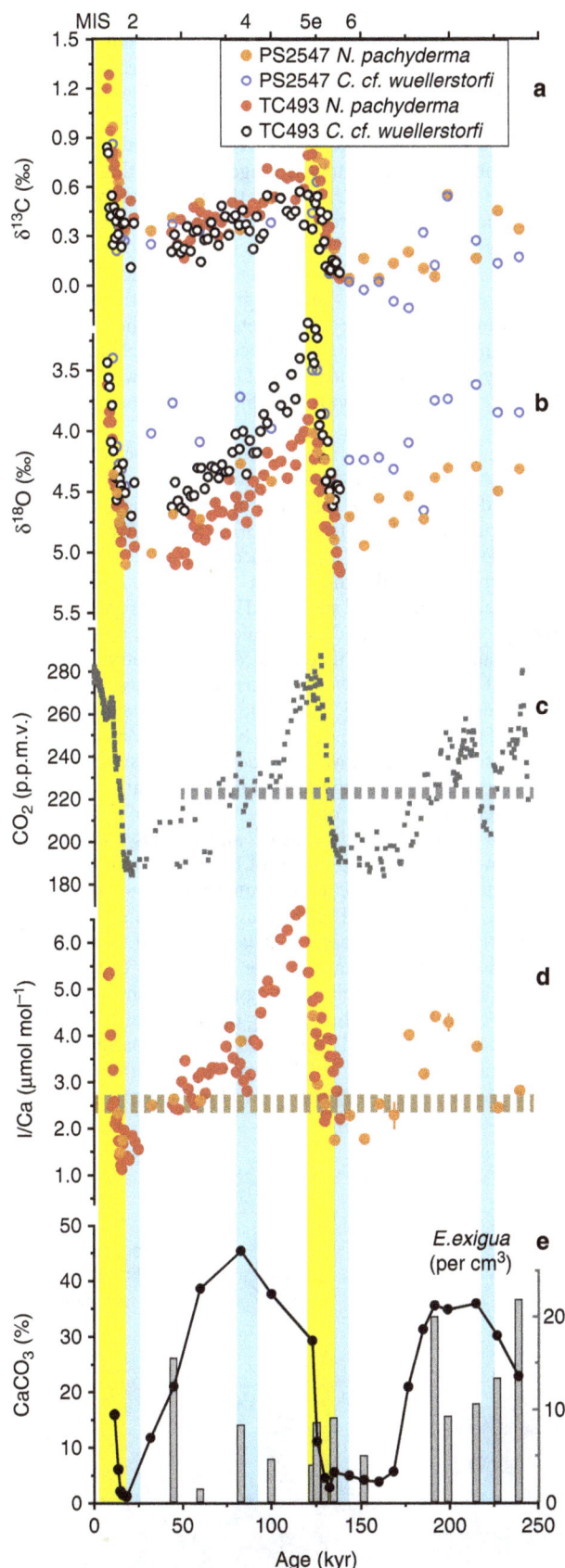

Figure 3 | Down-core records of studied sites. (a,b) Stable carbon and oxygen isotopes measured on benthic (*C.* cf. *wuellerstorfi*) and planktonic (*N. pachyderma*) foraminiferal tests[30]. It is well documented that δ13C of *N. pachyderma* (s) is offset by −1.0‰ south of the APF[49] and the values plotted here are uncorrected. **(c)** Atmospheric pCO2 record at EPICA Dome C is plotted for comparison, with dashed line indicating the long-term mean value of CO2 (50–270 ka) following Luethi *et al.*[56]. **(d)** I/Ca measured on *N. pachyderma* (s) tests from cores PS2547 and TC493. **(e)** Bulk sediment CaCO3 content from PS2547. Grey columns show the abundance of the benthic foraminifera *E. exigua* as number of tests per cm3 in core PS2547. Yellow shading highlights peak interglacial periods (including deglaciations), and blue shading marks glacial maxima and cooling intervals.

In contrast, recent planktonic foraminifera at sites with O_2 minima $> 100 \, \mu mol \, kg^{-1}$ have I/Ca $> 4 \, \mu mol \, mol^{-1}$, regardless of species (Fig. 2a). At site TC493/PS2547, the *N. pachyderma* (s) I/Ca ratio is high ($5–7 \, \mu mol \, mol^{-1}$) during the Holocene and MIS 5 relative to the lowest values ($< 2 \, \mu mol \, mol^{-1}$) during glacial MIS 2 and 6 (Fig. 3 and Supplementary Table 2).

Discussion

A tremendous amount of work has been devoted to developing foraminiferal proxies for temperature and pH, using global calibrations derived from core-top samples (for example, the Mg/Ca seawater temperature proxy[41]). Low I/Ca ratios of planktonic foraminifera unambiguously reveal the presence of low-oxygen waters, but a global calibration approach cannot establish planktonic foraminifera I/Ca as a linearly quantitative proxy for the continuum of dissolved O_2 concentration. Due to the stepwise nature of redox reactions[42], quantitative IO_3^- reduction does not occur before the dissolved oxygen is depleted to a certain threshold value, triggering nitrate reduction[43]. IO_3^- concentrations at water depths matching planktonic foraminiferal habitats are often not in equilibrium with the *in situ* O_2 concentrations, and O_2 contents which are sufficiently low to initiate major IO_3^- reduction may be detrimental to many species[44]. Instead, the I/Ca (recording the *in situ* IO_3^- concentration) is determined by the depth habitat of the foraminifera and the upper ocean IO_3^- mixing gradient. This mixing gradient is largely controlled by the surface water IO_3^- concentration and the depth of the IO_3^- reduction zone[28]. Nonetheless, a planktonic foraminifera proxy that semi-quantitatively approximates dissolved O_2 concentrations, indicative of the presence of an OMZ, can still be highly valuable for the paleoceanography community.

Before interpreting the down-core record from site TC493/PS2547, we identify the characteristic I/Ca signals for modern OMZs. IO_3^- depth profiles in the open ocean generally fall into two types (Fig. 2d): (1) the OMZ-type, with low surface water values and near-zero subsurface values in the OMZ; and (2) the normal open ocean type (for example, in a well-oxygenated water column), with relatively high surface water values and even higher subsurface values. A threshold O_2 concentration will cause complete IO_3^- reduction in the subsurface, and there may be a surface water IO_3^- threshold concentration below which complete IO_3^- reduction is likely to happen in the water column. Combined with modern water column IO_3^- and O_2 data, the I/Ca values measured on modern and late Holocene planktonic foraminifera consistently indicate that I/Ca $< 2.5 \, \mu mol \, mol^{-1}$ is equivalent to a surface water IO_3^- concentration of $< 0.25 \, \mu mol \, l^{-1}$, thus providing a marker for the presence of oxygen-depleted water with a subsurface O_2 concentration $< 20–70 \, \mu mol \, kg^{-1}$ (Fig. 2a–c).

Modern surface water IO_3^- concentrations are influenced by productivity and the presence of a subsurface OMZ[25,28]. To visualize this relationship, we compiled surface water IO_3^- concentrations from the literature and plotted them against the minimum O_2 concentrations in the subsurface water (Fig. 2b). The IO_3^- concentration broadly increases with the minimum O_2 concentration when the surface water IO_3^- concentration is $> 0.25 \, \mu mol \, l^{-1}$ (Fig. 2b). This correlation is likely a reflection of surface productivity versus subsurface respiration, because lower productivity leads to lower iodine uptake in surface water and less oxygen consumption by subsurface organic matter decomposition. In areas with a strong OMZ and near-zero O_2 values, the surface water IO_3^- concentrations are below $0.25 \, \mu mol \, l^{-1}$ (Fig. 2b). A partition coefficient K_d ($K_d = [I/Ca]/[IO_3^-]$ with units of $[\mu mol \, mol^{-1}]/[\mu mol \, l^{-1}]$) of ~ 10 was reported from abiological calcite synthesis experiments[17,20]. Using this K_d value, an IO_3^- concentration $< \sim 0.25 \, \mu mol \, l^{-1}$ results in I/Ca values $< \sim 2.5 \, \mu mol \, mol^{-1}$ in calcite. This estimate is consistent with modern I/Ca at OMZ Sites 658, 849 and 1242, as well as the last interglacial I/Ca value at Site 720 (Fig. 2a). Therefore, a surface water I/Ca value $< 2.5 \, \mu mol \, mol^{-1}$ indicates that a pronounced subsurface O_2 minimum exerted the dominant control on the upper ocean IO_3^- profile. This I/Ca threshold value does not seem to depend on foraminiferal species (Fig. 2a).

The O_2 threshold for maintaining an OMZ-type IO_3^- profile is useful for the paleoceanographic application of the planktonic I/Ca proxy. At O_2 concentrations $< 20 \, \mu mol \, kg^{-1}$, microbial processes become dominant[29], and IO_3^- likely would be completely reduced to I^- since the reaction is biologically mediated[45] (for example, ODP Sites 1242, 720 and 849 in Fig. 2a,c). ODP Site 658 is located at the northern edge of a shallow pocket of distinctively low-oxygen water with mean O_2 concentrations of $\sim 70 \, \mu mol \, kg^{-1}$ in the upper 200 m (ref. 46), which may be sufficiently low to generate an OMZ-type iodate profile. Three species of planktonic foraminifera analysed at ODP Site 1242 show exceptionally low I/Ca ratios around $0.5 \, \mu mol \, mol^{-1}$, corresponding to an IO_3^- concentration of $\sim 0.05 \, \mu mol \, l^{-1}$. Such a low IO_3^- concentration is comparable to that reported for a location where an extreme hypoxic event occurred[47]. Moreover, this low IO_3^- concentration implies that IO_3^- reduction should occur shallower than at Site 849 and at two sites with classic OMZ-type IO_3^- profiles (Eastern Equatorial

Pacific[28] and Arabian Sea[48]; Fig. 2c). A comparison of the O_2 profiles of these sites reveals that the O_2 threshold needs to be $> 50 \, \mu mol \, kg^{-1}$ to achieve a shallower IO_3^- reduction at Site 1242. Therefore, we suggest that I/Ca values lower than $\sim 2.5 \, \mu mol \, mol^{-1}$ indicate O_2 minima $< 20-70 \, \mu mol \, l^{-1}$. This O_2 range cannot be further narrowed down with the available information, and we refer to this range as the O_2 threshold for an OMZ-type IO_3^- profile. However, the threshold behaviour of IO_3^- reduction (relative to O_2) in subsurface waters does not necessarily lead to step changes in down-core records of planktonic I/Ca. This is because planktonic foraminifera typically record the IO_3^- mixing gradient in the top part of water column, above the O_2-depleted zone where rapid step changes in IO_3^- concentrations occur. Low planktonic I/Ca values may be driven by shoaling of O_2-depleted water, and/or by increasing productivity, both of which could change gradually over time.

The available data from modern and late Holocene planktonic foraminifera suggest that the I/Ca ratio acts as a robust (paleo-) proxy for determining the signature of O_2-depletion in the upper ocean (Fig. 2). At site TC493/PS2547, I/Ca was high ($5-7 \, \mu mol \, mol^{-1}$) during the Holocene and interglacial MIS 5 when compared with the lowest values ($< 2 \, \mu mol \, mol^{-1}$) characterizing peak glacial periods MIS 2 and 6 (Fig. 3). Changes in salinity, temperature and foraminiferal habitat, most likely, are not the main drivers for this record (Supplementary Discussion). The glacial I/Ca values of N. pachyderma (s) are best explained by the presence of a water mass with a dissolved O_2 content $< 70 \, \mu mol \, kg^{-1}$ close to, i.e., above or near, this site (Figs 2 and 3). We reiterate that the low I/Ca does not necessarily imply O_2-depleted seawater within the foraminiferal habitat.

At present, CDW wells up to a water depth of approximately 250–300 m in the Amundsen Sea[31] and has O_2 concentrations notably lower than the top 200 m of the water column (Fig. 2c). Although the interpretation of absolute values of planktonic $\delta^{13}C$ is far from straightforward in the seasonal ice zone (for example, disequilibrium from seawater[49]), it is reasonable to assume that CDW had a strong influence on the local water column during glacial periods, as its upwelling along the continental margin was probably responsible for the opening of the glacial polynyas. The CDW upwelling at site TC493/PS2547 today partly originates from Pacific Deep Water (PDW) moving southward from the equator, with a low-oxygen and high nutrient signature

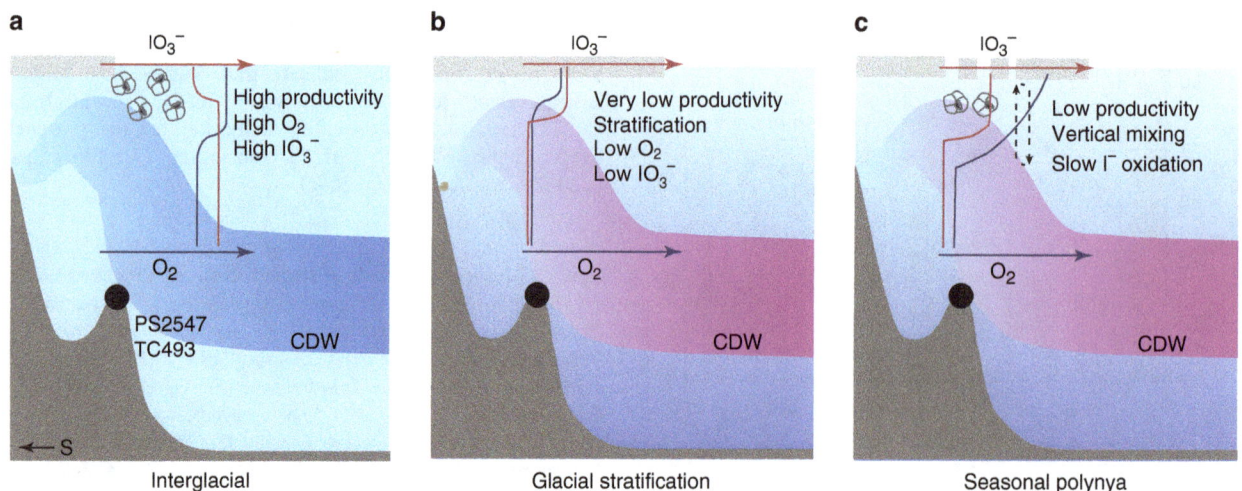

Figure 4 | Conceptual illustration of paleo-environmental changes. Upper ocean IO_3^- and O_2 profiles were influenced by circulation, productivity and polynyas over glacial cycles. (**a**) Well-oxygenated interglacial condition; (**b**) Relatively oxygen-depleted glacial conditions with expanded sea–ice cover; (**c**) Episodic polynya opening during glacials.

(Fig. 1)[50,51]. $\delta^{30}Si$ data from fossil diatoms and sponges indicate higher silicic acid concentrations in the Pacific sector of the Southern Ocean during the LGM, which further imply that either the southward transport of PDW was more efficient or PDW was less ventilated than today[52]. So glacial CDW was likely more O_2 depleted than during interglacials, and upwelling of this water contributed to the glacial I/Ca signal at site TC493/PS2547.

The oxidation of I^- to IO_3^- is thought to take from a few months up to 40 years[53]. Long-distance transport of well-oxygenated deep water with low IO_3^- concentrations ($<0.25\,\mu mol\,l^{-1}$) has not been documented in the modern ocean, but this scenario should be tested with further work on I^- oxidation kinetics. Today our site is bathed in CDW transported from a Pacific OMZ and the interglacial I/Ca values at site TC493/PS2547 do not show any remnant signal of the OMZ from the Pacific Ocean. On the basis of the knowledge about iodine speciation change in modern ocean, we interpret the observed glacial I/Ca values as a local signal, in principle, indicating the presence of a water mass with low O_2 and low IO_3^- vertically or horizontally close to the planktonic foraminiferal habitat.

In the setting of site TC493/PS2547 a coherent conceptual model for *N. pachyderma* (s) recording the presence/absence of O_2-depletion needs to integrate changes in productivity, sea–ice extent and the opening/closing of polynyas on time scales of glacial to seasonal cycles (Fig. 4). Although the polynyas complicate the interpretation of the proxy data, their presence arguably provides the only window for sufficient accumulation of planktonic microfossils to record upper ocean conditions during glacial periods at such high latitudes.

The modern O_2 profile at site TC493/PS2547 is defined by equilibration with the atmosphere at 0–250 m, and CDW influence below 250 m, as shown by the distinctively low O_2 concentrations (Fig. 2c). With O_2 above the threshold for complete IO_3^- reduction in the entire water column, the IO_3^- profile at site TC493/PS2547 should be similar to those at other high latitude locations, for example, site PS71/179–1 at 69°31′ S and 0°3′ W in the Weddell Sea[54] (Fig. 2d). An interglacial scenario of relatively high seasonal productivity, high O_2 and surface water IO_3^- ($>0.3\,\mu mol\,l^{-1}$) concentrations (Fig. 4a), is well described for the modern Atlantic sector of Southern Ocean[54].

Relative to the interglacial periods, the Southern Ocean experienced expanded sea–ice cover during glacial periods, and was less ventilated[9,36]. A more dynamic seasonal sea–ice cycle during ice ages would have increased water column stratification. Increased winter sea–ice formation (spatially and volumetrically) may have generated waters dense enough to sink ultimately to the bottom of the ocean[55]. On the other hand, melting of thicker sea ice during glacial-time summers in the seasonal sea–ice zone would have strengthened the halocline (not considering the influence of polynyas). So, the glacial seasonal stratification was likely stronger than today. These factors overall should have lowered the glacial O_2 concentrations in the Southern Ocean (Fig. 4b). At site TC493/PS2547, glacial I/Ca demonstrate that the IO_3^- profile was OMZ-like with complete IO_3^- reduction near the foraminiferal habitat (Fig. 2d). However, the dynamics of polynyas must be considered when interpreting the location of the low O_2 water mass, and the means by which the signal was recorded by *N. pachyderma* (s).

Without a polynya above site TC493/PS2547, glacial phytoplankton productivity under perennial sea–ice cover would have been relatively low due to the scarcity of light[34], and planktonic foraminifera depending on algae could not flourish. The water column would have been relatively poorly ventilated and strongly stratified during these times, creating the ideal environment for developing low O_2 conditions and an OMZ-type IO_3^- profile

(Fig. 4b). The episodic opening of a polynya re-established primary production (mainly by diatoms) and thus a planktonic foraminiferal habitat, vertical mixing and oxygenation in, at least, the uppermost part of the water column (Fig. 4c). While overall glacial-time production was reduced[30,34], the planktonic foraminifera preserved in the glacial sediments probably recorded transient I/Ca changes in the water column associated with polynya-induced peaks in glacial productivity. Modern open ocean productivity pulses do not lower IO_3^- concentrations to $<0.25\,\mu mol\,l^{-1}$ in oxygenated water (Supplementary Discussion)[54], thus the glacial I/Ca signal is most likely driven by changes in O_2 and not productivity.

The likely short-lived nature of glacial polynyas makes it difficult to envisage that very brief plankton blooms alone could produce a utilization-driven O_2 depletion in a cold, well-oxygenated Southern Ocean. For the same reason, it is difficult to imagine that the vertical mixing cells restricted by the size of the polynya could rapidly oxygenate voluminous nearby waters outside of the polynya, if most of the sea–ice covered areas were O_2-depleted. The more likely scenario is that the O_2 concentrations in the deep and abyssal Southern Ocean were generally lower during glacial periods than during interglacial periods. Upwelling of a more O_2-depleted CDW in the generally stratified upper ocean was mainly responsible for the IO_3^- reduction at site TC493/PS2547 (Fig. 4b), while the episodic opening of polynyas created habitable conditions for planktonic foraminifera to record the deoxygenation in the upper ocean (Fig. 4c). We suggest that the I/Ca proxy should be used as a local proxy, in principle. However, it is probably a reasonable speculation that this record (Fig. 3) shows oxygenation changes integrated over a regional volume of water (e.g. CDW).

The timing of glacial deoxygenation and deglacial reoxygenation at site PS2547 shows potential linkages to global climate changes (Fig. 3). The appearance of OMZ-type I/Ca values ($<\sim2.5\,\mu mol\,mol^{-1}$) during past glacial periods coincided with the lowering in atmospheric pCO_2 level below the long-term mean value[56]. Identical timing was reported for a strongly stratified Antarctic Zone coincident with pCO_2 decrease under the same threshold value (225 p.p.m.) in the Atlantic sector of the Southern Ocean[57]. Stronger stratification may be the common driving force for the productivity change (ODP Site 1094) and oxygenation change (PS2547/TC493) in the Antarctic zone. Furthermore, during the last interglacial period, the recovery of *N. pachyderma* (s) I/Ca values is offset from the $\delta^{18}O$ trend, with peak I/Ca occurring about 10 kyr after the peak $\delta^{18}O$ (Fig. 3), an observation worthy of future investigation.

Our I/Ca results build on other evidence[52,58,59] to make a stronger case for lower oxygen concentrations in CDW (and very likely PDW) during glacial periods. Altogether with the reconstructed O_2 content of deep waters in the glacial North Atlantic[10], these observations seem to allude to large scale deoxygenation in the glacial global ocean interior[60]. Future work providing quantitative reconstructions of bottom water O_2 concentrations in the Southern Ocean, especially south of the APF, and in other major ocean basins will shed new light on the mechanisms of sequestering atmospheric CO_2 during ice ages.

Methods

Foraminifera cleaning. Sediments were sampled from the split core sections and wet sieved. Approximately 40 tests of *N. pachyderma* sinistral were picked from the 200–250 μm size fraction of each sample. The cleaning procedure followed the Mg/Ca protocol of Barker *et al.*[61]. Cleaned glass slides were used to gently crack open all chambers. Clay particles were removed in an ultrasonic water bath. After adding NaOH-buffered 1% H_2O_2 solutions the samples were heated in boiling water for 10–20 min to remove organic matter. Calcareous microfossils were then thoroughly rinsed with de-ionized water. Reductive cleaning was not applied because contribution of iodine from Mn-oxides is deemed negligible[19].

ICP-MS measurements. The cleaned samples were dissolved in 3% nitric acid, and diluted to solutions with 50 p.p.m. Ca for analyses. Iodine calibration standards were freshly prepared also with 50 p.p.m. Ca. 0.5% tertiary amine solution (Spectrasol, CFA-C) was added to stabilize iodine within a few minutes after the sample dissolution. The measurements were performed immediately after that to minimize potential iodine loss. The sensitivity of iodine was tuned to above 80 kcps for a 1 p.p.b. standard. The precision for ^{127}I is typically better than 1%. The long-term accuracy is guaranteed by frequently repeated analyses of the reference material JCp-1 (ref. 17). The detection limit of I/Ca is on the order of 0.1 μmol mol^{-1}. The I/Ca measurements were performed using a quadrupole ICP-MS (Bruker M90) at Syracuse University.

References

1. Sigman, D. M., Hain, M. P. & Haug, G. H. The polar ocean and glacial cycles in atmospheric CO_2 concentration. *Nature* **466**, 47–55 (2010).
2. Marinov, I., Gnanadesikan, A., Toggweiler, J. R. & Sarmiento, J. L. The Southern Ocean biogeochemical divide. *Nature* **441**, 964–967 (2006).
3. Orsi, A. H., Whitworth, T. & Nowlin, W. D. On the meridional extent and fronts of the Antarctic Circumpolar Current. *Deep Sea Res. I* **42**, 641–673 (1995).
4. Schmidtko, S., Heywood, K. J., Thompson, A. F. & Aoki, S. Multidecadal warming of Antarctic waters. *Science* **346**, 1227–1231 (2014).
5. Martin, J. H. GLACIAL-interglacial CO_2 change: the iron hypothesis. *Paleoceanography* **5**, 1–13 (1990).
6. Martinez-Garcia, A. et al. Iron fertilization of the subantarctic ocean during the last ice age. *Science* **343**, 1347–1350 (2014).
7. Toggweiler, J. R. Variation of atmospheric CO2 by ventilation of the ocean's deepest water. *Paleoceanography* **14**, 571–588 (1999).
8. Stephens, B. B. & Keeling, R. F. The influence of Antarctic sea ice on glacial-interglacial CO_2 variations. *Nature* **404**, 171–174 (2000).
9. Skinner, L. C., Fallon, S., Waelbroeck, C., Michel, E. & Barker, S. Ventilation of the deep southern ocean and deglacial CO_2 rise. *Science* **328**, 1147–1151 (2010).
10. Hoogakker, B. A. A., Elderfield, H., Schmiedl, G., McCave, I. N. & Rickaby, R. E. M. Glacial-interglacial changes in bottom-water oxygen content on the Portuguese margin. *Nat. Geosci.* **8**, 40–43 (2015).
11. Knox, F. & McElroy, M. B. Changes in atmospheric CO_2 - influence of the marine biota at high-latitude. *J. Geophys. Res. Atmos.* **89**, 4629–4637 (1984).
12. Archer, D. E. et al. Model sensitivity in the effect of Antarctic sea ice and stratification on atmospheric pCO(2). *Paleoceanography* **18** (2003).
13. Jaccard, S. L. & Galbraith, E. D. Large climate-driven changes of oceanic oxygen concentrations during the last deglaciation. *Nat. Geosci.* **5**, 151–156 (2012).
14. Ceccaroni, L. et al. Late Quaternary fluctuations of biogenic component fluxes on the continental slope of the Ross Sea, Antarctica. *J. Mar. Syst.* **17**, 515–525 (1998).
15. Frank, M. et al. Similar glacial and interglacial export bioproductivity in the Atlantic sector of the Southern Ocean: Multiproxy evidence and implications for glacial atmospheric CO_2. *Paleoceanography* **15**, 642–658 (2000).
16. Hayes, C. T. et al. A stagnation event in the deep South Atlantic during the last interglacial period. *Science* **346**, 1514–1517 (2014).
17. Lu, Z., Jenkyns, H. C. & Rickaby, R. E. M. Iodine to calcium ratios in marine carbonate as a paleo-redox proxy during oceanic anoxic events. *Geology* **38**, 1107–1110 (2010).
18. Hardisty, D. S. et al. An iodine record of Paleoproterozoic surface ocean oxygenation. *Geology* **42**, 619–622 (2014).
19. Zhou, X. L., Thomas, E., Rickaby, R. E. M., Winguth, A. M. E. & Lu, Z. L. I/Ca evidence for upper ocean deoxygenation during the PETM. *Paleoceanography* **29**, 964–975 (2014).
20. Zhou, X. et al. Upper ocean oxygenation dynamics from I/Ca ratios during the Cenomanian-Turonian OAE 2. *Paleoceanography* **30** (2015).
21. Glock, N., Liebetrau, V. & Eisenhauer, A. I/Ca ratios in benthic foraminifera from the Peruvian oxygen minimum zone: analytical methodology and evaluation as a proxy for redox conditions. *Biogeosciences* **11**, 7077–7095 (2014).
22. Wong, G. T. F. The marine geochemistry of iodine. *Rev. Aquat. Sci.* **4**, 45–73 (1991).
23. Broecker, W. S. & Peng, H. T. *Tracers in the Sea* 690 (Lamont-Doherty Geological Observatory, 1982).
24. Moran, J. E., Oktay, S. D. & Santschi, P. H. Sources of iodine and iodine 129 in rivers. *Water Resources Research* **38** (2002).
25. Elderfield, H. & Truesdale, V. W. On the biophilic nature of iodine in seawater. *Earth Planet. Sci. Lett.* **50**, 105–114 (1980).
26. Nakayama, E., Kimoto, T., Isshiki, K., Sohrin, Y. & Okazaki, S. Determination and distribution of iodide-iodine and total-Iodine in the North Pacific Ocean - by using a new automated electrochemical method *Mar. Chem.* **27**, 105–116 (1989).
27. Waite, T. J., Truesdale, V. W. & Olafsson, J. The distribution of dissolved inorganic iodine in the seas around Iceland. *Mar. Chem.* **101**, 54–67 (2006).
28. Rue, E. L., Smith, G. J., Cutter, G. A. & Bruland, K. W. The response of trace element redox couples to suboxic conditions in the water column. *Deep Sea Res. I* **44**, 113–134 (1997).
29. Gilly, W. F., Beman, J. M., Litvin, S. Y. & Robison, B. H. in *Annual Review of Marine Science, Annual Review of Marine Science* Vol. 5 (eds Carlson, C. A. & Giovannoni, S. J.) 393–420 (Annual Reviews, 2013).
30. Hillenbrand, C. D., Futterer, D. K., Grobe, H. & Frederichs, T. No evidence for a Pleistocene collapse of the West Antarctic Ice Sheet from continental margin sediments recovered in the Amundsen Sea. *Geo-Mar. Lett.* **22**, 51–59 (2002).
31. Wahlin, A. K. et al. Some Implications of ekman layer dynamics for cross-shelf exchange in the amundsen sea. *J. Phys. Oceanogr.* **42**, 1461–1474 (2012).
32. Nakayama, Y., Schroder, M. & Hellmer, H. H. From circumpolar deep water to the glacial meltwater plume on the eastern Amundsen Shelf. *Deep Sea Res. I* **77**, 50–62 (2013).
33. Jacobs, S. et al. Getz Ice Shelf melting response to changes in ocean forcing. *J Geophys. Res. Oceans* **118**, 4152–4168 (2013).
34. Thatje, S., Hillenbrand, C. D., Mackensen, A. & Larter, R. Life hung by a thread: Endurance of antarctic fauna in glacial periods. *Ecology* **89**, 682–692 (2008).
35. Collins, L. G., Pike, J., Allen, C. S. & Hodgson, D. A. High-resolution reconstruction of southwest Atlantic sea-ice and its role in the carbon cycle during marine isotope stages 3 and 2. *Paleoceanography* **27** (2012).
36. Gersonde, R., Crosta, X., Abelmann, A. & Armand, L. Sea-surface temperature and sea ice distribution of the Southern Ocean at the EPILOG Last Glacial Maximum—a circum-Antarctic view based on siliceous microfossil records. *Quart. Sci. Rev.* **24**, 869–896 (2005).
37. Lisiecki, L. E. & Raymo, M. E. A Pliocene-Pleistocene stack of 57 globally distributed benthic delta O-18 records. *Paleoceanography* **20** (2005).
38. Bersch, M., Becker, G. A., Frey, H. & Koltermann, K. P. Topographic effects of the maud rise on the stratification and circulation of the weddell gyre. *Deep Sea Res. A* **39**, 303–331 (1992).
39. Martin, S. in *Encyclopedia of ocean sciences.* (eds Steele, H., Turekian, K. K. & Thorpe, S. A.) 2241–2247 (Academic Press, 2001).
40. Smith, J. A., Hillenbrand, C. D., Pudsey, C. J., Allen, C. S. & Graham, A. G. C. The presence of polynyas in the Weddell Sea during the Last Glacial Period with implications for the reconstruction of sea-ice limits and ice sheet history. *Earth Planet. Sci. Lett.* **296**, 287–298 (2010).
41. Elderfield, H. & Ganssen, G. Past temperature and delta O-18 of surface ocean waters inferred from foraminiferal Mg/Ca ratios. *Nature* **405**, 442–445 (2000).
42. Stumm, W. & Morgan, J. J. *Aquatic Chemistry*, 2nd edn 477 (Wiley, 1981).
43. Luther, G. W. et al. Simultaneous measurement of O-2, Mn, Fe, I-, and S(-II) in marine pore waters with a solid-state voltammetric microelectrode. *Limnol. Oceanogr.* **43**, 325–333 (1998).
44. Hull, P. M., Osborn, K. J., Norris, R. D. & Robison, B. H. Seasonality and depth distribution of a mesopelagic foraminifer, Hastigerinella digitata, in Monterey Bay, California. *Limnol. Oceanogr.* **56**, 562–576 (2011).
45. Kuepper, F. C. et al. Commemorating two centuries of iodine research: an interdisciplinary overview of current research. *Angew. Chem. Int. Ed.* **50**, 11598–11620 (2011).
46. Stramma, L., Huttl, S. & Schafstall, J. Water masses and currents in the upper tropical northeast Atlantic off northwest Africa. *J Geophys. Res. Oceans* **110** (2005).
47. Truesdale, V. W. & Bailey, G. W. Dissolved iodate and total iodine during an extreme hypoxic event in the Southern Benguela system. *Estuar. Coast. Shelf Sci.* **50**, 751–760 (2000).
48. Farrenkopf, A. M. & Luther, G. W. Iodine chemistry reflects productivity and denitrification in the Arabian Sea: evidence for flux of dissolved species from sediments of western India into the OMZ. *Deep Sea Res. II* **49**, 2303–2318 (2002).
49. Kohfeld, K. E., Anderson, R. F. & Lynch-Stieglitz, J. Carbon isotopic disequilibrium in polar planktonic foraminifera and its impact on modern and Last Glacial Maximum reconstructions. *Paleoceanography* **15**, 53–64 (2000).
50. Talley, L. D. Closure of the global overturning circulation through the indian, pacific, and southern oceans: schematics and transports. *Oceanography* **26**, 80–97 (2013).
51. McCave, I. N., Carter, L. & Hall, I. R. Glacial-interglacial changes in water mass structure and flow in the SW Pacific Ocean. *Quart. Sci. Rev.* **27**, 1886–1908 (2008).

52. Ellwood, M. J., Wille, M. & Maher, W. Glacial Silicic Acid Concentrations in the Southern Ocean. *Science* **330**, 1088–1091 (2010).

53. Chance, R., Baker, A. R., Carpenter, L. & Jickells, T. D. The distribution of iodide at the sea surface. *Environ. Sci. Process. Impacts* **16**, 1841–1859 (2014).

54. Bluhm, K., Croot, P. L., Huhn, O., Rohardt, G. & Lochte, K. Distribution of iodide and iodate in the Atlantic sector of the southern ocean during austral summer. *Deep Sea Res. II* **58**, 2733–2748 (2011).

55. Mackensen, A. Strong thermodynamic imprint on Recent bottom-water and epibenthic delta C-13 in the Weddell Sea revealed: Implications for glacial Southern Ocean ventilation. *Earth Planet. Sci. Lett.* **317**, 20–26 (2012).

56. Luthi, D. *et al.* High-resolution carbon dioxide concentration record 650,000-800,000 years before present. *Nature* **453**, 379–382 (2008).

57. Jaccard, S. L. *et al.* Two modes of change in southern ocean productivity over the past million years. *Science* **339**, 1419–1423 (2013).

58. Moy, A. D., Howard, W. R. & Gagan, M. K. Late Quaternary palaeoceanography of the Circumpolar Deep Water from the South Tasman Rise. *J. Quat. Sci.* **21**, 763–777 (2006).

59. Galbraith, E. D. & Jaccard, S. L. Deglacial weakening of the oceanic soft tissue pump: global constraints from sedimentary nitrogen isotopes and oxygenation proxies. *Quart. Sci. Rev.* **109**, 38–48 (2015).

60. Jaccard, S. L., Galbraith, E. D., Martínez-García, A. & Anderson, R. F. Covariation of deep Southern Ocean oxygenation and atmospheric CO_2 through the last ice age. *Nature* **530**, 207–210 (2016).

61. Barker, S., Greaves, M. & Elderfield, H. A study of cleaning procedures used for foraminiferal Mg/Ca paleothermometry. *Geochem. Geophys. Geosyst.* **4** (2003).

62. Comiso, J. C. *Large-Scale Characteristics and Variability of the Global Sea Ice Cover* 112–142 (Blackwell, 2003).

63. Olbers, D., Gouretski, V., Seiss, G. & Schroeter, J. *Hydrographic Atlas of the Southern Ocean* (Alfred-Wegener Institute (AWI), 1992).

64. Truesdale, V. W., Bale, A. J. & Woodward, E. M. S. The meridional distribution of dissolved iodine in near-surface waters of the Atlantic Ocean. *Progress Oceonagr.* **45**, 387–400 (2000).

65. Campos, M., Farrenkopf, A. M., Jickells, T. D. & Luther, G. W. A comparison of dissolved iodine cycling at the Bermuda Atlantic Time-Series station and Hawaii Ocean Time-Series Station. *Deep Sea Res. II* **43**, 455–466 (1996).

66. Wong, G. T. F. Dissolved iodine across the Gulf Stream Front and in the South Atlantic Bight. *Deep Sea Res. I* **42**, 2005–2023 (1995).

67. Farrenkopf, A. M., Dollhopf, M. E., NiChadhain, S., Luther, G. W. & Nealson, K. H. Reduction of iodate in seawater during Arabian Sea shipboard incubations and in laboratory cultures of the marine bacterium Shewanella putrefaciens strain MR-4. *Mar. Chem.* **57**, 347–354 (1997).

68. Smith, J. D., Butler, E. C. V., Airey, D. & Sandars, G. Chemical-properties of a low-oxygen water column in port-hacking (Australia) - arsenic, iodine and nutrients. *Mar. Chem.* **28**, 353–364 (1990).

69. Emerson, S., Cranston, R. E. & Liss, P. S. Redox species in a reducing fjord - equilibrium and kinetic considerations. *Deep Sea Res. A* **26**, 859–878 (1979).

Acknowledgements

Z.L. thanks NSF OCE 1232620. B.A.A.H. is supported by Natural Environment Research Council (NERC) grant NE/I020563/1. C.-D.H. is supported by NERC. Z.L. and R.E.M.R. were supported by NERC NE/E018432/1 during the initial development of the I/Ca proxy. We thank the captains, officers, crews and shipboard scientists of expeditions JR179 and ANT-XI/3 for recovering cores TC493 and PS2547, Professor David Hodell and Professor Andreas Mackensen for analysing stable isotopes on these cores, and Dr Hannes Grobe for providing samples from core PS2547. We acknowledge Professor Nick McCave for providing some of the core top materials and Mr Thomas Williams provided assistance for Fig. 1. This research used samples provided by the Ocean Drilling Program (ODP). ODP is sponsored by the US National Science Foundation and participating countries under management of the Joint Oceanographic Institutions (JOI). This manuscript greatly benefited from comments by Dr Nicolaas Glock and two anonymous reviewers.

Author contributions

Z.L., X.Z., K.M.G. and W.L. carried out the I/Ca analysis. C.-D.H. provided all samples from core TC493 and $CaCO_3$ and stable isotope data for cores TC493 and PS2547. B.A.A.H. and L.J. contributed the core top samples. E.T. identified *E. exigua*. All authors participated in data interpretation and manuscript preparation.

Additional information

18

The biogeography of red snow microbiomes and their role in melting arctic glaciers

Stefanie Lutz[1,2], Alexandre M. Anesio[3], Rob Raiswell[1], Arwyn Edwards[4,5], Rob J. Newton[1], Fiona Gill[1] & Liane G. Benning[1,2]

The Arctic is melting at an unprecedented rate and key drivers are changes in snow and ice albedo. Here we show that red snow, a common algal habitat blooming after the onset of melting, plays a crucial role in decreasing albedo. Our data reveal that red pigmented snow algae are cosmopolitan as well as independent of location-specific geochemical and mineralogical factors. The patterns for snow algal diversity, pigmentation and, consequently albedo, are ubiquitous across the Arctic and the reduction in albedo accelerates snow melt and increases the time and area of exposed bare ice. We estimated that the overall decrease in snow albedo by red pigmented snow algal blooms over the course of one melt season can be 13%. This will invariably result in higher melt rates. We argue that such a 'bio-albedo' effect has to be considered in climate models.

[1] Cohen Laboratories, School of Earth and Environment, University of Leeds, Leeds LS2 9JT, UK. [2] GFZ German Research Centre for Geosciences, Telegrafenberg, Potsdam 14473, Germany. [3] Bristol Glaciology Centre, School of Geographical Sciences, University of Bristol, Bristol BS8 1SS, UK. [4] Institute of Biological, Environmental and Rural Sciences (IBERS), Aberystwyth University, Aberystwyth SY23 3FL, UK. [5] Interdisciplinary Centre for Environmental Microbiology, Aberystwyth University, Aberystwyth SY23 3FL, UK. Correspondence and requests for materials should be addressed to S.L. (email: stlutz@gfz-potsdam.de) or to L.G.B. (email: benning@gfz-potsdam.de).

Glaciers are important components of Earth's climate and hydrologic system. The Arctic is being disproportionately affected by global warming, which in turn provides a strong feedback on the climate system[1]. One of the key parameters in the increase of glacial melt is albedo change[2]. The physical and chemical characteristics of snow and ice have been studied intensively; however, the field of glacial microbiology is still in its infancy. Snow and ice surfaces have been considered barren until recently, yet distinct habitats harbour species of all three domains of life[3]. So far, most attention has been paid to cryoconite holes[4–7], which are dominated by bacteria[8,9]. These are, however, only active once the long-lasting snow cover has melted away, and their coverage on glaciated areas usually reaches a maximum of only 10% (refs 3,4,8). In contrast, little is known about the diversity or function of snow algae, nor their global effect on albedo and hence glacial melting. This is despite the fact that coloured snow algal blooms have been known since Aristotle[10], and that they dominate primary production on snow and ice fields[11,12].

For most of the year, the largest proportion of the glacial surfaces in the Arctic is covered by snow. Moreover, permanent and seasonal snow can cover up to 35% of the entire Earth's surface[13]. We have recently shown that snow algae are critical players in glacial surface habitats and the dominating biomass immediately after the onset of melting[11]. Snow algae are prolific primary colonizers and producers that can form extensive blooms in spring and summer. Such snow algal blooms can substantially darken the surface of glaciers because of their red pigmentation (secondary carotenoids), which the algae produce as a protection mechanism (for example, from high levels of irradiation)[14,15]. We have shown that this phenomenon, known as 'red snow', can reduce the surface albedo locally by up to 20%, which in turn further increases melting rates of snow[11]. Previous studies have been unable to generalize this effect because of a lack of information on the distribution, and controls on red snow ecology and physiology. These studies have so far focussed on describing algae primarily through classical microbiological approaches[16–18] (for example, microscopy). In contrast, in the current study, we have employed high-throughput sequencing to characterize these cryophilic micro-eukaryotes and their associated microbiota, that is, bacteria and archaea. We evaluated the diversity and functionality of the red snow algal habitat in four geographically well-separated glacial systems across the Arctic, comprising of 40 red snow sites on 16 glaciers and snow fields. This way, we have produced the first large-scale biogeographical data set for red pigmented snow algae. Knowledge of the global distribution of species and their underlying spatial patterns and processes (that is, their biogeography) has long been assumed irrelevant for microbial communities. However, recently documented rapid changes in diversity across many ecosystems have led to an increased focus on biogeographical patterns and traits in microorganisms[19–21]. Identifying patterns can help to better understand their ecology within a specific ecosystem and make predictions about their role on a larger scale. We cross-correlated the marker gene data with geochemical and metabolic measurements. These parameters were then used to evaluate the environmental forcing factors on the snow microbial community composition.

Furthermore, recent snow-albedo models for Greenland[22] suggest that melting accelerates largely due to increased contributions from light-absorbing impurities, with impurities being primarily considered to be anthropogenic, forest fire-derived black carbon, Saharan or pro-glacial mineral dust[23]. However, the contribution of coloured algae to changing albedo and melt rates has not previously been considered[24].

Here, based on our albedo measurements on red snow and comparing with literature data for algae-free snow, we have estimated the reduction of albedo caused by microbial darkening of glacial surfaces (inferring higher melting rates). This will help to improve our understanding of the response of glacial systems to a warming climate.

Results

Cosmopolitan algal but local bacterial community structure. We have assessed the biogeographical patterns for red snow microbiomes across the Arctic by using high-throughput sequencing of the small subunit ribosomal RNA genes and characterized the species composition of 40 red snow sites in four well-separated and physico-chemically diverse Arctic settings (see Fig. 1 and Supplementary Table 1 for full details).

Our results show that, similar to recent studies of other habitats (for example, soil, marine)[25–27], the bacteria in our red snow samples inherited a strong geographical separation, despite their small cell size and therefore high potential for universal distribution. Bacteria were mostly represented by the phyla *Bacteriodetes*, *Proteobacteria* and *Cyanobacteria* (Supplementary Table 2). These bacterial phyla have previously been described in snow environments[13,28,29]. However, we found significant differences ($P < 0.05$) between locations for the major classes within these phyla (Supplementary Table 3). Samples clustered according to their geographic location (Fig. 2a), and the observed differences were derived from large variations in the relative abundance of *Sphingobacteria*, *Saprospirae*, *Alphaproteobacteria*, *Betaproteobacteria* and *Synechococcophycideae*. Among these, *Saprospirae*, *Cytophagia*, *Betaproteobacteria* and *Synechococcophycideae* were dominant in Svalbard; *Sphingobacteria* in Northern Sweden; *Sphingobacteria* and *Saprospirae* in Greenland; *Saprospirae*, *Betaproteobacteria* and *Alphaproteobacteria* in Iceland (Fig. 2b and Supplementary Table 2).

In contrast, these biogeographic patterns were not observed for the snow algae. Our results demonstrate that the snow algal community composition and their relative abundance in all studied Arctic sites was highly similar (Fig. 2c,d), despite the large distances, physico-chemical characteristics and associated bacterial composition differences between sites. We show that the snow algae are cosmopolitan. This is in contrast to recent molecular studies, which suggest that in other terrestrial habitats, and even within a specific habitat, micro-eukaryotes show strong biogeography[20,25–27]. Our data reveal a very low algal diversity. Six taxa make up $>99\%$ of the algal communities (Fig. 2c,d and Supplementary Tables 3 and 4) and all have similar relative abundance values across all samples. The uncultured *Chlamydomonadaceae* (2) was the most abundant species (39–75%, Fig. 2d), followed by *Chloromonas polyptera* (10–26%), *Chloromonas nivalis* (3–13%), *Chloromonas alpina* (0–1%), the uncultured *Chlamydomonadaceae* (1) and *Raphidonema sempervirens* (1–18%). The small variance in the algal data between sites (Fig. 2c) was mainly caused by changes in relative abundance of the uncultured *Chlamydomonadaceae* (2), *Chloromonas polyptera* and *Raphidonema sempervirens*. However, none of the samples clustered according to locations, and no significant differences were found between locations for most of the algal species. The exceptions were *Chloromonas nivalis*, which showed a higher relative abundance in samples from Greenland in comparison to Svalbard, and the uncultured *Chlamydomonadaceae* (2), which had a higher relative abundance in Svalbard in comparison to Iceland (full details of the eukaryotic and archaeal community compositions and diversity indexes can be found in Supplementary Tables 5–8).

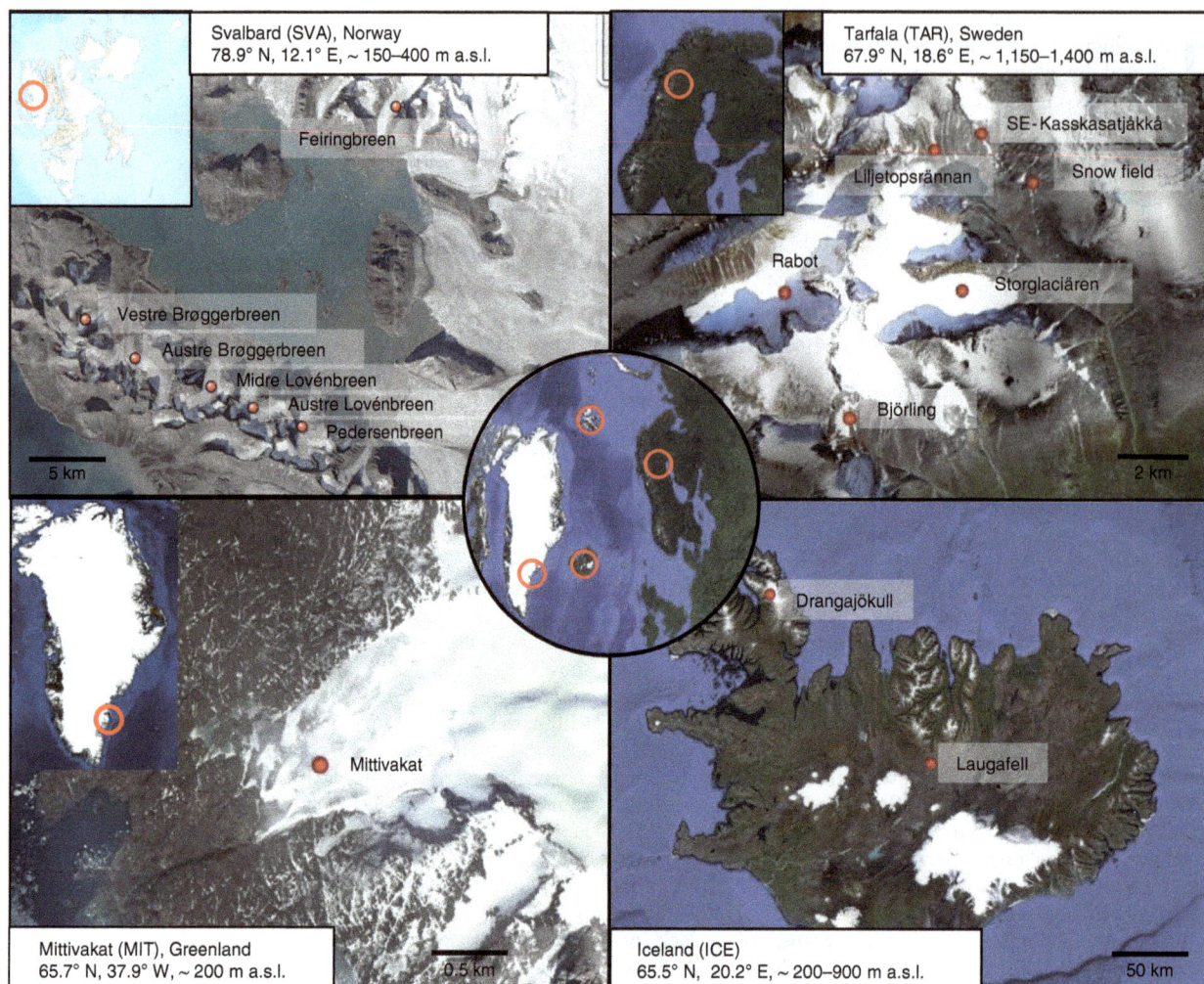

Figure 1 | Sample locations. Locations of the 16 glaciers and snow fields across the Arctic, where 40 sites of red snow were sampled: Svalbard ($n = 12$), Northern Sweden ($n = 24$), Greenland ($n = 2$) and Iceland ($n = 2$). These localities were chosen as they represent different geographical settings including low (67.9°N) versus high (78.9°N) latitude, low (150-400 m) versus high ($\sim$1,200-1,400 m) elevation, and maritime versus continental settings. Red dots represent sampling sites and several sampling events within one site (for full details, see Supplementary Table 1). Map data: Google, DigitalGlobe.

The homogeneous algal community composition described above was also mirrored in the similar composition of algal cell biomass, fatty acids and pigments with no significant differences between Svalbard and Northern Sweden (Fig. 3 and Supplementary Tables 9–11). On average, between 10^3 and 10^4 red pigmented algal cells per ml were present in our red snow samples. Despite the large variations in environmental parameters, no significant differences were found for cell numbers, cell sizes or total algal biomass (Supplementary Tables 3 and 9). Similarly, the fatty acid compositions in all analysed samples were similar with no statistically relevant differences between locations (Supplementary Tables 3 and 10). On average, $\sim$45–50% of all fatty acids were made up of polyunsaturated fatty acids, whereas saturated fatty acids comprised $\sim$30–40% and monounsaturated fatty acids were the least abundant ($\sim$10–15%; Fig. 3 and Supplementary Table 10). The high content of polyunsaturated fatty acids likely demonstrates their role as cryo-protectants, helping algal cells to maintain membrane fluidity and preventing intracellular ice crystal formation[30]. The production of fatty acids is often linked to pigments[14], which play the dominant role in changing the albedo. All samples were characterized by a high content of secondary carotenoids ($\sim$70–90%), which are synthesized by the snow algae as a protection mechanism from

high levels of irradiation, and with no significant differences between locations. The dominant secondary carotenoid was trans-astaxanthin (Supplementary Table 11). The remainder of the analysed pigments were typical primary carotenoids (up to 24%) or chlorophylls a and b (up to 55%; Supplementary Table 11).

Local environment affects bacteria but not algae. Changes in physico-chemical conditions are known to control variations in microbial diversity in the environment[21,31], yet for snow settings the importance or magnitude of these effects and whether they cause any biogeographical patterning are largely unknown. Our results show that the four chosen geographic locations differed substantially in the concentrations of essential nutrients, carbon species, trace elements (both dissolved and solid forms; Supplementary Tables 12 and 13) and mineralogy (Supplementary Table 14). Hence, they represent a good range of differing local snow environments across the Arctic. Dissolved organic carbon (DOC) concentrations varied significantly between locations (Supplementary Fig. 1 and Supplementary Tables 3 and 12), with up to five times higher values in snow from Northern Sweden in comparison to Svalbard. In contrast,

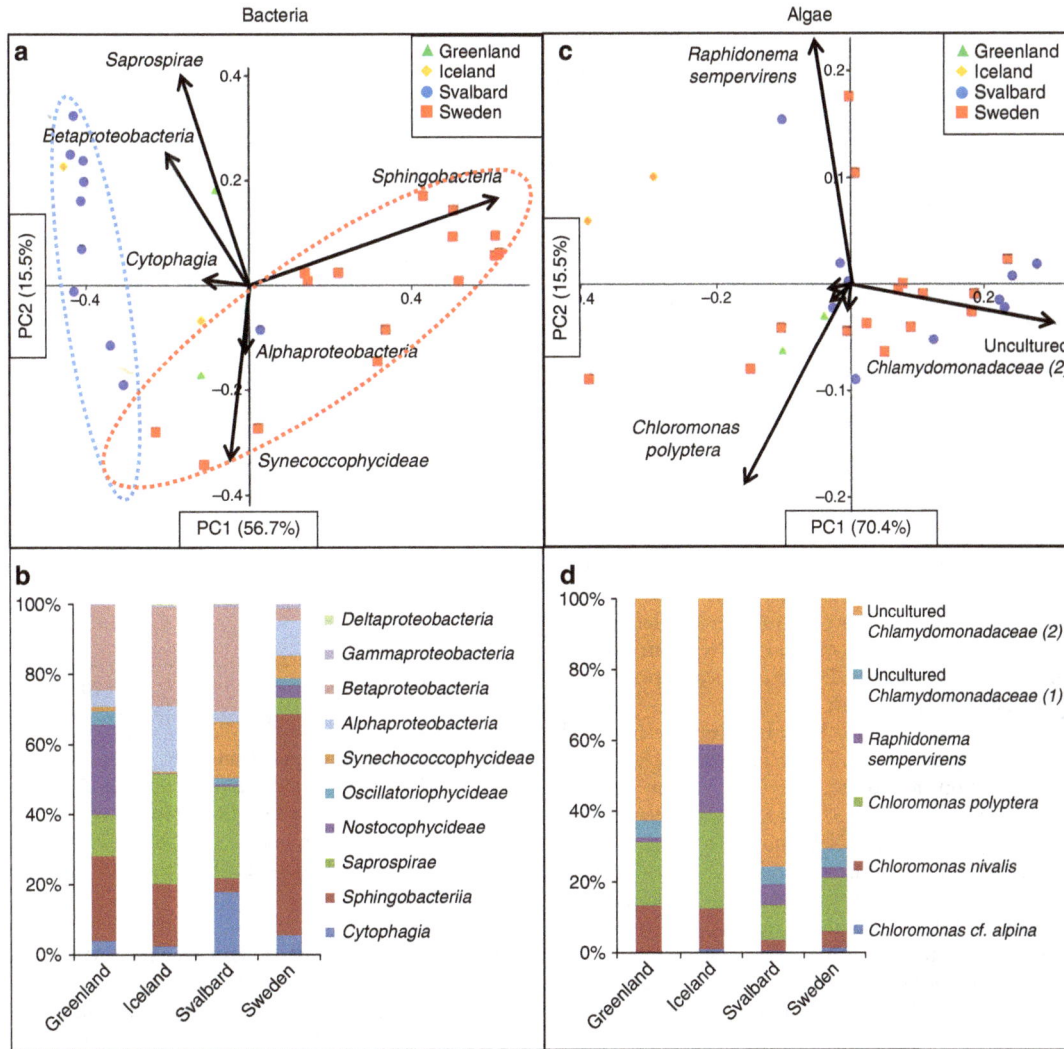

Figure 2 | Algal and bacterial community composition. Principal component analysis of bacterial classes (**a,b**) and algal species (**c,d**) revealing taxonomic distance between sampling sites and taxa causing separation. Algal species show homogenous community composition across all sites (**c**), whereas bacteria cluster according to locations, even on the higher taxonomic class level (**a**, dotted lines have been added to help guide the reader's eye). Bar charts show average community composition (Supplementary Tables 2 and 4 for individual values; Supplementary Table 3 for averages and P-values) for each location and confirm similar composition for algae (**d**) but large differences for bacteria (**b**).

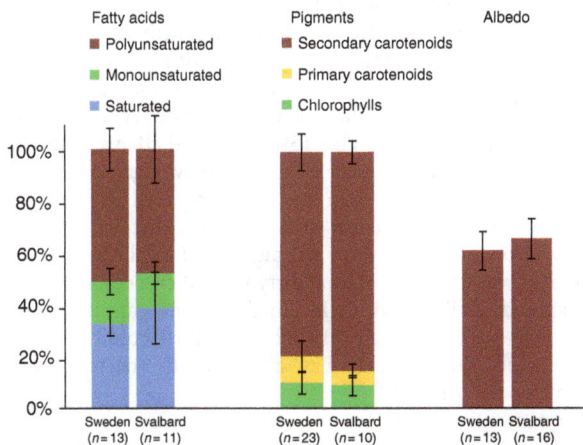

Figure 3 | Algal fatty acid and pigment composition and albedo values. Comparison between average fatty acid and pigment compositions with average surface albedo (all in % of total) for all Svalbard and Northern Sweden sites. Error bars are standard deviations (for full details see Supplementary Tables 1, 3 and 10).

concentrations of easily leachable elements (Ca, Cl, Mg, Mn, Na and K) were on average 10 times higher in red snow from Svalbard in comparison to Northern Sweden (Supplementary Table 12), whereas in Iceland the red snow samples contained up to 100 times higher iron concentrations than any of the other localities. These differences appear to be linked to the higher concentrations of more easily dissolvable mineral phases in Svalbard in comparison to Northern Sweden, and the higher Fe content in basaltic rocks from Iceland in comparison to the other sites (Supplementary Table 3). However, none of the essential nutrients (that is, NO_3, PO_4) showed any statistically relevant differences among locations (Supplementary Tables 3 and 12).

Mirroring the DOC trend, the total particulate carbon (TC) values as well as the total solid phase carbon to nitrogen (C/N) and carbon to phosphorous (C/P) ratios (Supplementary Table 13) were on average two or three times higher in Northern Sweden in comparison to Svalbard, whereas the $\delta^{13}C$ values of bulk organic matter were significantly lower in Svalbard in comparison to Northern Sweden. In both locations, the C/N ratios were below the Redfield ratio but C/P values below Redfield were only present in samples from Northern Sweden

Table 1 | Integrated albedo change.				
	Average	Minimum	Maximum	Reference
Dry (winter) clean snow	0.90	0.95	0.85	37
Wet clean snow	0.75	0.80	0.70	11
Red snow	0.65	0.77	0.53	11, Supplementary Table 1

Average, minimum and maximum albedo values (in terms of decrease in albedo) for dry clean snow, wet clean snow and red snow used to derive the integrated albedo change over 100 days (Supplementary Fig. 4).

(Supplementary Table 13). The high TC, DOC and $\delta^{13}C$ values in the snow samples from Northern Sweden likely document a higher amount of allochthonous carbon potentially derived from higher plants, and the large amounts of pine pollen blown onto the glaciers and snow fields from the lower parts of the Tarfala valley. However, all red snow samples, regardless of location, were characterized by similar, predominantly negative organic $\delta^{15}N$ values indicative of an atmospheric source (Supplementary Table 13).

Geochemical and mineralogical parameters varied dramatically between locations, yet no correlations between algal species distribution and these characteristics were found (Supplementary Fig. 2). This suggests that the uniform algal species composition remained unaffected by and independent of the local geochemical and mineralogical parameters in each site.

In contrast, our data show a clear links between the bacterial community composition and geochemical parameters, with the most positive correlation found between the carbon species (TC and DOC; Supplementary Fig. 3) and *Sphingobacteria*. This is not surprising as *Sphingobacteria* are known to be capable of degrading complex organic structures[32], and their abundance in Northern Sweden is consistent with the local high DOC and TC values (Supplementary Tables 12 and 13). In all samples and in total contrast to the algae, the bacteria are seemingly subjected to a much higher location-specific selection pressure and appeared more affected by the availability of allochthonous carbon and the local geology.

Algae decrease surface albedo. The above documented high algal biomass primarily made up of highly red pigmented algae, will invariably affect the amount of light that is reflected from the surface of snow fields. Our albedo measurements (Table 1 and Supplementary Table 1) showed a clear decrease in surface albedo in comparison to algal-free snow sites (0.90 ± 0.05; Table 1)[11,33]. The measured decrease where red pigmented algae were present was similar in all sites, independent of the local environment with albedo values reaching between ~0.50 and 0.75 (Supplementary Table 1). In addition, we found a significant ($P = 0.008$) negative correlation between algal biomass and surface albedo (Fig. 4), which clearly supports our assertion of the crucial role of red pigmented snow algae in decreasing surface albedo and increasing melting. This is also on par with the results by Painter *et al.*[34] and Aoki *et al.*[35], who showed that the strong light absorption is due to algal pigments in the 400–600 nm (carotenoids and chlorophylls) and 600–700 nm (chlorophylls) range, which is much stronger in comparison to absorption by mineral dust or black carbon if biomass is as high as in our documented red algal blooms.

The above described ubiquitous distribution, low diversity and similarity in snow algal community compositions and metabolic functions combined with the analogous values measured for the red snow algae induced albedo reduction (Supplementary Table 1), allow us to compare the impact that the red pigmented snow algae have on albedo in comparison to snow surfaces free of

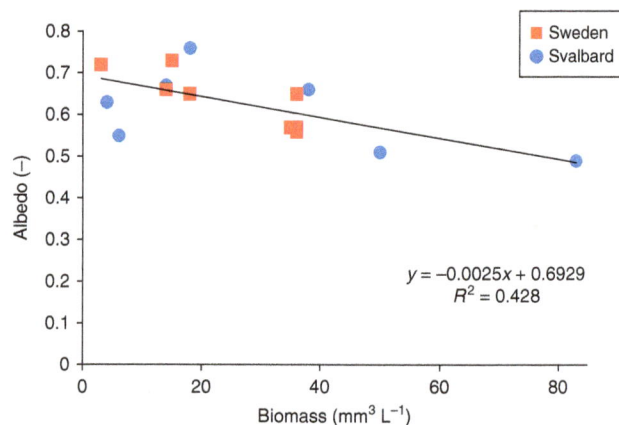

Figure 4 | Algal biomass and albedo. Plot shows a significant negative correlation (Pearson correlation factor: $r = -0.65$, $P = 0.008$) between algal biomass and surface albedo measured in red snow sites in Svalbard and Northern Sweden. This underpins the role of red pigmented snow algae in decreasing surface albedo and in turn melting.

algae over an estimated 100-day scenario. At the beginning of a melting season, we assume all glacier surfaces are covered by dry clean snow. Using values for albedo for such snow from the literature[2,22,36,37] (0.90 ± 0.05; Table 1) allowed us to linearly integrate the change in albedo of snow colonized by red pigmented algae versus algae-free snow surfaces for the season-long transition from dry clean snow to wet clean snow (0.75 ± 0.05) and to red snow (0.65 ± 0.12; Supplementary Fig. 4). We show a square root dependence between albedo values and time and applied a simple one-dimensional moving boundary approach to our data[38].

Our fit of the integrated albedo change with time (100 days) shows a 13% larger effect in the presence of red pigmented algal blooms in comparison to clean snow that has undergone a purely physical albedo change due to melting, and change of snow crystal size and structure. This 13% integrated change in albedo is an estimate for the overall effect of snow algal communities during an entire melt season and compares well with the single time-point albedo reduction of up to 20%, that we and others have previously measured on red pigmented algal snow sites[11,39]. Moreover, with further melting dirty ice and cryoconite holes will be exposed earlier and their albedo values can drop by an additional ~20% to 0.34 ± 0.15. This will likely culminate in even higher melt rates, which has also recently been shown in laboratory experiments[40].

A quantitative value for the area of Arctic glaciers and the Greenland Ice Sheet covered by snow algae during a melt season is still lacking. However, as we infer from our data, melting is one major driver for snow algal growth. Extreme melt events like that in 2012, when 97% of the entire Greenland Ice Sheet was affected by surface melting[41], are likely to re-occur with increasing frequency in the near future as a consequence of global warming[42]. Moreover, such extreme melting events are likely to even further intensify the effect of snow algae on surface albedo, and in turn melting rates[43].

With this work, we show that snow algae are ubiquitous and have little diversity across the Arctic, despite variations in environmental parameters that significantly impact the bacterial community composition. Although we may not have captured all environmental parameters, the patterns we observed occur on all studied glaciers regardless their geochemical and mineralogical compositions. Further investigations are needed to explore the validity of these findings for mid latitudes and the southern hemisphere. However, red snow is a ubiquitous phenomenon in Arctic sites (glaciers and permanent snow fields). The similarity in snow algal community composition, metabolic function and impact on albedo of snow habitats allows the upscaling of our findings and make predictions about the influence of snow algae on melt rates of glaciers across the Arctic. Our data show that the overall decrease in snow albedo by red pigmented snow algal blooms over an entire melt season can be 13%, likely leading to earlier exposure of dirty ice with an even lower surface albedo culminating in a further increase in melt rates. Our work paves the way for a universal model of algal–albedo interaction and a quantification of additional melting caused by algal blooms to be included in future climate models.

Methods

Field sites. A total of 40 red snow samples was collected from four well-separated Arctic locations on 16 glaciers and permanent snow fields: Svalbard (SVA, $n = 12$), Northern Sweden (TAR, $n = 24$), Greenland (MIT, $n = 2$) and Iceland (ICE, $n = 2$; Fig. 1; Supplementary Table 1). These localities were chosen as they represent different geographical settings including low (67.9°N) versus high (78.9 N) latitude, low (150–400 m) versus high ($\sim$1,200–1,400 m) elevation, and maritime versus continental settings. Vestre Brøggerbreen, Midtre Lovénbreen, Austre Brøggerbreen, Pedersenbreen, Austre Lovénbreen and Feiringbreen in Svalbard were sampled in July and August 2013. Samples from Storglaciären, Rabot, Liljetopsrännan, SE-Kasskasatjåkkå, Björling and nearby permanent snow fields in Northern Sweden were collected in July 2013 and July 2014. Mittivakkat glacier in Greenland and the glacier Drangajökull and permanent snow field Laugafell in Iceland were sampled in July 2012. Red snow samples were collected late in the melt season as those are the typical snow algal blooms that will have the largest effect on albedo.

Field sampling and measurements. All sampling, field measurements and most analyses have previously been described in full detail[11,33]. Here we summarize previously employed methods and give full details of new methods. At each sampling site, we measured pH, conductivity and temperature with a daily calibrated metre (Hanna instruments, HI 98129) before sampling. Photosynthetic active radiation, ultraviolet radiation and surface albedo (400–700 nm range) were measured using a radiometer (SolarLight, PMA2100) with specific photosynthetic active radiation (PMA2132), ultraviolet-A (PMA2110) and ultraviolet-B (PMA2106) sensors. Albedo was calculated by taking the ratio of reflected to incident radiation (400–700 nm range) and measuring the values always in the same position to the sun. The reading of the sensor was not affected by shading by the observer. Measurements were carried out with the sensors held at 30 cm above the snow surface (field of view 160°). At first, the sensor was pointed upwards (incident radiation) and then towards the snow surface (reflected radiation). Five measurements for incident and reflected radiation were acquired each, and the average was taken to avoid measuring bias. The standard deviations for each measurement set was below 10%. Data of the relative contribution of pigments, other light impurities (that is, mineral dust, black carbon) and snow metamorphism is lacking[42]. However, based on qualitative microscopic observations in the field, particularly mineral impurities (most often light coloured quartz and feldspars) were less important in changing albedo measurements in red snow surface samples in comparison to the pigment distribution. Moreover, Aoki et al.[35] and Painter et al.[34] showed that red snow has much higher light absorption below 600 nm because of the algal pigments in comparison to mineral dust or black carbon. Samples were collected in sterile centrifuge tubes or sterile Whirl-Pak bags and in pre-ashed glass jars (450 °C, >4 h) for organic analyses. All samples were slowly melted at room temperature within $\sim$6 h, and processed and preserved (for example, filtered, acidified) within $\sim$8 h after collection. Samples for DNA and organic analyses were flash-frozen in liquid nitrogen and stored at -80 °C until analysed. Inorganic samples were stored cold (4 °C) and in the dark. All analyses were carried out in the Cohen Laboratories at the University of Leeds unless stated otherwise.

Aqueous analyses. Aqueous analyses were carried out by Ion Chromatography (Dionex, anions), by inductively coupled plasma mass spectrometry (Agilent, cations, at the University of Sheffield), on a total organic carbon analyser (Shimadzu, for DOC contents, DOC, at the Plymouth University), and by segmented flow-injection analyses (AutoAnalyser3, Seal Analytical, dissolved phosphate).

Particulate analyses. Particulates in the samples were analysed for $\delta^{15}N$ and $\delta^{13}C$ by a Vario Pyro Cube elemental analyser (Elementar Inc) coupled to an Isoprime mass spectrometer. Samples were combusted in tin capsules at 1,150 °C, and gases were separated using temperature-controlled adsorption/desorption columns. Carbon analyses were calibrated with in-house C4-sucrose and urea standards assigned values of $-11.93‰$ and $-46.83‰$, respectively via calibration with the international standards LSVEC ($-46.479‰$), CH7 ($-31.83‰$), CH6 ($-10.45‰$) and CO-1 ($+2.48‰$). Nitrogen isotope values were calibrated using the international standards USGS-25 and USGS-26 with assigned values of $-30.4‰$ and $+53.7‰$, respectively. Total carbon, total nitrogen and total sulphur were derived from the thermal conductivity detector in the elemental analyser and calibrated using a sulphanilamide standard. Particulate phosphorus was extracted by ashing of the samples at 550 °C for 2 h and incubating in 1 M HCl for 16 h according to extraction step V in Ruttenberg et al.[44].

Algal biomass. Algal cells were imaged on a Leica DM750 microscope equipped with a $\times$ 63objective and counted with a haemocytometer in triplicate. For cell size analyses, 100 cell diameters per sample were measured in ImageJ. Cell volumes were calculated assuming a perfect spherical shape ($V = 4/3^*\pi^*r^3$). Total algal biomass was calculated using the average cell volume and cell abundance.

Pigment analysis. Carotenoid and chlorophyll contents in the samples were quantified by high-pressure liquid chromatography (HPLC) and a modified carotenoid/chlorophyll-specific extraction protocol[45]. Cells were disrupted by shock freezing in liquid nitrogen for 10 min followed by grinding with a Teflon mortar and pestle. The resulting powder was re-suspended in 1 ml of dimethylformamide and 1.0 mm glass beads and horizontally shaken on a laboratory shaker (MoBio Vortex Genie 2) at maximum speed (3,000 r.p.m.) for 10 min, followed by centrifugation for 5 min at 10,000 r.p.m. The supernatant was separated from the debris by filtering through a 0.45-μm Teflon filter and the filtrate was mixed with methanol (25 vol%).

Extracted samples were analysed immediately on an Agilent Technologies 1200 Infinity HPLC instrument with a gradient pump, an autosampler, a variable wavelength detector and ODS Hypersil column (250 $\times$ 4.6 mm^2; 5 μm particle size). Two solvents were used: solvent A consisted of a mixture of acetonitrile/water/methanol/hexane/tris buffer at ratios of 80:7:3:1:1, whereas solvent B was a mix of methanol and hexane at a ratio of 5:1. The HPLC was run at a flow rate of 1 ml min^{-1}, with an injection volume of 25 μl. Spectra were recorded from 200 to 800 nm. Chromatograms were quantified at 450 nm for carotenoids and 660 nm for chlorophyll a and b. Run time was 60 min. The protocol required a 15-min run with 100% of solvent A followed by a linear gradient from 100% solvent A, to 100% solvent B between 32 and 45 min, and finally with 15 min of column re-equilibration through a 5-min linear gradient from solvent B back to 100% solvent to A, followed by a further column conditioning with 100% solvent A for 10 min. The following commercially available standards were used for peak identification: chlorophyll a, chlorophyll b (Sigma), violaxanthin, neoxanthin, antheraxanthin, lutein, β-carotene, trans-astaxanthin and cis-astaxanthin (Caratenature).

Fatty acids analysis. Fatty acids were extracted according to the method described by Wacker and Martin-Creuzberg[46]. Briefly, 20 ng of internal standard (tricosanoic acid methyl ester) were added to each sample, followed by ultrasonic extraction using dichloromethane:methanol (2:1 (v:v)), and centrifugation to remove particulates and evaporation of solvent from the supernatant. Fatty acids were transesterified by adding methanolic HCl to the dried extract and heating at 60 ° C for 20 min. After cooling, fatty acid methyl esters were extracted in isohexane, the solvent was removed under nitrogen and the sample resuspended in isohexane for analysis.

Analysis of fatty acid methyl esters was carried out using a Trace 1300 gas chromatograph with flame ionization detector (Thermo Scientific), equipped with a non-polar-fused silica capillary column (CPSil-5CB, 50 m $\times$ 0.32 mm $\times$ 0.12 mm, Agilent Technologies). Samples (1 μl) were injected in splitless mode, with the injector maintained at 200 °C. Carrier gas was helium, with a constant flow rate of 1.5 ml min^{-1}. The following temperature programme was used: initial temperature 40 °C, rising to 140 °C at 20 °C min^{-1}, then rising to 240 °C at 4 °C min^{-1}, holding at 240 °C for 5 min. Fatty acid methyl esters were identified by comparison of retention time with those of reference compounds (Supelco) and by gas chromatography mass spectrometry (GC–MS). GC–MS was carried out using the gas chromatograph and column previously described, with identical operating conditions, coupled to an ISQ mass spectrometer (Thermo Scientific). The transfer line and the ion source were maintained at 300 °C. The emission current was set to 50 mA and the electron energy to 70 eV. The analyser was set to scan at m/z 50–650 with a scan cycle time of 0.6 s.

DNA sequencing. Total DNA was extracted from pelleted biomass using the PowerSoil DNA Isolation kit (MoBio Laboratories). 16S rRNA genes were amplified using bacterial primers 27F (5′-AGAGTTTGATCMTGGCTCAG-3′) and 357R (5′-CTGCTGCCTYCCGTA-3′; tagged with the Ion Torrent adapter sequences and MID barcode) spanning the V1-V2 hypervariable regions. 18S rRNA genes were amplified using the eukaryotic primers 528F (5′-GCGGTAA TTCCAGCTCCAA-3′) and 706R (5′-AATCCRAGAATTTCACCTCT-3′; Cheung et al., 2010 (ref. 48); tagged with the Ion Torrent adapter sequences and MID barcode) spanning the V4-V5 hypervariable region. PCRs were performed using Platinum PCR SuperMix High Fidelity according to the manufacturer's protocols. Initial denaturation at 95 °C for 5 min was followed by 30 cycles of denaturation at 95 °C for 30 s, annealing at 60 °C for 30 s and elongation at 72 °C for 30 s. Final elongation was at 72 °C for 7 min. Archaeal 16S rRNA genes were amplified following a nested PCR approach. The first PCR reaction was carried out using primers 20F (5′-TCCGGTTGATCCYGCCRG-3′) and 915R (5′-GTGCTCCCCCG CCAATTCCT-3′). Initial denaturation at 95 °C for 5 min was followed by 35 cycles of denaturation at 95 °C for 30 s, annealing at 62 °C for 30 s and elongation at 72 °C for 180 s. Final elongation was at 72 °C for 10 min. The PCR product was used as template for the second PCR reaction with primers 21F (5′-TCCGGTTGAT CCYGCCGG-3′) and 519R (5′-GWATTACCGCGGCKGCTG-3′; tagged with the Ion Torrent adapter sequences and MID barcode) spanning the V1-V2 hypervariable region. Initial denaturation at 95 °C for 5 min was followed by 30 cycles of denaturation at 95 °C for 30 s, annealing at 60 °C for 30 s and elongation at 72 °C for 30 s. Final elongation was at 72 °C for 7 min. Detailed information on the sequencing primers can be found in the Supplementary Information. All PCRs were carried out in triplicates to reduce amplification bias and in reaction volumes of 1×25 μl and 2×12.5 μl. All pre-amplification steps were done in a laminar flow hood with DNA-free certified plasticware and filter tips. The pooled amplicons were purified with AMPure XP beads (Agencourt) with a bead-to-DNA ratio of 0.6 to remove nucleotides, salts and primers. Quality, size and concentration were determined on the Agilent 2100 Bioanalyser (Agilent Technologies) with the High-Sensitivity DNA kit (Agilent Technologies). Sequencing was performed on an Ion Torrent Personal Genome Machine using the Ion Xpress Template Kit and the Ion 314 or Ion 316 chips following the manufacturer's protocols. All PCR amplifications and sequencing were carried out at the Aberystwyth University and the University of Bristol. The raw sequence data were processed in QIIME[47]. Barcodes and adapter sequences were removed from each sequence. Filtering of sequences was performed using an average cutoff of Q20 over the full sequence length (350 bp). Reads shorter than 200 bp were removed. Operational taxonomic units (OTUs) were picked de novo using a threshold of 97% identity. Taxonomic identities were assigned to representative sequences of each OTU using the reference databases Greengenes for bacteria and archaea. The Silva database (ref. 49; extended with additional 223 sequences of cryophilic algae kindly provided by Dr Thomas Leya from the CCCryo—Culture Collection of Cryophilic Algae, Fraunhofer IZI-BB) was used for eukaryotes. Data were aligned using PyNAST and a 0.80 confidence threshold. Singletons were excluded from the analysis. For bacterial sequence matching, plant plastids were removed from the data set before further analysis. For eukaryotic sequence matching, Chloroplastida were pulled out of the data set and stored in a separate OTU table. In order to focus upon algal diversity, sequences matching Embryophyta (for example, moss, fern) were removed from the data set. For archaea, sequences matching bacteria were removed. Finally, for further analyses, samples were rarefied to the minimum library size and Shannon indices were calculated in QIIME. All analyses were conducted at the 97% OTU level. A matrix of each OTU table representing relative abundance (raw data) was imported into PAST v3.06 (ref. 50) for multivariate statistical analyses (principal component analysis, canonical correspondence analysis) and Pearson correlations. One-way analysis of variance test was done in SPSS v19 (IBM).

Sequencing primers. Primers targeting the 18S rRNA gene were chosen because there are more sequences in the databases for green algae (that is, Chlorophyta, Charophyta) than for rbcL or internal transcribed spacer (ITS). Before sequencing, we carried out an in-silico investigation including 18S rNRA sequences from 218 snow and permafrost algae in order to make sure that the chosen primers are suitable for green algae and that there is enough variability in the chosen region (v4-v5) to distinguish between species.

Previous studies[51] have found that one primer pair is not sufficient to recover all eukaryotic groups in one sample. However, we chose our primer pair based on one group we were specifically targeting, that is, the green algae. We do not claim to have equally recovered all other groups among the micro-eukaryotes such as fungi or the 'SAR'-group. Furthermore, they found that libraries derived from different primer pairs grouped together for individual samples with no significant differences. Based on our in silico test of 218 snow and permafrost algae and the rarefaction curves (Supplementary Fig. 6), we are fairly confident that the choice of our primer pair has resulted in a good coverage of the algal diversity. However, we acknowledge that PCR-based approaches will always introduce a certain amount of bias.

This is similar for the archaea, which show no biogeographical patterns in our samples. The primers used are specific for archaea and since they are not the focus of this study and only the associated microbiome, we did not explore other primer possibilities. However, the results match what other studies have found before in cryo-environments[28,52].

Overall sampling design. All samples for DNA and aqueous analyses were analysed in exactly the same way for all samples from Greenland, Iceland, Svalbard and Sweden. Pigment and fatty acid data are only shown for the samples from Svalbard and Sweden because for Greenland and Iceland these data have previously been published[11,33]. The samples from Greenland have been excluded from the comparison here because the pigment and fatty acid data have been collected and quantified in a different way. The pigments were normalized to chlorophyll a, whereas in all other study area they were quantified with the appropriate pigment standards. The fatty acid components were analysed by GC–MS, whereas all samples from Svalbard and Sweden were also quantified by flame ionization detection. The pigment and fatty acid data from Iceland were also excluded from the comparison, as in all samples large amounts of moss (identified by microscopy and DNA sequences) that could not be separated from the algae before pigment and fatty acid extraction were present. This moss contribution would strongly 'skew' the data and thus these were excluded.

In addition, only selected samples in Greenland[11] and Iceland[33] were included in the comparison. This is because at both sites samples were collected at different stages in the melt season. The study in Greenland was conducted at the onset of melting and over a 3-week period when snow algae just started to bloom and a decrease in relative chlorophyll content and increase in carotenoid content could be observed. This led to our conclusion that there is a great heterogeneity in pigment composition both in space and time[11]. However, the few samples collected at the end of the study showed similar carotenoid contents. This end-of-season homogeneity in the red snow samples was the reason why in the current study we focused solely on samples from late in the melt season, which is the dominant red snow stage with the largest impact on albedo. Thus, we only included two DNA samples from Greenland. Similarly in Iceland[33], most of the samples were collected earlier in the year (June—July) and those samples were described as less 'typical' of red snow patches[33]. Off all samples from Iceland again only the two samples that were collected late in the melt season and therefore matched the conditions of the samples in the current study were used for comparisons.

Integrated albedo change. Using our mean, minimum and maximum measured albedo values for wet clean snow and red snow and literature data[22] for clean dry snow, we used a simple one-dimensional moving boundary approach that allows us to predict the effect of red pigmented snow algae on albedo. This approach is valid under the assumption that the snow and ice surfaces melt downwards relative to a fixed depth, and that at the same time such a change is accompanied by changes in albedo[38]. The parameters, equations and boundary conditions used are as follows:

Table 1 shows measured minimum, maximum and average albedo values for dry clean snow before the onset of melting[11], wet clean snow (no visual presence of algae) at the onset of melting and red snow (full red pigmented snow algal bloom). We used these values to derive linear regressions for albedo changes over a 100-day melt season (Supplementary Fig. 4). A conservative 100-day scenario was chosen, as this encompasses all our albedo measurements in the current and previous studies (June—August)[3,11,33]. In addition, this also corresponds to the number of days with mean air temperatures above 0 °C in the same period (Ny Alesund: 116 days in 2013 and 105 days in 2014, kindly provided by Dr Marion Maturilli and Siegrid Debatin, AWI; Storglaciären: 132 days in 2013 and 108 days in 2014; kindly provided by Dr Peter Jansson, Stockholm University; data are also publicly available at http://bolin.su.se/data/tarfala/). We compare a benchmark case of purely physical-driven albedo change (that is, changes in snow crystal sizes and increasing water content, scenario 1) with albedo change due to red pigmented algal growth (scenario 2).

Scenario 1 considers the transition over 25 days from clean snow to a wet melting surface without algal growth and with an albedo of 0.80 (a minimum value), which with continued melting results in an albedo of 0.75 (an average value) after 50 days and 0.70 (a maximum value) after 100 days (Table 1). Our benchmark case (scenario 1) shows albedo (α) changes with time and fits the equation:

$$\alpha = 0.8992 - 0.0203t^{0.5} \tag{1}$$

Scenario 2 considers the transition from clean snow to a surface where the growth of algae after 25 days produces light red snow with an albedo of 0.77 (a minimum value), and continued melting produces darker red snow with an albedo of 0.65 (an average value) after 50 days and 0.53 (a maximum value) after 100 days (Table 1). The albedo changes with time for this scenario fit the equation:

$$\alpha = 0.9177 - 0.0372t^{0.5} \tag{2}$$

These two equations can be integrated to obtain the cumulative effects of albedo (α) changes with time to give:

$$\int \alpha_{\text{scenario 1}}.dt = \int \left(0.8992 - 0.0203t^{-0.5}\right).dt$$
$$= 0.8992t + 0.0203t^{1.5}/1.5 = 0.8992t + 0.0135t^{1.5} \tag{3}$$

$$\int \alpha_{\text{scenario 2}}.dt = \int \left(0.9177 - 0.0372t^{-0.5}\right).dt$$
$$= 0.9177t + 0.0372t^{1.5}/1.5 = 0.9177t + 0.0248t^{1.5} \tag{4}$$

Subtracting equations (3) and (4) gives

$$\Delta\alpha = 0.0185t + 0.0113t^{1.5} \tag{5}$$

which represents the albedo changes attributable to algae growth alone. For a melt season of 100 days $\Delta\alpha = 1.85 + 11.3 = 13.15 \approx 13$.

In order to assess the error of our analysis, we carried out a sensitivity analysis using the data below (see also Supplementary Fig. 5 for details):

Drycleansnow (DCS) $0.95 - 0.90 - 0.85$

Wetcleansnow (WCS) $0.80 - 0.75 - 0.70$

Redsnow (RS) $0.77 - 0.65 - 0.53$

Comparing minimum and average albedo values:

$$\int \alpha_{min}.dt = \int \left(0.9458 - 0.0186 t^{-0.5}\right).dt$$
$$= 0.9458t + 0.0186t^{1.5}/1.5 = 0.9458t + 0.0124t^{1.5}$$

$$\int \alpha_{average}.dt = \int \left(0.9049 - 0.0243 t^{-0.5}\right).dt$$
$$= 0.9049t + 0.0243t^{1.5}/1.5 = 0.9049t + 0.0162t^{1.5}$$

Subtracting gives $\Delta\alpha = 0.0409t - 0.0038t^{1.5}$
So when $t = 100$, $\Delta\alpha = 4.09 - 0.0038 \times 1,000 = 4.09 - 3.8 = 0.29$

Comparing average and maximum values:

$$\int \alpha_{average}.dt = \int \left(0.9049 - 0.0243 t^{-0.5}\right).dt$$
$$= 0.9049t + 0.0243t^{1.5}/1.5 = 0.9049t + 0.0162t^{1.5}$$

$$\int \alpha_{max}.dt = \int \left(0.8641 - 0.0300 t^{-0.5}\right).dt$$
$$= 0.8641t + 0.0300t^{1.5}/1.5 = 0.8641t + 0.0200t^{1.5}$$

Subtracting gives $\Delta\alpha = 0.0408t - 0.0038t^{1.5}$
So when $t = 100$, $\Delta\alpha = 4.08 - 0.0038 \times 1,000 = 4.08 - 3.8 = 0.28$
So our sensitivity test is giving a crude range of ~ 0.3 about the mean.

References

1. Pachauri, R. K. et al. Climate Change 2014: Synthesis Report. *Contribution of Working Groups I. II and III to the Fifth Assessment Report of the Intergovernmental Panel on Climate Change* (2014).
2. Box, J. et al. Greenland ice sheet albedo feedback: thermodynamics and atmospheric drivers. *Cryosphere* **6**, 821–839 (2012).
3. Anesio, A. M. & Laybourn-Parry, J. Glaciers and ice sheets as a biome. *Trends Ecol. Evol.* **27**, 219–225 (2012).
4. Hodson, A. et al. Glacial ecosystems. *Ecol. Monogr.* **78**, 41–67 (2008).
5. Musilova, M., Tranter, M., Bennett, S. A., Wadham, J. & Anesio, A. M. Stable microbial community composition on the Greenland Ice Sheet. *Front. Microbiol.* **6**, 193 (2015).
6. Edwards, A. et al. Coupled cryoconite ecosystem structure-function relationships are revealed by comparing bacterial communities in alpine and Arctic glaciers. *FEMS Microbiol. Ecol.* **89**, 222–237 (2014).
7. Stibal, M., Šabacká, M. & Kaštovská, K. Microbial communities on glacier surfaces in Svalbard: impact of physical and chemical properties on abundance and structure of cyanobacteria and algae. *Microb. Ecol.* **52**, 644–654 (2006).
8. Anesio, A. M., Hodson, A. J., Fritz, A., Psenner, R. & Sattler, B. High microbial activity on glaciers: importance to the global carbon cycle. *Global Change Biol.* **15**, 955–960 (2009).
9. Edwards, A. et al. Possible interactions between bacterial diversity, microbial activity and supraglacial hydrology of cryoconite holes in Svalbard. *ISME J.* **5**, 150–160 (2011).
10. Gentz-Werner, P. *Roter Schnee: oder Die Suche nach dem färbenden Prinzip.* Vol. 28. (Akademie, 2007).
11. Lutz, S., Anesio, A. M., Villar, S. E. J. & Benning, L. G. Variations of algal communities cause darkening of a Greenland glacier. *FEMS Microbiol. Ecol.* **89**, 402–414 (2014).
12. Yallop, M. L. et al. Photophysiology and albedo-changing potential of the ice algal community on the surface of the Greenland ice sheet. *ISME J.* **6**, 2302–2313 (2012).
13. Hell, K. et al. The dynamic bacterial communities of a melting High Arctic glacier snowpack. *ISME J.* **7**, 1814–1826 (2013).
14. Remias, D., Lütz-Meindl, U. & Lütz, C. Photosynthesis, pigments and ultrastructure of the alpine snow alga Chlamydomonas nivalis. *Eur. J. Phycol.* **40**, 259–268 (2005).
15. Lutz, S., Anesio, A. M., Field, K. & Benning, L. G. Integrated 'omics', targeted metabolite and single-cell analyses of Arctic snow algae functionality and adaptability. *Front. Microbiol.* **6**, 1323 (2015).
16. Leya, T., Müller, T., Ling, H. U. & Fuhr, G. Snow algae from north-western Spitsbergen (Svalbard). *The Coastal Ecosystem of Kongsfjorden, Svalbard. Synopsis of Biological Research Performed at the Koldewey Station in the Years 1991-2003*, Vol. 492 (ed. Wiencke, C.) 46–54 (2004).
17. Hoham, R. & Duval, B. *Microbial Ecology of Snow and Freshwater Ice with Emphasis on Snow Algae. Snow Ecology: an Interdisciplinary Examination of Snow-covered Ecosystems*, 168–228 (Cambridge Univ., 2001).
18. Takeuchi, N. The altitudinal distribution of snow algae on an Alaska glacier (Gulkana Glacier in the Alaska Range). *Hydrol. Processes* **15**, 3447–3459 (2002).
19. O'Malley, M. A. The nineteenth century roots of 'everything is everywhere'. *Nat. Rev. Microbiol.* **5**, 647–651 (2007).
20. Ryšánek, D., Hrčková, K. & Škaloud, P. Global ubiquity and local endemism of free-living terrestrial protists: phylogeographic assessment of the streptophyte alga Klebsormidium. *Environ. Microbiol.* **17**, 689–698 (2014).
21. King, A. J. et al. Biogeography and habitat modelling of high-alpine bacteria. *Nat. Commun.* **1**, 53 (2010).
22. Tedesco, M. et al. The darkening of the Greenland ice sheet: trends, drivers, and projections (1981-2100). *Cryosph* **10**, 477–496 (2016).
23. Doherty, S., Warren, S., Grenfell, T., Clarke, A. & Brandt, R. Light-absorbing impurities in Arctic snow. *Atmospheric Chem. Phys.* **10**, 11647–11680 (2010).
24. Benning, L. G., Anesio, A. M., Lutz, S. & Tranter, M. Biological impact on Greenland's albedo. *Nat. Geosci.* **7**, 691–691 (2014).
25. Martiny, J. B. H. et al. Microbial biogeography: putting microorganisms on the map. *Nat. Rev. Microbiol.* **4**, 102–112 (2006).
26. Foissner, W. Biogeography and dispersal of micro-organisms: a review emphasizing protists. *Acta Protozoologica* **45**, 111–136 (2006).
27. Green, J. & Bohannan, B. J. Spatial scaling of microbial biodiversity. *Trends Ecol. Evol.* **21**, 501–507 (2006).
28. Cameron, K. A. et al. Diversity and potential sources of microbiota associated with snow on western portions of the Greenland Ice Sheet. *Environ. Microbiol.* **17**, 594–609 (2014).
29. Larose, C. et al. Microbial sequences retrieved from environmental samples from seasonal Arctic snow and meltwater from Svalbard, Norway. *Extremophiles* **14**, 205–212 (2010).
30. Spijkerman, E., Wacker, A., Weithoff, G. & Leya, T. Elemental and fatty acid composition of snow algae in Arctic habitats. *Frontiers in microbiology* **3**, 380 (2012).
31. Fierer, N. & Jackson, R. B. The diversity and biogeography of soil bacterial communities. *Proc. Natl Acad. Sci. USA* **103**, 626–631 (2006).
32. Thomas, F., Hehemann, J.-H., Rebuffet, E., Czjzek, M. & Michel, G. Environmental and gut bacteroidetes: the food connection. *Front. Microbiol.* **2**, 93 (2011).
33. Lutz, S., Anesio, A. M., Edwards, A. & Benning, L. G. Microbial diversity on Icelandic glaciers and ice caps. *Front. Microbiol.* **6**, 307 (2015).
34. Painter, T. H. et al. Detection and quantification of snow algae with an airborne imaging spectrometer. *Appl. Environ. Microbiol.* **67**, 5267–5272 (2001).
35. Aoki, T. et al. in *Radiation processes in the atmosphere and ocean (IRIS 2012): Proc. Int. Radiation Symp. (IRC/IAMAS)* 176–179 (AIP Publishing, 2016).
36. Tedesco, M. et al. Evidence and analysis of 2012 Greenland records from spaceborne observations, a regional climate model and reanalysis data. *Cryosphere Discussions (The)* **6**, 615–630 (2013).
37. Gardner, A. S. & Sharp, M. J. A review of snow and ice albedo and the development of a new physically based broadband albedo parameterization. *J. Geophys. Res. Earth Surface* **115**, F01009 (2010).
38. Zhang, Y. *Geochem. Kinetics* (Princeton Univ., 2008).
39. Thomas, W. H. & Duval, B. Sierra Nevada, California, USA, snow algae: snow albedo changes, algal-bacterial interrelationships, and ultraviolet radiation effects. *Arctic Alpine Res.* **27**, 389–399 (1995).
40. Musilova, M., Tranter, M., Bamber, J. L., Takeuchi, N. & Anesio, A. M. Experimental evidence that microbial activity lowers the albedo of glaciers. *Geochem. Perspect. Lett.* **2**, 106–116 (2016).
41. Nghiem, S. et al. The extreme melt across the Greenland ice sheet in 2012. *Geophys. Res. Lett.* **39**, L20502 (2012).
42. McGrath, D., Colgan, W., Bayou, N., Muto, A. & Steffen, K. Recent warming at Summit, Greenland: Global context and implications. *Geophys. Res. Lett.* **40**, 2091–2096 (2013).
43. Tedesco, M. et al. What Darkens the Greenland Ice Sheet? *EOS: Transactions, American Geophysical Union.* doi:10.1029/2015EO035773 (2015).
44. Ruttenberg, K. et al. Improved, high-throughput approach for phosphorus speciation in natural sediments via the SEDEX sequential extraction method. *Limnol. Oceanogr. Methods* **7**, 319–333 (2007).
45. Remias, D. & Lutz, C. Characterisation of esterified secondary carotenoids and of their isomers in green algae: a HPLC approach. *Algol. Studies* **124**, 85–94 (2007).
46. Wacker, A. & Martin-Creuzburg, D. Allocation of essential lipids in Daphnia magna during exposure to poor food quality. *Funct. Ecol.* **21**, 738–747 (2007).

47. Caporaso, J. G. *et al.* QIIME allows analysis of high-throughput community sequencing data. *Nat. Methods* **7**, 335–336 (2010).

48. Cheung, M. K., Au, C. H., Chu, K. H., Kwan, H. S. & Wong, C. K. Composition and genetic diversity of picoeukaryotes in subtropical coastal waters as revealed by 454 pyrosequencing. *ISME J.* **4**, 1053–1059 (2010).

49. Pruesse, E. *et al.* SILVA: a comprehensive online resource for quality checked and aligned ribosomal RNA sequence data compatible with ARB. *Nucl Acids Res.* **35**, 7188–7196 (2007).

50. Hammer, O., Harper, D. & Ryan, P. PAST: paleontological statistics software package for education and data analysis. *Paleontol. Electron.* **4**, 1–9 (2001).

51. Potvin, M. & Lovejoy, C. PCR-based diversity estimates of artificial and environmental 18S rRNA gene libraries. *J. Eukaryotic Microbiol.* **56**, 174–181 (2009).

52. Zarsky, J. D. *et al.* Large cryoconite aggregates on a Svalbard glacier support a diverse microbial community including ammonia-oxidizing archaea. *Environ. Res. Lett.* **8**, 035044 (2013).

Acknowledgements

We thank A. Detheridge (Aberystwyth University), and C. Waterfall and J. Coghill (University of Bristol) for help with the DNA sequencing and A. Stockdale (University of Leeds) for the phosphorus analysis. The research leading to these results was funded by a University of Leeds PhD Scholarship grant to S.L. and L.G.B., by a grant from the European Union Seventh Framework Program INTERACT (grant no 262693) to L.G.B. and by UK Natural Environment Research Council grants NE/J022365/1 to L.G.B. and NE/J02399X/1 to A.M.A. Financial support for SL's field and lab work through a Young Explorers grant from National Geographic (GEFNEY73-13) and a President's Fund for Research Visits from the Society for General Microbiology (PF13/16) are gratefully acknowledged. We also thank the scientific staff at the Tarfala and NERC Arctic research stations for great field support.

Author contributions

S.L. and L.G.B. designed the study. Field work was carried out by S.L., A.M.A. and L.G.B. All analyses were completed by S.L. with support from A.E. (DNA sequencing), R.J.N. (particulates) and F.G. (fatty acids). The integrated albedo change was developed by S.L., L.G.B. and R.R. All authors contributed to the discussion of the results. Manuscript was written by S.L. with major input from L.G.B. and contributions from A.M.A., R.R., A.E., R.J.N. and F.G.

Additional information

19

High potential for weathering and climate effects of non-vascular vegetation in the Late Ordovician

P. Porada[1,2], T.M. Lenton[3], A. Pohl[4], B. Weber[5], L. Mander[6], Y. Donnadieu[4,7], C. Beer[1,2], U. Pöschl[5] & A. Kleidon[8]

It has been hypothesized that predecessors of today's bryophytes significantly increased global chemical weathering in the Late Ordovician, thus reducing atmospheric CO_2 concentration and contributing to climate cooling and an interval of glaciations. Studies that try to quantify the enhancement of weathering by non-vascular vegetation, however, are usually limited to small areas and low numbers of species, which hampers extrapolating to the global scale and to past climatic conditions. Here we present a spatially explicit modelling approach to simulate global weathering by non-vascular vegetation in the Late Ordovician. We estimate a potential global weathering flux of 2.8 $(km^3$ rock$)$ yr^{-1}, defined here as volume of primary minerals affected by chemical transformation. This is around three times larger than today's global chemical weathering flux. Moreover, we find that simulated weathering is highly sensitive to atmospheric CO_2 concentration. This implies a strong negative feedback between weathering by non-vascular vegetation and Ordovician climate.

[1] Department of Environmental Science and Analytical Chemistry (ACES), Stockholm University, 10691 Stockholm, Sweden. [2] Bolin Centre for Climate Research, Stockholm University, 10691 Stockholm, Sweden. [3] Earth System Science Group, College of Life and Environmental Sciences, University of Exeter, Laver Building (Level 7), North Park Road, Exeter EX4 4QE, UK. [4] Laboratoire des Sciences du Climat et de l'Environnement, LSCE/IPSL, CEA-CNRS-UVSQ, Université Paris-Saclay, F-91191 Gif-sur-Yvette, France. [5] Max Planck Institute for Chemistry, PO Box 3060, 55020 Mainz, Germany. [6] Department of Environment, Earth and Ecosystems, The Open University, Milton Keynes, Buckinghamshire MK7 6AA, UK. [7] Aix-Marseille Université, CNRS, IRD, CEREGE UM34, 13545 Aix en Provence, France. [8] Max Planck Institute for Biogeochemistry, PO Box 10 01 64, 07701 Jena, Germany. Correspondence and requests for materials should be addressed to P.P. (email: philipp.porada@aces.su.se).

Lichens and bryophytes have been shown to significantly enhance weathering of the surface rocks on which they grow compared with abiotic conditions[1-3]. They increase both physical as well as chemical weathering rates: lichen rhizines and bryophyte rhizoids break up the rock surface, followed by dissolution of the minerals through different organic acids, alkalinolysis and chelating agents released by the organisms[4].

Measurements of weathering intensity below lichens indicate an enhancement by around one order of magnitude compared with abiotic conditions for cool mountainous regions, such as Norway[5], as well as for warm arid regions, for example, Lanzarote[6]. In moist tropical regions an enhancement of two orders of magnitude has been observed[7,8]. These measurements are complemented by studies based on mini-watersheds in the northeastern United States[9,10], which quantify directly the amount of chemical elements exported from the rock and which confirm the enhancement effect (factor 2–16). Bryophytes, too, have been proven to enhance weathering significantly under field conditions in Canada[1] as well as in microcosms[2,11].

At the global scale, however, it is difficult to assess the role that lichens and bryophytes play in weathering based on these local findings. The global distribution of rocks covered by lichens and bryophytes is poorly known[12]. Several studies have concluded that lichens and bryophytes could not significantly enhance the global weathering flux, since vascular plants show a considerably higher enhancement of weathering[2,13]. Other studies, however, suggest that enhancement of weathering by lichens and bryophytes is still sufficiently large for global effects[3,11].

The uncertainty regarding the relative importance of lichens and bryophytes for the global weathering flux limits climate reconstructions of the geological past[3]. On the basis of field observations, it has been concluded that the emergence of large vascular plants with rooting systems in the Devonian led to a large increase in global chemical weathering rates[14]. Including this effect in a global geochemical model[14] causes a decrease in atmospheric CO_2 based on the well-known silicate-weathering feedback[15].

Predecessors of today's bryophytes, however, could also have enhanced chemical weathering. Molecular phylogenies support the traditional view that non-vascular (bryophyte) plant lineages (liverworts, hornworts and mosses) all pre-date the origin of vascular plants[16]. The ancestors of modern bryophytes are thought to have evolved during the Mid-Late Ordovician period based on the microfossil record of spores[16] and improvements in molecular clock dating support an early origin[16]. The oldest lichen fossil found so far, exhibiting symbionts and anatomy similar to extant lichens, dates to the Early Devonian[17]; however, lichens could be much older, as shown by a potential symbiosis of fungal hyphae and cyanobacteria from the Late Precambrian[18]. Thus, early non-vascular plants, together with lichens, could also have led to a drawdown in atmospheric CO_2.

It has been hypothesized that the interval of glaciations at the end of the Late Ordovician (peaking in the Hirnantian Stage) could be explained by a decrease in atmospheric CO_2 because of the biotic enhancement of weathering[11]. Therefore, microcosms containing moss growing on mineral surfaces were used to estimate the enhancement of weathering compared with microcosms without moss. The results indicate a considerable increase in weathering for several elements, with highest values for iron (over two orders of magnitude) and phosphorus, with calcium and magnesium weathering amplified by factors of circa 1.5–5 across different silicate rock types.

In order to translate the increase in chemical weathering rates into a decrease in atmospheric CO_2, the results from the microcosm experiment were used to calibrate an enhancement factor in a global biogeochemical model[11]. Such an approach neglects the spatiotemporal variation of the global moss cover that results from climate patterns and that clearly controls weathering by moss. Moreover, the enhancement factor is based only on one species of moss, growing in the modern atmosphere, while weathering by lichens and bryophytes has been shown to be species-specific[1,19].

In this study, we present a complementary approach, which derives chemical weathering by lichens and bryophytes in the Ordovician from their net primary productivity (NPP), which is calculated by a novel, global process-based non-vascular vegetation model that simulates multiple, physiologically different, artificial species[20]. To translate NPP into a weathering flux it is assumed that lichens and bryophytes dissolve surface rocks to access phosphorus for the build-up of new biomass[21]. The amount of rock that is weathered can then be estimated from the phosphorus equivalent of NPP[22] and the phosphorus concentration of rocks. To account for the limited supply of unweathered rock material in shallow regions, we cap biotic weathering at the erosion rate. Furthermore, we constrain weathering by the transport capacity of runoff for dissolved weathered material. Since these limitations on weathering result in a lower nutrient supply from rocks, we reduce our estimate of NPP accordingly. By representing the species-specific, spatiotemporal dynamics of weathering by lichens and bryophytes in the Late Ordovician, we assess the significance of these organisms for Ordovician weathering and climate. Throughout the manuscript, we use the term chemical weathering to describe the partial dissolution and chemical transformation of a certain volume of primary minerals into solutes and secondary minerals. This includes all elements, such as calcium, phosphorus, iron and so on, in the primary minerals. We use volume as a basis to compare the rate of chemical weathering to the erosion rate.

Results

Weathering by lichens and bryophytes in the Ordovician. The global pattern of chemical weathering by Ordovician lichens and bryophytes, their NPP and their surface coverage as well as the pattern of the limiting factors on weathering are shown in Fig. 1. The estimates are obtained from a baseline simulation of the non-vascular vegetation model, run at eight preindustrial atmospheric level(PAL; 1 PAL = 280 p.p.m.) of atmospheric CO_2 and 14% of atmospheric O_2 for 600 years to reach steady state, with an initial number of 300 species (see Methods section for details).

Potential NPP of lichens and bryophytes depends only on climatic factors (Fig. 1a). It shows a latitudinal pattern, with maximum values at the equator and high values in the region between 30° and 60°. Between the equator and 30° N and 30° S, respectively, NPP is low and in some regions none of the simulated lichen and bryophyte species are able to survive. This distribution can be explained by water limitation of NPP in (sub)tropical regions and limitation by low temperature and light availability in polar regions (see Methods section). The global pattern of weathering by lichens and bryophytes based solely on their NPP is identical to that of potential NPP, since the conversion factors used to translate NPP into a weathering flux are assumed to be globally uniform (Fig. 1b).

Chemical weathering constrained by both the erosion rate and the transport capacity of runoff for dissolved weathered material is strongest in regions of pronounced relief and high rainfall (Fig. 1c). In areas of flat topography, weathering is limited by erosion, which is directly related to the mean surface elevation above the sea level. In arid regions, weathering is limited by low runoff. The global pattern of weathering thus results from the minimum of the limiting factors NPP, erosion and runoff

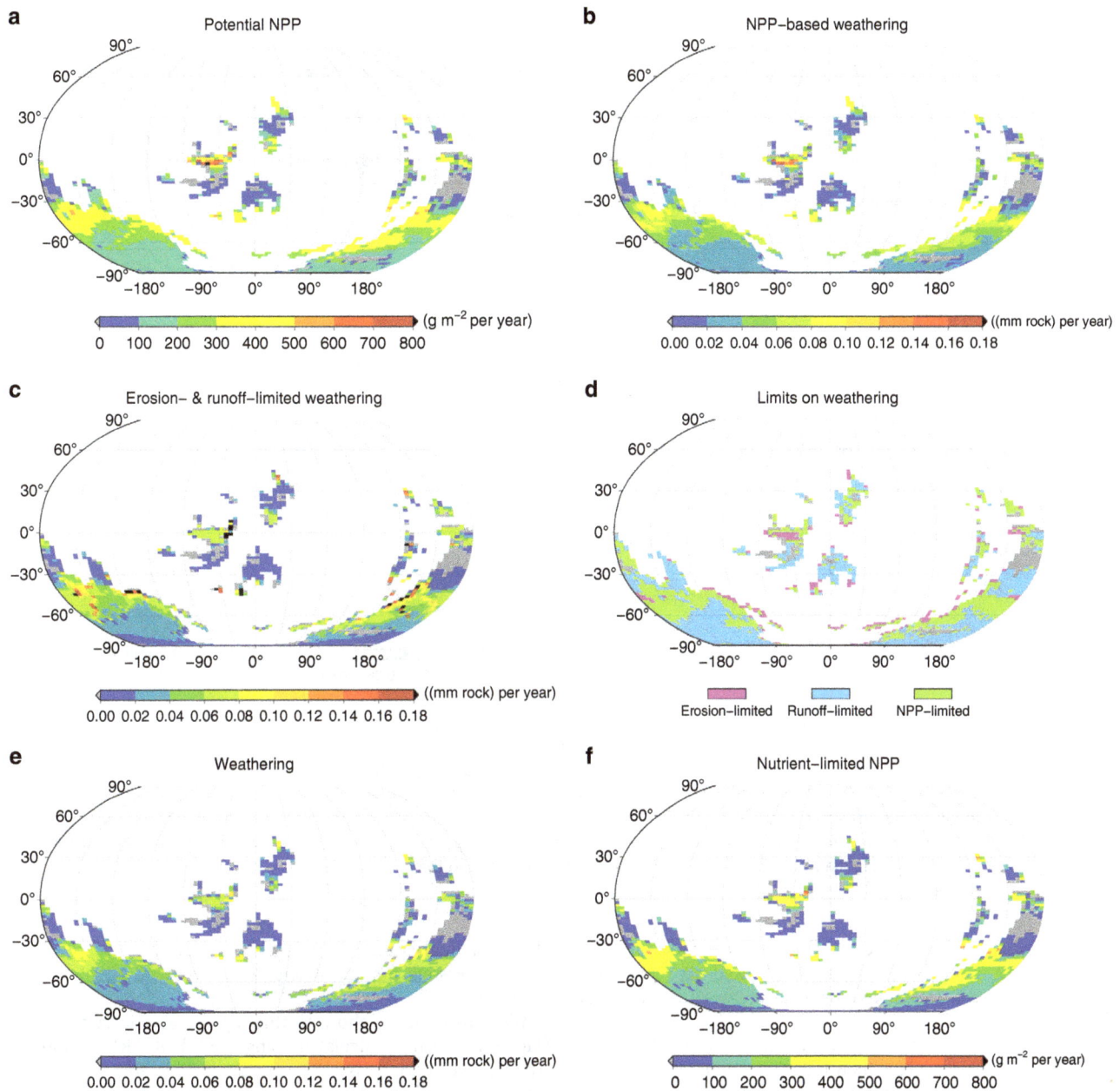

Figure 1 | Global patterns of NPP and weathering. (**a**) Potential NPP simulated by the lichen and bryophyte model for the Late Ordovician, 450 million years ago (450 Ma). (**b**) NPP-based weathering, (**c**) weathering based on the minimum of erosion and transport of solutes by runoff, (**d**) limiting factors on weathering, (**e**) weathering constrained by NPP, erosion and runoff and (**f**) NPP limited by nutrient supply from weathering. The maps show average values over the last 50 years of a 600-year simulation with 300 initial species. The atmospheric CO_2 concentration is 2,240 p.p.m. or 8 PAL. The grey areas denote regions where no species are able to survive. For the small grey area near the south pole the lichen and bryophyte model simulates a small ice sheet.

(Fig. 1d,e). In areas where NPP-based weathering exceeds erosion- and runoff-limited weathering, the supply of phosphorus from rocks is not sufficient to sustain potential NPP, which is consequently reduced (Fig. 1f). The strongest reduction of potential NPP occurs at the equator, because of low erosion rates, and in dry regions at 30° S and at high latitudes, because of low rainfall. At the global scale, NPP-based weathering of 3.6 (km³ rock) yr⁻¹ is reduced by 22% to a realized weathering of 2.8 (km³ rock) yr⁻¹. This is accompanied by a reduction in NPP because of phosphorus limitation in shallow or arid regions. Potential NPP of 18.7 (Gt C) yr⁻¹ is reduced by 23% to 14.4 (Gt C) yr⁻¹ of realized NPP.

Global annual values of lichen and bryophyte weathering, NPP, gross primary productivity (GPP), surface coverage and biomass are listed in Table 1 together with the corresponding values for today's climate in the presence of vascular plants. Today's values were calculated by forcing the lichen and bryophyte model with data fields for today's climate and other current environmental conditions[20]. In addition, shown in Table 1 are observation-based estimates of NPP, GPP, surface coverage and biomass of today's total terrestrial vegetation and total, that is, biotic plus abiotic, global chemical weathering.

Considering only lichens and bryophytes, all values corresponding to the Ordovician are considerably larger than those for

Table 1 | Global effects of lichens and bryophytes.

Simulation	NPP (Gt C) yr^{-1}	GPP (Gt C) yr^{-1}	Cover fraction	Biomass Gt C	Weathering (km^3 rock) yr^{-1}
Ordovician	14.4	30.1	0.44	133	2.8
Today					
Lichens and bryophytes	2.6	5.4	0.14	51	0.044
Total vegetation	60 (ref. 25)	123 (ref. 68)	0.74 (ref. 69)	442 (ref. 70)	0.85 (ref. 54)

GPP, gross primary productivity; Gt C, gigatons of carbon; NPP, net primary productivity.
Annual global values of NPP, GPP, surface coverage, biomass and weathering for the Ordovician baseline simulation (see Methods section) and for today's climate. Note that we converted the weathering estimate[54] by using an average density of surface rocks[22].

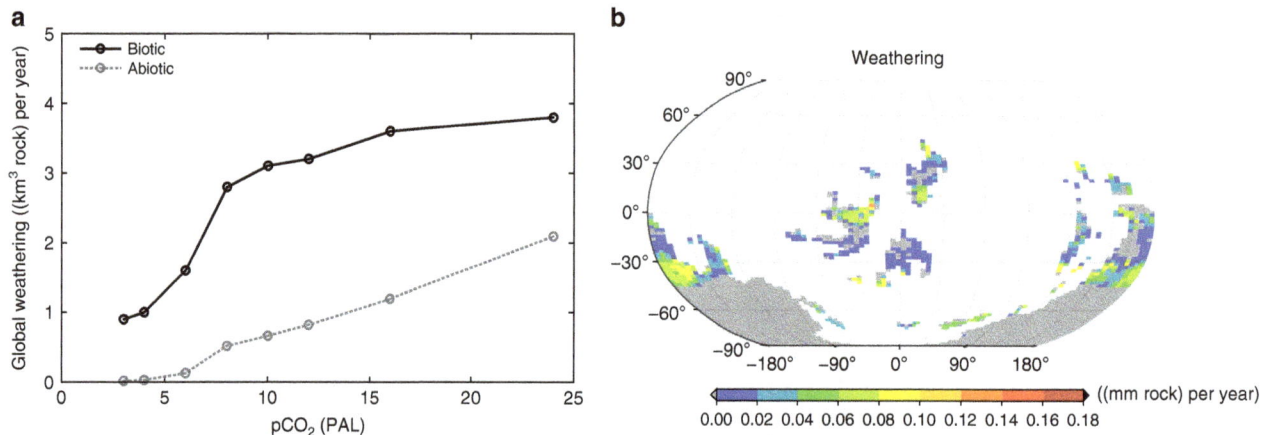

Figure 2 | Sensitivity of weathering to environmental factors. (a) Global chemical weathering as a function of atmospheric CO_2 concentration. One PAL corresponds to 280 p.p.m. Each dot represents a model simulation at the respective atmospheric CO_2 concentration. The black solid line shows weathering enhanced by non-vascular vegetation derived from a 600-year simulation with 300 initial species. The grey dashed line represents abiotic weathering, calculated by a GEOCARB approach. **(b)** Effect of a continental-sized ice sheet south of 30° S on the global pattern of weathering for the Late Ordovician (450 Ma), derived from NPP of lichens and bryophytes constrained by erosion and runoff. The map shows average values over the last 50 years of a 600-year simulation with 300 initial species. The atmospheric CO_2 concentration is 2,240 p.p.m. (8 PAL). The grey areas denote regions where no species are able to survive, notably including the ice sheet mask prescribed from the South Pole to the mid-latitudes.

today's climate. The relative increase differs between the variables. NPP and GPP increase by a factor of around six, cover and biomass by around three and weathering increases much more by a factor of ~ 60.

Chemical weathering by Ordovician lichens and bryophytes is more than three times larger than today's combined abiotic and biotic global chemical weathering. Values of NPP and GPP of Ordovician lichens and bryophytes, however, only amount to $\sim 25\%$ of today's total terrestrial vegetation, while Ordovician biomass and vegetation surface coverage are 30% and 60% of today's values, respectively. These results are explained below in the Discussion section.

In Table 1 we show today's combined abiotic and biotic global chemical weathering rate. It is estimated that vegetation enhances chemical weathering by a factor of ~ 4–7 (refs 14,23). Therefore, the combined rate should be a good approximation for weathering by vegetation alone, which includes vascular plants, bryophytes and lichens.

Sensitivity of weathering to environmental factors. Estimated global chemical weathering by lichens and bryophytes in the Ordovician shows a strong dependence on atmospheric CO_2 concentration, as illustrated in (Fig. 2a). The increase in weathering with increasing CO_2 results from the positive effect of high atmospheric CO_2 concentrations on NPP. While the relation looks exponential at values of CO_2 lower than 8 PAL, it seems to

change at high values to a logarithmic one. Below we discuss possible reasons for the strong dependence of NPP on atmospheric CO_2.

We, furthermore, show weathering for abiotic conditions in (Fig. 2a), which is calculated using the GEOCARB III model[14]. Abiotic weathering is in general considerably lower than weathering enhanced by non-vascular vegetation. However, abiotic weathering does not show saturation at very high values of CO_2, in contrast to biotic weathering.

To account for the existence of a continental-sized ice sheet south of 30° S in the Hirnantian Stage[24], an additional vegetation model run is performed with a corresponding glacier mask and updated climatic forcing fields (Fig. 2b). Although the ice sheet does not cover the regions of highest NPP and weathering at the equator and close to 30° S, it covers vast regions of Gondwana and, therefore, reduces weathering considerably from 2.8 to $1.5 \, km^3 \, yr^{-1}$ and NPP from 14.4 to 7.6 $Gt \, yr^{-1}$.

Sensitivity of weathering to number of artificial species. Figure 3 shows that global weathering by lichens and bryophytes increases with the number of initial artificial species used in a simulation and remains constant for larger numbers of initial species. This results from an undersampling of the very large space of possible combinations of physiological parameters. Species that are capable of high NPP and, consequently, strong weathering, are probably not generated if the number of initial species is too low. This means that, at the beginning of the colonization of the land surface by non-vascular plants, biotic weathering could have been lower because of very low diversity and productivity.

Discussion

In this study we estimate chemical weathering in the Late Ordovician by early forms of lichens and bryophytes. We quantify their NPP during this period with a process-based vegetation model and then translate the NPP into a weathering flux[22], additionally accounting for limitation of weathering by low erosion and runoff.

Our main finding is a huge weathering potential of Ordovician lichens and bryophytes, more than three times higher than today's combined abiotic and biotic global chemical weathering flux. This value results from a 60-fold increase in weathering by these organisms compared with today's value (Table 1), which could be explained by the following factors: first, lack of vascular plants in the Ordovician allows lichens and bryophytes to cover

the entire area available for growth of vegetation. Second, Ordovician lichens and bryophytes have increased access to mineral surfaces because of the absence of competing weathering by vascular plants. Finally, high Ordovician atmospheric CO_2 has a strong increasing effect on NPP of lichens and bryophytes and, consequently, on NPP-based weathering.

In the following, we discuss the plausibility of our estimates, starting with the large value of NPP for the Ordovician, which is $\sim 25\%$ of today's total terrestrial NPP[25]. The main factor for such a high productivity is the strong CO_2 sensitivity of simulated NPP, combined with high levels of atmospheric CO_2 in the Ordovician. The CO_2 response can be analysed in the model for individual artificial species, by varying ambient CO_2, but otherwise keeping constant boundary conditions, such as water content, temperature and radiation. To find appropriate species, we choose two regions with similar climate, one from a simulation with today's climatic and CO_2 forcing and another from an Ordovician simulation with a high atmospheric CO_2 value of 24 PAL. From these two regions we select the dominant species, shown in (Fig. 4a). It can be seen that for the high Ordovician CO_2 level, the model favours a species that has a higher photosynthetic capacity than the species dominant under today's CO_2, given otherwise similar climatic conditions. This seems plausible from the viewpoint of natural selection and leads to a higher NPP in the Ordovician simulation. Most dynamic vegetation models, however, do not implement such a large physiological flexibility, but rely on a few plant functional types with fixed photosynthesis parameters. They cannot adapt to strongly differing environmental conditions. Hence, we attribute the strong sensitivity of NPP to CO_2 to the physiological flexibility of our model.

Experiments confirm that the CO_2 sensitivity of lichens and bryophytes is species-specific (Fig. 4b). The species differ with regard to the slope of the CO_2 response as well as the ambient CO_2 concentration, at which photosynthesis saturates. We want to point out that Fig. 4 does not represent an evaluation of the model, it rather illustrates the large variety in CO_2 responses of different lichen and bryophyte species. This supports the possibility of high NPP of early lichens and bryophytes in the Late Ordovician.

Two of the experimental studies considered here measure NPP of bryophytes at CO_2 concentrations up to 8 PAL (see also Methods section). They find a two to threefold increase in NPP, whereas our artificial species from today shows a sixfold increase,

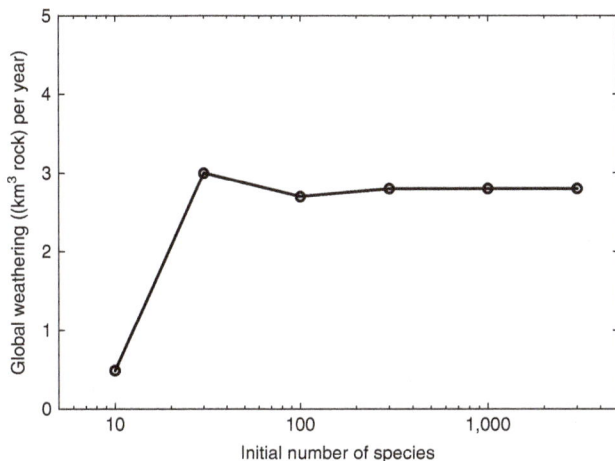

Figure 3 | Sensitivity of weathering to number of species. Global chemical weathering by lichens and bryophytes as a function of the initial number of species in a simulation conducted for the Late Ordovician at 8 PAL. Each dot represents a model simulation carried out at the corresponding initial species number.

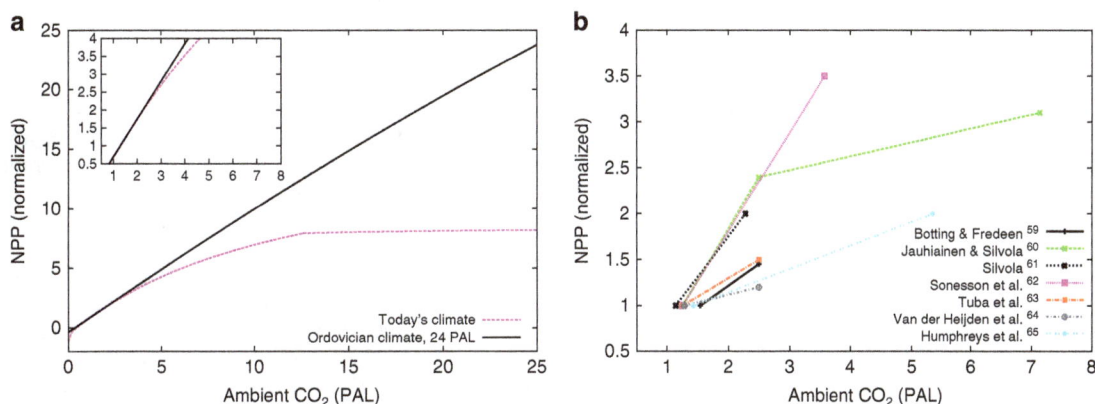

Figure 4 | Effect of elevated CO_2 on NPP. (a) NPP calculated by the model for constant light, temperature and water content. The blue solid line corresponds to a species adapted to Ordovician climate and an atmospheric CO_2 concentration of 24 PAL and the dashed magenta line corresponds to a species adapted to today's climate and CO_2 concentration. **(b)** NPP of lichens and bryophytes measured in field or laboratory experiments as a function of ambient CO_2. Each line stands for a different study and each dot represents a measurement of NPP at the corresponding CO_2 concentration. NPP is normalized in order to make the CO_2 response comparable between different species. The inset figure in **a** focuses on the range of ambient CO_2 shown in **b**.

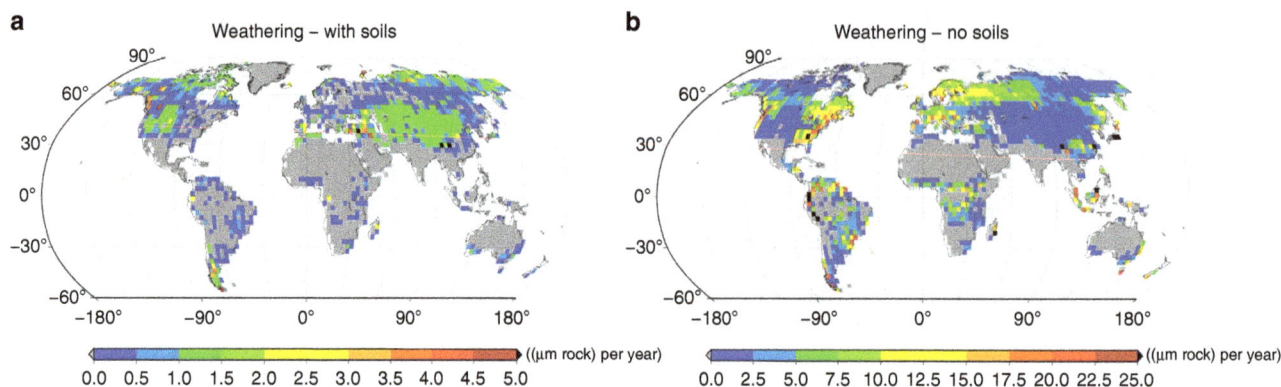

Figure 5 | Weathering for today's climate. Global pattern of chemical weathering by lichens and bryophytes for today's climate, constrained by their NPP, erosion and runoff. (**a**) NPP-based weathering is constrained in each grid cell to the fraction of bare rock surface. (**b**) NPP-based weathering may take place on the whole land surface area. The map shows average values over the last 50 years of a 600-year simulation with 300 initial species. Runoff is derived from annual average values of climate variables[20]. Note the differing ranges of the colour scales.

relative to NPP at 360 p.p.m. of CO_2. This may be explained by the fact that the artificial species has a lower photosynthetic capacity than the bryophytes from the experiments, maybe to avoid high respiration costs. As a consequence, NPP of the artificial species increases only slowly with rising ambient CO_2 leading to a much lower NPP at 360 p.p.m. of CO_2, compared with the bryophytes (see Methods section for details). For the same reason, however, NPP of the artificial species saturates at a higher CO_2 concentration than the bryophyte, leading to a larger CO_2 sensitivity. We do not want to exclude the possibility that our model is too flexible regarding the shape of the CO_2 response. We therefore test a minimum CO_2-sensitivity scenario, where potential NPP is reduced by 50%. Since weathering depends also on environmental limits in our approach, this translates into a 40% reduced estimate of global chemical weathering for the Ordovician of 1.6 (km³ rock) yr⁻¹.

Several studies substantiate the occurrence of high non-vascular NPP in the geological past. On the basis of estimated Ordovician soil CO_2 production, it has been concluded that the above-ground productivity fuelling this CO_2 production must have been similar to today's (ref. 26). Hence, non-vascular productivity could have been as high as the vascular one. The course of $\delta^{13}C$ through the Phanerozoic shows a marked increase in the Late Ordovician consistent with a new source of organic carbon on land and concomitant increase in global organic carbon burial[27]. Moreover, pre-vascular organic carbon deposits were already large, although they are not in the form of coal[3]. The occurence of coal before the Devonian is unlikely, albeit difficult to assess because of the sparse occurence of terrestrial sediments preserved from before the Devonian[16].

An alternative estimate of NPP for the Ordovician[28] has been made by multiplying today's global cryptogamic productivity on ground[12] with a value of 1.8, which is the ratio of today's total vegetated area to the area covered by cryptogams. The estimated value of 4.3 (Gt C) yr⁻¹ is markedly lower than our estimate, which can be explained by simplifying assumptions in this study, such as a linear relationship between area and productivity and no CO_2 fertilization.

Our estimate of realized NPP for the Ordovician is around six times higher than today's NPP of lichens and bryophytes. However, simulated chemical weathering by the organisms is enhanced by a factor of 60 in the Late Ordovician compared with today's value. We suggest that the absence of vascular plants in the Ordovician explains the remaining part of the difference in chemical weathering rates. Vascular plants are likely more

efficient in enhancing weathering rates than lichens and bryophytes[2,29]. As a consequence, lichens and bryophytes are usually separated from unweathered rock material by a layer of soil in ecosystems that are dominated by vascular plants, such as forests and grasslands. In these ecosystems, growth of lichens and bryophytes on bare rocks is confined to boulders and cliffs, which cover only a small fraction of the land surface. This strongly limits the contribution of lichens and bryophytes to total chemical weathering. In other ecosystems, such as cold and warm deserts, lichens and bryophytes often are the dominant vegetation. However, low runoff largely limits chemical weathering by the organisms in these regions. To assess the effect of soils on weathering by today's lichens and bryophytes, we assume that all simulated growth in the model may lead to weathering, not only the part that takes place on bare soil. This increases weathering by today's lichens and bryophytes by a factor of ~10, which, in combination with increased atmospheric CO_2, explains the strong increase in weathering from today to the Late Ordovician (Fig. 5).

Given the large value of today's total terrestrial NPP compared with the Ordovician, the current global chemical weathering flux appears to be quite small (Table 1). This implies that, for a given amount of biotic weathering, today's vascular vegetation exhibits a much higher productivity than Ordovician lichens and bryophytes at the global scale. We explain this finding by a high nutrient recycling ratio of modern ecosystems of ~50:1 (ref. 30), which results from the deep rooting systems of vascular plants and associated mycorrhiza. Although lichens and bryophytes are able to resorb phosphorus from senescing tissue, they most likely cannot access phosphorus that has been leached or released from decaying biomass and transported by water to deeper soil layers. In our model, we therefore assume a low recycling ratio of 5:1 for Late Ordovician ecosystems. This value is based on measurements of both weathering rates of lichens on rocks as well as their productivity in desert regions[20] (see Methods section). Our estimates of chemical weathering and NPP are sensitive to the recycling ratio. In the unlikely case that Ordovician ecosystems had the same high recycling ratio of 50:1 compared with modern ecosystems, weathering decreases to 0.34 (km³ rock) yr⁻¹ and NPP increases to 17.6 (Gt C) yr⁻¹. For a value of 1:1, meaning no recycling, weathering increases to 3.7 (km³ rock) yr⁻¹ because of the organisms' higher phosphorus requirement, while NPP is reduced to 3.9 (Gt C) yr⁻¹ owing to nutrient limitation.

It is likely that vascular plants are more efficient in enhancing weathering than lichens and bryophytes[2,29]. However, our estimate of global chemical weathering in the Late Ordovician is higher than today's weathering (Table 1). One explanation is that the environmental limits, erosion rate and runoff allowed for a much higher global chemical weathering rate in the Late Ordovician compared with today's value. Thus, weathering by vascular plants today may be largely limited by environmental factors, and not by the ability of the plants to enhance weathering. In parts of the Amazon basin, for instance, the weathering capacity of the vegetation is much higher than the weathering allowed by the low erosion rate, leading to thick, highly weathered soils, but a relatively low value of chemical weathering[31,32]. In contrast, chemical weathering in the Ordovician is limited by NPP in many regions (Fig. 1). This means that hypothetical vascular vegetation in the Ordovician could increase weathering to a value of 3.8 (km^3 rock) yr^{-1}, which is the maximum chemical weathering possible under erosion- and runoff limitation.

When early lichens and bryophytes colonized the land surface in the Ordovician, they could well have used their full potential to enhance chemical weathering, until weathering became limited by erosion in many regions and a certain amount of organic material including phosphorus had accumulated in some ecosystems, facilitating a certain degree of recycling. This rationale supports a scenario of strong weathering by lichens and bryophytes leading to a large, but temporally limited release of phosphorus into the oceans and thereby to a short period of cooling because of enhanced marine productivity[11]. The short duration of the phosphorus release is explained by an ecosystem shift to more efficient nutrient recycling caused by nutrient limitation.

Both the erosion- and the runoff-based limits to chemical weathering are maximum rates, which means that the true rate of chemical weathering may be lower for some reason, such as a strong kinetic control on the weathering reactions because of low residence time of water in the soil, for instance. Several observations indicate, however, that reaction kinetics are often not the limiting factor for chemical weathering at the global scale[33]. First, chemical weathering does not show a very clear dependence on temperature at large scale. Second, chemical weathering increases with runoff for low and intermediate values of runoff, which should not be the case if the weathering reaction is far from equilibrium. Finally, chemical weathering rates determined in laboratory experiments exceed those measured in the field by up to two orders of magnitude, which suggests that factors other than reaction kinetics may limit weathering rates in many regions of the world. However, in some areas, such as humid mountainous regions, kinetic limitation of weathering reactions may be important. This is also supported by our alternative estimate of today's global chemical weathering of 1.3 (km^3 rock) yr^{-1} based on the GEM-CO2 model, which is smaller than the 1.8 (km^3 rock) yr^{-1} from our limit-based approach. To account for this potential overestimation, we assume that the relations between runoff, lithology and chemical weathering established in GEM-CO2 for today are the same for the Ordovician. We then scale our estimate of Ordovician weathering by the ratio of today's estimates from GEM-CO2 and the limit-based approach. This would reduce global chemical weathering in the Ordovician by 28% to 2.0 (km^3 rock) yr^{-1}.

The weaker increase in weathering and NPP at high CO_2 values in (Fig. 2a) is because of the saturating Michaelis–Menten kinetics of the Farquhar scheme. Abiotic weathering, in contrast, does not show saturation. It increases almost linearly with atmospheric CO_2 because of an exponential temperature dependence, which is largely compensated by the logarithmic dependence of temperature on CO_2. The stronger than linear increase in biotic weathering and NPP for low CO_2 values is probably due to the retreat of continental ice sheets at rising CO_2 and a large increase in global land surface temperature. In addition, abiotic

weathering shows a nonlinear response to increases in atmospheric CO_2 below values of 8 PAL, since the relation between CO_2 and land surface temperature is not logarithmic anymore at low CO_2 values. The reason for this is a nonlinear increase in sea-surface temperatures with CO_2 because of the complex ocean dynamics in the Fast Ocean Atmosphere Model (FOAM)[34], which forces the Laboratoire de Météorologie Dynamique Zoom (LMDZ) atmospheric circulation model and, thereby, land surface temperature. The continental ice sheets result from a positive snow balance in the lichen and bryophyte model, which is calculated as a function of snowfall, surface temperature and a rate of lateral glacier movement. This snow scheme, however, is calibrated for the Holocene. For 8 PAL of CO_2, for instance, it estimates only a small ice sheet near the south pole (Fig. 1), while there are indications that the ice sheet was much larger[24]. Given the large uncertainty concerning the extent of the Ordovician ice sheet, we think that using the default snow scheme in the model is appropriate.

Our results indicate that weathering by lichens and bryophytes is sensitive to the diversity of plants growing on the land surface (Fig. 3), and this raises the possibility that fluctuations in the diversity of lichens and bryophytes through geological time could have major biogeochemical consequences. Owing to a lack of macroscopic plant remains in rocks of Ordovician age, plant diversity at this time can be examined using fossilized spores[35]. These fossils provide the first widely accepted evidence for land plants during the Dapingian-Darriwilian interval[36], and preserve evidence of an adaptive radiation of vascular plants in the Palaeozoic[37]. In light of our modelling results, shown in Fig. 3, we suggest that the magnitude of the weathering flux could have increased through the Ordovician as a consequence of the evolutionary radiation of primitive land plants. Thereby, our study highlights a possible link between the process of diversification and the biogeochemical processes that regulate Earth's climate.

Fossil cryptospores are not sufficiently abundant in the fossil record at this time to permit a detailed comparison between simulated vegetation and data on ancient vegetation derived from fossils. Nevertheless, recent work has demonstrated that cryptospores are found on all major Ordovician palaeocontinents and that, consequently, early land plants showed a global distribution before the Silurian[38]. Our simulations produce terrestrial vegetation that has such a global distribution (Fig. 1a), and in this respect our simulations are compatible with the available palaeobotanical data[37,38].

In summary, our estimates imply that the Ordovician non-vascular biosphere had a similarly high capacity for biotic weathering than today's biosphere. That should be taken into account for understanding and modelling palaeoclimatic evolution. The colonization of the land by early forms of lichens and bryophytes in the Mid-Late Ordovician[16,39] could have caused a large decrease in atmospheric CO_2 concentration at the end of the Ordovician through the enhancement of silicate weathering.

The results from this study could be used as a basis for further analyses of the feedback between weathering and climate in the geological past. By using a coupled biogeochemical model, the effect of weathering on atmospheric CO_2 concentration could be quantified. Subsequently, the CO_2 sensitivity of weathering by lichens and bryophytes derived here could be used to estimate the impact of changing CO_2 and climate on the biotic enhancement of weathering.

Methods

The non-vascular vegetation model. The process-based non-vascular vegetation model used in this study is a stand-alone vegetation model that uses gridded climatic fields to compute NPP as the difference between photosynthesis and

a

Mean annual total precipitation

b

Mean annual air temperature

c

Mean annual shortwave radiation

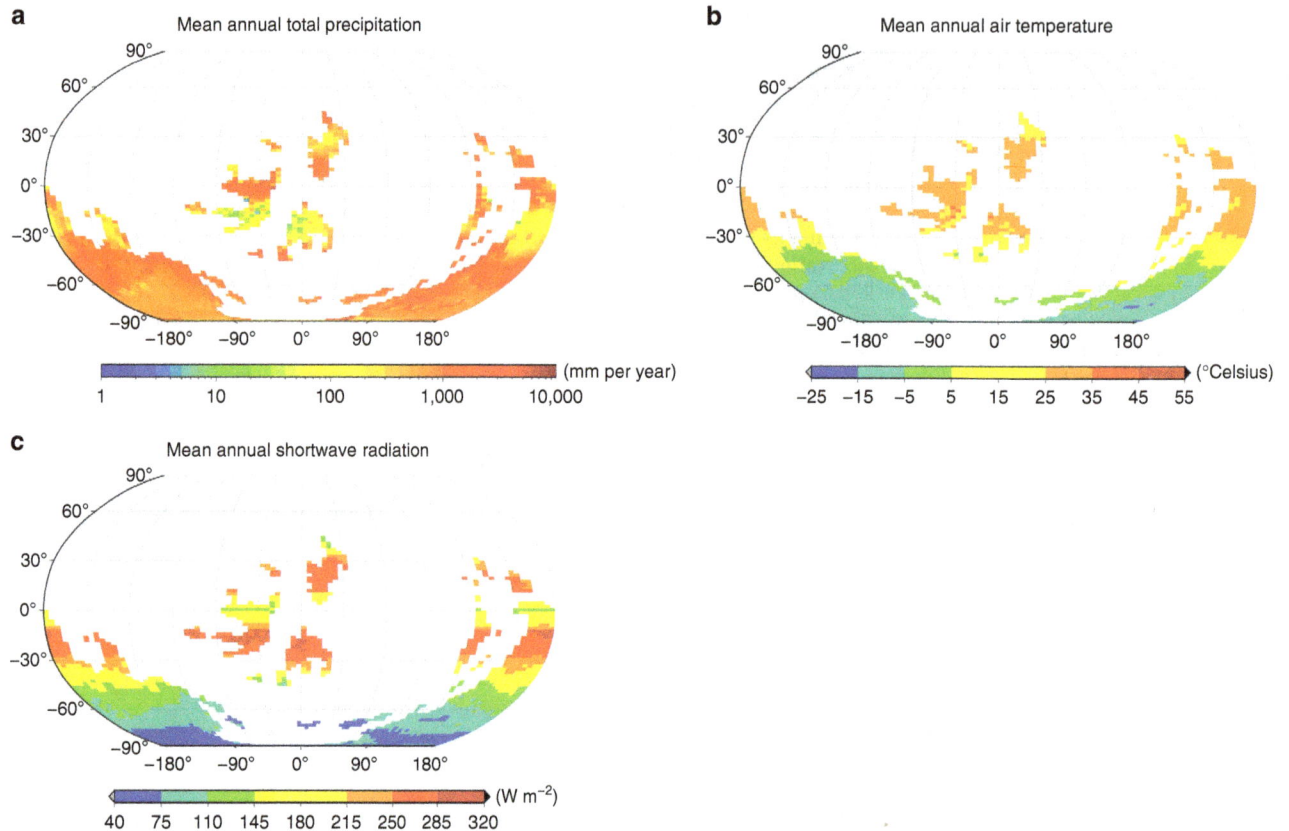

Figure 6 | Simulated Ordovician climate. Global patterns of (**a**) total precipitation, (**b**) near-surface air temperature and (**c**) shortwave radiation for the Late Ordovician (450 Ma). The atmospheric CO_2 concentration is 2,240 p.p.m. (8 PAL). The maps show average values over the whole time series of the interpolated climate variables that are derived from the monthly LMDZ output.

respiration[20]. Similar to many dynamical vegetation models, photosynthesis is calculated by the Farquhar scheme and respiration via a Q_{10} relationship. These approaches are combined with lichen/bryophyte-specific characteristics, for example, a passive control of the water status (poikilohydry) or the decrease in CO_2 diffusivity with increasing water content. In contrast to most global vegetation models, which use plant functional types, the lichen and bryophyte model explicitly represents physiological variation: the ranges of observed values of physiological parameters are sampled by a Monte-Carlo algorithm to generate artificial species. These species do not correspond directly to certain lichen and bryophyte species in the real world, but we assume that they represent the diverse physiological strategies of real lichens and bryophytes. This is a novel approach to include physiological variation in models of terrestrial vegetation, which is also used, for example, in the successful Jena Diversity-Dynamic Global Vegetation Model[40].

The model is run for all species at each grid point of the climatic fields and the performance of the different species is analysed. Those species that cannot maintain their biomass are assumed to die out at the respective grid point and the remaining ones are used to compute NPP. This is performed by weighting each surviving species by its NPP accumulated over the simulation.

Note that processes related to uptake and loss of nutrients are not explicitly modelled. Instead, it is assumed that the costs of weathering agents and nitrogen fixation are included in the respiration term, which is based on observed values.

Ordovician climate and environmental factors. The lichen and bryophyte model predicts surface coverage and NPP as a function of the following climate variables: downwelling shortwave and longwave radiation, rainfall, snowfall, air temperature, wind speed and relative humidity. These variables are calculated by the atmospheric general circulation model LMDZ[41] for the Late Ordovician, 450 million years ago.

The LMDZ model is an Intergovernmental Panel on Climate Change-class atmospheric general circulation model benefitting, as the atmospheric component of the state-of-the-art Institut Pierre Simon Laplace (IPSL) Earth system model, from the latest physical and dynamical refinements. A mid-resolution version of the model is used in this study, providing a resolution of $3.75° \times 1.875°$ with 39 vertical levels (21 in the troposphere) on a rectangular grid. LMDZ is forced with sea-surface temperatures from the Fast Ocean Atmosphere Model (FOAM[42]), also run at 450 Ma, to account for the effects of coupled ocean, atmosphere and sea ice. The palaeogeography for these simulations corresponds to the Blakey

configuration[24]. Further details about the FOAM set-up can be found in a previous study[34]. LMDZ is run for 20 years to reach the atmospheric steady state. The climate variables needed by the lichen and bryophyte model are provided in the form of monthly values averaged over the last 5 years of the LMDZ simulation, see Fig. 6 for examples. A comparison of the areas of simulated low rainfall and high temperature with the spatial distribution of indicators of arid climate[43] shows reasonable agreement.

The lichen and bryophyte model runs on an hourly time step to resolve fast processes such as dew formation, which are important for correctly simulating the organisms' NPP. Hence, to make the monthly climate variables from LMDZ usable for the lichen and bryophyte model, they are interpolated to hourly variables by a stochastic weather generator that was developed for this study (see Supplementary Fig. 1 and Supplementary Note 1).

In addition to the climatic fields, the lichen and bryophyte model requires information about the fraction of surface area that is available for growth, the frequency of disturbance events, such as fires, and the roughness length of the surface. The latter is needed to compute fluxes of heat and water between the lichen or bryophyte surface and the atmosphere. For the Ordovician, we assume that the whole land surface area is available for growth of lichens and bryophytes, except for regions covered by glaciers. We know little about the disturbance regime in the Ordovician, other than that there is no evidence of fires in the form of fossil charcoal, suggesting that oxygen was less than 17% of the atmosphere. Hence, we choose a relatively large disturbance interval of 100 years, which is a typical value for the boreal forest biome. The surface roughness length is set to a value of 0.01 m, corresponding to a desert or tundra ecosystem[20].

Model set-up. As baseline simulation for this study, the lichen and bryophyte model is run with the interpolated hourly climatic fields from LMDZ and an atmospheric CO_2 content of 8 PAL, which equals 2,240 p.p.m. This CO_2 level is close to the threshold required to trigger Late Ordovician glaciations in several climate models[44]. On the basis of the absence of fossil charcoal in the Ordovician and global geochemical modelling studies[45], atmospheric O_2 is set to 14% in our simulations. The initial number of artificial species is set to 300, since this amount of species sufficiently represents physiological variation (see Fig. 3). The model is run for 600 years to achieve a state where the number of surviving strategies remains constant for more than 100 years. We assume that this is a sufficient condition for a steady state. Since only 1 year of interpolated LMDZ output is available, we repeat this year 600 times. The estimates of the study are based on the averaged output of the last 50 years of the simulation.

Derivation of chemical weathering from NPP. To derive chemical weathering by lichens and bryophytes from their NPP, we use a method from a previous study[22]. First, the phosphorus requirement associated with a given NPP is calculated from the C:P ratio of lichen and bryophyte biomass. This phosphorus requirement is then translated into a potential uptake of phosphorus from surface rocks, thereby accounting for resorption of phosphorus from decaying biomass[46] and loss of phosphorus through leaching[47]. The values of C:P ratio, resorption and leaching vary considerably between different species of lichens and bryophytes. Here we use medium values of these properties[22], and we assume that today's lichens and bryophytes do not differ much from their Ordovician predecessors in this respect. Finally, the amount of rock weathered by lichens and bryophytes is estimated from their uptake of phosphorus divided by the phosphorus concentration of the rocks. Since we do not know the spatial distribution of rock types at the surface in the Ordovician, we assign a globally uniform phosphorus content to the land surface of 1,432 g per m^3 rock. This value corresponds to an area-weighted mean rock phosphorus concentration[22].

Our estimate of weathering would change proportionally for a different value of the mean global rock phosphorus concentration. The value of 1,432 g per m^3 used here is derived from a recent lithology map[48]. Older estimates of lithology[49] provide area fractions of rock types not only for today but also for the Ordovician, but probably overweight shallow phosphorus-poor sediments such as sandstones. Using these area fractions would result in 1,232 g phosphorus per m^3 rock for today and 1,374 g per m^3 for the Ordovician. Comparing these three values shows that the uncertainty in the mean rock phosphorus concentration is quite small.

Limits on chemical weathering. In areas where the land surface has a shallow relief and erosion is consequently low, weathering by lichens and bryophytes may become limited by a layer of highly weathered and phosphorus-depleted material blocking the access to unweathered rock substrate. This means that the rate of chemical weathering cannot exceed the rate of exposure of unweathered rock material at the surface, which is equal in the steady state to the erosion rate. We quantify the global distribution of erosion rates as a function of surface elevation above the sea level. We use a method[32] that relates erosion to the mean relief, and we set the mean relief proportional to the mean surface elevation[50]. Elevation data for the Ordovician are taken from a global palaeoclimate modelling study[34].

Our method to estimate chemical weathering implies that a corresponding amount of physical weathering takes place to provide sufficient surface area for chemical weathering. We represent this potential limitation of chemical weathering implicitly by calculating the erosion rate. Only material that has been subjected to physical weathering to some extent can be transported by erosion. Hence, the generation rate of fresh rock material by physical weathering, which equals the erosion rate in steady state, sets the upper limit for the rate of chemical weathering in our approach.

Weathering reactions can only occur at mineral surfaces if water that is saturated with the products of weathering is exchanged by precipitation and runoff. Thus, chemical weathering is limited in steady state to the rate of export of dissolved weathering products. We estimate this export of dissolved material by calculating the calcium flux from the soil as a product of runoff and soil calcium concentration, and then divide by the calcium concentration of rocks to obtain the flux of total rock material. To determine soil calcium concentration, we apply an equilibrium approach[51]. The maximum concentration of calcium in soil water is approximated by the equilibrium with respect to calcite. This means that higher concentrations of calcium in the soil solution are prevented by precipitation of calcite. The calcite equilibrium depends on the partial pressure of CO_2 in the soil and on the mineral composition of the soil or regolith. The latter influences calcium concentration through the relative activity of minerals with and without calcium and the effects of cations other than Ca^{2+} on the charge balance with HCO_3^-. Owing to the uncertain spatial distribution of Ordovician surface rocks, we use a globally uniform mineral composition, based on the area-weighted mean for today's lithology classes[51]. In addition, our estimate of rock calcium concentration is based on this mean mineral composition, assuming a calcium content of 40% for calcite and 14% for calcium-feldspar. Since we lack information about soil CO_2 concentration in the Ordovician, we assign a globally uniform value of 10,000 p.p.m., which is a realistic value for today's soils at intermediate depths[52]. Note that CO_2 concentration in the uppermost part of the soil was at least 2,240 p.p.m. in the Ordovician, assuming an atmospheric CO_2 concentration of 8 PAL.

Runoff is generated in the lichen and bryophyte model when water input by rainfall, dew or snowmelt exceeds the water-storage capacity of the vegetation. We only consider runoff from lichen- and bryophyte-covered areas to contribute to chemical weathering, which leads to underestimation of weathering for regions with sparse vegetation cover. However, we do not expect this to influence strongly our estimates, since abiotic chemical weathering on bare areas is probably several times lower than biotic weathering[14,23]. By assuming that only runoff is in contact with mineral surfaces, we potentially further underestimate chemical weathering, since also mineral dust deposited on top of lichens and bryophytes and wetted by rainfall or dewfall may be weathered by the organisms. In the Ordovician,

atmospheric dust consisted most probably of unweathered rock particles because of the lack of organisms other than lichens and bryophytes, which were capable of transforming significant amounts of rock phosphorus into organic phosphorus. The dust could therefore represent a source of phosphorus in arid regions, where no runoff is generated and dewfall is the only source of water for lichens and bryophytes.

Evaluation of erosion- and runoff-limited chemical weathering. We apply our limit-based approach to estimate today's global chemical weathering flux. To be consistent with our estimate for the Ordovician, we use the same globally uniform mean values of soil mineral composition and soil CO_2 concentration. We determine runoff for each grid cell from the Budyko curve, where we calculate potential evaporation by the equilibrium evaporation approach[53]. Annual mean values of radiation, temperature and precipitation are computed from today's climatic fields[20]. The simulated potential global calcium flux amounts to 10.7e12 mol yr^{-1}, which compares well with the 9.3e12 mol yr^{-1} estimated by a more complex model that represents the spatial distribution of lithological classes and includes a process-based scheme for soil CO_2 (ref. 51). Subsequently, we take the minimum of the runoff-based calcium flux and the flux supported by erosion, where we convert the rock material flux associated with erosion to a potential calcium flux using average mineral properties. This reduces the estimated calcium flux to 7.7e12 mol yr^{-1}, which corresponds to a global chemical weathering of 1.8 (km^3 rock) yr^{-1}. This value is around twice the value based on measurements in the world's rivers[54], converted to rock volume using an average rock density of around 2,500 kg m^{-3}. We explain this by two reasons: First, the measurements likely underestimate the rock volume affected by chemical weathering, since some rock types form insoluble clay minerals during weathering, which are not captured by the measurement method (total dissolved solids). Second, kinetic limitation of weathering reactions may be important in some regions, such as mountain areas in the humid tropics, for instance, leading to an overestimation of chemical weathering by the limit-based approach.

To test these potential explanations we apply the GEM-CO2 model[55] that predicts chemical weathering as a linear function of runoff for different lithology classes. These classes are the same as those used in our approach. On the basis of average mineral properties we estimate a global calcium flux of 5.5e12 mol yr^{-1}, which translates into a rock volume of 1.3 (km^3 rock) yr^{-1}. This suggests that clay formation and kinetic limitation contribute roughly equally to the overestimation of chemical weathering by our approach compared with measurements from the world's rivers. Keeping these uncertainties in mind we believe that the erosion- and runoff-based limits on chemical weathering represent a reasonable first approximation to the actual rate of global chemical weathering and should be suitable to constrain our estimate of chemical weathering in the Late Ordovician to realistical values.

To distinguish between biotic and abiotic contributions to the global chemical weathering flux, we quantify Ordovician abiotic weathering using the GEOCARB III model[14]. This means that we scale today's global calcium flux to the Ordovician calcium flux, thereby assuming abiotic conditions in the Ordovician. As a first step, we calculate today's calcium fluxes associated with chemical weathering of silicates as well as carbonates with a spatially explicit version of the GEM-CO2 model, since GEOCARB uses different equations for the two fluxes. We estimate 2.1e12 mol yr^{-1} of carbonate weathering and 3.4e12 mol yr^{-1} of silicate weathering, which matches well with the value of 5.5e12 mol yr^{-1} for the total calcium flux based on average lithology and mineral properties stated above.

The scaling relationship in GEOCARB III consists of several factors, which describe how changes in temperature, runoff and atmospheric CO_2 between today and the Ordovician influence rates of weathering reactions. We use the standard values from GEOCARB III parameterize these factors and we adopt the derivation of runoff from temperature. We do not, however, use the GEOCARB approach to derive Ordovician surface temperature from atmospheric CO_2, but instead we use the average land surface temperature estimated by the LMDZ atmospheric circulation model, which is also used to force the non-vascular vegetation model. In this way estimates of both biotic and abiotic weathering are based on the same CO_2 and temperature forcing. Furthermore, GEOCARB III includes the effects of changes in uplift, land surface area and biotic enhancement of weathering. For uplift, we use the GEOCARB III value for the late Ordovician. The change in land surface area is derived from our forcing data to be consistent with the estimate of biotic weathering. GEOCARB III assumes that the lack of large vascular land plants before the Devonian reduced the global weathering rate to a fraction of 0.25. GEOCARB does not distinguish between non-vascular vegetation and abiotic conditions. Hence, the factor of 0.25 integrates studies that compare weathering below vascular plants to weathering in bare soil as well as studies that compare vascular plants to lichens and bryophytes. For bare soil the factor ranges from 0.25 to 0.1 and, for lichens and bryophytes it amounts to 0.25 based on one study. Since we do not want to underestimate abiotic weathering, we adopt the GEOCARB III standard value of 0.25 for the Ordovician. Finally, we convert the calcium fluxes to rock volume using average rock properties of carbonate and silicate minerals.

We use the GEOCARB approach for estimating abiotic weathering since our erosion- and runoff-limited approach does not contain an explicit description of the influence of biota on weathering. While the original, more complex version of the limit-based approach includes the impact of root respiration on soil CO_2, the authors point out that many other factors that are important for enhancing the weathering rate are not considered[51]. Among these are release of organic acids, retention of water at the rock surface, break up of rock surface by rhizines and rhizoids and supply of carbon to soil microorganisms that perform weathering. Instead of representing the factors explicitly, the limit-based approach rests on the assumption that the factors reduce the kinetic limitation of the weathering reactions to a large extent. As a consequence, chemical weathering is rather limited by climatic and topographic constraints. The fact that the limit-based approach provides a reasonable approximation of realized chemical weathering suggests that kinetic limitation of weathering is indeed not a decisive factor in large parts of the world, such as flat or dry areas. However, without biota, the kinetic limitation of the weathering reactions probably becomes much more important. For this reason, we think it is more appropriate to use a kinetic approach for abiotic weathering, such as the GEOCARB III model.

Phosphorus limitation of NPP. In some regions the phosphorus requirement of lichens and bryophytes based on their potential NPP may not be met by the amount of phosphorus available from the substrate, since biotic weathering is limited by low erosion or runoff. In these areas, NPP is phosphorus-limited and we consequently reduce our estimate of NPP by converting the flux of available phosphorus into realized NPP via the C:P ratio of biomass. However, the availability of phosphorus from the substrate may be higher than the amount released by weathering because of additional release of phosphorus by microorganisms that decompose organic matter, or phosphorus remaining in the substrate after leaching. If the residence time of water in the substrate is sufficiently long, this additional phosphorus may be taken up by the organisms, resulting in recycling of phosphorus in addition to the effects of resorption and leaching described above, which almost compensate each other. The availability of phosphorus in the substrate probably depends on spatially varying hydrological conditions: rainfall is the main driver for leaching of phosphorus from the organisms, while runoff is necessary to export phosphorus from the substrate. Therefore, phosphorus availability should be lower in humid regions. However, water is necessary for the decomposition of organic matter, which releases phosphorus, leading to lower phosphorus availability in arid regions. In today's ecosystems, these effects are probably strongly modulated by the effect of roots of vascular plants on soil water-residence time. For simplicity, we assume here a global uniform value of the recycling of phosphorus in the substrate.

The recycling ratio for phosphorus can be estimated by dividing the phosphorus uptake associated with NPP by the phosphorus release from weathering. Since we want to estimate the recycling ratio for the Late Ordovician, we consider those of today's ecosystems, which are dominated by lichens and bryophytes, such as deserts. Typical values for NPP of lichens and bryophytes in deserts are ~5 (g C) m^{-2} yr^{-1}, although they show large spatial variation[20]. Weathering by lichens and bryophytes on rock surfaces[9] ranges from 0.0004 to 0.01 mm yr^{-1}. Assuming a typical value of 0.002 mm yr^{-1}, the mean values for C:P ratio and rock phosphorus concentration, we calculate a recycling ratio of ~5:1.

We run the lichen and bryophyte model with and without nutrient limitation for today's climate. We assume a recycling ratio of 5:1 for bare soil, but for regions that are dominated by vascular vegetation we use a recycling ratio of 50:1 (ref. 30). This high recycling ratio lowers the phosphorus requirement of lichens and bryophytes below the amount provided by erosion- or runoff-limited chemical weathering; the reduction of potential NPP thus mostly occurs in desert regions. At the global scale, NPP is reduced by 16% from 3.1 to 2.6 (Gt C) yr^{-1}. In dry areas, the reduction of NPP reduces the average value for the desert biome from 13.7 to 6.2 (g C) m^{-2} yr^{-1}, which improves the model fit to the typical observed value of 5 (g C) m^{-2} yr^{-1} estimated in a previous study[20]. Note that simulated chemical weathering rates of today's lichens and bryophytes in desert regions (Fig. 5a) compare well with observed rates on bare rocks[9].

Sensitivity analyses. We perform a sensitivity analysis to assess the relevance of several uncertainties, which are associated with our modelling approach. To account for the considerable uncertainty regarding the value of Ordovician atmospheric CO_2 content[3], we force the lichen and bryophyte model with additional LMDZ climatic fields corresponding to levels of atmospheric CO_2 of 3, 4, 6, 10, 12, 16 and 24 PAL (see Fig. 2a).

We further analyse the sensitivity of simulated NPP to ambient CO_2 concentrations by running the model for individual artificial species for a range of ambient CO_2 concentrations assuming constant light, temperature and water content. We test one species well adapted to today's climate and another one well adapted to Ordovician climate with an atmospheric CO_2 concentration of 24 PAL (Fig. 4a). To find well-adapted species, we select two regions with similar climatic conditions, one for the Ordovician at 24 PAL and one for today's climate, and use the parameters of the dominant species in these regions. NPP is calculated in the model as the difference of GPP and respiration. GPP is simulated by the well-established Farquhar scheme, which quantifies photosynthesis as a function of light, temperature and CO_2 concentration within the chloroplasts. The latter depends on diffusion of CO_2 from the atmosphere into the thallus, which is controlled in the model by the limiting effect of water content on CO_2 diffusivity. The Farquhar scheme depends on several physiological parameters, such as the carboxylation rate of Rubisco, for instance, which vary considerably between the

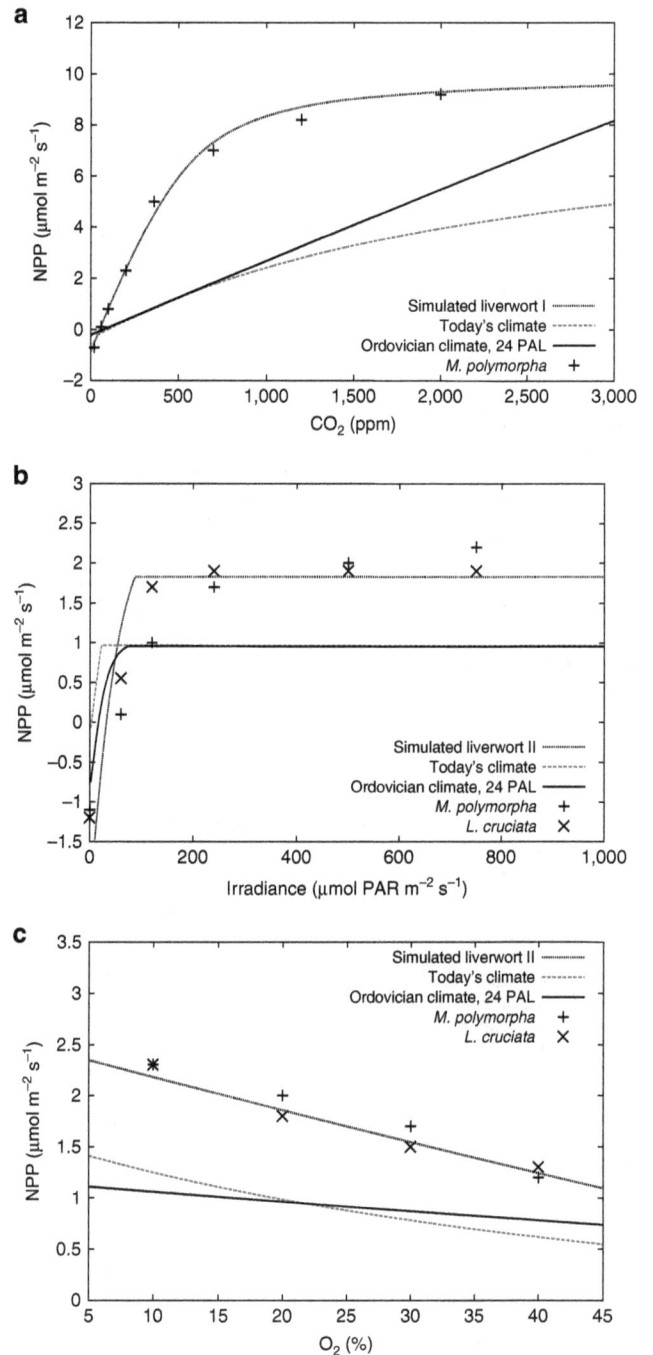

Figure 7 | Response of NPP to environmental conditions. NPP calculated by the model for saturated water content and a temperature of 20 °C at varying levels of CO_2, light and O_2 compared with observational data from a study on liverworts[66]. The black dotted line correspond to an artificial species parameterized to reproduce the observational data. Two versions of this species, I and II, are used to take into account that the liverworts on which the data points are based differ in photosynthetic capacity. The blue solid line corresponds to an artificial species adapted to Ordovician climate and an atmospheric CO_2 concentration of 24 PAL, the dashed magenta line corresponds to an artificial species adapted to today's climate and CO_2 concentration. (**a**) CO_2 response of our model for an irradiance of 250 μmol m^{-2} s^{-1} of photosynthetically active radiation and an ambient O_2 concentration of 21%. (**b**) Light response of the model for ambient concentrations of 400 p.p.m. of CO_2 and 21% of O_2. (**c**) O_2 response at an irradiance of 250 μmol m^{-2} s^{-1} and a CO_2 concentration of 400 p.p.m.

Table 2 | Sensitivity to model parameters.

Scenario	NPP ((Gt C) yr^{-1})	GPP ((Gt C) yr^{-1})	Cover fraction	Biomass (Gt C)	Weathering ((km^3 rock) yr^{-1})
Φ_{RR} low	10	23	0.4	102	2.3
$D_{CO_2,sat}$ low	12	25	0.4	124	2.5
η_{CCM} Low	13	27	0.4	126	2.6
τ_D 250 yr	14	31	0.5	170	2.8
Baseline	14	30	0.4	133	2.8
η_{CCM} high	15	31	0.4	135	2.9
$D_{CO_2,sat}$ High	15	31	0.4	134	2.9
τ_D 10 yr	15	27	0.2	40	3.0
Φ_{RR} High	19	38	0.4	153	3.4

GPP, gross primary productivity; Gt C, gigatons of carbon; NPP, net primary productivity.
Global annual values of NPP, GPP, surface coverage, biomass and weathering for the Ordovician baseline simulation (see Methods section) and for upper and lower bounds of four key model parameters.
See text for a description of the parameters and the associated upper and lower bounds.

different artificial species simulated by the model. In addition, the relation between thallus CO_2 diffusivity and water content is species-specific. At constant temperature and water content, respiration is constant in the model and it does not depend on CO_2. Therefore, the response of NPP to CO_2 is driven by GPP. In addition to atmospheric CO_2, weak photorespiration because of low atmospheric O_2 may lead to high GPP in Ordovician climate conditions.

The strong CO_2 sensitivity of the lichen and bryophyte model begs the question: is there experimental evidence for a large increase in NPP because of CO_2 fertilization? We reviewed studies where lichens and bryophytes were treated with elevated ambient CO_2 levels and their growth was measured. While some studies did not find any significant effects[56–58], others found partly strong increases in NPP at elevated CO_2 (refs 59–65). Figure 4b shows a summary of those studies.

In some cases where NPP did not increase at elevated levels of ambient CO_2 or decreased again after a longer period of fertilization, phosphorus limitation was suggested as an explanation[56,57]. These studies were, however, conducted with moss growing in peatlands or similar environments, where the access to rock substrates is limited. Lichens and bryophytes growing on bare rock surfaces are capable of taking up elements from the rocks[1]. In this case, enough phosphorus could be available to alleviate nutrient limitation of NPP.

To further test model performance, we compare simulated sensitivity of NPP to varying levels of ambient CO_2, light and O_2 to observations from a study on liverworts[66]. In Fig. 7 we show that the model is able to reproduce the CO_2, light and O_2 responses of the liverworts. The observational data for the light and O_2 response curves were obtained from two species, *Marchantia polymorpha* and *Lunularia cruciata*, whereas the CO_2 response includes only *M. polymorpha*, Furthermore, the observed NPP values in Fig. 7a are based on liverworts grown under different conditions than the liverworts of which NPP measurements are shown in Fig. 7b,c. While laboratory-grown *M. polymorpha* has an NPP of $\sim 5\,\mu$mol m^{-2} s^{-1} at 400 p.p.m. and 250 μmol m^{-2} s^{-1} of photosynthetically active radiation, *M. polymorpha* grown outdoors has an NPP of $\sim 2\,\mu$mol m^{-2} s^{-1} for the same CO_2 and light forcing (compare Fig. 7a,b). Hence, we use two different parameterizations for our model, which mainly differ in photosynthetic capacity, to reproduce the observational data in Fig. 7. Moreover, we include in Fig. 7 the CO_2, light and O_2 response of the two artificial species from Fig. 4a for comparison. The artificial species have a lower NPP than the liverworts at the CO_2 level of 400 p.p.m., but their relative increase in NPP from 400 p.p.m. of CO_2 to 1,000 p.p.m. is larger than the liverworts' (Fig. 7a). This can be explained largely by the lower photosynthetic capacity of the two artificial species from Fig. 4a compared with the simulated liverworts; their NPP saturates only at higher levels of ambient CO_2. In addition, other parameters have an effect, such as thallus diffusivity for CO_2 and parameters, which control the temperature response of photosynthesis. The latter plays a role since the regions from which the two artificial species have been selected exhibit different temperature and light conditions than the environment in which the liverworts were grown.

The existence of a continental-sized ice sheet south of 30° S in the Hirnantian Stage is debated[24], which would strongly reduce the available area for growth of lichens and bryophytes. Hence, we run a second baseline simulation with a prescribed glacier mask and updated climatic fields accounting for the presence of the ice sheet. The outcome of this sensitivity analysis is shown in (Fig. 2b).

We test the sensitivity of our estimates to the chosen globally uniform mean values of soil CO_2 concentration and mineral composition. For a soil CO_2 concentration of 3,000 p.p.m., weathering decreases by 18% to 2.3 (km^3 rock) yr^{-1} because of reduced runoff-limited chemical weathering and NPP decreases equally by 17% to 11.9 (Gt C) yr^{-1} as a result of phosphorus limitation. Conversely, at 30,000 p.p.m. of soil CO_2, weathering increases by 7% to 3.0 (km^3 rock) yr^{-1} and NPP increases equally by 8% to 15.5 (Gt C) yr^{-1}. Mineral composition controls the equilibrium calcium concentration in the soil as well as the calcium content of the minerals. Consequently, it modulates the relation between runoff and the chemical weathering flux. We assign extreme values for mineral composition based on a map of today's lithology[48], which contains six lithological classes: sandstone,

limestone, shale, granite, rhyolite and basalt. Chemical weathering at a given runoff and soil CO_2 concentration is maximal for granite; therefore, weathering and NPP increase by ~ 15% to 3.2 (km^3 rock) yr^{-1} and 16.6 (Gt C) yr^{-1}, respectively. For shale, chemical weathering is minimal, which leads to a reduction of weathering and NPP by ~ 40% to 1.6 (km^3 rock) yr^{-1} and 8.3 (Gt C) yr^{-1}, respectively.

In addition to environmental factors such as atmospheric CO_2, ice sheets and substrate properties, we test the sensitivity of the model with regard to the following model parameters, which have been shown to considerably influence simulated NPP[20] and which might have changed from the Ordovician to today: disturbance interval, diffusivity of the water-saturated lichen or bryophyte thallus for CO_2, ratio of Rubisco content to maintenance respiration at a given temperature and efficiency of the carbon concentration mechanism (CCM; Table 2).

We know very little about the disturbance regime in the Ordovician, hence we chose strongly different disturbance intervals τ_D, from 10 years to 250 years. The effect on NPP and, consequently, weathering, however, is not large. Only cover and biomass are significantly reduced by frequent disturbances. The ratio of Rubisco content to maintenance respiration at a given temperature Φ_{RR} controls how much organisms have to pay in form of maintenance respiration to achieve a certain photosynthetic capacity, which is determined by the Rubisco content. High photosynthetic capacity represents a competitive advantage in high-light environments because of high potential growth, but under low-light conditions it may have negative effects because of the associated high respiration. For a doubling of Φ_{RR}, weathering and NPP increase by ~ 20% and 30%, respectively, whereas halving Φ_{RR} leads to a decrease of 17 and 26%. Φ_{RR} has the strongest impact on the model. The efficiency of the CCM η_{CCM} determines by which factor the organisms can increase their internal CO_2 concentration above the ambient level through channelling off electrons from the light-dependent reactions of photosynthesis. While the baseline scenario uses a value of 8, which leads to realistic patterns of NPP for today's lichens and bryophytes, η_{CCM} is varied from a low factor of 2 to the theoretical factor of 45 based on a study in cyanobacteria[67]. It has a weaker impact on NPP than Φ_{RR}. Φ_{RR} and η_{CCM} have the same value for each species since they represent universal physiological constraints[20]. The diffusivity of the water-saturated lichen or bryophyte thallus for CO_2, $D_{CO_2,sat}$, however, does not have a uniform value but is sampled for each species from the observed range of values. $D_{CO_2,sat}$ does not represent a physiological constraint by itself, but it is part of a tradeoff between the influx of CO_2 into the thallus and resistance to water loss. A dense thallus prevents rapid water loss by evaporation, but it also hampers high influx of CO_2. Thus, we extend the range once by doubling the maximum $D_{CO_2,sat}$ and once by halving the minimum $D_{CO_2,sat}$. The effect on NPP is similar to that of η_{CCM}. To summarize, the estimates seem to be robust over a wide range of physiological parameter values.

At low numbers of artificial species, the lichen and bryophyte model may underestimate NPP since the random sample may be too small to contain combinations of physiological parameter values leading to high NPP. We run simulations with initial numbers of 10, 30, 100, 300, 1,000 and 3,000 artificial species to determine at which number of species physiological variation is represented to a sufficient degree, so that NPP and weathering remain constant thereafter. The number of 300 initial species chosen for the baseline scenario seems to be sufficient to represent the physiological variation of the organisms (Fig. 3).

Code availability. The non-vascular vegetation model used in this study is integrated into an interface for parallel computing, which was developed at the Max Planck Institute for Biogeochemistry, Jena, Germany. In addition to the model, a post-processing script is necessary to derive the results presented in this study from the model output. The model, the interface and the script are freely available as long as the copyright holders and a disclaimer are distributed along with the code in source or binary form. The code is available from the corresponding author upon request.

References

1. Jackson, T. Weathering, secondary mineral genesis, and soil formation caused by lichens and mosses growing on granitic gneiss in a boreal forest environment. *Geoderma* 251-252, 78–91 (2015).
2. Quirk, J. et al. Constraining the role of early land plants in Palaeozoic weathering and global cooling. *Proc. R Soc. B* 282, 20151115 (2015).
3. Boucot, A. & Gray, J. A critique of Phanerozoic climatic models involving changes in the CO_2 content of the atmosphere. *Earth-Sci. Rev.* 56, 1–159 (2001).
4. Chen, J., Blume, H.-P. & Beyer, L. Weathering of rocks induced by lichen colonization - a review. *Catena* 39, 121–146 (2000).
5. McCarroll, D. & Viles, H. Rock-weathering by the lichen Lecidea auriculata in an arctic alpine environment. *Earth Surf. Processes Landforms* 20, 199–206 (1995).
6. Stretch, R. & Viles, H. The nature and rate of weathering by lichens on lava flows on Lanzarote. *Geomorphology* 47, 87–94 (2002).
7. Jackson, T. & Keller, W. A comparative study of the role of lichens and 'inorganic' processes in the chemical weathering of recent Hawaiian lava flows. *Am. J. Sci.* 269, 446–466 (1970).
8. Brady, P. et al. Direct measurement of the combined effects of lichen, rainfall, and temperature on silicate weathering. *Geochim. Cosmochim. Acta* 63, 3293–3300 (1999).
9. Aghamiri, R. & Schwartzman, D. Weathering rates of bedrock by lichens: a mini watershed study. *Chem. Geol.* 188, 249–259 (2002).
10. Zambell, C., Adams, J., Gorring, M. & Schwartzman, D. Effect of lichen colonization on chemical weathering of hornblende granite as estimated by aqueous elemental flux. *Chem. Geol.* 291, 166–174 (2012).
11. Lenton, T., Crouch, M., Johnson, M., Pires, N. & Dolan, L. First plants cooled the Ordovician. *Nat. Geosci.* 5, 86–89 (2012).
12. Elbert, W. et al. Contribution of cryptogamic covers to the global cycles of carbon and nitrogen. *Nat. Geosci.* 5, 459–462 (2012).
13. Cochran, M. & Berner, R. Promotion of chemical weathering by higher plants: field observations on Hawaiian basalts. *Chem. Geol.* 132, 71–77 (1996).
14. Berner, R. & Kothavala, Z. Geocarb III: a revised model of atmospheric CO_2 over Phanerozoic time. *Am. J. Sci.* 301, 182–204 (2001).
15. Berner, R., Lasaga, A. & Garrels, R. Carbonate-silicate geochemical cycle and its effect on atmospheric carbon dioxide over the past 100 million years. *Am. J. Sci.* 283, 641–683 (1983).
16. Kenrick, P., Wellman, C., Schneider, H. & Edgecombe, G. A timeline for terrestrialization: consequences for the carbon cycle in the Palaeozoic. *Phil. Trans. R Soc. Lond. B* 367, 519–536 (2012).
17. Honegger, R., Edwards, D. & Axe, L. The earliest records of internally stratified cyanobacterial and algal lichens from the Lower Devonian of the Welsh Borderland. *New Phytol.* 197, 264–275 (2013).
18. Yuan, X., Xiao, S. & Taylor, T. Lichen-like symbiosis 600 million years ago. *Science* 308, 1017–1020 (2005).
19. Weber, B., Scherr, C., Bicker, F., Friedl, T. & Budel, B. Respiration-induced weathering patterns of two endolithically growing lichens. *Geobiology* 9, 34–43 (2011).
20. Porada, P., Weber, B., Elbert, W., Pöschl, U. & Kleidon, A. Estimating global carbon uptake by lichens and bryophytes with a process-based model. *Biogeosciences* 10, 6989–7033 (2013).
21. Sharma, K. Inorganic phosphate solubilization by fungi isolated from agriculture soil. *J. Phytol.* 3, 11–12 (2011).
22. Porada, P., Weber, B., Elbert, W., Pöschl, U. & Kleidon, A. Estimating impacts of lichens and bryophytes on global biogeochemical cycles of nitrogen and phosphorus and on chemical weathering. *Global Biogeochem. Cycles* 28, 71–85 (2014).
23. Bergman, N., Lenton, T. & Watson, A. COPSE: a new model of biogeochemical cycling over Phanerozoic time. *Am. J. Sci* 304, 397–437 (2004).
24. Blakey, R. *Carboniferous-Permian Paleogeography of the Assembly of Pangaea in Proceedings of the XVth International Congress on Carboniferous and Permian Stratigraphy, Utrecht, 10-16 August 2003.* (ed. Wong, T.) 443–456 (Royal Dutch Academy of Arts and Sciences, 2007).
25. Ito, A. A historical meta-analysis of global terrestrial net primary productivity: are estimates converging? *Glob. Change Biol.* 17, 3161–3175 (2011).
26. Yapp, C. & Poths, H. Productivity of pre-vascular continental biota inferred from the Fe(CO3)OH content of goethite. *Nature* 368, 49–51 (1994).
27. Veizer, J. et al. $^{87}Sr/^{86}Sr$, $\delta^{13}C$ and $\delta^{18}O$ evolution of Phanerozoic seawater. *Chem. Geol.* 161, 59–88 (1999).
28. Edwards, D., Cherns, L. & Raven, J. A. Could land-based early photosynthesizing ecosystems have bioengineered the planet in Mid-Paleozoic times? *Paleontology* 58, 1–35 (2015).
29. Moulton, K., West, J. & Berner, R. Solute flux and mineral mass balance approaches to the quantification of plant effects on silicate weathering. *Am. J. Sci.* 300, 539–570 (2000).
30. *Gaia's Body.* (ed. Volk, T.) 1-269 (Springer Science & Business Media, 1998).
31. Buendía, C. et al. On the potential vegetation feedbacks that enhance phosphorus availability - insights from a process-based model linking geological and ecological timescales. *Biogeosciences* 11, 3661–3683 (2014).
32. Arens, S. & Kleidon, A. Eco-hydrological versus supply-limited weathering regimes and the potential enhancement of weathering at the global scale. *Appl. Geochem.* 26, 274–278 (2011).
33. Kump, L., Brantley, S. & Arthur, M. Chemical weathering, atmospheric CO_2, and climate. *Annul. Rev. Earth Planet. Sci.* 28, 611–667 (2000).
34. Pohl, A., Donnadieu, Y., Le Hir, G., Buoncristiani, J.-F. & Vennin, E. Effect of the Ordovician paleogeography on the (in)stability of the climate. *Clim. Past* 10, 2053–2066 (2014).
35. Strother, P., Traverse, A. & Vecoli, M. Cryptospores from the Hanadir Shale member of the Qasim formation, Ordovician (Darriwilian) of Saudi Arabia: taxonomy and systematics. *Rev. Palaeobot. Palynol.* 212, 97–110 (2015).
36. Wellman, C. & Strother, P. The terrestrial biota prior to the origin of land plants (embryophytes): a review of the evidence. *Palaeontology* 58, 601–627 (2015).
37. Wellman, C., Steemans, P. & Vecoli, M. in *Early Palaeozoic Biogeography and Palaeogeography.* (eds Harper, D. & Servais, T.) 461–476 (Geological Society, 2013).
38. Raevskaya, E., Dronov, A., Servais, T. & Wellman, C. Cryptospores from the Katian (Upper Ordovician) of the Tungus basin: the first evidence for early land plants from the Siberian paleocontinent. *Rev. Palaeobot. Palynol.* 224, 4–13 (2016).
39. Schwartzman, D. in *Biology of Lichens, Bibliotheca Lichenologica* (eds Nash III, T. et al.) 191–196 (J.Cramer, 2010).
40. Pavlick, R., Drewry, D., Bohn, K., Reu, B. & Kleidon, A. The Jena Diversity-Dynamic Global Vegetation Model (JeDi-DGVM): a diverse approach to representing terrestrial biogeography and biogeochemistry based on plant functional trade-offs. *Biogeosciences* 10, 4137–4177 (2013).
41. Hourdin, F. et al. Impact of the LMDZ atmospheric grid configuration on the climate and sensitivity of the IPSL-CM5A coupled model. *Clim. Dyn.* 40, 2167–2192 (2013).
42. Jacob, R. *Low Frequency Variability in a Simulated Atmosphere Ocean System.* PhD thesis, Univ. Wisconsin (1997).
43. Boucot, A., Xu, C. & Scotese, C. *Phanerozoic Paleoclimate: an atlas of lithologic indicators of climate* (Society of Economic Paleontologists and Mineralogists, 2013)
44. Herrmann, A., Patzkowsky, M. & Pollard, D. Obliquity forcing with 8-12 times preindustrial levels of atmospheric pCO2 during the Late Ordovician glaciation. *Geology* 31, 485–488 (2003).
45. Berner, R. Modeling atmospheric O2 over Phanerozoic time. *Geochim. Cosmochim. Acta* 65, 685–694 (2001).
46. Eckstein, R., Karlsson, P. & Weih, M. Leaf life span and nutrient resorption as determinants of plant nutrient conservation in temperate-arctic regions. *New Phytol.* 143, 177–189 (1999).
47. Belnap, J. Nitrogen fixation in biological soil crusts from southeast Utah, USA. *Biol. Fertil. Soils* 35, 128–135 (2002).
48. Amiotte Suchet, P., Probst, J.-L. & Ludwig, W. Worldwide distribution of continental rock lithology: implications for the atmospheric/soil CO_2 uptake by continental weathering and alkalinity river transport to the oceans. *Global Biogeochem. Cycles* 17, 1038 (2003).
49. Bluth, G. & Kump, L. Phanerozoic paleogeology. *Am. J. Sci.* 291, 284–308 (1991).
50. Pitman, W. & Golovchenko, X. The effect of sea level changes on the morphology of mountain belts. *J. Geophys. Res.* 96, 6879–6891 (1991).
51. Arens, S. *Global Limits on Silicate Weathering and Implications for the Silicate Weathering Feedback.* PhD thesis, Friedrich-Schiller-Universität (2013).
52. Moncrieff, J. & Fang, C. A model for soil CO_2 production and transport 2: application to a Florida Pinus elliotte plantation. *Agr. Forest Meteorol.* 95, 237–256 (1999).
53. McNaughton, K. G. & Jarvis, P. G. in *Water Deficits and Plant Growth* (ed Kozlowski, T. L.) 1–47 (Academic Press, 1983).
54. Gaillardet, J., Dupré, B., Louvat, P. & Allègre, C. Global silicate weathering and CO_2 consumption rates deduced from the chemistry of large rivers. *Chem. Geol.* 159, 3–30 (1999).
55. Suchet, P. A. & Probst, J. A global model for present-day atmospheric/soil CO_2 consumption by chemical erosion of continental rocks (GEM-CO2). *Tellus B* 47, 273–280 (1995).
56. Heijmans, M., Klees, H. & Berendse, F. Competition between Sphagnum magellanicum and Eriophorum angustifolium as affected by raised CO_2 and increased N deposition. *Oikos* 97, 415–425 (2002).
57. Toet, S. et al. in *Plants and Climate Change* (eds Rozema, J., Aerts, R. & Cornelissen, H.) 27–42 (Springer, 2006).

58. Field, K. et al. Contrasting arbuscular mycorrhizal responses of vascular and non-vascular plants to a simulated Palaeozoic CO_2 decline. *Nat. Commun.* **3**, 835 (2012).

59. Botting, R. & Fredeen, A. Net ecosystem CO_2 exchange for moss and lichen dominated forest floors of old-growth sub-boreal spruce forests in central British Columbia, Canada. *Forest Ecol. Manage.* **235**, 240–251 (2006).

60. Jauhiainen, J. & Silvola, J. Photosynthesis of Sphagnum fuscum at long-term raised CO_2 concentrations. *Ann. Bot. Fenn.* **36**, 11–19 (1999).

61. Silvola, J. CO_2 dependence of photosynthesis in certain forest and peat mosses and simulated photosynthesis at various actual and hypothetical CO_2 concentrations. *Lindbergia* **11**, 86–93 (1985).

62. Sonesson, M., Gehrke, C. & Tjus, M. CO_2 environment, microclimate and photosynthetic characteristics of the moss Hylocomium splendens in a subarctic habitat. *Oecologia* **92**, 23–29 (1992).

63. Tuba, Z., Csintalan, Z., Szente, K., Nagy, Z. & Grace, J. Carbon gains by desiccation-tolerant plants at elevated CO_2. *Funct. Ecol.* **12**, 39–44 (1998).

64. Van Der Heijden, E., Verbeek, S. & Kuiper, P. Elevated atmospheric CO_2 and increased nitrogen deposition: effects on C and N metabolism and growth of the peat moss Sphagnum recurvum P. Beauv. var. mucronatum (Russ.) Warnst. *Global Change Biol.* **6**, 201–212 (2000).

65. Humphreys, C. et al. Mutualistic mycorrhiza-like symbiosis in the most ancient group of land plants. *Nat. Commun.* **1**, 103 (2010).

66. Fletcher, B., Brentnall, S., Quick, W. & Beerling, D. BRYOCARB: a process-based model of thallose liverwort carbon isotope fractionation in response to CO_2, O_2, light and temperature. *Geochim. Cosmochim. Acta* **70**, 5676–5691 (2006).

67. Reinhold, L., Zviman, M. & Kaplan, A. A quantitative model for inorganic carbon fluxes and photosynthesis in cyanobacteria. *Plant Physiol. Biochem.* **27**, 945–954 (1989).

68. Beer, C. et al. Terrestrial gross carbon dioxide uptake: global distribution and covariation with climate. *Science* **329**, 834–838 (2010).

69. Bartholomé, A. E. Belward. GLC2000: a new approach to global land cover mapping from Earth observation data. *Int. J. Remote Sensing* **26**, 1959–1977 (2005).

70. Carvalhais, N. et al. Global covariation of carbon turnover times with climate in terrestrial ecosystems. *Nature* **514**, 213–217 (2014).

Acknowledgements

The Bolin Centre for Climate Research is thanked for financial support. The Max Planck Institute for Biogeochemistry provided computational resources and the Laboratoire des Sciences du Climat et de l'Environnement provided data resources. P.P. acknowledges funding from the European Union FP7-ENV project PAGE21 under contract number GA282700. T.M.L. was supported by the Leverhulme Trust (RPG-2013-106) and by a Royal Society Wolfson Research Merit Award. A.P. and Y.D. thank the CEA/CCRT for providing access to the HPC resources of TGCC under the allocation 2015012212 made by GENCI. B.W. gratefully acknowledges support by the Max Planck Society (Nobel Laureate Fellowship) and the German Research Foundation (project WE2393/2). The authors thank four anonymous reviewers for their thorough and helpful comments.

Author contributions

P.P., A.K. and T.M.L. designed the study. P.P. did the modelling analyses with help and input from A.P. and Y.D.; P.P. wrote the paper with input from T.M.L., A.K., B.W., L.M., C.B. and U.P. in form of literature research and pieces of manuscript text.

Additional information

20

Global pulses of organic carbon burial in deep-sea sediments during glacial maxima

Olivier Cartapanis[1,2], Daniele Bianchi[2,3,4], Samuel L. Jaccard[1] & Eric D. Galbraith[2,5,6]

The burial of organic carbon in marine sediments removes carbon dioxide from the ocean–atmosphere pool, provides energy to the deep biosphere, and on geological timescales drives the oxygenation of the atmosphere. Here we quantify natural variations in the burial of organic carbon in deep-sea sediments over the last glacial cycle. Using a new data compilation of hundreds of sediment cores, we show that the accumulation rate of organic carbon in the deep sea was consistently higher (50%) during glacial maxima than during interglacials. The spatial pattern and temporal progression of the changes suggest that enhanced nutrient supply to parts of the surface ocean contributed to the glacial burial pulses, with likely additional contributions from more efficient transfer of organic matter to the deep sea and better preservation of organic matter due to reduced oxygen exposure. These results demonstrate a pronounced climate sensitivity for this global carbon cycle sink.

[1] Institute of Geological Sciences and Oeschger Centre for Climate Change Research, University of Bern, 3012 Bern, Switzerland. [2] Department of Earth and Planetary Sciences, McGill University, Montreal, Canada H3A 2A7. [3] School of Oceanography, University of Washington, Seattle, Washington 98105, USA. [4] Department of Atmospheric and Oceanic Sciences, University of California Los Angeles, Los Angeles, California 90095-1565, USA. [5] Institució Catalana de Recerca i Estudis Avancats (ICREA), 08010 Barcelona, Spain. [6] Institut de Ciència i Tecnologia Ambientals and Department of Mathematics, Universitat Autonoma de Barcelona, 08193 Barcelona, Spain. Correspondence and requests for materials should be addressed to O.C. (email: olivier.cartapanis@geo.unibe.ch).

The climate history of the Earth offers a wide range of timescales on which the sensitivity of organic carbon burial to climate might be tested, including the relatively well-documented glacial–interglacial cycles triggered by insolation variations over the quaternary[1]. Much research has focused on changes in the partitioning of carbon between the ocean and atmosphere over these cycles, amplifying orbital forcing by transferring CO_2 from the atmosphere to the ocean during glacial times, and back to the atmosphere during interglacial periods[1–3]. However, interactions between the long-term storage of organic carbon in oceanic sediments and global climate variations on glacial–interglacial timescales have received little attention.

More than 60 years ago, the discovery of high concentrations of organic carbon in deep-sea sediments near the Galapagos Islands led Arrhenius to propose that exposure of the Galapagos Plateau during the low sea level of the last ice age had caused the resuspension and downslope transport of organic-rich coastal sediments[4]. Later, enhanced burial of organic carbon in glacial-age sediments from the nearby equatorial Pacific was interpreted as reflecting increased biological productivity accompanying intensified wind-driven upwelling[5], and more recently, reinterpreted as better preservation under reduced oxygenation[6]. Enhanced burial of organic carbon in glacial sediments was also observed in the equatorial Atlantic over several glacial cycles[7], and at low latitudes for the last glacial cycle[8], and interpreted as the response of primary productivity to either orbitally forced changes in ocean circulation or to changes in trade winds[7,8].

Here we make a quantitative analysis of several hundreds of high-quality sediment records to extend these pioneering observations to the global scale. We focus our analysis on organic carbon burial in deep-sea sediments, given that the deposition of sediments on continental shelves is complicated by pronounced

spatial heterogeneity in biological production and sedimentation (see Methods). Our results suggest that organic carbon burial in deep-sea sediments increased during peak glacial conditions, demonstrating the sensitivity of this component of the global carbon cycle to climatic variation.

Results

Organic carbon burial rates during the Last Glacial Maximum.

We estimate global variations of total organic carbon (TOC) mass accumulation rate (MAR) over the past 150 kyr by combining the modern organic carbon burial distribution (see Methods and Supplementary Fig. 1) with time series of TOC MAR from a global compilation of 561 sediment cores extending from the present to the Marine Isotopic Stage 6 (MIS6; Methods). This analysis reveals the existence of robust geographical patterns (Fig. 1a). TOC burial was higher during the Last Glacial Maximum (LGM) in most provinces (Fig. 1a, see Methods and Supplementary Figs 2 and 3), including the tropical and subtropical Atlantic, the eastern and southern Pacific, and the Arabian Sea. The only provinces in which sedimentary TOC burial was lower during the LGM are the high latitudes of the Arctic and Antarctic oceans, the Bering and Okhotsk seas, the California Current, and the Caribbean region. Our reconstruction constrains the global deep-sea TOC MAR during the LGM to between 118 and 171% of the Holocene MAR (Table 1), with a mean estimate of $147 \pm 18\%$. Despite the uncertainties, none of the scenarios that we considered (Table 1, see Methods) allows lower burial during the LGM compared with the Holocene. On the basis of the downcore sediment MAR changes and modern burial map, significantly increased glacial burial occurred in the tropical regions of the Atlantic and eastern Pacific, and the Subantarctic (Fig. 1b). The elevated TOC burial during the LGM

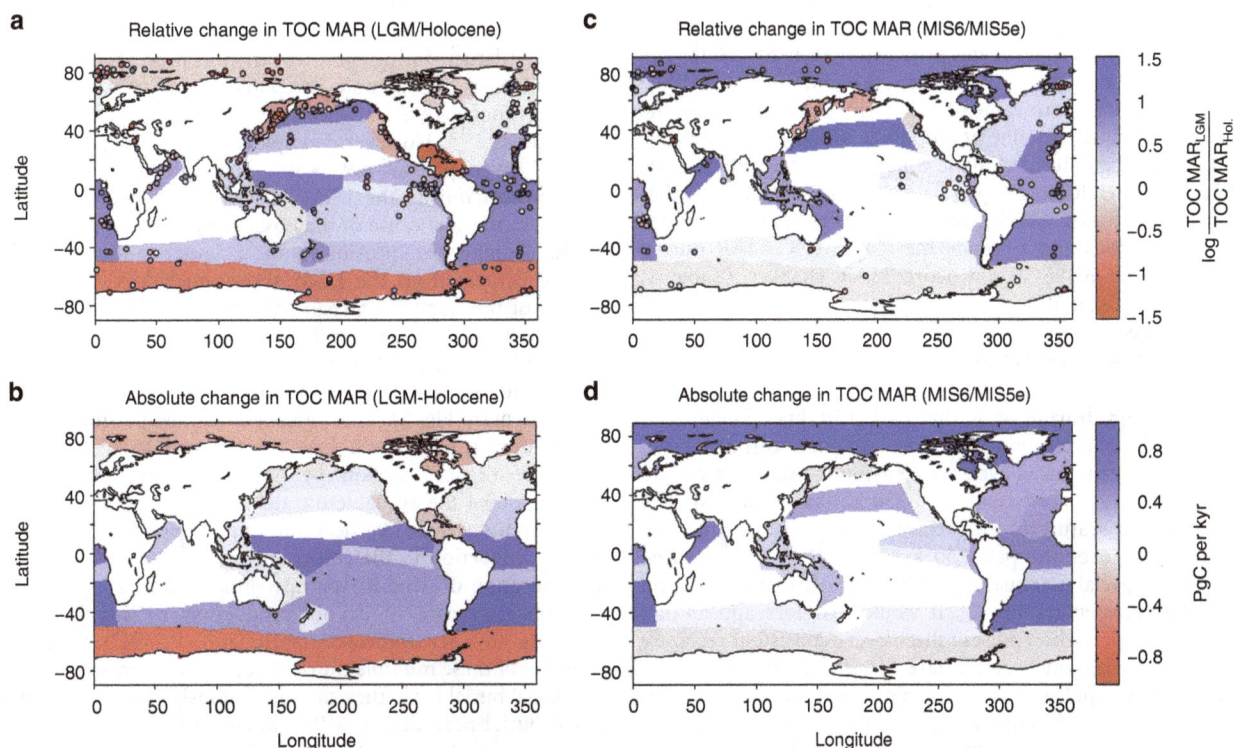

Figure 1 | Relative and absolute changes in deep-sea burial of organic carbon for the two last deglacial transition. Relative (**a,c**) and absolute (**b,d**; PgC per kyr) changes in deep-sea burial of organic carbon from MIS2 to Holocene transition (**a,b**) and MIS6 to MIS5e transition (**c,d**). Individual sedimentary ratios are shown as coloured circles in **a,c**. Shadings correspond to the mean ratio in each province (**a,c**) and to the absolute changes in **b,d**. Note that (**b,d**) show total burial changes in each province, and the absolute changes are not area-normalized.

Table 1 | Changes in organic carbon burial over the two last glacial–interglacial transitions.

Province map	Holocene MAR (PgC per kyr)	LGM/ Holocene (%)	LGM MAR (PgC per kyr)	MIS6/MIS5e (%)	MIS6 MAR (PgC per kyr)	Glacial excess burial (PgC)	Number of provinces
Ocean	17.1	117.7	20.1	152.0	26.0	196	7
Seas	17.1	134.7	23.0	153.2	26.2	242	101
Longhurst (L.)	16.8	159.3	26.8	148.2	24.9	338	56
Modified L. 1	17.1	141.0	24.1	155.6	26.6	262	30
Modified L. 1 + depth.	17.1	147.0	25.1	167.3	28.6	281	60
Modified L. 2	17.1	158.6	27.1	176.0	30.1	435	15
Modified L. 2 + depth.	17.1	170.7	29.2	183.0	31.3	501	30
Mean		147.00	25.06	162.19	27.67	322	
s.d.		17.75	2.99	13.38	2.37	110	

Holocene and LGM deep-ocean TOC burial (MAR) estimated for the different province maps (shown in Supplementary Fig. 2). A subdivision of the provinces following the 1,500 m isobath was added for province maps 5 and 7. The glacial excess burial was calculated between 80 ka, when global burial diverged from MIS5 values, and 10 ka, when global burial reached low Holocene values (Fig. 1).

is due, in similar proportions, to higher sedimentary organic matter concentrations ($126 \pm 8\%$ of interglacial value), and to higher sedimentation rates ($125 \pm 15\%$ of interglacial value, see Methods).

Organic carbon burial rates during MIS6. A range of factors could potentially bias our LGM/Holocene TOC MAR estimates, including sediment compaction in cores for which density measurements are not available (which would tend to overestimate Holocene MARs) and diagenetic remineralization of TOC (which would tend to underestimate LGM TOC). These biases should be minimized for the penultimate glacial termination, which is generally found at significantly greater depths below the seafloor, where vertical gradients in compaction and labile TOC are far more subdued. Thus, we performed the same analyses for 135 available records in our database that include the MIS6–MIS5e transition (between 135–143 ka and 119–127 ka). Depending on the province map used (Methods), MIS6 burial corresponded to between 148 and 183% of MIS5e TOC MAR, with a mean value of $162 \pm 13\%$ (Table 1). Thus, the MIS6/MIS5e ratio shows similar amplitude to the LGM/Holocene ratio. Considering that the climatic conditions during interglacials (MIS5e and the Holocene) were relatively similar, we assume that carbon burial during MIS5e was the same as that during the modern conditions and calculate the global TOC MAR during MIS6 using the MIS6/5e downcore MAR changes (Table 1, Fig. 1c,d). The distribution of changes is very similar to the LGM/ Holocene transition, except for the Arctic (Fig. 1c,d.), where age models are arguably poorly constrained[9].

Organic carbon burial over the past 150 kyr. Finally, we reconstructed a time series of global organic carbon burial throughout the last glacial cycle by applying the same procedure each 1 kyr time step between 0 and 150 ka (see Methods). The most prominent feature of the reconstructed global organic carbon burial rate over the past 150 kyr (Fig. 2) is the increased burial during glacial maxima, regardless of the province selection strategy (Supplementary Fig. 2). It would therefore appear that the global organic carbon pulses reflect a very consistent response to glacial maxima, which could have resulted from any combination of more rapid export of organic matter from the surface mixed layer, more efficient transfer of organic matter from the upper ocean and continental margins to the seafloor, and better preservation of organic matter in the sediment. We estimate the 'excess' removal of C from the system that would have occurred during the last glacial period, above the baseline of interglacial C burial rates, by integrating the C burial between 80 and 10 ka, and

subtracting the baseline interglacial burial. Our estimate suggests that excess burial in the deep sea, that is, that which would have exceeded a constant interglacial burial flux, removed between 200 and 500 PgC from the ocean–atmosphere system (Table 1).

Discussion

In general, higher glacial organic carbon burial occurred in regions where a previous qualitative reconstruction, summarizing diverse proxies[10], inferred higher export production from the surface ocean. The potential importance of increased export is further suggested by similar geographic and temporal patterns between TOC and opal burial (see Methods and Supplementary Fig. 4), which also suggests some degree of increased silicic acid supply to the low latitudes[11], particularly in the Atlantic. Export production is limited by the supply of nitrogen to the mixed layer over most of the ocean, and by iron and/or other factors limiting growth in nitrate-rich regions[12]. The global nitrate inventory may have been larger during the LGM than at present, due to slower denitrification rates, and a potential fertilization by N_2 fixing cyanobacteria enhanced by the supply of iron from glacial dust[13–15]. However, isotopic constraints suggest that the fixed N inventory was not >50% and likely not >30% larger than present during the LGM[16,17], which makes it unlikely to have been the sole cause of the observed >50% increase in TOC MAR. Despite the possibility of a larger nitrate inventory, reconstructions from the LGM show much less nitrate at the surface of the Southern Ocean and subarctic Pacific[18], with low TOC MAR in the coldest parts of these regions, while the high TOC MAR in the Subantarctic is consistent with accelerated export due to dust-borne iron inputs in this region[19,20] (see also Supplementary Fig. 5). Thus, increased export production due to higher dust supply could have contributed to the accelerated burial of organic carbon in presently iron-limited regions, drawing down the available nitrate, while expanded summer sea ice cover and reduced vertical nutrient resupply[21] likely throttled export production in the coldest oceanic realms. Meanwhile, an intensification of wind-driven upwelling could have provided an additional increase of export production by supplying more nutrients to the tropical Atlantic and Pacific[5]. Despite the potential of these multiple nutrient supply mechanisms to explain the glacial burial peak, they cannot obviously explain the lack of a global burial peak during MIS4, when dust flux was high[20] and many other features of the global climate were similar to MIS2 (Supplementary Fig. 5).

The transfer of organic matter from the sunlit surface and continental margins to the ocean floor could also have varied over glacial cycles. The transfer efficiency of organic detritus through

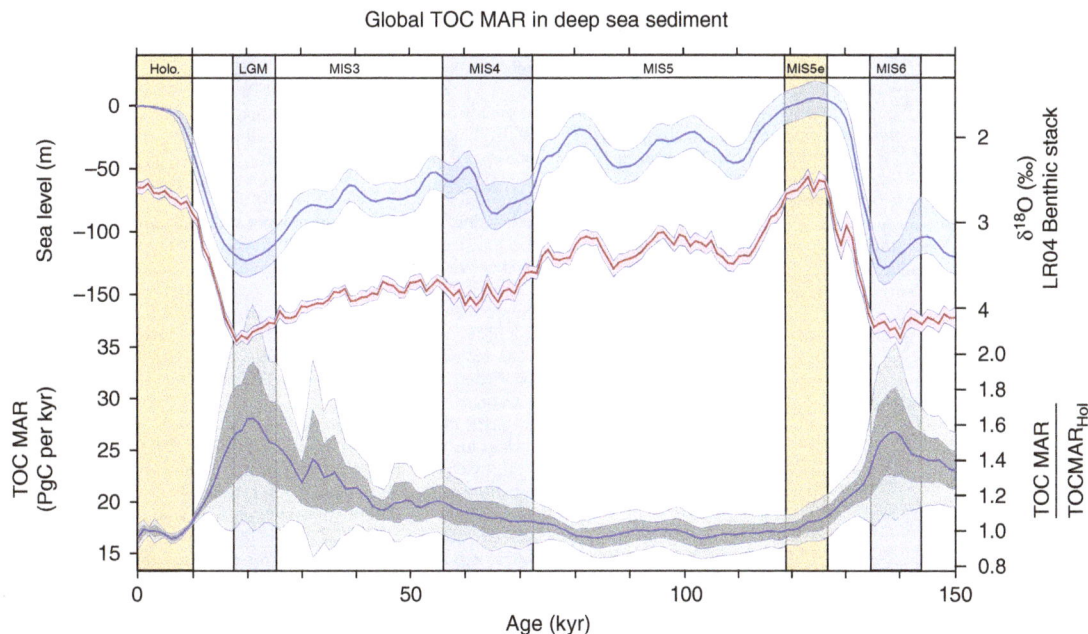

Figure 2 | Global TOC burial over the past 150 kyr. Sea-level reconstruction (and associated confidence interval[50]; LR04 benthic foraminifera $\delta^{18}O$ stack[51] ($\pm$ 2 s.e.), and reconstructed global organic carbon burial in deep-sea sediment over the past 150 kyr (PgC per kyr, $\pm1\sigma$, $\pm2\sigma$ and mean scenario based on the different province scenarios, see Methods, Supplementary Fig. 2 and Supplementary Data 1). The right axis shows relative changes in TOC MAR as compared with Holocene values. The mean organic carbon burial is significantly correlated to both the sea-level reconstruction and the benthic stack ($R = 0.85$, $P \ll 0.05$ and $R = 0.77$, $P \ll 0.05$, respectively). Yellow vertical bands correspond to interglacial periods (Holocene and Marine Isotopic Stage (MIS) 5e), blue vertical bands correspond to glacial condition (Last Glacial Maximum (LGM), MIS4 and MIS6).

the water column depends on the types of organic particles produced by the phytoplankton community, the abundance and behaviour of the heterotrophic community that feeds on the organic matter, and the presence of ballast minerals that protect organic matter from bacterial respiration and/or increase sinking velocities of aggregates[22]. Changes in the transfer efficiency could be consistent with the similar spatial and temporal patterns observed in TOC and opal burial (Supplementary Fig. 4). Colder water temperatures may also have increased the transfer efficiency by reducing the metabolic rates of heterotrophs[11]. In addition, both marine and terrestrial organic matter can be transferred from the coastal zone to the deep sea through downslope transport and nepheloid layers. The exposure of the continental shelf during glacial times favoured direct transfer of coastal and riverine sediment to the deep ocean, bypassing temporary storage and partial remineralisation on shelves, and increasing burial efficiencies[23]. Indeed, terrestrial organic carbon deposited at the mouth of the Amazon River under present-day conditions was transferred almost entirely to the deep-sea fan during the LGM[24,25]. Increased downslope transport of fine-grained lithogenic material during the glacial sea-level lowstand could also help to explain the increased sedimentation rates evidenced in this study (Methods), and over most major deep-sea fan systems[26]. However, the glacial burial peaks do not appear to have been amplified closer to coasts, as might be expected were downslope transport of shelf sediments the main driver of the global burial pulses (see Methods, and Supplementary Figs 6, 7 and 8).

The third candidate explanation involves enhanced preservation of organic matter in glacial sediment. Increased transport of fine-grained material to the deep sea could have favoured organic matter preservation, which is enhanced with increasing surface area of lithogenic sediment[23]. In addition, organic matter preservation on the seafloor is related to oxygen exposure time[27], a function of sedimentation rates and bottom water oxygenation. Higher bulk accumulation rates during glacial

maxima, potentially due to enhanced dust flux, continental erosion by ice sheets and/or erosion of exposed shelves, would have reduced the oxygen exposure time. Furthermore, proxy reconstructions have shown that the global deep ocean was less oxygenated during the LGM[28,29]. A postulated decrease in the proportion of well-ventilated North Atlantic Deep water[30,31] could have contributed to reduced oxygenation of the deep Atlantic[28] (see Supplementary Fig. 5), and enhanced the burial of organic matter during glacial intervals[32]. However, changes in oxygen exposure time cannot explain covariations between opal and TOC burial (see Supplementary Fig. 4), and the non-linear relationship between oxygen exposure time and burial efficiency[27] suggests that this process may have less leverage in the deep sea, where sedimentation rates are low. It is possible that both enhanced TOC and opal burial reflect better preservation due to increased bulk sedimentation rates[33,34].

The glacial burial pulses documented here, and their correlation with sea level and global ice volume (Fig. 2), are evidence of a striking sensitivity of deep-sea organic carbon burial to climate. The excess burial of 200–500 PgC of organic carbon in deep-sea sediment during glacials is comparable to the excess carbon stored below ice sheets[35], or within soil, permafrost and peat deposits[36] during glacials. While our results constrain the deep-sea organic carbon sink during glacials, a closure of the long-term carbon budget in the ocean will require an estimate of deep-sea calcium carbonate burial, as well as burial on shelves. In addition, the fact that a single major deep-sea fan off of the Amazon basin accounts for a significant part of glacial organic matter burial points to the need for studies specifically targeting deep-sea fans and slopes.

Three general factors have been presented here that could have contributed to the burial pulses, acting individually or together. The apparent lack of an organic carbon burial peak during MIS4, an interval in many respects similar to MIS2 and MIS6, is an intriguing observation that deserves further attention. Disentangling the interwoven causes behind the burial pulses provides an important challenge for understanding the global carbon cycle, and its variations through Earth history.

Methods

Analytical strategy overview. Our estimate for carbon burial involves three consecutive steps. First, we use a global map of organic carbon burial in modern (core-top) sediments, based on ref. 37 (Supplementary Fig. 1). Second, in order to infer regional changes in burial patterns from individual sediment cores while accounting for the geographic and bathymetric variability in global organic carbon burial, we subdivide the world ocean into different sets of geographical provinces. Given that any such division will introduce inherent biases, we used a total of seven different subdivision strategies (Supplementary Fig. 2), and compare the results, as a test of the degree to which the results depend on the selection of provinces, and to evaluate uncertainties. Third, we use quality-controlled sediment records to reconstruct relative changes, as compared with the Holocene, of the burial rate in each province. These composite time series are then used, in combination with modern estimates of burial rates, to calculate absolute changes in the burial rates in each province.

Modern TOC MAR and associated uncertainties. The modern organic carbon MAR map (Supplementary Fig. 1) was generated by multiplying the TOC concentrations reported for surface sediments (core tops) by the corresponding bulk sediment accumulation rate. The TOC content map was generated by combining the map reported by ref. 38, augmented with unpublished data, while the MAR map was determined based on the geometric mean of two pre-existing maps (ref. 39 and other unpublished data, see details in ref. 37). It is important to note that both the MAR and the TOC maps have insufficient spatial resolution to adequately resolve continental margins. Even though a substantial portion of organic carbon burial may occur in coastal environments (up to 90% (ref. 40)), our maps with a 1×1 degree spatial resolution cannot resolve small-scale features such as shorelines, shelves and deep-sea fans. Uncertainties related to the modern burial estimates may have two impacts on our results:

(1) The absolute values of the modern burial conditions and the burial values inferred for the past. The modern map that we use is consistent with the most commonly cited value for deep-ocean organic carbon burial[41], but other studies have reported values that diverge by orders of magnitude (see details and references in ref. 42). For this reason, we reported relative changes in Table 1, based on the downcore records only. Similarly, the right y axis on the bottom panel of Fig. 2 shows relative changes as compared with the Holocene, rather than absolute values.

(2) A different spatial distribution of the modern burial would change the relative contribution of the provinces to the global budget, altering both the pattern and absolute value of the reconstructed global TOC burial. To test the impact of a modern map with TOC burial focused on continental slopes and coastal environments, we used the global map of modelled organic carbon flux to the seafloor from ref. 43. This map shows a similar pattern in the deep sea, but orders of magnitude higher flux in coastal regions compared with the map used in our study (note that the flux of TOC to the seafloor differs from burial, because of organic carbon remineralization on the seafloor; Supplementary Fig. 9, blue axis). Following exactly the same approach, but using this alternative map as modern reference, did not substantially modify the shape and amplitude of the relative global TOC burial changes (Supplementary Fig. 9, blue lines), as compared with the reference maps (Supplementary Fig. 9, red lines).

We further tested the influence of the modern burial map on our results, using an updated TOC content map, and only the Jahnke bulk MAR map[39] (Supplementary Fig. 9, green lines). Once again, the relative changes in global burial obtained were very similar to the one presented in this study, suggesting that our reconstruction is robust and relatively independent on our starting assumptions in terms of variability.

Because paleoceanographic studies favour using sedimentary archives that are continuous, and hence not perturbed by abrupt sedimentary processes such as mass flows or turbidity currents, or by hiatuses in sedimentation rates, the coring sites are generally expected to have been selected to avoid such processes. However, sedimentation over slopes and deep-sea fans probably occurs mostly as massive and sporadic, yet very localized events. Meanwhile, some portions of the ocean floor probably see no sediment deposition at all, or even erosion. These types of sedimentary environments are likely under sampled by the sediment cores used in this study.

Thus, our reconstruction applies mainly for deep-sea sediment, poorly resolving coastal sediment deposits. No MAR map is available for the Mediterranean Sea to our knowledge, and as a consequence this region was not considered further.

Province definition strategies. The first two strategies subdivide the global ocean into the major ocean basins (Supplementary Fig. 2a), as well as finer subdivisions of the oceans into seas (Supplementary Fig. 2b), based on ref. 44. Next, based on the assumption that the large-scale ocean features driving biogeochemical cycles in the modern ocean have remained relatively stationary over time, we used the annual climatology of the ocean biogeochemical provinces based on ref. 45, later updated by ref. 46. This map defines 56 coherent provinces from a biogeochemical

perspective (Supplementary Fig. 2c). To adapt the subdivision of the ocean to our sample distribution, the Longhurst map was redefined and simplified into two different maps with, respectively, 30 (Supplementary Fig. 2d) and 15 different provinces (Supplementary Fig. 2e). To investigate potential influences of the depth of the record, these two simplified maps were used to create two new sets of maps by dividing each province into a shallow and a deep component using the 1,500 m isobaths (Supplementary Fig. 2f,g).

Extracting burial rate changes from sediment core database. First, we created a database comprising surface and downcore sediment composition, retrieving available data from the NOAA (ftp://ftp.ncdc.noaa.gov/pub/data/paleo/paleocean/sediment_files/) and PANGAEA (http://www.pangaea.de) online repositories (Supplementary Data 2). Thus, any data set used in this study is available online in one of these repositories. All the data sets that contained TOC concentrations, TOC accumulation rates, age models, sediment density values, along other parameters were taken into consideration. Given that some records are reported with their original depths, while others can be reported with composite or corrected depth scales (from drilling disturbance, voids, or instantaneous deposits such as turbidites or tephras), we first verified the internal consistency of the depth scale, by comparing similar proxies from different records from the same core. When more than one combination between TOC content, age model and density was possible, the best one was selected. The objective criterion used to select the best combination was as follows: outliers detected visually from the cloud of different solutions were removed. More recent versions of age models were favoured over older versions. Age models with a high number of tie points and calibrated ^{14}C measurements were preferred, and a composite age scale was created for some of the records, using least square splines.

230-Thorium normalized sediment accumulation rates were used whenever available. If they were not available (the majority of records), sedimentation rates were inferred at the depth of each TOC measurement by calculating the linear sedimentation rate implied by the age model. We used measured dry bulk densities, when available, interpolated onto the TOC record sampling depth, to convert linear accumulation rate to MAR. When dry bulk densities were not directly available, we assumed constant values equal to $0.9 \, \mathrm{g \, cm^{-3}}$, corresponding to the mean dry bulk density for compacted marine sediments in our database.

Each sedimentary record of organic carbon accumulation was expressed as a ratio to the mean Holocene value of the same record. The mean Holocene to LGM ratio of sedimentary records was calculated for each province (Fig. 1a), and used to estimate LGM TOC burial by multiplying the province ratio with the corresponding province modern burial value (Fig. 1b, Table 1). The LGM global burial rate was then obtained by summing the inferred burial in each province (Table 1). Note that we used the geometric, rather than the arithmetic mean of LGM to Holocene ratios, given that the values are log-normally distributed.

Our database includes 561 total TOC MAR time series. Of these, 260 were of sufficient resolution and length for both the Holocene (defined as 0–10 ka; 454 records), and the LGM (18–25 ka; 303 records) to calculate average LGM to Holocene TOC MAR ratios (coloured circles, Fig. 1a). The provinces for which no downcore records are available account for a small fraction (0 to 23%, depending on the province strategy) of modern global burial. Significant variations of sedimentary LGM/Holocene ratios occur within a single province, indicating local variability in the factors affecting the production, transfer and sedimentation of organic matter (Supplementary Fig. 3).

The same procedure was used to calculate global TOC burial rate over the past 150 kyr with a 1 kyr time step. We used interpolated downcore MAR records with a 1 kyr time step, expressed as a fraction of Holocene MAR (0–10 kyr). Any time interval within a sediment record in which no measurement was present over a period >10 kyr was excluded from the calculation of the regional stacks. When no record was available for a specific province, and/or for a specific period, the province flux was assumed to be constant and equal to modern values.

To evaluate the relative influence of organic carbon content and sedimentation rate variations on the results, we also calculated global burial assuming (1) constant sedimentation rates and (2) constant TOC content. This test indicates that the more elevated TOC burial during the LGM is due, in similar proportions, to higher sedimentary organic matter concentrations ($126 \pm 8\%$ of interglacial value) and to higher sedimentation rates ($125 \pm 15\%$ of interglacial value).

Given that the deglacial trend in organic matter burial in the Arctic is uncertain, and shows opposite behaviour from MIS6 to MIS5e, and LGM to Holocene transitions, we excluded the Arctic Ocean. As only a small proportion of global burial currently occurs in the Arctic Ocean (<8%), this exclusion has only a very minor impact on the results in terms of timing and amplitude.

Long-term trend in global organic matter burial. The mean of the different scenarios shows a pronounced increasing trend towards the present (Supplementary Fig. 10). There are different explanations to account for that trend, which are not related to actual burial changes. Given that density measurements were not always available, we performed our calculation assuming a constant density for the entire core. The consequence of this assumption is that the calculated MARs are slightly overestimated for the most recent sediments, for which the density is expected to be lower than for older sediments, because of sediment compaction. This can also partly explain the increasing trend during the late Holocene (Supplementary Figs 5 and 10).

In addition, the slow long-term diagenesis of refractory organic carbon contributes to the gradual decrease of apparent burial fluxes with age.

Another source of uncertainty arises from the use of raw ^{14}C ages to derive some sediment core-age models. Records for which only raw ^{14}C ages were available should slightly overestimate the most recent calculated TOC MAR, due to the difference between calendar and ^{14}C ages. Thus, these records can slightly reduce the LGM to Holocene amplitude, but cannot explain the decreasing trend. This is confirmed by similar patterns between MIS6 and MIS5, out of the range covered by ^{14}C-dating.

Finally, apparent sedimentation rates decrease as a power law function of the intervals between two age control[47], indicating that sedimentation rates could be overestimated during the Holocene and the deglaciation, when the age model resolution is generally higher, as compared with older periods.

To correct for these biases, we removed the long-term trend from each scenario (Supplementary Fig. 10) calculated using a least-squares spline modelling tool (MATLAB Shape Language Modeling toolbox, D' Errico, 2009, MATLAB Central File Exchange, retrieved online on February 2012). It is important to note that this correction implicitly assumes similar burial rates for the Holocene and MIS5.

Biogenic opal burial variations. To evaluate the changes in opal burial over the past 150 kyr, we applied the same method described in this paper for TOC burial to biogenic opal records available in our database. The number of sedimentary records for opal burial was lower than for TOC. However, we obtained remarkably similar results from a spatial and temporal point of view, in provinces with a sufficient number of records (Supplementary Fig. 4).

Impact of sea-level change on sediment redistribution. As sea level dropped during glacial periods, continental shelves were exposed and eroded, activating submarine canyons and rerouting coastal deposits directly to the deep sea[48]. Indeed, the present-day deposition of terrestrial organic carbon at the mouth of the Amazon River was transferred almost entirely to the deep-sea fan during the LGM, representing 3.7 PgC per kyr[24], or >13% of global LGM deep-sea burial rate (Table 1). Although this particular hot spot in marine accumulation rate is not resolved in our analysis, it is possible that downslope transport of organic matter does contribute to some continental slope records included in our database. Despite its importance, this mechanism is unlikely to explain more than a fraction of our reconstructed changes in TOC MAR, given that many of the records in our database are far from continental slopes. It is worth noting that erosion of coastal deposits during sea-level lowstands may have enhanced nutrient availability and productivity of the ocean[48,49], while increased organic matter reaching the seafloor may have contributed to reduced oxygenation of the deep sea[48].

To further test the potential influence of coastal deposit remobilization during sea-level lowstands, we performed additional analyses, following exactly the same procedure as outline above, using newly designed province maps, for which we distinguished coastal and open ocean regions. Assuming that glacial–interglacial sediment remobilization/relocation was more important near shorelines, we used the ocean and seas maps, as well as the two simplified Longhurst maps (Supplementary Fig. 6) and split each province into a coastal and an open ocean province. Coastal regions were defined using distance thresholds of 500, 1,000 and 1,500 km from the closest point on the coastline (Supplementary Fig. 6). Changes of global TOC burial based on these nine new province maps were similar to the previous analyses, in terms of shape and absolute value (Supplementary Figs 6 and 7). Moreover, the burial in open ocean and in coastal regions displays similar patterns to the whole deep ocean regardless of the width of the coastal provinces, suggesting similar temporal patterns in both open ocean and coastal regions (Supplementary Fig. 7). Finally, we calculated the distance from the shelf, here defined as the 150 m isobath, for each individual sediment core (Supplementary Fig. 8). The absence of correlations between the MAR change between the Holocene and the LGM, and the distance from the shelf, further suggests that regional patterns of burial dominate over continental influences.

References

1. Jouzel, J. et al. Orbital and millennial Antarctic climate variability over the past 800,000 years. Science 317, 793–796 (2007).
2. Broecker, W. S. Ocean chemistry during glacial time. Geochim. Cosmochim. Acta 46, 1689–1705 (1982).
3. Petit, J. R. et al. Climate and atmospheric history of the past 420,000 years from the Vostok ice core, Antarctica. Nature 399, 429–436 (1999).
4. Arrhenius, G., Nyberg, A., Blomqvist, N. & Svenska, D. Sediment cores from the East Pacific (Elanders boktryckeri aktiebolag, 1952).
5. Pedersen, T. F. Increased productivity in the eastern equatorial Pacific during the last glacial maximum (19,000 to 14,000 yr B.P). Geology 11, 16–19 (1983).
6. Bradtmiller, L. I., Anderson, R. F., Sachs, J. P. & Fleisher, M. Q. A deeper respired carbon pool in the glacial equatorial Pacific Ocean. Earth Planet Sci. Lett. 299, 417–425 (2010).
7. Lyle, M. Climatically forced organic carbon burial in equatorial Atlantic and Pacific Oceans. Nature 335, 529–532 (1988).
8. Sarnthein, M., Winn, K., Duplessy, J.-C. & Fontugne, M. R. Global variations of surface ocean productivity in low and mid latitudes: Influence on CO₂

9. Sundby, B., Lecroart, P., Anschutz, P., Katsev, S. & Mucci, A. When deep diagenesis in Arctic Ocean sediments compromises manganese-based geochronology. Mar. Geol. 366, 62–68 (2015).
10. Kohfeld, K. E., Quéré, C. L., Harrison, S. P. & Anderson, R. F. Role of marine biology in glacial-interglacial CO₂ cycles. Science 308, 74–78 (2005).
11. Matsumoto, K. Biology-mediated temperature control on atmospheric pCO₂ and ocean biogeochemistry. Geophys. Res. Lett. 34, L20605 (2007).
12. Moore, C. M. et al. Processes and patterns of oceanic nutrient limitation. Nat. Geosci. 6, 701–710 (2013).
13. Galbraith, E. D., Kienast, M., Albuquerque, A. L., Altabet, M. A. & Batista, F. et al. The acceleration of oceanic denitrification during deglacial warming. Nat. Geosci. 6, 579–584 (2013).
14. Ganeshram, R. S., Pedersen, T. F., Calvert, S. E. & Murray, J. W. Large changes in oceanic nutrient inventories from glacial to interglacial periods. Nature 376, 755–758 (1995).
15. Falkowski, P. G. Evolution of the nitrogen cycle and its influence on the biological sequestration of CO₂ in the ocean. Nature 387, 272–275 (1997).
16. Deutsch, C., Sigman, D. M., Thunell, R. C., Meckler, A. N. & Haug, G. H. Isotopic constraints on glacial/interglacial changes in the oceanic nitrogen budget. Global. Biogeochem. Cycles. 18, GB4012 (2004).
17. Eugster, O., Gruber, N., Deutsch, C., Jaccard, S. L. & Payne, M. R. The dynamics of the marine nitrogen cycle across the last deglaciation. Paleoceanography 28, 116–129 (2013).
18. Galbraith, E. D. & Jaccard, S. L. Deglacial weakening of the oceanic soft tissue pump: global constraints from sedimentary nitrogen isotopes and oxygenation proxies. Quat. Sci. Rev. 109, 38–48 (2015).
19. Martínez García, A. et al. Southern Ocean dust-climate coupling over the past four million years. Nature 476, 312–315 (2011).
20. Winckler, G., Anderson, R. F., Fleisher, M. Q., McGee, D. & Mahowald, N. Covarying Glacial-Interglacial Dust Fluxes in the Equatorial Pacific and Antarctica. Science 320, 93–96 (2008).
21. Sigman, D. M., Hain, M. P. & Haug, G. H. The polar ocean and glacial cycles in atmospheric CO₂ concentration. Nature 466, 47–55 (2010).
22. Klaas, C. & Archer, D. E. Association of sinking organic matter with various types of mineral ballast in the deep sea: Implications for the rain ratio. Global. Biogeochem. Cycles. 16, 1116 (2002).
23. Keil, R., Tsamakis, E., Wolf, N., Hedges, J. & Goñi, M. in Proceedings of the Ocean Drilling Program. Scientific results (1997).
24. Schlünz, B., Schneider, R. R., Müller, P. J., Showers, W. J. & Wefer, G. Terrestrial organic carbon accumulation on the Amazon deep sea fan during the last glacial sea level low stand. Chem. Geol. 159, 263–281 (1999).
25. Goñi, M. A. in Proceedings of the Ocean Drilling Program, Scientific Results (1997).
26. Covault, J. A. & Graham, S. A. Submarine fans at all sea-level stands: Tectono-morphologic and climatic controls on terrigenous sediment delivery to the deep sea. Geology 38, 939–942 (2010).
27. Hartnett, H. E., Keil, R. G., Hedges, J. I. & Devol, A. H. Influence of oxygen exposure time on organic carbon preservation in continental margin sediments. Nature 391, 572–574 (1998).
28. Hoogakker, B. A. A., Elderfield, H., Schmiedl, G., McCave, I. N. & Rickaby, R. E. M. Glacial-interglacial changes in bottom-water oxygen content on the Portuguese margin. Nat. Geosci. 8, 40–43 (2015).
29. Jaccard, S. L. & Galbraith, E. D. Large climate-driven changes of oceanic oxygen concentrations during the last deglaciation. Nat. Geosci. 5, 151–156 (2012).
30. Böhm, E. et al. Strong and deep Atlantic meridional overturning circulation during the last glacial cycle. Nature 517, 73–76 (2015).
31. Jonkers, L. et al. Deep circulation changes in the central South Atlantic during the past 145 kyrs reflected in a combined ²³¹Pa/²³⁰Th, Neodymium isotope and benthic ¹³C record. Earth Planet Sci. Lett. 419, 14–21 (2015).
32. Koho, K. A. et al. Microbial bioavailability regulates organic matter preservation in marine sediments. Biogeosciences 10, 1131–1141 (2013).
33. Ragueneau, O. et al. A review of the Si cycle in the modern ocean: recent progress and missing gaps in the application of biogenic opal as a paleoproductivity proxy. Global Planet. Change 26, 317–365 (2000).
34. Sayles, F. L., Martin, W. R., Chase, Z. & Anderson, R. F. Benthic remineralization and burial of biogenic SiO₂, CaCO₃, organic carbon, and detrital material in the Southern Ocean along a transect at 170°West. Deep-Sea Res. Part II 48, 4323–4383 (2001).
35. Zeng, N. Glacial-interglacial atmospheric CO₂ change —The glacial burial hypothesis. Adv. Atmos. Sci. 20, 677–693 (2003).
36. Ciais, P. et al. Large inert carbon pool in the terrestrial biosphere during the Last Glacial Maximum. Nat. Geosci. 5, 74–79 (2012).
37. Dunne, J. P., Sarmiento, J. L. & Gnanadesikan, A. A synthesis of global particle export from the surface ocean and cycling through the ocean interior and on the seafloor. Global Biogeochem. Cycles 21, GB4006 (2007).

38. Seiter, K., Hensen, C., Schröter, J. & Zabel, M. Organic carbon content in surface sediments—defining regional provinces. *Deep-Sea Res. Part I* **51**, 2001–2026 (2004).

39. Jahnke, R. A. The global ocean flux of particulate organic carbon: Areal distribution and magnitude. *Global Biogeochem. Cycles* **10**, 71–88 (1996).

40. Sarmiento, J. L. & Gruber, N. *Ocean Biogeochemical Dynamics* (Princeton Univ. Press, 2006).

41. Burdige, D. J. Burial of terrestrial organic matter in marine sediments: A re-assessment. *Global Biogeochem. Cycles* **19**, 4, Gb4011 (2005).

42. Burdige, D. J. Preservation of organic matter in marine sediments: controls, mechanisms, and an imbalance in sediment organic carbon budgets? *Chem. Rev.* **107**, 467–485 (2007).

43. Dunne, J. P., Hales, B. & Toggweiler, J. R. Global calcite cycling constrained by sediment preservation controls. *Global Biogeochem. Cycles* **26**, GB3023 (2012).

44. International Hydrographic Organization. *Limits of Oceans and Seas.* Special Publication 23 (International Hydrographic Organization, 1953).

45. Longhurst, A. Seasonal cycles of pelagic production and consumption. *Prog. Oceanogr.* **36**, 77–167 (1995).

46. Reygondeau, G. *et al.* Dynamic biogeochemical provinces in the global ocean. *Global Biogeochem. Cycles* **27**, 1046–1058 (2013).

47. Schumer, R. & Jerolmack, D. J. Real and apparent changes in sediment deposition rates through time. *J Geophys Res.* **114**, F00A06 (2009).

48. Tsandev, I., Rabouille, C., Slomp, C. P. & Van Cappellen, P. Shelf erosion and submarine river canyons: implications for deep-sea oxygenation and ocean productivity during glaciation. *Biogeosciences* **7**, 1973–1982 (2010).

49. Menviel, L., Joos, F. & Ritz, S. P. Simulating atmospheric CO_2, ^{13}C and the marine carbon cycle during the Last Glacial–Interglacial cycle: possible role for a deepening of the mean remineralization depth and an increase in the oceanic nutrient inventory. *Quat. Sci. Rev.* **56**, 46–68 (2012).

50. Waelbroeck, C. *et al.* Sea-level and deep water temperature changes derived from benthic foraminifera isotopic records. *Quat. Sci. Rev.* **21**, 295–305 (2002).

51. Lisiecki, L. E. & Raymo, M. E. A Pliocene-Pleistocene stack of 57 globally distributed benthic $\delta^{18}O$ records. *Paleoceanography* **20**, PA1003 (2005).

Acknowledgements

We thank John Dunne for providing the modern burial flux maps, and three anonymous reviewers for insightful comments, which helped improve the quality of the manuscript. O.C. and S.L.J. were funded by the Swiss National Science Foundation (SNF grant PP00P2_144811). The Canadian Institute for Advanced Research (CIFAR), the Canadian Foundation for Innovation (CFI), and the Natural Sciences and Engineering Research Council (NSERC) supported O.C., D.B. and E.D.G. D.B. acknowledges funding from University of Washington.

Author contributions

O.C., D.B. and E.D.G. designed the proxy database and the data analysis. O.C., D.B., E.D.G. and S.L.J. wrote the manuscript. All authors contributed to the interpretation of the results.

Additional information

Competing financial interests: The authors declare no competing financial interest.

21

Differences in the efficacy of climate forcings explained by variations in atmospheric boundary layer depth

Richard Davy[1] & Igor Esau[1]

The Earth has warmed in the last century and a large component of that warming has been attributed to increased anthropogenic greenhouse gases. There are also numerous processes that introduce strong, regionalized variations to the overall warming trend. However, the ability of a forcing to change the surface air temperature depends on its spatial and temporal distribution. Here we show that the efficacy of a forcing is determined by the effective heat capacity of the atmosphere, which in cold and dry climates is defined by the depth of the planetary boundary layer. This can vary by an order of magnitude on different temporal and spatial scales, and so we get a strongly amplified temperature response in shallow boundary layers. This must be accounted for to assess the efficacy of a climate forcing, and also implies that multiple climate forcings cannot be linearly combined to determine the temperature response.

[1] Nansen Environmental and Remote Sensing Center and Bjerknes Centre for Climate Research, Thormøhlensgt. 47, 5006 Bergen, Norway. Correspondence and requests for materials should be addressed to R.D. (email: Richard.davy@nersc.no).

The detection of processes that affect the surface climate is one of the fundamental tools in our understanding of climate change. The assessment of climate-forcing processes is essentially a statistical, signal-to-noise problem. The most famous example is the detection of warmer surface air temperatures (SATs) attributed to the human-induced enhanced concentration of atmospheric carbon dioxide[1,2]. One of the challenges in the detection of climate forcing signals has been the strong spatial correlation between the strength of the surface temperature trends and the strength of natural variability[3]. This relation leads to an inherently low signal-to-noise ratio, regardless of the strength of temperature trends. Indeed, despite the well-established rapid warming in the Arctic[4-6], the polar regions were the last part of the globe for which there was a successful detection and attribution of the recent warming to anthropogenic-enhanced greenhouse gas (GHG) concentrations[7].

In addition to the detection of the influence of enhanced GHGs on surface temperatures, there have been numerous studies assessing how clouds[8-12], precipitation[8,12,13] and soil moisture[8,12] may have introduced some of the observed temporal and regional variations to the overall warming trend that has been seen in the latter half of the twentieth century. In all these studies, the authors have adopted the commonly accepted linear regression model for establishing the relationship between a change in a given property and the temperature response[14]. However, it has been established from energy-budget models of the climate that temperature changes are linearly related to changes in the surface heating, but also inversely related to the effective heat capacity of the system[15,16]. Thus, although a linear-model approximation may work well under constrained conditions, such as focusing on a given region or season, in more wide-reaching studies it becomes necessary to account for variations in the effective heat capacity of the atmosphere. In recent times, it has also been recognized that the atmospheric convective mixing may have a strong effect on the observed and simulated climate through modulation of the low cloud and lapse-rate feedbacks[17].

As the effective heat capacity is defined by the volume of air through which the heat is mixed, it can be related to the depth of the atmospheric boundary layer[18]. If we consider the one-dimensional energy budget of the lower atmosphere, then we define the change in SAT as being linearly related to the forcing and feedbacks in the climate system, and inversely related to the effective heat capacity of the system[19], such that:

$$\frac{d\theta_v}{dt} = \frac{Q}{\rho c_p h} \tag{1}$$

where Q ($\mathrm{W\,m^{-2}}$) is the heat flux divergence across a boundary layer of depth h (m), ρ ($\mathrm{kg\,m^{-3}}$) is the air density, c_p ($\mathrm{J\,kg^{-1}\,K^{-1}}$) is the heat capacity at constant pressure and θ_v (K) is the virtual potential temperature that is representative of the boundary layer. Henceforth, θ_v is taken at a height of 2 m above the ground. If differences in the effective heat capacity of the system are relatively small, we can linearly relate any changes to the heating to a change in θ_v. However, if there are large variations in the effective heat capacity, we need to account for the dependency of the temperature response on the effective heat capacity. This is especially important when the effective heat capacity is small, as this can strongly magnify the strength of the SAT response to a perturbation in the surface energy budget[3].

The hypothesis we put forward here presumes that there is essentially a decoupling of the planetary boundary layer (PBL) from the rest of the atmosphere. Such a decoupling may not be intuitive, but it is a rather natural concept, supported by numerous direct observations[20,21], and indeed it is embedded in the PBL schemes of many (if not all) climate models[22].

Consider the case of the urban PBL. The urban PBL is better mixed than the PBL in the rural background due to high surface roughness in the urban environment. Nevertheless, even the urban PBL shows a rather clear top boundary, separating it from the rest of the atmosphere[23,24], which can be visualized by water vapour (Fig. 1). The well-known urban heat island effect[25], which in northern cities may reach up to 10 K (ref. 26), demonstrates how heat may be trapped inside the urban boundary layer.

Now consider in this urban context, which can be so clearly visualized, the presumption that this decoupling is valid on climatological time scales. The urban heat island is highly variable in time and may appear for only a few hours in a given day. However, as the PBL effect is strongly selective and nonlinear, the average over climatological time scales is accumulating. We cannot state that the maximum effect will be observed over the time scale of 30–40 years, as addressed in this study, but over this period of time it is sufficiently large as well. Figure 2 illustrates the urban heat island, that is, the additional anthropogenic heat

Figure 1 | Visualization of the separation of the stable boundary layer. Water vapour in the planetary boundary layer over Bergen, Norway, highlights the trapping of air in the urban boundary layer. Courtesy, T. Wolf, NERSC.

boundary layer depth and the magnitude of trends in θ_v ($R = 0.35$, $P < 0.05$ and $R = 0.32$, $P < 0.05$, respectively). This process where heat gets trapped in a shallow layer near the surface by stable stratification in the boundary layer has been shown to be one of the dominant causes of Arctic amplification[30,31], but it also has a crucial implication for how we assess the efficacy of different forcings in affecting the SAT. Given the large differences between the climatology of the PBL depth in re-analyses and the global climate models shown here, and the controlling influence the boundary layer depth has on the surface climate[32], this is clearly an area that requires more attention, especially in future model development.

Variability in the strength of the boundary layer effect. Owing to its varied climatology (Supplementary Fig. 2), there are some geographical differences as to when the boundary layer depth becomes important in determining the strength of temperature trends. For deep boundary layers the relationship between boundary layer depth and temperature trends is expected to be small, and it is the strength of the local forcing factors themselves, which will principally determine the variations in the rate of warming. However, in shallower boundary layers the strength of temperature trends may be expected to become increasingly dependent on the boundary layer depth. This can be seen in the correlation between the magnitude of temperature trends and the inverse boundary layer depth for different geographical regions (Fig. 4). In high-latitude continental regions such as North America, North Asia and Antarctica, where we frequently get very shallow boundary layers in autumn and winter, there is a strong correlation between inverse boundary layer depth and θ_v trends. Whereas in more tropical regions such as Africa, South Asia and South America, where cases of shallow boundary layers are less frequent, there is no evidence of this amplification effect.

There is also a strong seasonal variation in the PBL amplification effect: in the boreal spring and summer, the boundary layer is relatively deep and so the amplification effect is relatively weak. However, during the boreal autumn and winter, we can see that the PBL depth is very small over land (Supplementary Fig. 2); thus, during these periods the amplification effect of the PBL depth can become important and should be taken into account. This is why studies that have chosen to focus on the mid-latitudes during the summer seasons have had some success in demonstrating a relationship between a given forcing process and changes to the SAT[9], whereas more global analysis of the same processes have shown weaker relationships[8]. From the ERA-Interim results, we can see that the amplification effect becomes very apparent for boundary layers less than a few hundred metres (Fig. 3). This is quite common, with PBL depths <400 m occurring $>46\%$ of the time in ERA-Interim. This is a good indication of the fraction of time that the PBL depth becomes important in determining the strength of temperature trends.

Including the boundary layer effect in signal detection. One way to account for the PBL depth in the analysis of climate processes is by considering the integrated temperature response within a co-variability framework (Fig. 5). In this framework, the net temperature change is proportional to the time integration of the product of the perturbations in the forcing, dQ, with the inverse boundary layer depth, h^{-1}. In this regard, the conventional methodology is a limit of the co-variability framework, when the variations in the heat capacity can be neglected and we can directly relate perturbations in the forcing to perturbations in the surface temperature (case A, Fig. 5). This is a reasonable approximation only if that forcing is applied solely to deep or

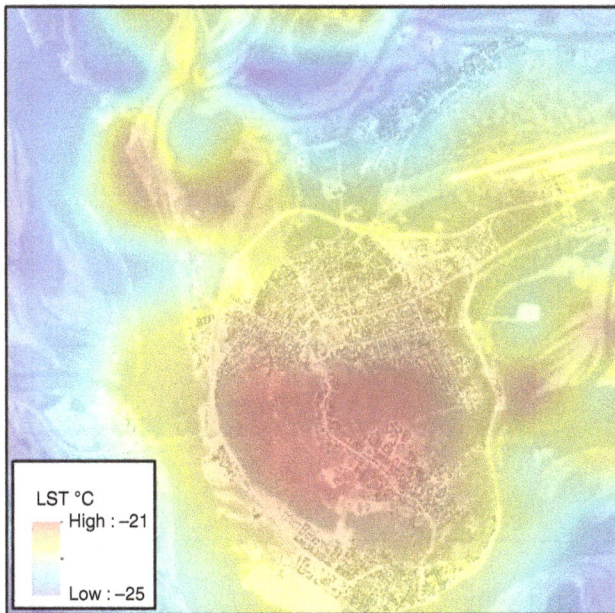

Figure 2 | Satellite observation of the urban heat island effect. The urban heat island effect, as seen in 15-year climatological mean winter-time land surface temperatures (LST) over Khanty-Mansiysk, Russia, from MODIS observations. Courtesy V. Miles, NERSC.

trapped in by the shallow PBL in Khanty–Mansiysk, according to MODIS satellite data analysis for 2000–2014. Thus, from numerous local studies, it has been firmly established that the PBL depth effect does exist. Our study here is the first to quantify this effect globally, using available data sets and models.

Different climate forcing processes such as variations in solar forcing, GHGs and aerosols have different efficacies in affecting the SAT[27–29]. In this work we have demonstrated that it is the variations in the effective heat capacity of the atmosphere, defined by the PBL depth, which can explain these differences in climate forcing efficacy. We must therefore question the assumption that different climate forcings are linearly additive in nature. This work highlights the pressing need to obtain a robust physical climatology of the boundary layer depth from the observational network and to be able to simulate this climatology with global climate models, to better constrain our estimates of surface climate response to climate forcings.

Results
Temperature trend amplification in shallow boundary layers. Here we have identified the relationship between the boundary layer depth and the trend in θ_v during the recent warming period, as seen in different re-analyses, and in two state-of-the-art global climate models (Fig. 3). There is a distinctly inverse relationship, where the strongest warming trends are found in shallow boundary layers, and correspondingly low atmospheric heat capacity. The strength of the trends decrease rapidly as we go towards deeper boundary layers and then remain relatively constant across a wide range of boundary layer depths. This amplified temperature trend in the shallowest boundary layers can also be seen in climate models with very different climatologies of the PBL depth. The Norwegian Earth System Model and Geophysical Fluid Dynamic Laboratories Coupled Model 3 have very different climatologies of the PBL depth compared with the re-analyses (Supplementary Fig. 1); both have biases towards deeper PBLs, but they still show a significant correlation between inverse

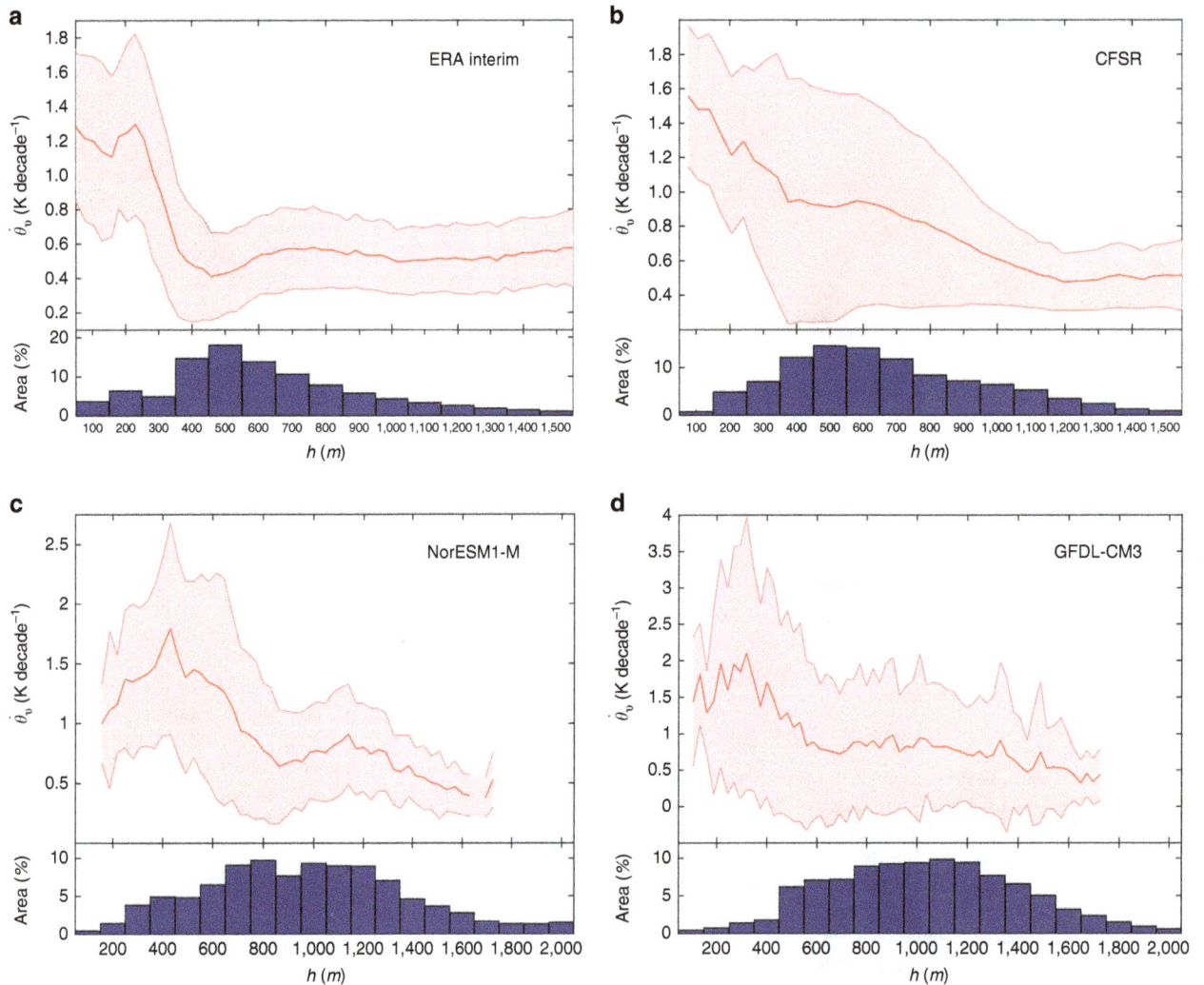

Figure 3 | The amplification of temperature trends in shallow boundary layers. The bin mean and s.d. of the inter-annual trend in the virtual potential temperature at a height of 2 m above the ground, as a function of the climatological monthly mean planetary boundary layer depth for (**a**) ERA-Interim reanalysis, (**b**) CFSR reanalysis, (**c**) the Geophysical Fluid Dynamic Laboratories Coupled Model 3 (GFDL-CM3) and (**d**) the Norwegian Earth System Model (NorESM1-M) climate model simulations. The ERA-Interim and CFSR data are taken over the period 1979–2014 and the climate model data from 1979–2005.

weakly varying boundary layers such as the tropical marine boundary layer[17]. The other limit of the co-variability framework occurs when we have a uniformly applied perturbation to the climate forcing (case B, Fig. 5). In this limit, it is the climatology of the boundary layer depth that principally determines the pattern of warming/cooling in response to a perturbation in the surface heating. The enhanced concentration of GHGs is one such example of a near-uniform perturbation in the surface heating and, as such, there is a strong relationship between the inverse boundary layer depth and the strength of temperature trends in re-analysis (Fig. 3), and both within and between global climate models[32].

However, in most cases it is both the PBL depth and the strength of the forcing that will be important in determining the spatial and temporal variations of climate change. In these cases, it is necessary to account for the nonlinear amplification effect of the PBL depth. Let us take the example of the influence of cloud cover on surface temperatures. We expect that an increase in cloud cover during the day will damp incoming solar radiation and thus decrease surface temperatures[9]. However, an increase in cloud cover at night is expected to reduce longwave cooling and hence result in warmer

surface temperatures. Thus, the net effect of changes to the cloud cover on the surface climate is determined by the balance between the cooling effect of damped shortwave radiation and the warming effect of reduced longwave cooling. The cooling effect principally applies during conditions with strong surface heating (when the surface energy balance is dominated by shortwave radiation) and applies to deep PBLs, compared with the warming effect that dominates when there is a net longwave cooling and relatively shallow PBLs. Therefore, when we consider the effect of changes in cloud cover on the atmospheric heat content, rather than on the surface temperature, we expect the cooling effect to become more apparent. This can be seen in the regressions of the cloud cover anomalies, $\hat{C}$, against surface temperature anomalies, $\hat{T}$, and against normalized atmospheric heat content anomalies, $h\hat{T}$ (Fig. 6). If we look at the sensitivity of surface temperature to cloud cover we can see that in the high latitudes the strong winter-time warming effect of increased cloud cover dominates on the inter-annual scale and we get a strong positive relationship, whereas when we consider the effect of cloud cover on heat anomalies we find a more widespread cooling effect of increased cloud cover, even in these high-latitude continental interiors.

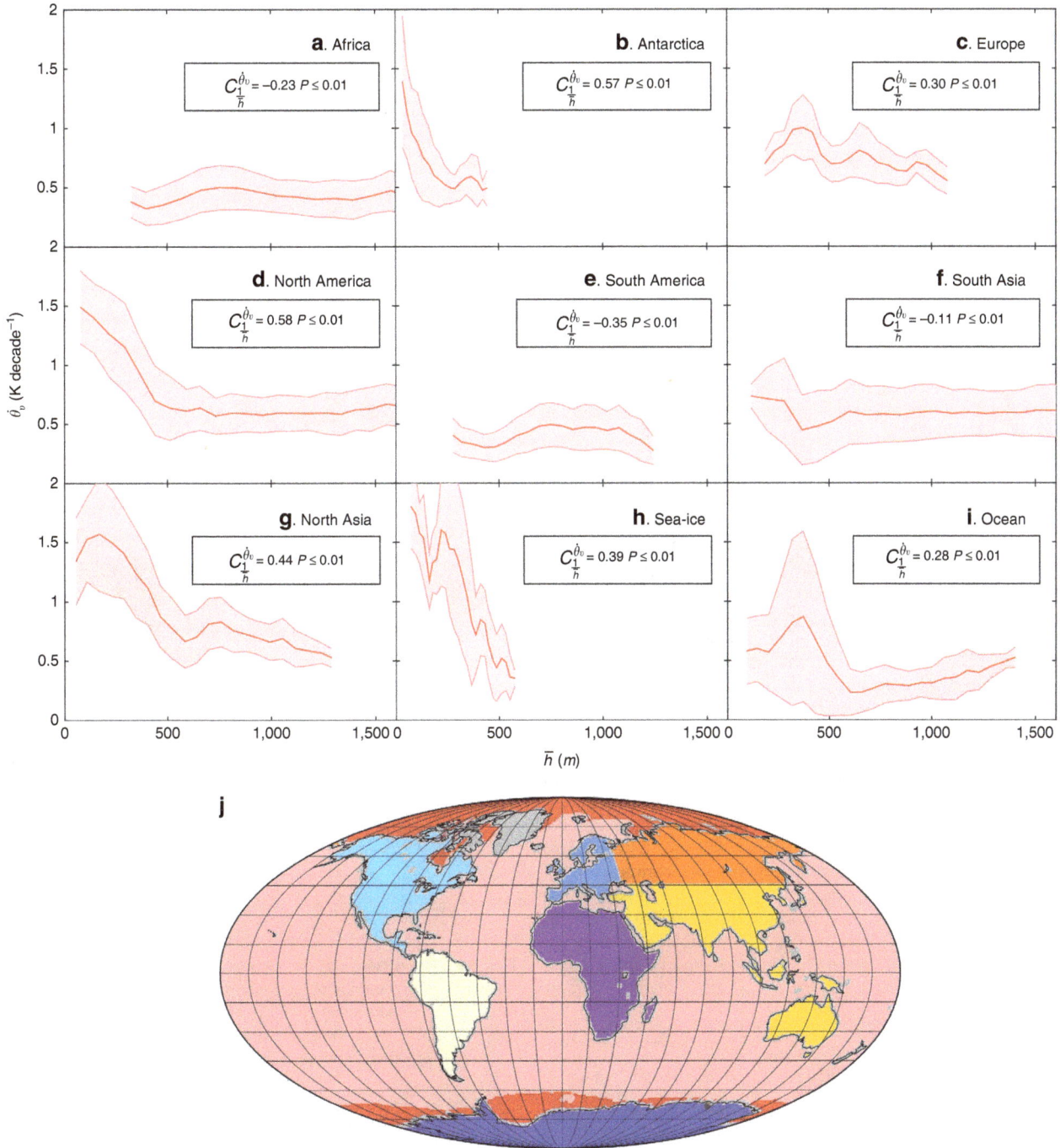

Figure 4 | Geographical variations to the planetary boundary layer effect. The inter-annual trend in the virtual potential temperature at a height of 2 m above the ground as a function of the climatological monthly mean planetary boundary layer depth is shown for nine different regions: (**a**) Africa, (**b**) Antarctica, (**c**) Europe, (**d**) North America, (**e**) South America, (**f**) South Asia, (**g**) North Asia, (**h**) Sea-ice and (**i**) Ocean, as illustrated on the (**j**) map of the Earth. The thick red line indicates the bin-mean and the shaded area shows the region of 1 s.d. The correlation between the magnitude of the temperature trends and the inverse boundary layer depth is given for each region.

This may be expected, as the warming effect of positive cloud cover perturbations on the surface temperature that occurs during the winter months only has a small impact on the atmospheric heat content compared with the cooling that occurs in deep PBLs in the summer months. Thus, when we account for variations in the effective heat capacity, we get a significantly stronger damping of atmospheric heat content from increased cloud cover than we found when assessing surface temperatures: the globally averaged overland temperature sensitivity to cloud cover is $-12\,(\pm 17)\times 10^{-3}\,\mathrm{K}\,\%^{-1}$, compared with a sensitivity of norm-

alized heat content to cloud cover of $-32\,(\pm 16)\times 10^{-3}\,\mathrm{K}\,\%^{-1}$. This marks a much clearer signal of an overall cooling effect of increased cloud cover on the surface climate.

Discussion

What is proposed here is essentially a way of accounting for the fact that processes such as changes to cloud cover, soil moisture and so on introduce perturbations to the heating, but what we measure is the temperature perturbation. It has been shown that

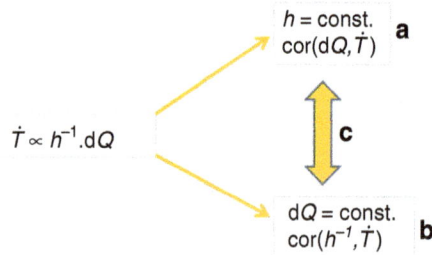

Figure 5 | Relationship of proposed co-variability method to existing methodology. Schematic of the proposed co-variability method and the relation to: (**a**) current methodology where variations in the PBL depth, h, are neglected and we directly relate surface temperature trends, $\dot{T}$, with climate perturbations, dQ; (**b**) a uniform climate-forcing perturbation where it is the climatology of the PBL depth, which is the best predictor of temperature trends; and (**c**) intermediary conditions where both the PBL depth and the perturbation in the forcing are significant in determining the temperature response.

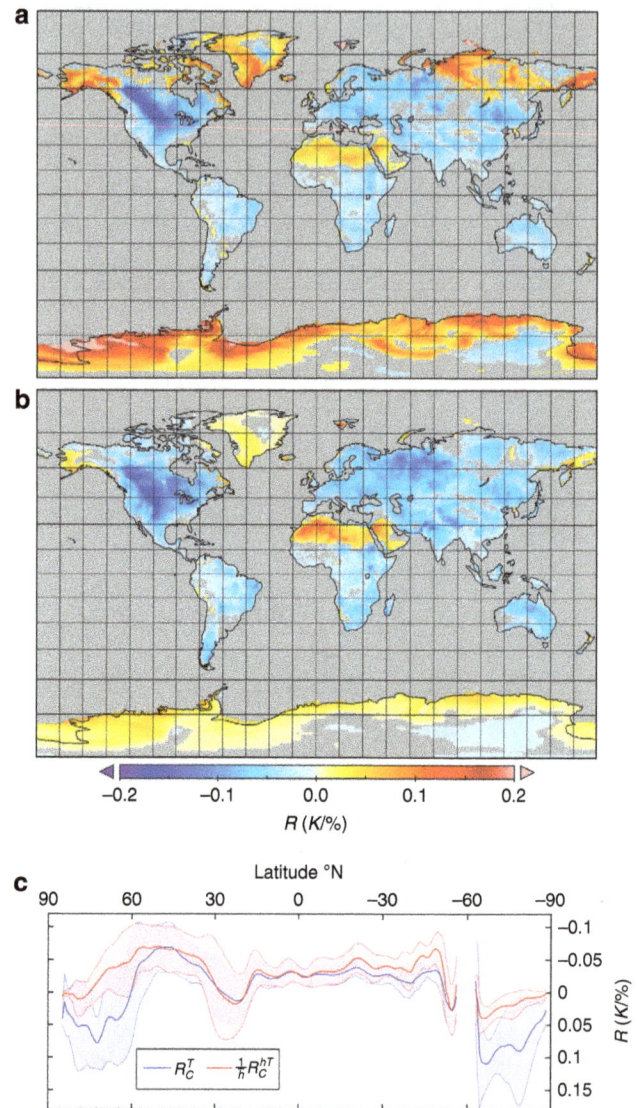

Figure 6 | The climate effect of clouds on temperature and heat content. The maps show the slope of the linear regression, R, between anomalies in cloud cover and (**a**) anomalies in SAT, and (**b**) anomalies in normalized heat content in K%$^{-1}$ change in cloud cover. We also show (**c**) the bin mean and s.d. of the regression coefficients as a function of latitude.

we can better understand influences on the climate system by considering the heat content, rather than the temperature, of the components of the climate system[33]; however, the importance of spatial and temporal variations in the heat capacity have largely been neglected. This can be especially important within the diurnal cycle, when the shallow boundary layer depths at night drive strong changes to the nocturnal temperatures[12], which have a strong impact on vegetation[34].

Here we have shown that by accounting for the variations in the effective heat capacity of the atmosphere we can more accurately assess how a given process influences the surface climate. This allows us to directly compare, in an apples-to-apples manner, the relative importance of different processes in determining the overall response of the climate system to perturbations in the climate forcing. This is also critical for model-model and model-observation comparisons of climate forcing processes. There is a very varied climatology of the PBL in different climate models and re-analysis (Fig. 3) (refs 32,35,36) and so these models can exhibit very different temperature responses, given the same change in the climate forcing.

Methods

Processing of re-analysis data. We obtained the monthly mean time series of air and dew-point temperature at a height of 2 m above the ground, boundary layer depth and sea-ice concentration from the European Centre for Medium-range Weather Forecasts website for the full period of available data, 1979–2014. The air and dew-point temperature were then used to calculate the virtual potential temperature at a height of 2 m above the ground, θ_v. The boundary layer depth in the ERA-Interim model is calculated using an iterative bulk-Richardson method, which scans upwards from the lowest model level and interpolates between model levels to find the height at which the bulk Richardson number first exceeds the critical Richardson number, Ri_{cr}, taken to be 0.25. Although the bulk-Richardson method has been shown to work well in both stable and unstable stratification[35], this definition does create some problems in tropical regions, including parts of South America, Africa and South Asia. The estimation of the boundary layer depth from different methods shows great variation in these regions of high surface humidity and strong convection, and so the ERA-Interim PBL depth is not necessarily reflective of the vertical extent of turbulent mixing[36]. This is likely to be the reason for the lack of a clear signal of shallow boundary layer amplification of temperature trends in these regions (Fig. 4).

The models used in both re-analysis and GCMs also tend to be biased towards producing deeper-than-observed boundary layers, especially under stably stratified conditions[35,37]. This is probably due to our limited understanding of geophysical turbulence under stable stratification and the limited applicability of current turbulence parameterization schemes to strongly, stably stratified conditions[37]. This bias has been shown to lead to an underestimation of the SAT response to climate forcing[32].

The sea-ice concentration was used to define the region 'sea ice' in Fig. 4. The requirement for being considered over sea ice was that the minimum concentration for a location in a given month be $>80\%$ in all years.

The monthly mean time series of air temperature and specific humidity at a height of 2 m above the ground, the boundary layer depth and sea-ice concentration were obtained from the National Center for Atmospheric Research website. The boundary layer depth in this model is calculated in a similar way as to that in ERA-Interim, using the bulk-Richardson formulation and the same critical Richardson number.

Analysis of climate model data. To calculate the boundary layer depth in these models, we used a bulk Richardson method. This was chosen due to the limits of data availability for other methods, for example, no information on local gradients was available, which would be required for a flux-Richardson method, and the vertical resolution in these models is too coarse for any of the profile-based methods. The bulk-Richardson method has also been shown to be the most robust across a wide range of thermal stability and is consistent with the method used in the re-analyses.

First, we obtained the 6-hourly resolution three-dimensional fields of the wind components, humidity and temperature, and the 3-hourly resolution surface temperature, pressure and humidity for the period 1979–2005 for Norwegian Earth System Model and Geophysical Fluid Dynamic Laboratories Coupled Model 3 from the British Atmospheric Data Centre archive. The temperature and humidity fields were used to calculate the virtual potential temperature at each model level and at the surface using the definition $\theta_v = \theta(1 + 0.61M)$, where θ is the potential temperature and M is the water–vapour mixing ratio. The thickness of each model level, Z, was calculated using a hydrostatic assumption, such that $Z_i = \frac{R_d}{2g}(p_i - p_{i-1})(\frac{\theta_{vi}}{p_i} + \frac{\theta_{vi-1}}{p_{i-1}})$, where i is the index of the model level with the surface being $i = 0$, θ_v is the virtual potential temperature, p the pressure, R_d the dry gas constant ($287.06\,\mathrm{J\,kg^{-1}\,K^{-1}}$) and g the surface gravity ($9.807\,\mathrm{m\,s^{-2}}$). The geometric height of each level was then calculated by the summation of the thickness of the lower levels. The bulk Richardson number at each pressure level was calculated using the difference between that level and the surface: $Ri = \frac{gz}{\theta_{vs}}\frac{\theta_v - \theta_{vs}}{U^2}$, where g is the surface gravity, z is the height above the surface, θ_v is the virtual potential temperature at height z, θ_{vs} is the virtual potential temperature at a height of 2 m above the surface and U is the wind speed at height z. We then scanned upwards from the lowest level above the surface and linearly interpolated between levels, to find the first height at which the bulk Richardson number exceeded the critical value, Ri_{cr}. We also applied a requirement that the PBL depth should be > 10 m and < 4 km: as this was an automated method, these constraints were necessary to avoid cases where the routine returned unphysical PBL depths.

Statistical methodology. The correlations given in Fig. 4 are the area-weighted spatial correlations between the monthly mean inverse boundary layer depth and the inter-annual trend in the monthly mean virtual potential temperature at a height of 2 m above the ground. The P-values were computed to test against the null hypothesis of zero correlation using a Student's t distribution for a transformation of the correlation.

The regression coefficients in Fig. 6 were determined from a least-square, best-fit linear regression between the two variables under consideration. These were the cloud-cover anomalies against the anomalies in θ_v and the cloud cover anomalies against the normalized heat anomalies, $\frac{1}{\bar{h}}h\theta_v$, where $\bar{h}$ is the area-weighted, climatological mean boundary layer depth. Anomalies were calculated by removing the climatological mean of each month from the time series using the full period of the data, 1979–2014.

Code availability. The Matlab code used to generate the PBL depth and virtual potential temperature data sets discussed in this work have been archived by the authors and are available on request from the corresponding author, R.D. (email: Richard.davy@nersc.no).

References

1. Barnett, T. P. *et al.* Detection and attribution of recent climate change: A status report. *Bull. Am. Meteorol. Soc.* **80**, 2631–2659 (1999).
2. IPCC. in *Climate Change 2013: The Physical Science Basis. Contribution of Working Group I to the Fifth Assessment Report of the Intergovernmental Panel on Climate Change* (eds Stocker, T. F. *et al.*) 1535 (Cambridge Univ. Press, 2013).
3. Esau, I., Davy, R. & Outten, S. Complementary explanation of temperature response in the lower atmosphere. *Environ. Res. Lett.* **7**, 044026 (2012).
4. Graverson, R. G., Mauritsen, T., Tjernstrom, M., Kallen, E. & Svensson, G. Vertical structure of recent Arctic warming. *Nature* **451**, 53–56 (2008).
5. Simon, C., Arris, L. & Heal, B. *Arctic Climate Impact Assessment* (Cambridge Univ. Press, 2005).
6. Johannessen, O. *et al.* Arctic climate change: observed and modeled temperature and sea-ice variability. *Tellus A* **56**, 328–341 (2004).
7. Gillet, N. P. *et al.* Attribution of polar warming to human influence. *Nat. Geosci.* **1**, 750–754 (2008).
8. Dai, A., Trenberth, K. E. & Karl, T. R. Effects of clouds, soil moisture, precipitation, and water vapor on diurnal temperature range. *J. Climate* **12**, 2451–2473 (1999).
9. Tang, Q. & Leng, G. Damped summer warming accompanied with cloud cover increase over Eurasia from 1982 to 2009. *Eniron. Res. Lett.* **7**, 014004 (2012).
10. Crook, J. A., Forster, P. M. & Stuber, N. Spatial patterns of modeled climate feedback and contributions to temperature response and polar amplification. *J. Climate* **24**, 3575–3592 (2011).
11. Sun, B., Groisman, P. Y., Bradley, R. S. & Keimig, F. T. Temporal changes in the observed relationship between cloud cover and surface air temperature. *J. Climate* **13**, 4341–4357 (2000).
12. Davy, R., Esau, I., Chernokulsky, A., Outten, S. & Zilitinkevich, S. Diurnal asymmetry to the observed global warming. *Int. J. Climatol.* doi:10.1002/joc.4688 (2016).
13. Zhou, L. *et al.* Spatial dependence of diurnal temperature range trends on precipitation from 1950 to 2004. *Clim. Dyn.* **32**, 429–440 (2009).
14. Stone, D. A. *et al.* The detection and attribution of human influence on climate. *Annu. Rev. Environ. Resour.* **34**, 1–16 (2009).
15. Hasselmann, K. Stochastic climate models Part 1. Theory. *Tellus* **28**, 473–485 (1976).
16. Kim, K.-Y. & North, G. R. Surface temperature fluctuations in a stochastic climate model. *J. Geophys. Res.* **96**, 18573–18580 (1991).
17. Sherwood, S. C., Bony, S. & Dufresne, J.-L. Spread in model climate sensitivity traced to atmospheric convective mixing. *Nature* **505**, 37–42 (2014).
18. Esau, I. & Zilitinkevich, S. On the role of the planetary boundary layer in the climate system. *Adv. Sci. Res.* **4**, 63–69 (2010).
19. Hansen, J. *et al.* Climate impact of increasing atmospheric carbon dioxide. *Science* **213**, 957–966 (1981).
20. Wood, R. & Bretherton, C. S. Boundary layer depth, entrainment, and decoupling in the cloud-capped subtropical and tropical marine boundary layer. *J. Climate* **17**, 3576–3588 (2004).
21. Petäjä, T. *et al.* Enhanced air pollution via aerosol-boundary layer feedback in China. *Sci. Rep.* **6**, 18998 (2016).
22. Mauritsen, T. *et al.* Tuning the climate of a global model. *J. Adv. Model. Earth Syst.* **4**, M00A01 (2012).
23. Barlow, J. F. *et al.* Boundary layer dynamics over London, UK, as observed using Doppler lidar during REPARTEE-II. *Atmos. Chem. Phys.* **11**, 2111–2125 (2011).
24. Wolf, T., Esau, I. & Reuder, J. Analysis of the vertical temperature structure in the Bergen valley, Norway, and its connection to pollution episodes. *J. Geophys. Res. Atmos.* **119**, 10645–10662 (2014).
25. Zhao, L., Lee, X., Smith, R. B. & Oleson, K. Strong contributions of local background climate to urban heat islands. *Nature* **511**, 214–219 (2014).
26. Peng, S. *et al.* Surface urban heat island across 419 global big cities. *Environ. Sci. Technol.* **46**, 796–803 (2012).
27. Marvel, K., Schmidt, G. A., Miller, R. L. & Nazarenko, L. S. Implications for climate sensitivity from the response to individual forcings. *Nat. Clim. Change* **6**, 386–389 (2016).
28. Shindell, T. Inhomogenous forcing and transient climate sensitivity. *Nat. Clim. Change* **4**, 274–277 (2014).
29. Hansen, J. *et al.* Efficacy of climate forcings. *J. Geophys. Res.* **110**, D18104 (2005).
30. Pithan, F. & Mauritsen, T. Arctic amplification dominated by temperature feedbacks in contemporary climate models. *Nat. Geosci.* **7**, 181–184 (2014).
31. Serreze, M. C., Barrett, A. P., Stroeve, J. C., Kindig, D. N. & Holland, M. M. The emergence of surface-based Arctic amplification. *The Cryosphere* **3**, 11–19 (2009).
32. Davy, R. & Esau, I. Global climate models' bias in surface temperature trends and variability. *Environ. Res. Lett.* **9**, 114024 (2014).
33. Levitus, S. *et al.* Anthropogenic warming of Earth s climate system. *Science* **292**, 267–270 (2001).
34. Peng, S. *et al.* Asymmetric effects of daytime and night-time warming on Northern Hemisphere vegetation. *Nature* **501**, 88–93 (2013).
35. Seidel, D. J. *et al.* Climatology of the planetary boundary layer over the continental United States and Europe. *J. Geophys. Res.* **117**, D17106 (2012).
36. Von Engeln, A. & Teixeira, J. A planetary boundary layer height climatology derived from ECMWF re-analysis data. *J. Climate* **26**, 6575–6590 (2013).
37. Holtslag, A. A. M. *et al.* Stable atmospheric boundary layers and diurnal cycles: challenges for weather and climate models. *Bull. Am. Meteor. Soc.* **94**, 1691–1706 (2013).

Acknowledgements

Support for this study was provided by the Belmont Forum project Anthropogenic Heat Islands in the Arctic—Windows to the Future of the Regional Climates, Ecosystems and Society and by the Bjerknes Centre for Climate Research project, BASIC. We thank T. Wolf and V. Miles for supporting materials. We also thank the Environmental Modeling Center (EMC), National Centers for Environmental Prediction (NCEP) for providing the CFSR data, the European Centre for Medium Range Weather Forecasting for the ERA-Interim, the NorESM and GFDL modelling groups for producing model data and the Earth System Grid Foundation for archiving that data.

Author contributions

R.D. and I.E were jointly responsible for discussion of theory and methodology. R.D. performed the data analysis and prepared the manuscript.

Additional information

Competing financial interests: The authors declare no competing financial interests.

22

Deep-reaching thermocline mixing in the equatorial pacific cold tongue

Chuanyu Liu[1,2,3], Armin Köhl[1], Zhiyu Liu[4], Fan Wang[2,3] & Detlef Stammer[1]

Vertical mixing is an important factor in determining the temperature, sharpness and depth of the equatorial Pacific thermocline, which are critical to the development of El Ninõ and Southern Oscillation (ENSO). Yet, properties, dynamical causes and large-scale impacts of vertical mixing in the thermocline are much less understood than that nearer the surface. Here, based on Argo float and the Tropical Ocean and Atmosphere (TAO) mooring measurements, we identify a large number of thermocline mixing events occurring down to the lower half of the thermocline and the lower flank of the Equatorial Undercurrent (EUC), in particular in summer to winter. The deep-reaching mixing events occur more often and much deeper during periods with tropical instability waves (TIWs) than those without and under La Niña than under El Niño conditions. We demonstrate that the mixing events are caused by lower Richardson numbers resulting from shear of both TIWs and the EUC.

[1] Institute of Oceanography, Center for Earth System Research and Sustainability (CEN), University of Hamburg (UHH), Hamburg 20146, Germany. [2] Key Lab of Ocean Circulation and Waves (KLOCAW), Institute of Oceanology, Chinese Academy of Sciences (IOCAS), Nanhai Road 7, Qingdao 266071, China. [3] Function Laboratory for Ocean and Climate Dynamics, Qingdao National Laboratory for Marine Science and Technology (QNLM), Qingdao 266237, China. [4] State Key Laboratory of Marine Environmental Science (MEL) and Department of Physical Oceanography, College of Ocean and Earth Sciences, Xiamen University, Xiamen 361102, China. Correspondence and requests for materials should be addressed to C.L. (email: chuanyu.liu@qdio.ac.cn).

The maintenance of the equatorial Pacific thermocline relies either on high-latitude buoyancy forcing or on extra-tropical wind and buoyancy forcing[1,2] at annual to inter-annual time scales, but is modulated by local Kelvin waves[3] and wind stress curl[4] at intra-seasonal to seasonal time scales. Numerical experiments suggest that the sharpness and depth of the thermocline is also determined by vertical mixing within it[5,6]. Measurements and model studies suggests that turbulence and mixing below the mixed layer base of the equatorial Pacific are attributed to the vertical velocity gradient (shear) between the eastward flowing Equatorial Undercurrent (EUC) and the westward flowing South Equatorial Current[7-15], which is likely to be further modulated by the wind stress[16-18]. Mixing or instabilities in layers further below, ranging from the upper[13] to the lower[14,19] parts of the thermocline, are also observed from limited measurements. The instabilities in the lower part of the thermocline may be caused by absorption and saturation of wave energy at critical levels[19], whereas the mixing in the upper part of the thermocline is found to be related to baroclinic inertial-gravity waves[20], Kelvin waves[14] and, in particular, the tropical instability waves (TIWs)[13].

TIWs refer to energetic meanders frequently emerging in the middle and eastern equatorial ocean. They have long been proposed to be a combination of a Yanai(-like) wave on the Equator and a first-meridional-mode Rossby wave just north of the Equator, with periods of 12–40 days and wavelengths of 700–1,600 km (refs 21–26). Alternatively, TIWs are also suggested to be manifestations of tropical vortices or highly nonlinear waves[27,28].

A prominent feature of TIWs is the large meridional velocity ranging from the surface to the core of the EUC, providing the potential for vigorous interactions with the already energetic equatorial current system. Turbulence measurements taken by a Lagrangian float encountering a TIW[29] and modelling studies of the impact of TIWs[30] in the eastern/middle equatorial Pacific both found strong vertical mixing at the base of the surface mixed layer, which induces intensive cooling of the sea surface[30,31]. The measurements[29] suggest that the strong mixing can be explained by the enhancement of shear modulated by the TIW[29]. Direct turbulence measurements at 0°, 140° W encountering a TIW further confirmed the enhancement of mixing by TIW both in and below the surface mixed layer; in particular, the measurements also revealed a tenfold increase in turbulent heat flux in the upper half of the thermocline[13]. The resulting mixing was accompanied with a significant temperature change in the upper 150 m within a cycle of the TIW[32]. The vigorous deep-reaching mixing are also attributed to additional shear provided by the meridional velocity of the TIW above the EUC core[13].

If this identified relationship between TIW and enhanced deep thermocline mixing is largely representative, it implies that after a long duration of TIWs the associated thermocline mixing may have the potential to alter the structure of the thermocline and the subsurface temperature of the Pacific cold tongue, which may further have an impact on the large-scale oceanic-atmospheric dynamics, such as El Niño and Southern Oscillation (ENSO)[6] and the global climate at large[33].

However, observational evidence for the link between the TIWs and enhanced thermocline mixing is far from adequate. To date, direct turbulence measurements were confined to a few specific locations and covered only short time spans. Whether the thermocline mixing is organized in seasonal or longer-period cycles that are mechanistically related to variations of TIWs at the same periods and to what depths the TIWs may have an impact on the vertical mixing need to be explored.

Two databases could be employed to investigate the both issues. One is the Argo float database[34]. More than 3,000 freely drifting Argo floats continuously provide millions of profiles of temperature and salinity in the upper ∼2,000 m ocean. The measurements offer a great opportunity to shed light on vertical mixing, because many profiles possess fine resolution, that is, small enough sample spacing ($O(1\,m)$), to resolve turbulent mixing processes in the ocean interior. Mapping the global distribution of vertical mixing based on the Argo observations[35] with a fine-scale parameterization method[36] has demonstrated the usefulness of Argo measurements in ocean mixing studies. However, such estimation so far has been restricted to extra-equatorial regions below the thermocline due to limitations of the employed method[36,37]. Alternatively, the Thorpe method (see Methods and ref. 38) is suitable in the thermocline and could be applied to fine resolution Argo profiles, to detect mixing events in the equatorial thermocline.

The second database is the Tropical Atmosphere and Ocean (TAO) mooring observations[39]. The TAO array has provided continuous and high-quality oceanographic data including velocity, temperature and salinity in the upper 500 m over the last two decades. The method of linear stability analysis (LSA; see Methods and refs 18,40–42) is applied to the hourly profiles of density and velocity at a location in the middle equatorial Pacific. This method enables to detect potential instabilities occurring in the thermocline.

From both databases, we obtained large amount of possible mixing events (featured as density overturns and potential instabilities). We show that the mixing events occurred not only in the upper part of the thermocline but also deep down to the centre and lower part of it. We also show that the mixing events occurred more often and much deeper during periods of TIWs and of La Niña conditions because of stronger shear instabilities.

Results

Deep-reaching density overturns in Argo float measurements. The equatorial Pacific is a region with accumulated Argo float observations. Among all observed profiles, there exist ∼20,000 fine resolution profiles that are with a sample spacing of no more than 2 m (see Methods) in the upper 200 m and covering 10° S to 10° N and 180 to 80° W over the period of January 2000 to June 2014 (Supplementary Fig. 1). The Thorpe method (see Methods) is applied to the fine resolution profiles and eventually ∼800 density overturns are identified. Among them, a large portion of the overturns occurred after 2008, a period when most of the fine-resolution profiles exist. The horizontal distributions of the detected overturns are shown in Fig. 1a. In particular, in the region between 160° and 100° W, most overturns are confined to the equatorial band, ranging from 3° S to 6° N (about 400 overturns are found between 160°–110° W and 3° S–6° N); east of 100° W, overturns extend meridionally to ± 10°. Away from these regions, fewer overturns are detected.

Overlapped by the overturns in Fig. 1a is the occurrence probability of overturns calculated in 2° × 2° bins. The occurrence probability refers to the ratio of the number of Argo profiles that contain overturn(s) to the number of total qualified Argo profiles (see Methods) in a given area. Here, the time span is January 2000–June 2014. The occurrence probability ranges from 1 to 20% and peaks at 3%. It is noteworthy that the small values may not reflect the real occurrence probability of turbulent overturns, because the real overturns may have sizes of 10 cm to several metres, whereas here only the overturns larger than the sample spacing of the Argo profiles (2 m) were detected. Despite the scale selection, the results are still indicative for inferring the relationship between mixing events and TIWs. For example, a prominent feature of the horizontal distribution of the overturns is that they are concentrated in a band across the Equator but display a meridional asymmetry: 3° S–6° N.

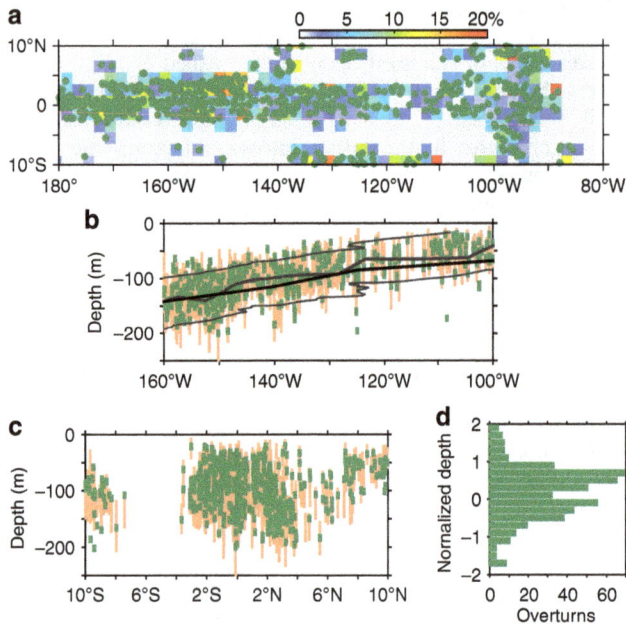

Figure 1 | Spatial distribution of detected density overturns in the equatorial Pacific cold tongue. (**a**) Occurrence probability (colour) and horizontal locations of overturns (dark green dots). (**b**) Depth and sizes (in metres) of the overturns (dark green bars) occurred between 3° S and 6° N, and the corresponding PLs (light orange bars) from a latitudinal view. The blue curve denotes the mean depth of the EUC core (averaged over ± 1°; data are obtained over the 1990s (ref. 70)). The thick black curve and thin black curves denote the centre (depth of maximum N^2; N^2_{max}) and bounds (depths of half N^2_{max}) of the mean pycnocline. N^2 is calculated from sample mean density that is meridionally averaged from all fine-resolution Argo profiles over ± 1°. (**c**) The same as in **b** but assembled from data between 160 and 100° W (curves of average variables are not added due to large zonal variation). (**d**) Histogram of overturns at the referenced and normalized vertical coordinate.

This band is within the region of the South Equatorial Current and EUC, and matches well the region of the TIWs, which are concentrated at the Equator and have two centres of high temperature variability at 5° N and 2° S (ref. 21).

We further graph the overturns with their corresponding instantaneous pycnocline layers (PLs) against both longitudes (Fig. 1b) and latitudes (Fig. 1c). Here, the centre of a PL is defined as the depth of the maximum N^2 (hereafter N^2_{max}, $N^2 = -g\rho_z/\rho_0$ is the buoyancy frequency squared, $\rho = \rho(z)$ is monotonically sorted potential density and is smoothed by a 40-m running mean, to remove influences of noises or intermittent internal waves, $\rho_0 = 1{,}000\,kg\,m^{-3}$ is the reference potential density and g is the gravitational acceleration). The upper and lower bounds of a PL are defined as the depths where $N^2 = 0.5 \times N^2_{max}$ above and below the PL centre (but is additionally bounded by the depth of $N^2 = 0.625 \times 10^{-4}\,s^{-2}$). The size of the most detected overturns is 6 m (Supplementary Fig. 2), whereas the thicknesses of instantaneous PL vary from 50 to 100 m and are generally larger in the western than in the eastern equatorial Pacific (Fig. 1b). The synoptic overturns are confined to the temporally averaged PL (Fig. 1b; thin black curves) of the Equator. It is noticeable that a large fraction occurred below the centre of the average pycnocline (Fig. 1b; the thick black curve); they reached as deep as ~ -200 and $\sim -100\,m$ in the western and eastern equatorial Pacific, respectively.

We emphasize that the deep-reaching overturns in the pycnocline in the meantime also reached to the lower flank of the EUC. This can be inferred from Fig. 1b, where the average

core of the EUC (the depth of maximum eastward velocity; Fig. 1b; the blue curve) is more or less coincident with the centre of the pycnocline. This result is confirmed in the following by LSA examinations.

Figure 1c shows the meridional distribution of the detected overturns. It confirms the feature that the equatorial overturns are confined between 3° S and 6° N, the regime of the TIWs. Overturns outside this region are mainly found in the region east of 100° W.

The overturns and PLs shown on the physical depths (Fig. 1b,c) are subject to spatial and temporal variations. To provide an overview of the vertical distribution of the overturns relative to their corresponding pycnocline, we redistribute the overturns with respect to a transformed and normalized vertical coordinate (Fig. 1d). This coordinate is referred to the depth of N^2_{max}, and normalized in the upper (lower) half of the pycnocline by the thicknesses of the upper (lower) half of each PL. As such, in this coordinate, 0 represents the PL centre, while 1 and -1 represent the upper and lower bounds of the PL, respectively.

It shows that overturns occur not only in the upper part of the PL but also below the centre of the PL. Most overturns occur in the upper three quarters of the PL (between -0.5 and 1). Although in Fig. 1d the overturns peak at ~ 0.7, that is, near the upper bound of the pycnocline, it may not mean that the overturns in the ocean really peak here. This is because the prescribed cutoff buoyancy frequency (minimum of $N^2 = 0.5 \times N^2_{max}$ and $N^2 = 0.625 \times 10^{-4}\,s^{-2}$) in the Thorpe method may have omitted overturns in weak-stratification layers, including the mixed layer and the layer just below. Nevertheless, the overturns peaking at ~ 0.7 needs an interpretation. Taking the location 0°, 140° W for reference, the centre and upper bound of the temporally averaged pycnocline are at -100 and $-60\,m$, respectively (Fig. 1b); in consequence, depth 0.7 of the normalized coordinate corresponds to 28 m above the pycnocline centre, that is, at the physical depth of $-72\,m$. According to direct turbulence measurements[13,32], this depth mostly belongs to the upper core layer, which refers to a layer that is located above the EUC core and accompanied with strong TIW-induced turbulence. In the depths above 0.7, the overturns may come from the deep cycle layer[7,8,10,11,14], which refers to a layer several tens of metres below the base of the surface mixed layer that undergoes a nighttime enhancement of turbulence; this layer is dynamically related to the diurnal varying surface buoyancy and wind forcing. It is noteworthy that the TIW-related upper core layer is seemingly separated from the surface-driven deep cycle layer[32]. Between the depths -0.5 and 0.7, more than a half overturns as those at depth 0.7 are found, indicating that intensive turbulence extends into the deep pycnocline.

Relationship between the deep-reaching overturns and TIWs. In general, TIWs are active from boreal summer to winter, while inactive in boreal spring[26]. To investigate whether the occurrence of overturns is also organized in such a seasonal cycle, the monthly occurrence probability of overturns in the region of active TIWs, 3° S and 6° N, and 160° W and 100° W, is calculated over the period between January 2005 and December 2013 (Fig. 2b). It is shown that the occurrence probability is indeed subject to similar seasonal variation: they peak in August and December, and have minimum values in boreal spring (April to June) and October. This seasonality is statistically significant. The maximum in August is twice the minima in October and April; the secondary maximum in December is >50% larger than the minima. From direct turbulence measurements over a 6-year span at 0°, 140° W, the vertical heat flux in the subsurface layers (-60 to $-20\,m$) is found to be largest in boreal August, second largest in December, and least in spring and second least in

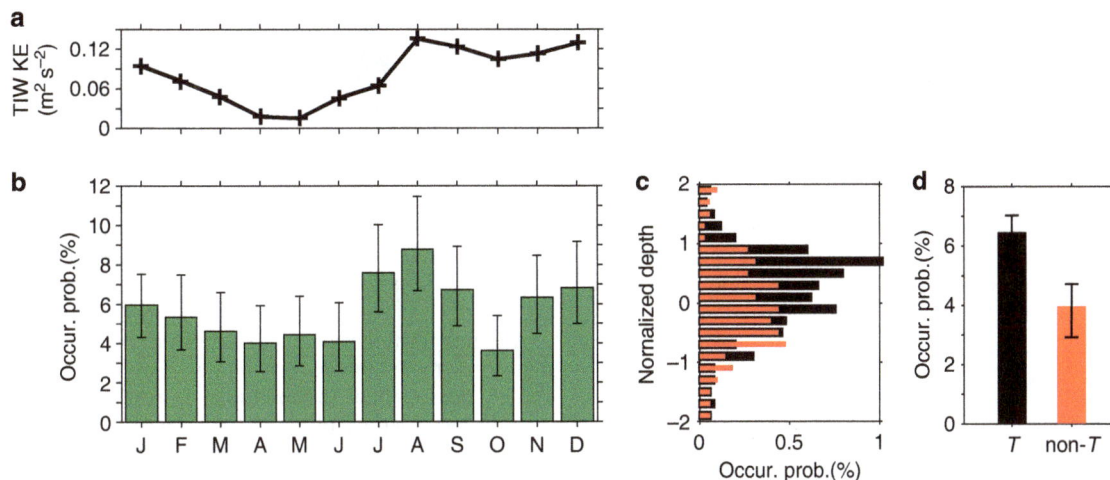

Figure 2 | Occurrence probability of detected overturns and its relation to tropical instability waves. (**a**) Monthly climatology of TIWs KE (averaged over years 2005–2011). (**b**) Monthly occurrence probability of overturns between 3° S and 6° N over 160 and 270° W; error bars are 95% bootstrap CIs. Peak at August is significantly different from surrounding troughs in June and October at the 95% bootstrap confidence level; the peak in December is significantly different from the trough at October at the 95% bootstrap confidence level and from troughs in April and June at the 90% bootstrap confidence level. The correlation coefficient between monthly TIWs KE in **a** and occurrence probability in **b** is $r = 0.66$, with P-value $= 0.020$ and 95% CI $= (0.14, 0.89)$, that is, statistically significant. (**c**) Histogram of the occurrence probability in the normalized coordinate for periods of TIW (blue) and non-TIW (red) (see text). (**d**) Total occurrence probability of overturns during TIW (6.34%, blue) and non-TIW (3.76%, red) periods; error bars are 95% bootstrap CIs (blue: (5.67%, 7.07%) and red: (2.92%, 4.71%)). In **b,c** and **d**, the occurrence probability is calculated over the years 2005–2013.

October and November[43], consistent with the occurrence probability of overturns shown here. Specifically, the monthly occurrence probability is significantly correlated with the multi-year (2005–2010) averaged monthly TIW kinetic energy (KE) at 0°, 140° W (Fig. 2a). (The TIW KE is calculated as $(\langle \widehat{u} \rangle^2_{30-70} + \langle \widehat{v} \rangle^2_{30-70})$, where u and v are the eastward and northward components, respectively, of velocity observed by TAO moorings, $\widehat{u}$ and $\widehat{v}$ are the 12–40 days band-pass filtered, $\langle \rangle_{30-70}$ denotes the vertical mean over -70 to ~ -30 m and $\overline{(\ )}$ denotes a 40-day low-pass filtering.)

The results strongly indicate the modulating effect of TIWs on the occurrence of deep-reaching overturning in the pycnocline. Given that overturning and mixing usually accompany with each other, the result not only confirms the notion that TIWs lead to enhanced turbulence and mixing in the upper part of the thermocline[13] but also implies that the modulation effects of TIWs on mixing can reach deeper depths of the pycnocline. (As density here is dominated by temperature[32], in the remainder of the study we focus on the thermocline instead of the pycnocline.)

In the following, we will further verify the impact of TIWs in enhancing the occurrence of overturns by comparing overturn properties of TIW periods with those of non-TIW periods. To this end, we followed ref. 44 and defined TIW periods and non-TIW periods based on meridional sea surface temperature (SST) gradient. The reason why we adopted this strategy is because a large portion of the fine resolution Argo profiles and overturns are found between the years 2011 and 2014, whereas the TAO velocity measurement and hence the TIW KE index are not available since 2011. SST[45] (see Methods for data source) is first averaged in longitude spanning 12° centred at two latitudes of 140° W, 4.5° N and 0.5°N, and then the averaged SSTs are 140-day low-pass filtered; finally, the meridional gradient of the filtered SSTs (SST_y) is calculated. TIW periods are defined as the periods when the SST_y is $> 0.25 \times 10^{-2}$ °C km^{-1}; other periods are defined as non-TIW periods (Fig. 3). This proxy matches the TIW KE index well (Fig. 3).

Overall, about two-thirds of the total time periods belong to TIW periods and one-third of them belong to non-TIW periods (Fig. 3). The numbers of overturns within 160°–110° W and 3° S–6° N over the years 2008–2013 are 314 and 67, whereas the numbers of fine-resolution Argo profiles are 4,952 and 1,783, in TIW and non-TIW periods, respectively. This leads to the occurrence probability of 6.34% for TIW periods (the 95% bootstrap confidence interval (95% CI) = (5.67%, 7.07%)) and of 3.76% for non-TIW periods (the 95% bootstrap CI = (2.92%, 4.71%)). The former is 69% larger than the latter (Fig. 2d); in addition, the occurrence probability for TIW periods is larger at almost every depth than non-TIW periods within the upper and centre of the thermocline (depths -0.5 to ~ 1; Fig. 2c). The results demonstrate again that TIWs are associated with a higher occurrence of overturns.

Link the overturns with TIWs via shear instability. The observed higher occurrence of deep-reaching overturns during TIWs, so far established in the seasonal and period-to-period cycles, calls for a physical interpretation. The overturns are indicative of breaking of internal waves and/or turbulence generated by shear instability. Two ways may be employed to demonstrate this physical interpretation. One is the LSA, which can determine the potential instabilities of an observed flow by providing locations and other detailed properties of the exponentially growing unstable modes (see Methods and refs 18,40–42). The LSA is applied to $\sim 8 \times 10^4$ hourly TAO profiles of years 2000–2010 at 0°, 140° W (this site locates meridionally at the centre of Pacific TIWs and thus is representative for TIW studies[21]).

The monthly counts of the potential instabilities (in terms of the critical levels of the detected unstable modes; see Methods) is shown on physical depth in Fig. 4a and on referenced depths in Fig. 4b,c. In Fig. 4b, the depth is referenced to the thermocline centre of each profile, which is defined as the depth of maximum vertical temperature gradient (before calculation, temperature is 40-m running smoothed, to remove effects of noises and intermittent waves). In Fig. 4c, the depth is referenced to the

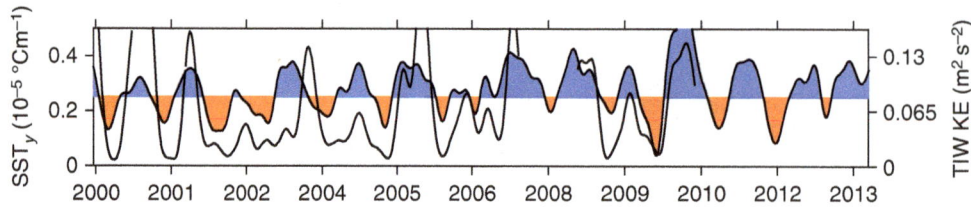

Figure 3 | Separating TIW and non-TIW periods with meridional SST gradient. The blue curve denotes the meridional SST gradient (SST_y). The TIW (non-TIW) periods are depicted with blue (red) shading. The black curve denotes the 140-day low-pass-filtered TIW KE. The correlation coefficient between the filtered TIW KE and the SST_y over 2000–2010 is $r = 0.47$, with $P < 0.001$ and 95% CI = (0.44, 0.49).

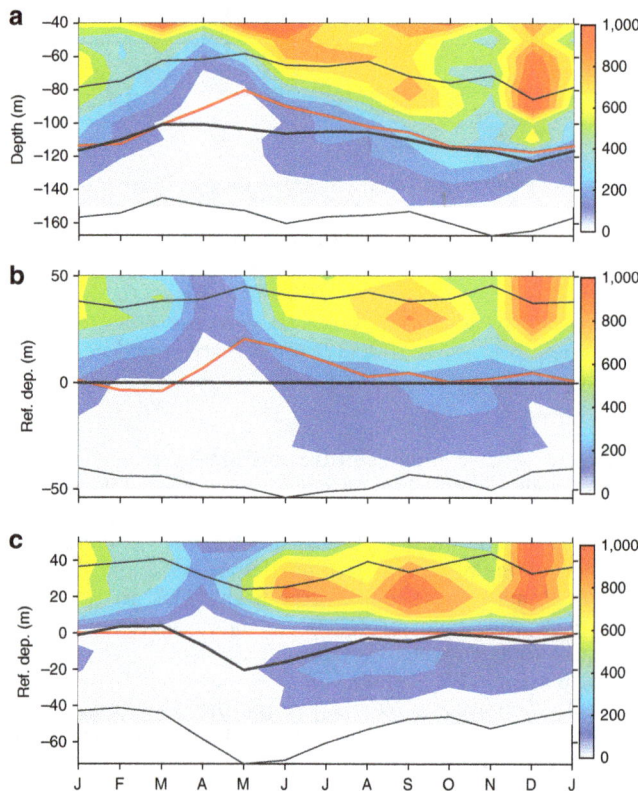

Figure 4 | Monthly climatology of count of critical levels at 140° W of the Equator. (**a**) On physical depths. (**b**) On depths that is referenced to hourly centres of the thermocline (defined as the depth of maximum vertical gradient of 40-m running-averaged temperature). (**c**) On depth that is referenced to hourly centres of the EUC (defined as the depth of maximum eastward velocity). Shown are counts in 10-m bins. In each panel, the red curve denotes the average depth of EUC core, the black thick curve denotes the average depth of the thermocline centre and the two black thin curves denote the average upper and lower bounds of the thermocline.

EUC core of each profile, which is defined as the depth of maximum eastward velocity. In addition, the long-term averaged monthly depths of the EUC core, the thermocline centre and the upper and lower thermocline bounds (defined as the depths of half the maximum vertical temperature gradient), are overlaid on the three panels of Fig. 4.

A prominent feature is the seasonal variation of the potential instabilities (Fig. 4a). Within the thermocline, more potential instabilities were found in boreal summer to winter, while relatively few potential instabilities were found in boreal spring. There were also fewer potential instabilities occurring during October and November. This feature is consistent with the seasonal variations of both TIW KE and occurrence probability of Argo-determined overturns (Fig. 2a,b).

Another distinguished characteristic is the deep-reaching nature of the potential instabilities. From boreal summer to winter, the potential instabilities may reach down to -120 m, with deepest depths of -150 m in fall. Potential instabilities occurring below the upper thermocline bound are as many as those above it. As the depth of the upper thermocline bound varies from -80 to -60 m and roughly coincides with the top of the observed upper core layer[13], it demonstrates that the upper core layer mixing is a remarkably persistent phenomenon in the study site. Moreover, $\sim 15\%$ of all the determined instabilities locate well below the average centres of both the EUC and the thermocline, which coincide with each other in boreal June to March (Fig. 4a). By contrast, in spring, instabilities can only occur within the upper 75 m, which is ~ 25 m above the thermocline centre.

As the numbers of hourly profiles in each month and at each depth are nearly the same, the occurrence probability of the instabilities (Fig. 5a) displays a similar pattern as the counts of potential instabilities shown in Fig. 4a.

When all the potential instabilities are redistributed on the depth that is referenced to the thermocline centre of each profile (Fig. 4b), a centre of potential instabilities emerges ~ 26 m above the thermocline centre (the averaged thickness of the upper flank of the thermocline is ~ 40 m), in particular in summer to fall. The centre coincides well with the peak at 0.7 on the normalized coordinate of the Argo-detected overturns (Fig. 1d). In addition, potential instabilities occur also in and below the centres of the thermocline. Alternatively, when the potential instabilities are redistributed on the vertical coordinate that is referenced to the EUC core of each profile (Fig. 4c), a striking feature is clearly observed: in addition to those in the upper core layer, an isolated region of potential instabilities stand out in the lower flank of the EUC. These potential instabilities occur only in summer to winter and accounts for $\sim 20\%$ of those in the upper core layer. Instabilities peaked ~ 20 m above and below the EUC core, but no instabilities were found in the EUC core.

The relation of the deep-reaching potential instabilities to TIWs is further illustrated from a period-to-period point of view in Fig. 5b,c. In these two panels, we show the occurrence probability of potential instabilities in the TIWs and non-TIWs periods, respectively, on the depth referenced to the EUC core. Here, the TIW periods are defined as periods when the 140-day low-passed TIW KE (Fig. 3) is larger than $4 \times 10^{-2}\,\mathrm{m^2\,s^{-1}}$ (corresponding to a characteristic horizontal velocity of $20\,\mathrm{cm\,s^{-1}}$); the other periods are defined as non-TIW periods. These newly defined periods of TIWs and non-TIWs are consistent with those defined based on the SST gradient (Fig. 3).

The occurrence probability is ~ 50 to $\sim 100\%$ larger, almost at every depth during TIW periods than non-TIW periods in the upper core layer (except in February). In particular, the instabilities of the lower flank of the EUC can only occur with the existence of TIWs. Consequently, the results clearly demonstrate the enhancement effect of TIWs on the occurrence of potential instabilities in both the upper and lower flanks of the EUC.

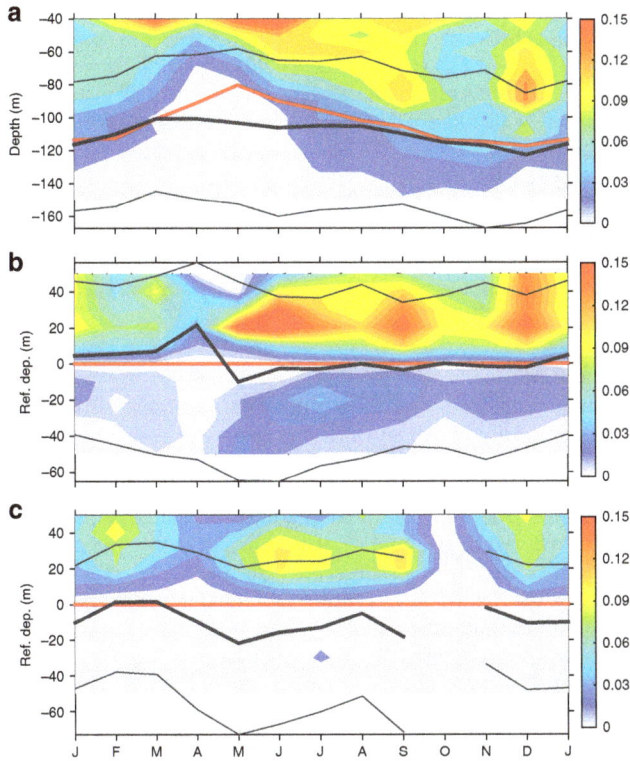

Figure 5 | Monthly climatology of occurrence probability of critical levels at 140° W of the Equator. (**a**) On physical depths. (**b**) For periods of TIWs but on the depth that is referenced to instantons EUC cores (see caption of Fig. 4). (**c**) The same as in **b**, but for periods of non-TIWs. TIW (non-TIW) periods are defined when the TIW KE is larger (less) than $0.04 \, m^2 \, s^{-2}$. The occurrence probability is defined as the ratio of the number of unstable modes over the number of profiles in 10-m bins. The red curve denotes the average depth of EUC core, the black thick curve denotes the average depth of the thermocline centre and the black thin curves denote the average depths of the thermocline bounds.

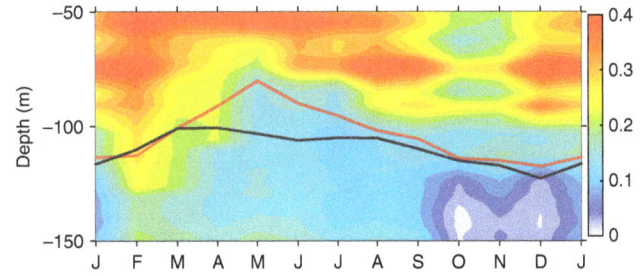

Figure 6 | Monthly occurrence frequency of low Richardson number. The occurrence frequency is calculated as the ratio of numbers of $Ri \leq 0.35$ over numbers of all Ri in 10-m bins. The red curve denotes the average depth of EUC core and the black curve denotes the average depth of the thermocline centre.

Link the instabilities to low Richardson numbers. The other way to link the shear instability to the deep-reaching feature of the detected overturns in the Argo profiles (as well as the TAO-determined potential instabilities) is to examine the Richardson number, Ri ($= N^2/S^2$, where $S^2 = |\partial u/\partial z|^2 + |\partial v/\partial z|^2$ is the shear squared). The shear instability (in particular of the Kelvin–Helmholtz type) is dynamically related to the local Richardson number, a critical value of which Ri_c is ~ 0.25. $Ri = Ri_c$ is an equilibrium state for turbulence in a stratified shear flow. When $Ri < Ri_c$, turbulence may be initiated or continue to grow due to shear instabilities. When $Ri > Ri_c$, the flow is dynamically stable and any turbulence will decay[46,47]; however, when Ri is close to Ri_c, the flow may lie in the regime subject to marginal instability[48]. For example, turbulence can persist up to a Ri value typically near 1/3 (ref. 49). The marginal instability is well identified in the upper layer (upper ~ 75 m) of TIW periods at 0°, 140° W[50]. Accordingly, based on the hourly TAO measurements over the years 2000–2010, the occurrence frequency of $Ri \leq 0.35$ was computed to represent the possibility of instabilities (Fig. 6).

The high occurrence frequency of $Ri \leq 0.35$ is roughly associated with the high-occurrence probability of potential instabilities as shown in Fig. 5a, although they match well only in their main structure, rather than in details. The pattern of higher occurrence frequency (say ≥ 0.25) includes a deep extension to ~ -100 m in winter and summer months, and a subsurface centre (at -50 m) from February to September. The lower bound

of the higher-occurrence frequency is generally confined to the centres of the thermocline and the EUC. This feature is consistent with the occurrence of the potential instabilities.

The inconsistence in detailed structures between the occurrence frequency of $Ri \leq 0.35$ and the occurrence probability of potential instabilities is explainable. In particular, from February to June, relatively high occurrence frequency of $Ri \leq 0.35$ is found below -100 m, where fewer potential instabilities were determined here (Fig. 5a). This may be because the shear in the depths is weak (Fig. 7). Hence, although a large portion of Ri is small (resulted from weak stratification), there was not enough KE available in the mean flow to drive unstable modes that have high-enough growth rate[18] that could pass the growth rate criterion used in the LSA (see Methods).

As mentioned, the annual cycle of the occurrence frequency of low Ri should have resulted from not only the shear of EUC and the TIWs, but also the thermal structure of the upper ocean. All the processes and properties are ultimately also related to wind stresses and exhibit seasonal variations. For example, in boreal spring the wind reduces, the shear weakens and the water warms with increased stratification in the subsurface layers; since late summer, the wind stress increases, the shear strengthens and the surface water cools down with decreased stratification.

Nevertheless, the contribution of TIWs to the low Ri and therefore the generation of potential instabilities could be roughly isolated from the EUC. This was done by separating the individual shear they induce. The shear squared induced by the background EUC is calculated as $S_0^2 = |\partial \bar{u}/\partial z|^2 + |\partial \bar{v}/\partial z|^2$, where $\bar{u}, \bar{v}$ are the 40-day low-pass-filtered velocities, representing the background flows, whereas the shear squared associated with the TIWs is estimated as the difference between the original and the background shear squared: $S_{tiw}^2 = S^2 - S_0^2 = (|\partial u/\partial z|^2 + |\partial v/\partial z|^2)$-$(|\partial \bar{u}/\partial z|^2 + |\partial \bar{v}/\partial z|^2)$ (Fig. 7a,b). In general, the EUC is associated with stronger shear squared, which centres ~ 20–50 m above the seasonally varying EUC core (Fig. 7a), whereas the TIWs are associated with weaker shear (Fig. 7b).

However, the magnitude of TIW-induced shear squared could reach half of that induced by the EUC in a thick layer. In addition, as a prominent feature, the TIW-induced shear is centred just above the EUC core and covers both the upper core layer and the layers immediately below the EUC core, in particular during TIW seasons (boreal summer to winter). The TIW shear covering the EUC core adds to the EUC-induced shear and provides the conditions favourable for instability; besides, the strong velocity of TIWs provides necessary KE for the instability to grow fast. This explains the occurrence of potential instabilities occurring below the centres of both the EUC and thermocline (Fig. 4).

The portion of the TIW-induced shear is calculated as S_{tiw}^2/S^2 (Fig. 7c). The TIW-induced shear accounts for $30 \sim 50\%$ for most

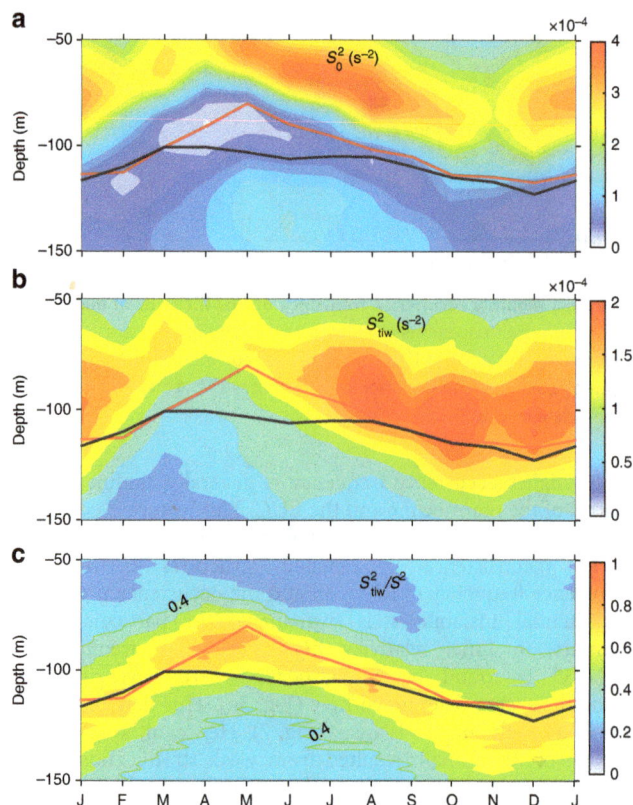

Figure 7 | Shear of background flows and TIWs. (a) Shear squared induced by the background flow, S_0^2. (b) Shear squared associated with TIWs, S_{tiw}^2. (c) The proportion of the shear squared associated with TIWs, S_{tiw}^2/S^2. In **a,b** and **c**, the red curve denotes the average depth of the EUC core and the black curve denotes the average depth of the thermocline centre. In **c**, contour of 0.4 is highlighted for reference. The different colour scales in **a** and **b** are noteworthy.

of the upper layer. This percentage is consistent with the direct measurements that shows ~30% larger of shear induced by TIW[13]. In particular, it accounts for 60~80% just above and below the EUC core. These results indicate that the TIWs provide a modulating effect on the generation of unstable disturbances.

TIWs and instabilities at ENSO timescales. In Fig. 8a,d we show the monthly TIW KE and the occurrence probability of unstable modes within the thermocline (-50 to ~ -150 m) for the years 2000–2010. In years of stronger TIWs, larger occurrence probability of unstable modes are observed. The high correlation between them (correlation coefficient $r = 0.71$, P-value < 0.001, 95% CI = (0.62, 0.79)) further demonstrates that TIWs are associated with higher thermocline instability occurrence also at the inter-annual timescale.

Previous studies, based on modelling results and a TIW proxy in terms of the SST variance, found that the activity of TIWs is larger under La Niña conditions and smaller under El Niño conditions, because the former are associated with stronger latitudinal gradient of SST immediately north of the Equator and thus more occurrence of baroclinic instability[51]. Using the monthly TIW KE and the Oceanic Niño Index (ONI) calculated from the monthly Optimum Interpolation Sea Surface Temperature (ref. 52) (Fig. 8a,b), we confirmed such a significantly negative correlation (correlation coefficient $r = -0.69$, with P-value < 0.001 and 95% CI = (-0.77, -0.59)).

The implication of the relation is that the inter-annual variation of occurrence probability of instabilities could also be related to El Niño and La Niña conditions. The correlation coefficient between the ONI and the occurrence probability (Fig. 8d) is -0.58, with P-value < 0.001 and 95% CI = (-0.68, -0.45). It implies that there were more potential instabilities, associated with more TIWs, under La Niña than under El Niño conditions.

Moreover, the extension range of potential instabilities differs between two conditions (Fig. 8c). Under El Niño conditions, the potential instabilities are mainly confined to the upper flank of the thermocline, except for stronger TIWs. By contrast, under La Niña conditions, the potential instabilities mostly can reach to the lower flank of the thermocline.

Discussion

In the present study, we show the existence of overturns in the deep depths of the thermocline. We also show that the potential instabilities are organized in a physically quite reasonable structure. Given the good coincidence of the determined potential instabilities and the measured mixing during November 2008 (Supplementary Figs 3 and 4), it is anticipated that the potential instabilities during other time are also associated with mixing, although the mixing intensity and accompanying heat fluxes can not be correctly estimated yet.

In the cold tongue of the equatorial Pacific, maintaining cool SSTs in the presence of intense solar heating requires a combination of subsurface mixing and vertical advection to transport surface heat downward[43,53–56]. Analyses of direct turbulence measurements have demonstrated that the subsurface mixing (over -60 to ~ -20 m) reduces SST during a particular season—boreal summer[43]. If mixing is indeed associated with the detected overturns and potential instabilities in the deep depths, it could also blend water between the upper part and middle/lower part of the thermocline, resulting in cooling of the upper thermocline, and further cooling of the surface.

The TIW-related mixing during La Niña conditions may have rich implications for ENSO dynamics. It has been found that incorporating TIWs in the ocean–atmosphere coupled models results in a significant asymmetric negative feedback to ENSO[57–59] (anomalously heating the Equator under La Niña conditions and cooling it under El Niño conditions via horizontal advection). Accordingly, the asymmetric negative feedback is argued to explain the observed asymmetric feature of a stronger-amplitude El Niño and weaker-amplitude La Niña relative to the models. However, the cooling effect via vertical mixing associated with TIWs, in particular during La Niña conditions, was missed or under-represented by the numerical models due to underestimation by the parameterizations[60]. Therefore, the effects of TIWs on ENSO development requires to be re-examined.

To best simulate the oceans, numerical models need to reproduce or properly represent the TIWs and the associated turbulence. Although the main structure of TIWs can be reproduced in some coarse resolution ocean general circulation models (OGCMs)[30], the small-scale structures of the frontal areas of TIWs, which are key regions of turbulence generation[28,29,61,62], remain unresolved by coarse resolution OGCMs and ocean-atmospheric coupled models. This shortage may lead to underestimates of thermocline mixing by the oversimplified vertical mixing parameterizations incorporated in coarse resolution OGCMs[60] and hence to model-data deviations not only in the equatorial ocean but also in mid-latitudes[63,64].

Methods

Data processing. The Argo data (see below) covers the period from 2000 till June 2014 and only profiles with vertical sample resolution of at least 2 m and with maximal sampling depth deeper than -200 m are used to detect overturns. Both

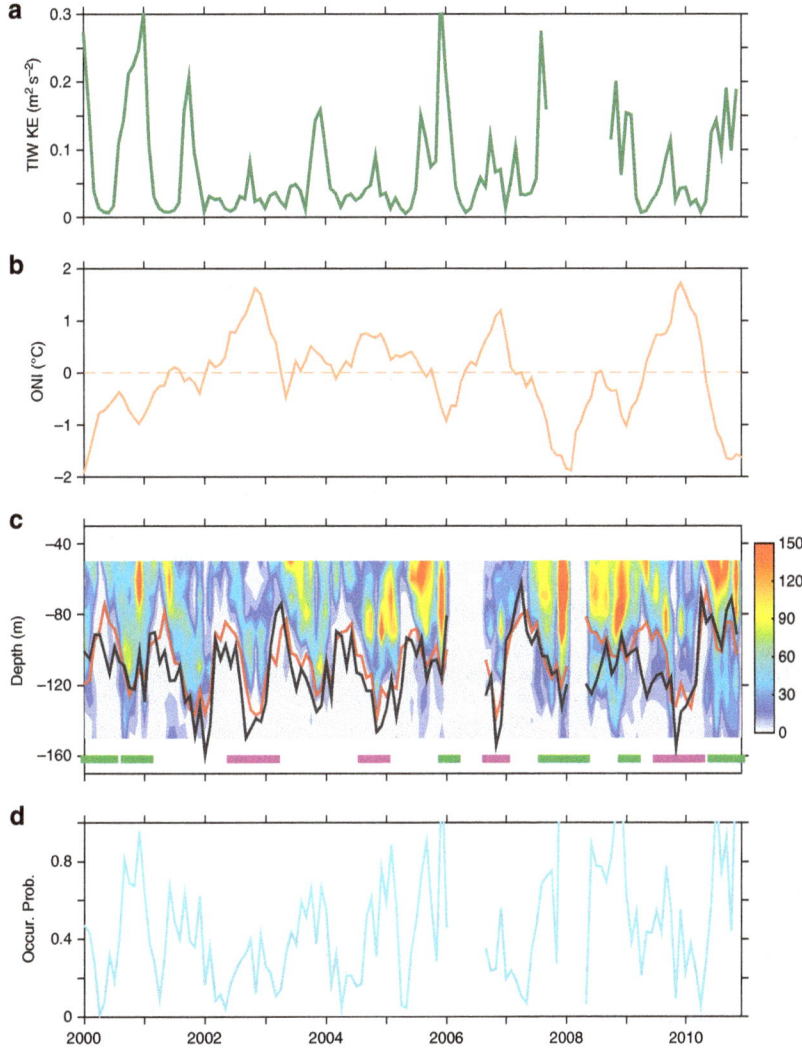

Figure 8 | Relation among instabilities, TIWs and large-scale processes on the inter-annual time scale. (**a**) Monthly TIW KE. (**b**) ONI, showing the Niño 3.4 (5° S–5° N, 170–120° W) SST anomaly (1981–2010 mean removed) calculated from v2 of the Optimum Interpolation Sea Surface Temperature (OISST). The dashed line denotes zero. (**c**) Monthly count of unstable modes in 10-m bins. The green and magenta bars on the bottom denote periods of the La Niña and El Niño conditions, defined when the ONI is $\leq -0.5\,°C$ and $\geq 0.5\,°C$, respectively. The red curve denotes the average depth of EUC core and the black curve denotes the average depth of the thermocline centre. (**d**) Monthly occurrence probability of unstable modes, defined as the ratio of counts of critical levels occurring between -50 and -150 m of a month over the number of profiles of the given month.

hourly temperature and velocity from TAO mooring at 0°, 140° W is interpolated (extrapolated if needed) at 1 m spacing grids for both Ri calculation and the LSA (see below). As salinity observations of TAO are not sufficient and their contribution to density in this region is minor, salinity needed for real-time density calculation was often replaced by its temporal average[50]; here we use the salinity climatology averaged from 251 Argo and CTD (CTD is obtained from the same source of Argo) profiles falling into the range of 0° ± 0.5°, 140° W ± 0.5°.

The Thorpe method. The Thorpe method[38] is commonly used for estimating dissipation rate and vertical turbulence diffusivities, which is based on the size of detected density overturn patch and the stratification intensity over the patch, of a measured potential density profile. In this work the Thorpe method is applied only for overturn detection, rather than diffusivity estimation (because the resolution of the Argo profiles is still too low for such estimation). The Thorpe method is suitable to be applied in the thermocline where the stratification is strong; in contrast, care should be taken in layers of low stratification due to its sensitivity to noise[65]; therefore, both the upper and lower layers of low stratifications defined by $N^2 < 0.625 \times 10^{-4}\,s^2$ are omitted from our analysis. The size of any overturn should not be smaller than three times the profile resolution, a minimum criteria for the overturn size[66]. Profiles with large spikes from any of the properties (temperature, salinity and pressure) were removed from the data set. With the above criteria, few unreasonably large overturns are still found (see Supplementary Fig. 2). Therefore, each detected overturn and the corresponding density profile were carefully

examined afterwards, to guarantee that it is physically sensible. In particular, the detected overturns that have sizes > 30 m were removed.

Linear stability analysis. The LSA is designed to detect instabilities potentially occurring in an observed flow. The stability of an inviscid, incompressible, stratified, unidirectional shear flow to small disturbances is determined by the solutions of the Taylor–Goldstein equation:

$$\mathrm{d}^2\varphi/\mathrm{d}z^2 + \{N^2/(U-c)^2 - k^2 - \mathrm{d}^2U/\mathrm{d}z^2/(U-c)\}\varphi = 0, 0 \leq z \leq h \quad (1)$$

where $U(z)$ and $N(z)$ are the profiles of z-dependent mean velocities and buoyancy frequency, respectively, and $\varphi(z) = \varphi_0(z)\exp[ik(l-ct)]$ is the z-dependent stream function of a disturbance with real horizontal wavenumber k and complex phase speed $c = c_r + ic_i$. Here, $\varphi_0(z)$ is the amplitude of the stream function, t is time and l is along the direction of the perturbation wave vector. For non-parallel flows, the stability can be examined by taking U as the velocity component in the direction (α) of the disturbance wave vector: $U = u\cos(\alpha) + v\sin(\alpha)$, where (u, v) is the measured eastward and northward components of the velocity vector.

The Taylor–Goldstein equation (equation (1)) is a linear eigenvalue problem that can be solved numerically using matrix method[40–42,67] subject to prescribed boundary conditions. In this study, zero condition is applied at both surface ($z = 0$) and the lower boundary $z = -200$ m. Unstable modes should at least have the property $kc_i > 0$, to ensure that the disturbance grows exponentially. Useful quantities derived through solving the problem include the following: the wave vector direction of the instability perturbation wave, α; the perturbation wave length, $2\pi/k$; the horizontal phase speed of the perturbation wave, c_r; the growth rate of the perturbation wave, kc_i and the critical level of it, z_c, which is defined as the depth where $U(z_c) - c_r = 0$; the critical level is also considered as the position of the unstable mode.

Before applying the LSA, the data are carefully processed. The hourly TAO velocities are interpolated to a 1-m grid using cubic splines from surface to -200 m. Specifically, in the upper layers where velocity data are not available (the upper -25 m or -37.5 m, depending on data quality), the profiles of u and v are extrapolated using a polynomial fit, which should meet the requirement that the first derivative is continuous between -200 m and the surface, and approaches zero at the top two grids (-1 m and 0).

The hourly temperature of TAO observations is inter/extrapolated into the same 1-m spacing grids. As the temperature sample spacing is sparse (see Supplementary Fig. 3), a special strategy is applied. First, the raw data are interpolated into 5-m (above -50 m), 10-m (between -50 and -150 m) and 50-m (between -150 and -500 m) grids using a linear method. Second, the inter/extrapolated data are further inter/extrapolated into the 1-m resolution grids using a cubic spline method (other grid spacing and inter/extrapolated methods are tested; Supplementary Table 1). If data at the lowest sample grids (-300 m and -500 m; sometimes also at -180 m) are not available, their first derivatives are required to smoothly approach climatologic values, a similar manner as used in dealing with (u, v) on the top grids. We found that if the unavailable temperature data in the lowest sample grids are not constrained by prescribed values (such as by nudging their first derivatives to climatology), extrapolation may produce unphysically extreme values out of the data ranges and lead to obviously artificial unstable modes. Apart from this, the LSA results seem only weakly sensitive to grid spacing and inter/extrapolation methods (see Supplementary Fig. 3 and Supplementary Table 1). Nevertheless, all detected unstable modes that are below -150 m are rejected for accuracy.

Salinity measurements are sparse, so a temporal and spatial mean that is averaged into the 1-m grid over all fine resolution Argo measurements over the region of $0° \pm 0.5°$, $140° W \pm 0.5°$ is employed instead. This substitution of salinity does not induce problems in N^2 calculation and in the LSA, because the effect of salinity on density is small[32]. A similar treatment is adopted in ref. 50 at the same location.

As no prior information of the perturbation waves exist, the LSA scans both the wave vector directions and the wave numbers for each observed flow. For computational efficiency, the disturbance wave vector direction α is scanned from $0°$ to $180°$ with interval $15°$ (direction 0 represents east) in this study (direction 0 to $\sim -180°$ is symmetry to 0 to $\sim 180°$ and need not be scanned); the hourly mean component of TAO velocities (u, v) at α direction, that is, U, is calculated subsequently. The wave number k of the perturbation wave is scanned over 85 values ranging from $2\pi/60$ to $2\pi/1{,}000$ m^{-1}. For a given wave vector (k, α), the number of unstable modes, which are determined based on criteria of ref. 18 (see below), ranges from 0 to ~ 10. Furthermore, all potential unstable modes of the flow (in terms of critical levels) constitute several mode families.

The idea of mode family is based on the nature that the critical level z_c of the flow is relatively consistent, even though most mode properties profile vary with k and α. A histogram of z_c obtained from all (k, α) is constructed and peaks identified. All modes close to a given peak (that is, having critical levels between the adjacent minima of the histogram) are considered part of the same mode family, which usually focus in different depth ranges[16,18] (Supplementary Fig. 3). For each mode family, the fastest growing mode (should satisfy additional criteria; see below) is defined as the unstable modes of the flow (Supplementary Table 1 and Supplementary Fig. 4).

The criteria (see ref. 18) adopted to determine all possible unstable modes (which are in terms of critical levels) and reasonable mode families include cutoff depths, cutoff growth rate of instability, cutoff wavelengths, Ri criteria, critical layer criteria and others, which are described in details below.

The cutoff depths are -40 m and -150 m in the present study. This criterion rejects potentially unphysical modes that are induced by extrapolation near the boundaries as mentioned above. Previous sensitivity studies also demonstrated that the location of detected unstable mode near boundary may depart to some extent from its real position[42].

The cutoff growth rate of instability is $1 h^{-1}$. This criterion guarantees that the instability grows faster than hourly variation of the mean flow.

The cutoff wavelengths depend on the sampling spacing of the temperature measurements. Instabilities of inviscid, non-diffusive, stratified shear layers typically have wavelength around 2π times the thickness of the shear layer[18,41,68]. Based on this, modes with wavelength 65 m are likely to grow from layers of thickness 10 m, the Nyquist wavelength of the ~ 5-m vertical bins of TAO temperature data in the above -60 m depth; similarly, modes with wavelength 250 m (500 m) are likely to grow from the ~ 20-m (~ 40cm) vertical bins of TAO temperature data over -60 to ~ -140-m (-140 to ~ -200) depths. We remove modes of wavelengths < 65 m above -60 m depth and of wavelengths < 250 m between -60 and -140 m, and of wavelengths smaller than 500 m between -140 and -200 m, to avoid possibly unphysical modes that are resulted from interpolation. The bins of TAO velocity data are constantly 5 m and thus do not require extra limitations of wavelengths.

The lowest Ri of the profile should be < 0.25. Moreover, in the vicinity ($\pm 1/14$ wavelength) of the critical level of the potential unstable mode, lowest Ri is required to be < 0.25, to assure typical Kelvin–Helmholtz instability of this unstable mode[17].

Any mode family must include at least one resolved mode with a larger wavelength than the fastest growing mode and at least one with smaller wavelength. This effectively rejects modes whose true maxima lie outside the range of wavelengths tested.

In addition, the critical level(s) is determined as the depth(s) where $|U(z) - c_r| \leq 0.01$ (m s^{-1}) in the present study. Why we added such a criterion is because the 1-m spacing, inter/extrapolated velocity profile is still discrete so that it may not guarantee a depth that meets the restrict definition of critical level: $U(z) - c_r = 0$. Under such a criterion, there may exist more than one critical level of a (k, α, c_r) vector that satisfy the above criteria. All are retained for further analysis.

(It is noteworthy that the LSA performed here differs in physics from ref. 18 in that we did not include effect of eddy viscosity in equation (1), while the referred work did. In ref. 18, the authors demonstrated that the addition of eddy viscosity to equation (1) damped the generation of instabilities mainly at night when turbulence is strongest. However, the effect of eddy viscosity could be subtle under different conditions, that is, it may also destabilize a stratified shear flow[69].)

Supplementary Figs 3 and 4 show detailed results of LSA that is applied to an example profile and to consecutive profiles over a period of ~ 8 days, respectively. Based on the flow shown on Supplementary Fig. 3, we also discuss the sensitivity of the LSA to the inter/extrapolation method (Supplementary Table 1). In Supplementary Note 1 we describe the details of both analyses. In summary, the sensitivity study suggests that the unstable modes occur in vicinity of low Ri, and as long as this region of low Ri is accurately solved and not close to the boundary, reasonable unstable mode can be detected. By Supplementary Fig. 4, we demonstrate the usefulness of LSA via showing the coincidence of the detected unstable modes with the direct turbulence measurements.

References

1. Shin, S. I. & Liu, Z. Y. Response of the equatorial thermocline to extratropical buoyancy forcing. *J. Phys. Oceanogr.* **30**, 2883–2905 (2000).
2. Huang, R. X. & Pedlosky, J. Climate varialblity of the equatorial thermocline inferred from a two-moving-layer model of the ventilated thermocline. *J. Phys. Oceanogr.* **30**, 2610–2626 (2000).
3. Kessler, W. S., Mcphaden, M. J. & Weickmann, K. M. Forcing of intraseasonal Kelvin waves in the equatorial Pacific. *J. Geophys. Res. Oceans* **100**, 10613–10631 (1995).
4. Wang, B., Wu, R. G. & Lukas, R. Annual adjustment of the thermocline in the tropical Pacific Ocean. *J. Climate* **13**, 596–616 (2000).
5. Li, X. J., Chao, Y., McWilliams, J. C. & Fu, L. L. A comparison of two vertical-mixing schemes in a Pacific Ocean general circulation model. *J. Climate* **14**, 1377–1398 (2001).
6. Meehl, G. A. *et al.* Factors that affect the amplitude of El Nino in global coupled climate models. *Clim. Dynam.* **17**, 515–526 (2001).
7. Gregg, M. C., Peters, H., Wesson, J. C., Oakey, N. S. & Shay, T. J. Intensive measurements of turbulence and shear in the equatorial undercurrent. *Nature* **318**, 140–144 (1985).
8. Moum, J. N. & Caldwell, D. R. Local influences on shear-flow turbulence in the equatorial ocean. *Science* **230**, 315–316 (1985).
9. Peters, H., Gregg, M. C. & Toole, J. M. On the parameterization of equatorial turbulence. *J. Geophys. Res. Oceans* **93**, 1199–1218 (1988).
10. Peters, H., Gregg, M. C. & Sanford, T. B. The diurnal cycle of the upper equatorial ocean - turbulence, fine-scale shear, and mean shear. *J. Geophys. Res. Oceans* **99**, 7707–7723 (1994).
11. Moum, J. N., Caldwell, D. R. & Paulson, C. A. Mixing in the equatorial surface-layer and thermocline. *J. Geophys. Res. Oceans* **94**, 2005–2021 (1989).
12. Moum, J. N. & Nash, J. D. Mixing measurements on an equatorial ocean mooring. *J. Atmos. Ocean Technol.* **26**, 317–336 (2009).
13. Moum, J. N. *et al.* Sea surface cooling at the Equator by subsurface mixing in tropical instability waves. *Nat. Geosci.* **2**, 761–765 (2009).
14. Lien, R. C., Caldwell, D. R., Gregg, M. C. & Moum, J. N. Turbulence variability at the equator in the Central Pacific at the beginning of the 1991-1993 El-Nino. *J. Geophys. Res. Oceans* **100**, 6881–6898 (1995).
15. Wang, D. L. & Muller, P. Effects of equatorial undercurrent shear on upper-ocean mixing and internal waves. *J. Phys. Oceanogr.* **32**, 1041–1057 (2002).

16. Sun, C. J., Smyth, W. D. & Moum, J. N. Dynamic instability of stratified shear flow in the upper equatorial Pacific. *J. Geophys. Res. Oceans* **103**, 10323–10337 (1998).

17. Moum, J. N., Nash, J. D. & Smyth, W. D. Narrowband oscillations in the upper equatorial ocean. Part I: interpretation as shear instabilities. *J. Phys. Oceanogr.* **41**, 397–411 (2011).

18. Smyth, W. D., Moum, J. N., Li, L. & Thorpe, S. A. Diurnal shear instability, the descent of the surface shear layer, and the deep cycle of equatorial turbulence. *J. Phys. Oceanogr.* **43**, 2432–2455 (2013).

19. Smyth, W. D. & Moum, J. N. Shear instability and gravity wave saturation in an asymmetrically stratified jet. *Dynam. Atmos. Oceans* **35**, 265–294 (2002).

20. Peters, H., Gregg, M. C. & Sanford, T. B. Equatorial and off-equatorial fine-scale and large-scale shear variability at 140-degrees-W. *J. Geophys. Res. Oceans* **96**, 16913–16928 (1991).

21. Lyman, J. M., Johnson, G. C. & Kessler, W. S. Distinct 17-and 33-day tropical instability waves in subsurface observations. *J. Phys. Oceanogr.* **37**, 855–872 (2007).

22. Legeckis, R. Long waves in eastern equatorial Pacific ocean - view from a geostationary satellite. *Science* **197**, 1179–1181 (1977).

23. Miller, L., Watts, D. R. & Wimbush, M. Oscillations of dynamic topography in the eastern equatorial Pacific. *J. Phys. Oceanogr.* **15**, 1759–1770 (1985).

24. Strutton, P. G., Ryan, J. P. & Chavez, F. P. Enhanced chlorophyll associated with tropical instability waves in the equatorial Pacific. *Geophys. Res. Lett.* **28**, 2005–2008 (2001).

25. McPhaden, M. J. Monthly period oscillations in the Pacific North equatorial countercurrent. *J. Geophys. Res. Oceans* **101**, 6337–6359 (1996).

26. Halpern, D., Knox, R. A. & Luther, D. S. Observations of 20-day period meridional current oscillations in the upper ocean along the Pacific Equator. *J. Phys. Oceanogr.* **18**, 1514–1534 (1988).

27. Lyman, J. M., Chelton, D. B., deSzoeke, R. A. & Samelson, R. M. Tropical instability waves as a resonance between equatorial Rossby waves. *J. Phys. Oceanogr.* **35**, 232–254 (2005).

28. Kennan, S. C. & Flament, P. J. Observations of a tropical instability vortex. *J. Phys. Oceanogr.* **30**, 2277–2301 (2000).

29. Lien, R. C., D'Asaro, E. A. & Menkes, C. E. Modulation of equatorial turbulence by tropical instability waves. *Geophys. Res. Lett.* **35**, L24607 (2008).

30. Menkes, C. E. R., Vialard, J. G., Kennan, S. C., Boulanger, J. P. & Madec, G. V. A modeling study of the impact of tropical instability waves on the heat budget of the eastern equatorial Pacific. *J. Phys. Oceanogr.* **36**, 847–865 (2006).

31. Jochum, M. & Murtugudde, R. Temperature advection by tropical instability waves. *J. Phys. Oceanogr.* **36**, 592–605 (2006).

32. Inoue, R., Lien, R. C. & Moum, J. N. Modulation of equatorial turbulence by a tropical instability wave. *J. Geophys. Res. Oceans* **117**, C1009 (2012).

33. Xie, S. P. Climate science unequal equinoxes. *Nature* **500**, 33–34 (2013).

34. Argo. Argo float data and metadata from Global Data Assembly Centre (Argo GDAC). *Ifremer* (2000).

35. Whalen, C. B., Talley, L. D. & MacKinnon, J. A. Spatial and temporal variability of global ocean mixing inferred from Argo profiles. *Geophys. Res. Lett.* **39**, L18612 (2012).

36. Kunze, E., Firing, E., Hummon, J. M., Chereskin, T. K. & Thurnherr, A. M. Global abyssal mixing inferred from lowered ADCP shear and CTD strain profiles. *J. Phys. Oceanogr.* **36**, 1553–1576 (2006).

37. Gregg, M. C., Sanford, T. B. & Winkel, D. P. Reduced mixing from the breaking of internal waves in equatorial waters. *Nature* **422**, 513–515 (2003).

38. Thorpe, S. A. Turbulence and mixing in a Scottish Loch. *Philos. Trans. R. Soc. Lond. A Math. Phys. Sci.* **286**, 125–181 (1977).

39. McPhaden, M. J. The tropical atmosphere ocean array is completed. *Bull. Am. Meteorol. Soc.* **76**, 739–741 (1995).

40. Moum, J. N., Farmer, D. M., Smyth, W. D., Armi, L. & Vagle, S. Structure and generation of turbulence at interfaces strained by internal solitary waves propagating shoreward over the continental shelf. *J. Phys. Oceanogr.* **33**, 2093–2112 (2003).

41. Smyth, W. D., Moum, J. N. & Nash, J. D. Narrowband oscillations in the upper equatorial ocean. Part II: properties of shear instabilities. *J. Phys. Oceanogr.* **41**, 412–428 (2011).

42. Liu, Z. Y. Instability of Baroclinic tidal flow in a stratified Fjord. *J. Phys. Oceanogr.* **40**, 139–154 (2010).

43. Moum, J. N., Perlin, A., Nash, J. D. & McPhaden, M. J. Seasonal sea surface cooling in the equatorial Pacific cold tongue controlled by ocean mixing. *Nature* **500**, 64–67 (2013).

44. Contreras, R. F. Long-term observations of tropical instability waves. *J. Phys. Oceanogr.* **32**, 2715–2722 (2002).

45. Reynolds, R. W. *et al.* Daily high-resolution-blended analyses for sea surface temperature. *J. Climate* **20**, 5473–5496 (2007).

46. Miles, J. W. On the stability of heterogeneous shear flows. *J. Fluid Mech.* **10**, 496–508 (1961).

47. Rohr, J. J., Itsweire, E. C., Helland, K. N. & Vanatta, C. W. Growth and decay of turbulence in a stably stratified shear-flow. *J. Fluid Mech.* **195**, 77–111 (1988).

48. Thorpe, S. A. & Liu, Z. Y. Marginal instability? *J. Phys. Oceanogr.* **39**, 2373–2381 (2009).

49. Smyth, W. D. & Moum, J. N. Length scales of turbulence in stably stratified mixing layers. *Phys. Fluids* **12**, 1327–1342 (2000).

50. Smyth, W. D. & Moum, J. N. Marginal instability and deep cycle turbulence in the eastern equatorial Pacific Ocean. *Geophys. Res. Lett.* **40**, 6181–6185 (2013).

51. Yu, J. Y. & Lui, W. T. A linear relationship between ENSO intensity and tropical instability wave activity in the eastern Pacific Ocean. *Geophys. Res. Lett.* **30**, 1735 (2003).

52. Reynolds, R. W., Rayner, N. A., Smith, T. M., Stokes, D. C. & Wang, W. Q. An improved in situ and satellite SST analysis for climate. *J. Climate* **15**, 1609–1625 (2002).

53. Wang, W. M. & McPhaden, M. J. The surface-layer heat balance in the equatorial Pacific Ocean. Part I: mean seasonal cycle. *J. Phys. Oceanogr.* **29**, 1812–1831 (1999).

54. Wang, B. & Fu, X. H. Processes determining the rapid reestablishment of the equatorial Pacific cold tongue/ITCZ complex. *J. Climate* **14**, 2250–2265 (2001).

55. Jouanno, J., Marin, F., du Penhoat, Y., Sheinbaum, J. & Molines, J. M. Seasonal heat balance in the upper 100 m of the equatorial Atlantic Ocean. *J. Geophys. Res. Oceans* **116**, C09003 (2011).

56. Mcphaden, M. J., Cronin, M. F. & Mcclurg, D. C. Meridional structure of the seasonally varying mixed layer temperature balance in the eastern tropical Pacific. *J. Climate* **21**, 3240–3260 (2008).

57. Ham, Y. G. & Kang, I. S. Improvement of seasonal forecasts with inclusion of tropical instability waves on initial conditions. *Clim. Dynam.* **36**, 1277–1290 (2011).

58. Imada, Y. & Kimoto, M. Parameterization of tropical instability waves and examination of their impact on ENSO characteristics. *J. Climate* **25**, 4568–4581 (2012).

59. An, S. I. Interannual variations of the Tropical Ocean instability wave and ENSO. *J. Climate* **21**, 3680–3686 (2008).

60. Zaron, E. D. & Moum, J. N. A new look at Richardson number mixing schemes for equatorial ocean modeling. *J. Phys. Oceanogr.* **39**, 2652–2664 (2009).

61. Johnson, E. S. A convergent instability wave front in the central tropical Pacific. *Deep Sea Res. Pt II* **43**, 753–778 (1996).

62. Flament, P. J., Kennan, S. C., Knox, R. A., Niiler, P. P. & Bernstein, R. L. The three-dimensional structure of an upper ocean vortex in the tropical Pacific Ocean. *Nature* **383**, 610–613 (1996).

63. Furue, R. *et al.* Impacts of regional mixing on the temperature structure of the equatorial Pacific Ocean. Part 1: vertically uniform vertical diffusion. *Ocean Model* **91**, 91–111 (2015).

64. Jia, Y. L., Furue, R. & McCreary, J. P. Impacts of regional mixing on the temperature structure of the equatorial Pacific Ocean. Part 2: depth-dependent vertical diffusion. *Ocean Model* **91**, 112–127 (2015).

65. Gargett, A. & Garner, T. Determining Thorpe scales from ship-lowered CTD density profiles. *J. Atmos. Ocean Technol.* **25**, 1657–1670 (2008).

66. Galbraith, P. S. & Kelley, D. E. Identifying overturns in CTD profiles. *J. Atmos. Ocean Technol.* **13**, 688–702 (1996).

67. Liu, Z., Thorpe, S. A. & Smyth, W. D. Instability and hydraulics of turbulent stratified shear flows. *J. Fluid Mech.* **695**, 235–256 (2012).

68. Hazel, P. Numerical studies of stability of inviscid stratified shear flows. *J. Fluid Mech.* **51**, 39–61 (1972).

69. Li, L., Smyth, W. D. & Thorpe, S. A. Destabilization of a stratified shear layer by ambient turbulence. *J. Fluid Mech.* **771**, 1–15 (2015).

70. Johnson, G. C., Sloyan, B. M., Kessler, W. S. & McTaggart, K. E. Direct measurements of upper ocean currents and water properties across the tropical Pacific during the 1990s. *Prog. Oceanogr.* **52**, 31–61 (2002).

Acknowledgements

C.L., A.K. and D.S. acknowledge funding by the German Federal Ministry for Education and Research via the project RACE (FZ 03F0651A). Contribution to the DFG funded CliSAP Excellence initiative of the University of Hamburg. C.L. was also supported by the Knowledge Innovation Program of the Chinese Academy of Sciences (Y62114101Q). Z.L. was funded by the National Basic Research Program of China (2012CB417402), the National Natural Science Foundation of China (NSFC) (41476006) and the Natural Science Foundation of Fujian Province of China (2015J06010). F.W. was funded by the Strategic Priority Research Program of the Chinese Academy of Sciences (XDA11010201), the NSFC Innovative Group Grant (41421005) and the NSFC-Shandong Joint Fund for Marine Science Research Centers (U1406401). We are grateful to three anonymous reviewers who provided instructive suggestions that greatly improved the manuscript. We thank all the data providers. We thank M. Carson for proofreading.

Author contributions

C.L. and A.K. designed the research and conducted data analysis. Z.L. proposed and C.L. conducted the LSA. C.L. and A.K. wrote the first draft of the paper with all the authors' contribution to the revisions. The idea, analysis and manuscript were motivated, performed and written in UHH. The first and later revisions were made in UHH and IOCAS, respectively.

Additional information

Competing financial interests: The authors declare no competing financial interests.

23

Evidence for the stability of the West Antarctic Ice Sheet divide for 1.4 million years

Andrew S. Hein[1], John Woodward[2], Shasta M. Marrero[1], Stuart A. Dunning[2,3], Eric J. Steig[1,4], Stewart P.H.T. Freeman[5], Finlay M. Stuart[5], Kate Winter[2], Matthew J. Westoby[2] & David E. Sugden[1]

Past fluctuations of the West Antarctic Ice Sheet (WAIS) are of fundamental interest because of the possibility of WAIS collapse in the future and a consequent rise in global sea level. However, the configuration and stability of the ice sheet during past interglacial periods remains uncertain. Here we present geomorphological evidence and multiple cosmogenic nuclide data from the southern Ellsworth Mountains to suggest that the divide of the WAIS has fluctuated only modestly in location and thickness for at least the last 1.4 million years. Fluctuations during glacial–interglacial cycles appear superimposed on a long-term trajectory of ice-surface lowering relative to the mountains. This implies that as a minimum, a regional ice sheet centred on the Ellsworth-Whitmore uplands may have survived Pleistocene warm periods. If so, it constrains the WAIS contribution to global sea level rise during interglacials to about 3.3 m above present.

[1]School of GeoSciences, University of Edinburgh, Drummond Street, Edinburgh EH8 9XP, UK. [2]Department of Geography, Northumbria University, Ellison Place, Newcastle upon Tyne NE1 8ST, UK. [3]Department of Geography, School of Geography, Politics and Sociology, Newcastle University, Newcastle upon Tyne NE1 7RU, UK. [4]Quaternary Research Center and Department of Earth and Space Sciences, University of Washington, Seattle, Washington 98195, USA. [5]Scottish Universities Environmental Research Centre, Rankine Avenue, East Kilbride G75 0QF, UK. Correspondence and requests for materials should be addressed to A.S.H. (email: Andy.Hein@ed.ac.uk).

The West Antarctic Ice Sheet (WAIS) is pinned on an archipelago with its central dome situated over subglacial uplands and bedrock basins, the latter more than 1,500 m below sea level (Fig. 1). For over four decades there has been a fear that this topography could lead to marine instability, since retreat of the ice margin into the basins would enhance ice calving and ice-mass loss, leading to loss of the WAIS and a rapid rise in global sea level of 3–5 m (refs 1–3). Recent studies have suggested that such a collapse may already be underway in the Pacific-facing sector of the ice sheet[4,5]. Constraining the past ice behaviour would allow a more confident assessment of its potential contribution to past and future sea level change. Marine biological evidence based on diatoms and the similarity of octopus and *Bryozoa* between the Pacific and Atlantic sectors suggests that much of the ice sheet disappeared during interglacials, creating an open seaway between these sectors[6–8]. Such a conclusion is reinforced by estimates of higher-than-present global sea levels during interglacials[9,10]. Efforts to constrain the minimum configuration of the ice sheet in the past have relied on numerical ice-sheet models, each with its own set of assumptions on boundary conditions, internal dynamics and external forcing (for example, climate and sea level)[3,11,12]. The models suggest that most upland areas could have remained glaciated even during the warmest interglacials, but whether as individual mountain glaciers or as larger regional ice sheets remains uncertain, and there is no direct evidence from the continent to constrain this.

The Heritage Range, situated in the heart of the Weddell Sea embayment, lies within 50 km of the interface between the grounded ice sheet and the floating Filchner-Ronne Ice Shelf in Hercules Inlet (Fig. 1). Two component massifs, Patriot Hills and Marble Hills are summits of a 15-km-wide upland bounded by troughs excavated to below sea level[13]. At present, ice from the central WAIS flows around and between these mountains to the grounding line. The WAIS divide forms a broad saddle between the main dome 300 km to the west and another 200 km to the northwest. Katabatic winds flow down the ice slope from the divide towards Hercules Inlet crossing the mountains and creating blue-ice areas in their lee (Fig. 1). The winds cause ablation of surface ice that in turn causes a compensating upward flow of ice that brings basal debris to the ice-surface as blue-ice moraines[14,15]. This ice-marginal, basally derived material is deposited higher on the mountain flanks and records past changes in ice thickness.

The use of cosmogenic nuclide dating on bedrock and glacially transported material on nunataks in Antarctica has provided much quantitative data on the history of ice thickness changes over time[16–23]. Initial work in the Heritage Range revealed a scatter of ages of up to 400 ka in elevated blue-ice moraines[24,25]. These data led to the untested hypothesis that the spread in ages represented the continuous presence of an ice sheet that fluctuated in thickness in response to glacial–interglacial cycles[14]. The range of ages reflects preservation of some erratics and deposition of others during successive glaciations. An alternative possibility is that the moraines represent composite features formed by multiple ice-sheet inundations interspersed with periods of local mountain glaciation or deglaciation.

The combination of geomorphological analysis of landforms and measurement of multiple cosmogenic nuclides can provide rare insight into ice-sheet history. The advantage of measuring multiple cosmogenic nuclides in single samples is that both the age and exposure history can be constrained[22]. For example, if a previously exposed clast is buried by ice long enough for the shorter lived of two nuclides to decay preferentially, the signal will be observed in the isotopic ratio[26]. In the case of cosmogenic ²⁶Al/¹⁰Be, it takes several tens of thousands of years for the burial

signal to become evident. By measuring multiple isotopes in three adjacent erratics at each specific sampling site, the degree of scatter and extent to which the erratics have shared the same history of exposure can be determined. Thus, one can gain information on the age of deposition and possible subsequent overriding and disturbance by ice.

Here we use geomorphological analysis of landforms and deposits supported by in situ cosmogenic ²⁶Al, ¹⁰Be and ²¹Ne from newly collected, quartz-bearing erratics to investigate elevated blue-ice moraines. Our evidence reveals several relict ice-marginal blue-ice moraine deposits as old as 1.4 Ma. The isotopic evidence indicates that the highest deposits have not been disturbed by ice since deposition, but lower deposits have experienced subsequent burial. All geomorphic and cosmogenic nuclide data are consistent with an ice sheet that thickened and thinned in response to quaternary glacial–interglacial cycles. We find no evidence to suggest a change in glaciological conditions that would accompany the loss of the entire ice sheet and the build-up of individual mountain glaciers. We interpret this consistency as evidence for continuous ice-sheet conditions in this part of the Weddell Sea sector. The minimum configuration that maintains strong katabatic winds is a regional ice sheet centred on the Ellsworth-Whitmore block. This interpretation, where the WAIS shows dynamic equilibrium about a continuous ice divide, supports numerical models that indicate a maximum WAIS contribution to sea level of about 3.3 m (refs 3,11), consistent with low-end estimates of global sea level during past interglacial periods[9]. Such an interpretation is also consistent with marine biological evidence indicating an open seaway in West Antarctica during some interglacials[6–8].

Results

Geomorphology. The geomorphological analysis of landforms and deposits reveals currently active blue-ice moraines at the edge of glaciers at the eastern foot of the mountains (Supplementary Figs 1–4). Striated, basally derived clasts occur in the moraines and in folded debris bands in the adjacent glacier surface. Airborne radio echo sounding (RES) data reveals that the debris sequences originate ∼800 m below at the glacier base at a depth close to present sea level (Fig. 2). Above the ice margin are two formerly glaciated zones marked by an upper erosional trimline (Fig. 1d), the latter recognized throughout the Ellsworth Mountains[27]. The upper weathered zone occurs up to 650 m above the present ice-surface and is covered by iron-stained, quartz-rich erratics and till patches on an ice-eroded limestone or marble bedrock surface. Lower down in this zone, the weathered erratics and till have been disturbed by eastward flowing ice. This is demonstrated by erratics preferentially trapped in irregularities in the bedrock and the preservation of till patches in basins and on the eastern side of bedrock bumps, leaving the western slopes and summits relatively free of debris. The weathered deposits represent former ice-marginal blue-ice moraines. This conclusion is borne out by the location and concentration of till patches at the foot of a mountain escarpment athwart katabatic winds, their proximity to a former ice margin and the lack of erratics above the trimline (Supplementary Fig. 4). Moreover, the shape and lithology of the erratics is the same as the quartz-rich lithologies in the moraines at the current ice edge. The lower unweathered zone is characterized by fresh erratics, perched boulders and ice-cored tills; it is thought to reflect deposition by ice during the Last Glacial Maximum[24] and is not considered here.

Cosmogenic nuclide data. We measured in situ cosmogenic ²⁶Al, ¹⁰Be and ²¹Ne on newly collected quartz-bearing erratics (see Methods and Supplementary Tables 1–3). The exposure ages

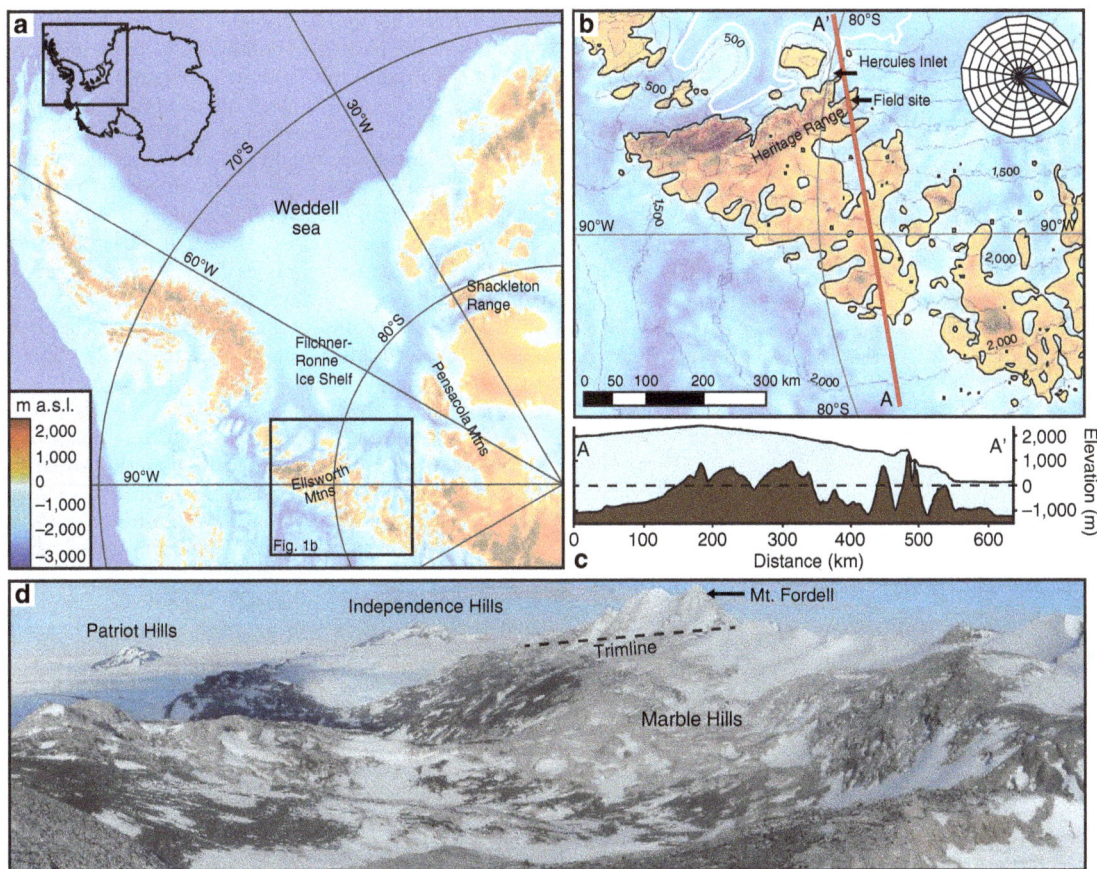

Figure 1 | The Heritage Range field site. (**a**) Subglacial topography of Antarctica[57] showing the field location in the Ellsworth Mountains and prominent geographical features within the Weddell Sea embayment; inset shows wider Antarctic setting. (**b**) Subglacial topography of the wider Ellsworth block and the present-day ice-surface contours (250 m) of the WAIS. White line indicates the grounding line[58]; black lines are bed elevation contours at 0 m elevation (WGS84). The red line A-A' shows the profile line used in **c**; it runs between the main dome of the WAIS and Hercules inlet. The rose diagram shows persistent katabatic winds from the south-southwest recorded at the Patriot Hills blue-ice aircraft runway over two months in the austral summer of 2008 (ref. 59). (**c**) Profile of the bed and ice-sheet surface from the WAIS divide to Hercules Inlet, showing the deep troughs excavated below sea level surrounding the Patriot and Marble Hills. (**d**) Photograph showing dark-coloured erratics scattered across ice-scoured limestone bedrock of the Marble Hills. Wind-drift glaciers can be seen along the summit ridge to the right of Mt. Fordell (see Supplementary Fig. 2 for a geomorphic map).

of the weathered erratics decline with decreasing elevation towards the glacier surface (Fig. 3). Three adjacent erratics from each of two sites in the upper weathered zone in the Marble Hills have [10]Be exposure ages of 1.2–1.4 Ma, and at a slightly lower elevation above the ice, 0.6–0.7 Ma. The samples at both sites yield tightly clustered ages for each isotope with ratios that do not indicate prolonged burial (see Methods and Supplementary Fig. 5). Lower down, erratics have younger [10]Be exposure ages of 0.5–0.6 Ma with some isotopic evidence of burial. Comparable erratics from a patch of weathered till in the Patriot Hills have similarly clustered [10]Be exposure ages of 0.4–0.5 Ma and isotopic ratios indicating more than 300 ka burial (Supplementary Fig. 6). Seven samples emerging from the ice today in front of both the Marble and Patriot Hills have exposure ages of <1.5 ka and thus total inheritance is low. The striking feature of the data from the high elevation samples is that they reflect a shared origin and exposure history. Rather than the scatter of ages one might expect with repeated episodes of burial by ice, there is a consistent pattern of decreasing age of exposure and increasing degree of burial towards the present glacier margin. This suggests that any subsequent burial at the different sites was by cold-based ice that did not move existing material or deposit new material in the process.

Discussion

The simplest explanation of the pattern of cosmogenic nuclide data is that an ice margin fluctuated in elevation on the mountain flank (Fig. 4). The highest erratics are exposed for the longest time, while progressively lower erratics are exposed for increasingly shorter periods of time. This explains both the younging trend and evidence of increased burial with decreased altitude. The implication of exposure ages of up to 1.4 Ma at higher elevations is that ice thickening and blue-ice moraine formation also occurred during earlier glacial cycles in the Pleistocene. Given the mountains are situated near the grounding line of today, increases in ice thickness near the mountains would accompany any seaward migration of the grounding line as ocean temperature cooled and global sea level fell[28]. Over millions of years one would expect glacial erosion to lower the ice-sheet surface relative to the mountains[29,30] and thus the cyclic changes in ice thickness would be superimposed on a trajectory of lowering relative to the mountains. This scenario is consistent with the great ages and minimal burial of the highest erratics.

It could be argued that the scenario above should produce a scatter of exposure ages of up to 1.4 Ma, rather than clustering at certain ages. Indeed, such scatter has been measured in tills

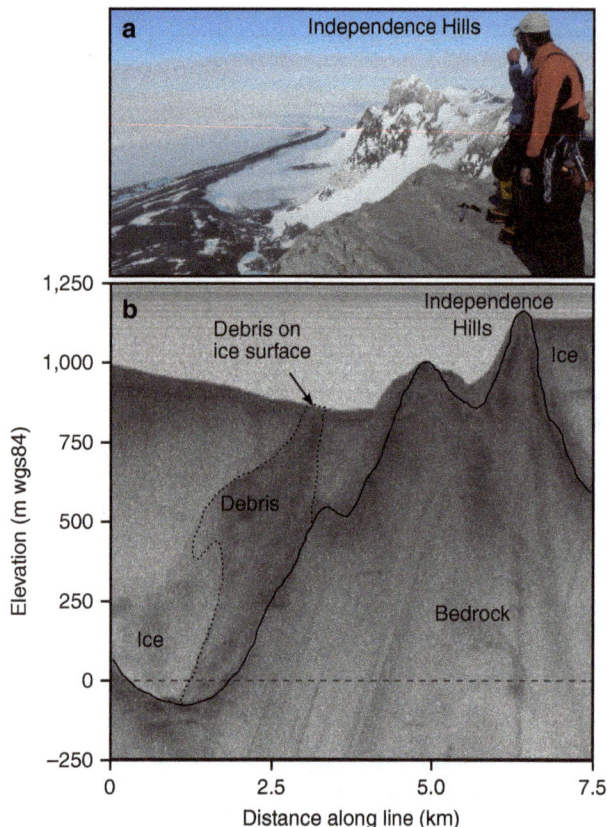

Figure 2 | The source of blue-ice moraine debris. (a) The blue ice moraine at the foot of Independence Hills. **(b)** RES transect across blue-ice moraines adjacent to Independence Hills showing how debris underlies the folded surface moraines. The origin of the debris (dotted line), much of it locally derived, is from deep within the glacier trough, indeed close to present sea level. In addition to the lateral flow there is a limited longitudinal component of flow to the east (into the page). The 150 MHz pulse-processed radar data was collected by the British Antarctic Survey Polarimetric-radar Airborne Science Instrument ice-sounding radar in 2010/2011.

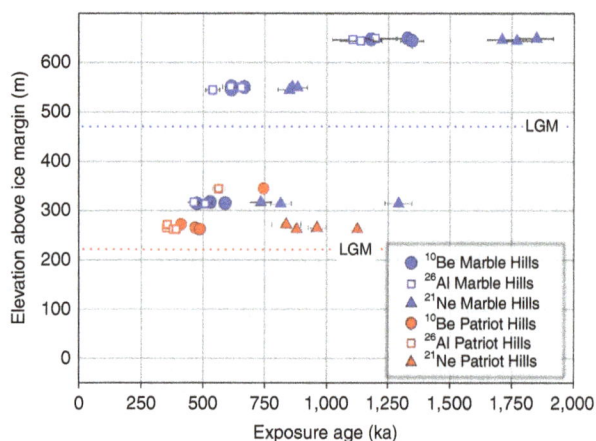

Figure 3 | The cosmogenic nuclide data. The apparent exposure ages of weathered rock samples from the Marble Hills (blue) and Patriot Hills (red) plotted against elevation above the ice margin. Error bars (1σ) reflect analytical uncertainties only. There is a clear decline in exposure ages with decreasing elevation at the Marble Hills. Three Marble Hills samples with [10]Be ages of 500–600 ka survived beneath the ice cover of the Last Glacial Maximum[24] (blue dotted line). A similar exposure history involving burial is inferred for samples above the Last Glacial Maximum limit[24] in the Patriot Hills (Supplementary Figs 5–6).

within the lower unweathered zone in the Marble Hills, which ranged from 15 to 250 ka (ref. 24). The clustering may reflect episodes in the past when conditions were particularly favourable for the formation of blue-ice moraines at those particular locations, just as the modern concentration of blue-ice material varies with local topography and ice-surface elevation. In addition, clustering would be an expected artefact of the sampling, which was concentrated on sites at specific altitudes.

The implication of the evidence above is that the WAIS divide and associated katabatic winds have also been present for at least 1.4 Ma. The blue-ice moraines form because of strong katabatic winds and these in turn are strongest and most consistent when they flow downslope for hundreds of km. The loss of the ice divide would diminish both katabatic winds and blue-ice moraine formation.

Could such an assemblage of deposits survive loss of the WAIS? The recorded ages leave adequate intervals of time for the ice divide to disappear. If the ice sheet disappeared, ice caps and glaciers would likely build up on mountain massifs in a fjord landscape. Each massif would have a different and locally radial pattern of flow depending on the type and scale of the topography (Fig. 4c). We found none of the features characteristic of such a scenario. Rather than radial flow from the mountain axis, Marble and the Patriot Hills bear geomorphologic evidence of eastward ice flow (Supplementary Fig. 4). In other fjord areas of the world, glacial deposits typically include marine traces, such as diatoms, shells and glaciomarine muds[31]. Examination of the Heritage Range tills, that RES shows are sourced from deep within the glacier troughs (Fig. 2), revealed no traces of marine diatoms or other biogenic silica in either present-day or elevated blue-ice tills (R. Scherer, personal communication, 2014). Local glaciation typically produces deposits associated with local corrie glaciers, as in the Asgard and Olympus ranges in the Transantarctic Mountains[32]. Rather than corrie glacier deposits, concentrations of material with a local origin in the study area are restricted to wind-drift glaciers that merged local rocks with exotic material in blue-ice moraines. Indeed, the very existence of former wind-drift glaciers supports the existence of the ice sheet and associated katabatic winds.

The lack of evidence of marine and local glaciation cannot on its own rule out short periods of complete deglaciation. The cosmogenic nuclide data alone are not a direct test of this hypothesis. It is possible that some evidence may remain preserved beneath the ice sheet or that the characteristic geomorphology is missing or poorly developed. Corrie and wind-drift glaciers could produce geomorphology that may be indistinguishable, while cold-based glaciers may leave no mark at all. Furthermore, interglacial periods are relatively short lived.

While we recognize the above possibility of complete deglaciation, there are arguments in favour of persistent glaciation. Recent atmospheric modelling of the Antarctic climate response to a collapse of the WAIS indicates significant warming would occur in the Atlantic sector of the WAIS[33]. Any such warming in an Antarctic maritime environment would cause an increase in snowfall and the growth of mountain glaciers. The re-glaciation of the WAIS would begin on upland areas such as the Heritage Range. Moreover, modelling suggests the increased temperature and accumulation relative to today would favour the formation of warm-based local glaciers that are efficient at removing sediment from their beds[12]. However, even if re-glaciation involved cold-based glaciers, which can move sediment selectively[32,34], one would expect more scatter in the exposure age results, especially in the highest samples on Mt Fordell; these samples are situated high on the mountain and within 40 m of the present wind-drift glacier margin. Instead of scatter, the exposure ages are tightly clustered. In summary, while we acknowledge the limitations of our

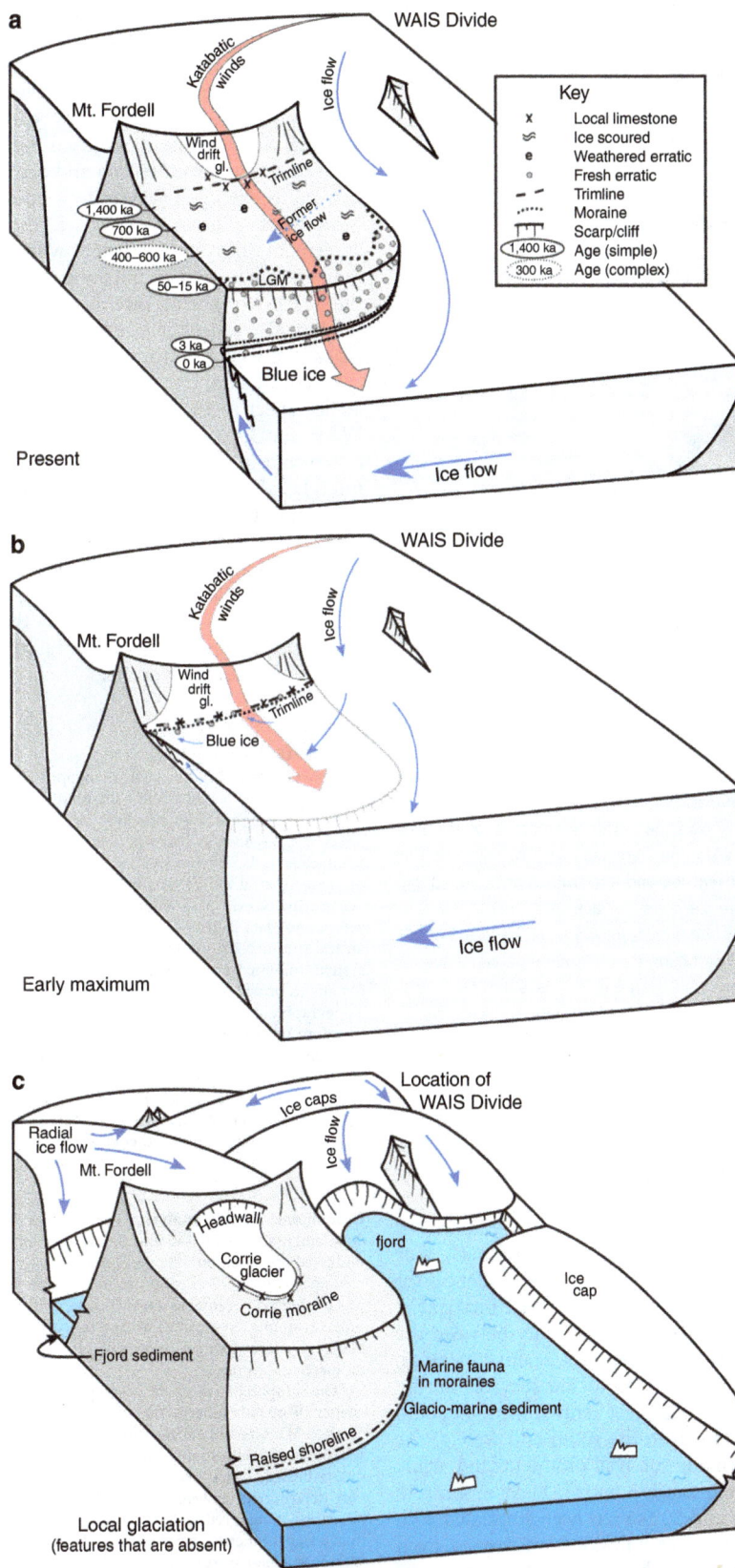

Figure 4 | A conceptual reconstruction of different ice-sheet configurations surrounding the Marble Hills massif. The reconstruction (looking west) shows (**a**) the present minimum, (**b**) past maximum blue-ice relationships and (**c**) features that would accompany local glaciation. The ages of weathered quartz-rich erratics on the ice-scoured bedrock upland decline and their burial history becomes more complex at lower elevations, the latter reflecting longer periods of burial by ice. The Last Glacial Maximum limit is often the lowest on the upland plateau. The evidence is consistent with oscillations in ice thickness related to Pleistocene sea level fluctuations. There is no evidence of local glaciation radiating from the massifs.

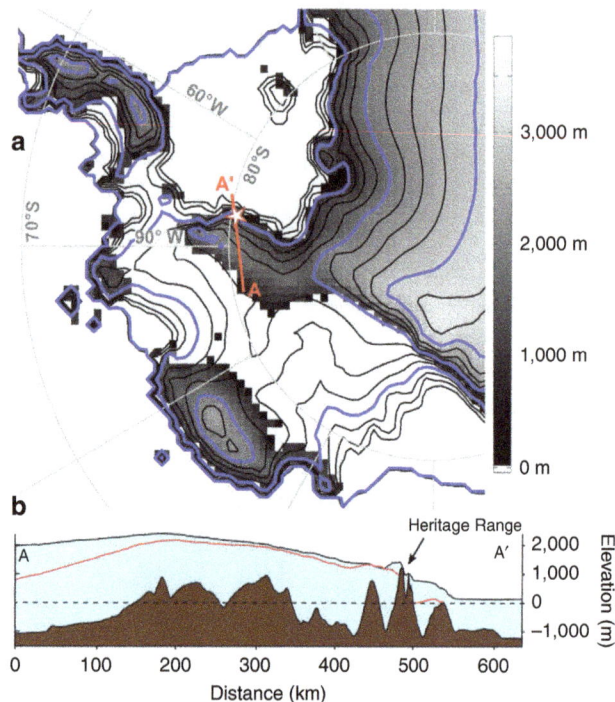

Figure 5 | The minimum ice-sheet configuration. (**a**) Comparison of the modelled minimum Pleistocene extent of the WAIS[11] in grey shading with the present ice-sheet surface represented by black contours (250 m); the blue contours indicate 1,000 m increments. The star indicates the Heritage Range. The catchment area for katabatic winds affecting the Heritage Range is comparable. The red line A-A' shows the profile line used in **b**; it runs between the main dome of the WAIS and Hercules inlet. (**b**) cross-section along A-A' showing the same present ice and bed topography[57] used in Fig. 1c and the modelled minimum ice-sheet surface (red line)[11]. The regional ice-sheet surface remains at a similar elevation as the WAIS today. The latter surface was generated by taking the difference between the modelled present-day and minimum (205 ka) ice thickness and then subtracting this value from the present-day ice-sheet surface elevation. Small isostatic changes in bedrock elevation are accounted[11] but are not shown.

evidence, it seems significant to draw attention to the nature of the cosmogenic and geomorphological evidence from the inner reaches of the Weddell Sea embayment, which point to ice-sheet glaciation for 1.4 Ma.

One possible explanation for persistent ice-sheet conditions for at least 1.4 Ma is that the whole WAIS survived intact. Our evidence is compatible with this view, but the evidence for substantial increases in eustatic sea level during past interglacial periods[9,35] and the evidence of marine connectivity between the Pacific and Atlantic marine sectors[6-8] argues against the idea. Instead, we suggest our evidence points to the survival during interglacials of a smaller regional ice sheet centred on the upland of the Ellsworth-Whitworth massif to the south and west of the Ellsworth Mountains; an ice divide survived on the upland, while ice was lost from surrounding marine basins. Such a scenario during interglacials at 1.2 Ma and 205 ka has been simulated with an ice-sheet model, based on the assumption that ocean-driven melting at the ice-sheet margins is the primary determinant of changes in the WAIS[11] (Fig. 5). The catchment area and fetch for katabatic winds crossing the Heritage Range remains approximately the same. Further, because the regional ice-sheet centres on a topographic high, the ice-surface altitude at the divide decreases by just a few hundred metres during interglacials and remains essentially the same in the vicinity of the Heritage

Range (Fig. 5b). The preservation of a high WAIS divide would help explain the ~7 Ma record of continuous high polar conditions in the adjacent Transantarctic Mountains, inferred from some of the lowest bedrock erosion rates in Antarctica[20]. This reconstruction also permits marine interaction between the Weddell Sea and the Amundsen Sea as suggested based on the analysis of diatoms, *Bryozoa* and octopus[6-8]; indeed, it physically constrains the open seaway to a position north of the Ellsworth Mountains. There are two further implications. First, the agreement of our evidence with models driven by ocean temperature supports the long held view of the sensitivity of the WAIS to marine forcing. Second, our field results support numerical ice-sheet models that imply a WAIS contribution to global sea level of no more than ~3.3 m above present[3,11].

In conclusion, our results point to the continuous presence of an ice sheet in the southern Ellsworth Mountains sector of the WAIS for 1.4 Ma. The ice divide adjusted to accommodate the loss of marine-based portions of the main ice sheet during some interglacials, but a regional ice sheet based on the Ellsworth-Whitmore uplands was sufficient in size and altitude to maintain katabatic winds and form blue-ice moraines in the Heritage Range throughout the Pleistocene. Here, near the grounding line, the ice sheet has experienced a long-term trajectory of lowering relative to the mountains, one that is marked by modest fluctuations in thickness as it responded to glacial–interglacial climate and sea level cycles.

Methods

Geomorphology. Landforms were mapped in the field using differential global positioning satellite and satellite imagery. This formed the basis for detailed work on sediment morphology, lithology, weathering and cosmogenic isotope analysis. Selected areas, such as complex debris accumulations, were mapped with a laser scanner as well as high-resolution vertical aerial photographs taken from an unmanned aerial vehicle. The British Antarctic Survey Polarimetric-radar Airborne Science Instrument ice-sounding radar was used to image deeper englacial reflections (data collected during an airborne survey of the Institute Ice Stream, austral summer 2010/2011 (ref. 36)). A differential global positioning satellite system, mounted on a snowmobile, traversed along the ice margin of both massifs to provide a reference surface from which to normalise sample elevations. Topographic control on the ice margin introduces variability in elevation and therefore uncertainties in the normalization process are estimated at ± 15 m. Supplementary Fig. 1 shows the general ice flow configuration around the southern Heritage Range. Supplementary Fig. 2 shows a geomorphologic map of the Marble Hills massif. Supplementary Fig. 3 shows sample locations from the Patriot Hills. Supplementary Fig. 4 illustrates the nature of the glaciated upland surface and the type and distribution of weathered debris.

Cosmogenic nuclide analysis. The cosmogenic nuclide data are presented in Supplementary Tables 1–3 and Supplementary Figs 5–6. All exposure ages discussed are based on [10]Be ages because the production rate is better constrained; [26]Al and [21]Ne are used to constrain exposure histories. The measurement of [21]Ne was completed because, as a stable isotope, it gives a measure of the total exposure time (assuming no erosion) irrespective of subsequent burial, and because it can potentially record longer periods of exposure than is generally possible with radioactive isotopes.

The sampling strategy for cosmogenic nuclides was designed to reduce the chance of nuclide inheritance, and exclude the possibility of nuclide loss through erosion. We targeted subglacially derived clasts with striated surfaces and subangular to subrounded shapes. We sampled the freshest appearing, quartz-bearing, brick-sized clasts resting on flat bedrock to minimize problems of post-depositional movement and self-shielding. It is crucial to be convinced that the exposure ages reflect the time since deposition of a freshly exposed clast rather than a signal inherited from the past. To test whether clasts emerging on the glacier surface in blue-ice areas have no inherited cosmogenic nuclides, we analysed seven clasts on the present ice margin. All had negligible amounts of both [10]Be and [26]Al, implying that they were first exposed to cosmic rays when they emerged at the ice-surface. In view of the similar lithologies, and thus origin, of quartz-rich erratics at higher elevations, it is reasonable to argue that they too were deposited with no significant pre-exposure. This is reinforced by the clustering of multi-isotope exposure ages from each sampled site. Thus, we conclude that the cosmogenic nuclide concentration in the erratics accurately reflects their exposure history since deposition.

Laboratory and analytical techniques. Whole-rock samples were crushed and sieved to obtain the 250–710 μm fraction. Be and Al were selectively extracted from the quartz component of the whole-rock sample at the University of Edinburgh's Cosmogenic Nuclide Laboratory following established methods[37,38]. $^{10}Be/^{9}Be$ and $^{26}Al/^{27}Al$ ratios were measured in 20–30 g of quartz at the Scottish Universities Environmental Research Centre Accelerator Mass Spectrometry (AMS) Laboratory in East Kilbride, UK. Measurements are normalized to the NIST SRM-4325 Be standard material with a revised[39] nominal $^{10}Be/^{9}Be$ of 2.79×10^{-11} and half-life of 1.387 Ma (refs 40,41), and the Purdue Z92-0222 Al standard material with a nominal $^{26}Al/^{27}Al$ of 4.11×10^{-11}, which agrees with the Al standard material of Nishiizumi[42] with half-life of 0.705 Ma (ref. 43). Scottish Universities Environmental Research Centre ^{10}Be-AMS is insensitive to ^{10}B interference[44] and the interferences to ^{26}Al detection are well characterized[45]. Process blanks ($n=6$) were spiked with 250 μg ^{9}Be carrier (Scharlau Be carrier, 1,000 mg l^{-1}, density 1.02 g ml^{-1}) and 1.5 mg ^{27}Al carrier (Fischer Al carrier, 1,000 p.p.m.). Samples were spiked with 250 μg ^{9}Be carrier and up to 1.5 mg ^{27}Al carrier (the latter value varied depending on the native Al-content of the sample). Blanks range from 3.3 to 9.3×10^{-15} [$^{10}Be/^{9}Be$] (<1% of total ^{10}Be atoms in all but the ice-margin samples); and 1.6–7.5×10^{-15} [$^{26}Al/^{27}Al$] (<1% of total ^{26}Al atoms in all but the ice margin samples). Concentrations in Supplementary Table 1 are corrected for process blanks; uncertainties include propagated AMS sample/lab-blank uncertainty and a 2% carrier mass uncertainty and a 3% stable ^{27}Al measurement inductively coupled plasma optical emission spectrometry uncertainty.

Neon isotopes were measured in ~250 mg of leached quartz (250–500 μm). Samples were wrapped in aluminium foil and loaded into a Monax glass tree and evacuated to $<10^{-8}$ torr for 48 h before analysis. Samples were successively heated for 20 min to 1,200 °C in a double-vacuum resistance furnace with a tungsten heating element and a molybdenum crucible. The extracted gas was cleaned on two hot SAES TiZr getters. The heavy noble gases (Ar, Kr and Xe) were absorbed onto a charcoal trap cooled with liquid nitrogen. Neon was then absorbed on to a charcoal trap at −228 °C for 20 min, and the residual He was removed by a turbomolecular pump. The Ne was released from the charcoal at −173 °C, and the isotopic composition analysed using a MAP 215-50 noble gas mass spectrometer. All Ne isotopes were measured in 11 peak jumping cycles using a Burle channeltron electron multiplier operated in pulse-counting mode. Neon abundances were determined by peak height comparison with Ne from 95.2 ± 0.5 μcc STP air. The reproducibility of Ne abundances was better than ±1.5%, and isotopic ratios of replicate calibrations were better than ±0.5%. Interference corrections and detailed analytical procedure is presented elsewhere[46]. The ^{20}Ne blank at 1,200 °C was typically $\sim 1 \times 10^{-11}$ ccSTP and were indistinguishable from the atmospheric isotopic composition after correction for interfering species. Consequently no blank correction is made to the data in Supplementary Table 2. The consistency of procedures is demonstrated by the reproducibility of the cosmogenic ^{21}Ne concentration in replicate analyses of MH12-27 (Supplementary Table 2). In all samples Ne isotope compositions are consistent with binary mixture of air and cosmogenic Ne. The ^{21}Ne concentrations in Supplementary Table 2 include a correction for nucleogenic ^{21}Ne (that is, non-cosmogenic ^{21}Ne) of $7.7 \pm 2.4 \times 10^{6}$ at g^{-1}. This is the value estimated by Middleton *et al.*[47] for Beacon Sandstone; we use this value as a best estimate based on the similar lithology and thermal history of the rocks. This value is close to the mode of the range of nucleogenic ^{21}Ne measured in Antarctic rocks (see Balco and Shuster[48] for a review). However, there is likely variability in the nucleogenic ^{21}Ne concentrations that could impact the youngest samples. The conclusions are insensitive to uncertainty in burial time.

Exposure age calculations. For exposure age calculations we used default settings in Version 2.0 of the CRONUScalc programme[49]. This is the product of the CRONUS-Earth collaboration that allows for all commonly used nuclides to be calculated using the same underlying framework, resulting in internally consistent cross-nuclide calculations for exposure ages, erosion rates and calibrations. The CRONUS-Earth production rates[50] with the nuclide-dependent scaling of Lifton-Sato-Dunai[51] were used to calculate the ages presented in the paper. Sea level and high latitude production rates are 3.92 ± 0.31 atoms g^{-1} a^{-1} for ^{10}Be and 28.5 ± 3.1 atoms g^{-1} a^{-1} for ^{26}Al. However, the use of Lal/Stone[26,52] scaling does not change the conclusions of the paper despite the ~3 and 8% older exposure ages for ^{10}Be and ^{26}Al, respectively. Rock density is 2.7 g cm^{-3} and the attenuation length used is 153 ± 10 g cm^{-2}. No corrections are made for rock surface erosion or snow cover and thus exposure ages are minima. Finally, we make no attempt to account for production rate variations caused by elevation changes associated with glacial isostatic adjustment of the massif through time[53]. This is justified because the samples have been exposed for multiple glacial cycles and thus any variations in elevation associated with ice loading and unloading, which has been of similar magnitude (maximum elevation difference 170 m), are likely to have been averaged out to the point of being smaller than other sources of uncertainty.

The CRONUScalc code for ^{21}Ne was modelled after the existing code for ^{3}He and only includes spallation production. The ^{21}Ne production rate is tied to the total CRONUScalc ^{10}Be production rate (assuming 1.5% production from muons[49]) with a $^{21}Ne/^{10}Be$ ratio of 4.08 ± 0.37 (ref. 48), resulting in a ^{21}Ne production rate of 16.26 ± 1.96 atoms g^{-1} a^{-1} at sea level, high latitude scaled according to nuclide-dependent Lifton-Sato-Dunai[48,50,51]. There are several other alternative ^{21}Ne production rates (all converted to be consistent with Lifton-Sato-Dunai scaling): 14.5 (Amidon *et al.*[54], 18.0 (Vermeesch *et al.*[55]) and 18.9 (Niedermann *et al.*[56]). The Balco and Shuster[48] rate was used because it is based on ratios tied to ^{10}Be instead of ^{26}Al, it uses a relatively large dataset compared with other ^{21}Ne studies, it was performed using Antarctic samples, and the resulting rate falls in the middle of the production rate range. The differences in age using the other production rates given above range from 12% older to 14% younger than those given in the paper. While these changes are significant, the exposure ages are consistently similar or older than the corresponding ^{10}Be and ^{26}Al ages so the exact choice of ^{21}Ne production rate does not affect the conclusions presented in the paper. For comparison, Lal/Stone[26,52] scaling in CRONUScalc was used in conjunction with the production rate from Balco and Shuster[48]and produced ^{21}Ne exposure ages that were approximately 3% younger than those produced using the nuclide-dependent Lifton-Sato-Dunai scaling scheme with the Balco and Shuster[48] production rate.

Supplementary Figs 5 and 6 show plots of the isotopic ratios of $^{26}Al/^{10}Be$ and $^{21}Ne/^{10}Be$. Samples should plot within the erosion island if they have been continuously exposed and eroding, and within the complex zone if they have been buried for a significant period of time, long enough for the shorter lived nuclide to preferentially decay. The $^{26}Al/^{10}Be$ system should be more sensitive to recent burial than the $^{21}Ne/^{10}Be$ system because of the shorter half-life of ^{26}Al (0.705 ka). In our samples, the burial signal implied by the $^{21}Ne/^{10}Be$ ratios is greater than that implied by $^{26}Al/^{10}Be$ ratios. There are a few possible explanations. First, this may partly reflect the uncertainties on ^{21}Ne production rates as discussed above. Second, it is possible that the samples contain additional nucleogenic ^{21}Ne that has not been corrected for. A final explanation is that ^{21}Ne, which is stable, is recording a period of exposure that is not evident in the $^{26}Al/^{10}Be$ system. At present it is not possible to discriminate between the above scenarios. In any case, our conclusions are not sensitive to these minor discrepancies.

Till analysis for marine traces. Scherer (R. Scherer, personal communications, 2014) examined four till samples from both current and elevated blue-ice moraines and found no evidence of diatoms or biogenic silica.

References

1. Joughin, I. & Alley, R. B. Stability of the West Antarctic ice sheet in a warming world. *Nat. Geosci.* **4,** 506–513 (2011).
2. Mercer, J. H. West Antarctic Ice Sheet and CO_2 greenhouse effect - threat of disaster. *Nature* **271,** 321–325 (1978).
3. Bamber, J. L., Riva, R. E. M., Vermeersen, B. L. A. & LeBrocq, A. M. Reassessment of the potential sea-level rise from a collapse of the West Antarctic Ice Sheet. *Science* **324,** 901–903 (2009).
4. Joughin, I., Smith, B. E. & Medley, B. marine ice sheet collapse potentially under way for the Thwaites Glacier Basin, West Antarctica. *Science* **344,** 735–738 (2014).
5. Rignot, E., Mouginot, J., Morlighem, M., Seroussi, H. & Scheuchl, B. Widespread, rapid grounding line retreat of Pine Island, Thwaites, Smith, and Kohler glaciers, West Antarctica, from 1992 to 2011. *Geophys. Res. Lett.* **41,** 3502–3509 (2014).
6. Scherer, R. P. *et al.* Pleistocene collapse of the West Antarctic ice sheet. *Science* **281,** 82–85 (1998).
7. Strugnell, J. M., Watts, P. C., Smith, P. J. & Allcock, A. L. Persistent genetic signatures of historic climatic events in an Antarctic octopus. *Mol. Ecol.* **21,** 2775–2787 (2012).
8. Barnes, D. K. A. & Hillenbrand, C.-D. Faunal evidence for a late quaternary trans-Antarctic seaway. *Global Change Biol.* **16,** 3297–3303 (2010).
9. Raymo, M. E. & Mitrovica, J. X. Collapse of polar ice sheets during the stage 11 interglacial. *Nature* **483,** 453–456 (2012).
10. Kopp, R. E., Simons, F. J., Mitrovica, J. X., Maloof, A. C. & Oppenheimer, M. Probabilistic assessment of sea level during the last interglacial stage. *Nature* **462,** 863–U851 (2009).
11. Pollard, D. & DeConto, R. M. Modelling West Antarctic ice sheet growth and collapse through the past five million years. *Nature* **458,** 329–U389 (2009).
12. Jamieson, S. S. R., Sugden, D. E. & Hulton, N. R. J. The evolution of the subglacial landscape of Antarctica. *Earth Planet. Sci. Lett.* **293,** 1–27 (2010).
13. Ross, N. *et al.* The Ellsworth subglacial highlands: inception and retreat of the West Antarctic Ice Sheet. *Geol. Soc. Am. Bull.* **126,** 3–15 (2014).
14. Fogwill, C. J., Hein, A. S., Bentley, M. J. & Sugden, D. E. Do blue-ice moraines in the Heritage Range show the West Antarctic ice sheet survived the last interglacial? *Palaeogeogr. Palaeoclimatol. Palaeoecol.* **335,** 61–70 (2012).
15. Bintanja, R. On the glaciological, meteorological, and climatological significance of Antarctic blue ice areas. *Rev. Geophys.* **37,** 337–359 (1999).
16. Stone, J. O. *et al.* Holocene deglaciation of Marie Byrd Land, West Antarctica. *Science* **299,** 99–102 (2003).
17. Ackert, R. P. *et al.* Measurements of past ice sheet elevations in interior West Antarctica. *Science* **286,** 276–280 (1999).
18. Mackintosh, A. *et al.* Exposure ages from mountain dipsticks in Mac. Robertson Land, East Antarctica, indicate little change in ice-sheet thickness since the Last Glacial Maximum. *Geology* **35,** 551–554 (2007).

19. Ackert, Jr R. P. *et al.* Controls on interior West Antarctic Ice Sheet elevations: inferences from geologic constraints and ice sheet modeling. *Quat. Sci. Rev.* **65**, 26–38 (2013).

20. Mukhopadhyay, S., Ackert, R. P., Pope, A. E., Pollard, D. & DeConto, R. M. Miocene to recent ice elevation variations from the interior of the West Antarctic ice sheet: constraints from geologic observations, cosmogenic nuclides and ice sheet modeling. *Earth Planet. Sci. Lett.* **337**, 243–251 (2012).

21. Joy, K., Fink, D., Storey, B. & Atkins, C. A 2 million year glacial chronology of the Hatherton Glacier, Antarctica and implications for the size of the East Antarctic Ice Sheet at the Last Glacial Maximum. *Quat. Sci. Rev.* **83**, 46–57 (2014).

22. Balco, G., Stone, J. O. H., Sliwinski, M. G. & Todd, C. Features of the glacial history of the Transantarctic Mountains inferred from cosmogenic Al-26, Be-10 and Ne-21 concentrations in bedrock surfaces. *Antarctic Sci.* **26**, 708–723 (2014).

23. Lilly, K., Fink, D., Fabel, D. & Lambeck, K. Pleistocene dynamics of the interior East Antarctic ice sheet. *Geology* **38**, 703–706 (2010).

24. Bentley, M. J. *et al.* Deglacial history of the West Antarctic Ice Sheet in the Weddell Sea embayment: constraints on past ice volume change. *Geology* **38**, 411–414 (2010).

25. Todd, C. & Stone, J. O. Deglaciation of the southern Ellsworth Mountains, Weddell Sea sector of the West Antarctic Ice Sheet. in *Proceedings, 11th Annual WAIS Workshop* (Sterling, Virginia, West Antarctic Ice Sheet Initiative, 2004).

26. Lal, D. Cosmic-ray labeling of erosion surfaces—*in situ* nuclide production-rates and erosion models. *Earth Planet. Sci. Lett.* **104**, 424–439 (1991).

27. Denton, G. H., Bockheim, J. G., Rutford, R. H. & Andersen, B. G. in *Geology and Palaeontology of the Ellsworth mountains, West Antarctica* Vol. Memoir 170 (eds Webers, G. F., Craddock, C. & Splettstoesser, J. F.) (Geological Society of America, 1992).

28. Alley, R. B., Anandakrishnan, S., Dupont, T. K., Parizek, B. R. & Pollard, D. Effect of sedimentation on ice-sheet grounding-line stability. *Science* **315**, 1838–1841 (2007).

29. Sugden, D. E. *et al.* Emergence of the Shackleton Range from beneath the Antarctic Ice Sheet due to glacial erosion. *Geomorphology* **208**, 190–199 (2014).

30. Stern, T. A., Baxter, A. K. & Barrett, P. J. Isostatic rebound due to glacial erosion within the Transantarctic Mountains. *Geology* **33**, 221–224 (2005).

31. Finlayson, A., Fabel, D., Bradwell, T. & Sugden, D. Growth and decay of a marine terminating sector of the last British-Irish Ice Sheet: a geomorphological reconstruction. *Quat. Sci. Rev.* **83**, 28–45 (2014).

32. Lewis, A. R., Marchant, D. R., Ashworth, A. C., Hemming, S. R. & Machlus, M. L. Major middle Miocene global climate change: evidence from East Antarctica and the Transantarctic Mountains. *Geol. Soc. Am. Bull* **119**, 1449–1461 (2007).

33. Steig, E. J. *et al.* Influence of West Antarctic Ice Sheet collapse on Antarctic surface climate. *Geophys. Res. Lett.* **42**, 4862–4868 (2015).

34. Atkins, C. B., Barrett, P. J. & Hicock, S. R. Cold glaciers erode and deposit: evidence from Allan Hills, Antarctica. *Geology* **30**, 659–662 (2002).

35. Church, J. A. & Clark, P. U. *Sea Level Change* 1137–1216 (Cambridge University Press, 2013).

36. Ross, N. *et al.* Steep reverse bed slope at the grounding line of the Weddell Sea sector in West Antarctica. *Nat. Geosci.* **5**, 393–396 (2012).

37. Bierman, P. R. *et al.* in *Reviews in Mineralogy and Geochemistry* Vol. 50 147−205(*MineralogicalSocAmerica*, 2002).

38. Kohl, C. P. & Nishiizumi, K. Chemical isolation of quartz for measurement of in situ-produced cosmogenic nuclides. *Geochim. Cosmochim. Acta.* **56**, 3583–3587 (1992).

39. Nishiizumi, K. *et al.* Absolute calibration of Be-10 AMS standards. *Nucl. Instrum. Methods Phys. Res. B* **258**, 403–413 (2007).

40. Chmeleff, J., von Blanckenburg, F., Kossert, K. & Jakob, D. Determination of the Be-10 half-life by multicollector ICP-MS and liquid scintillation counting. *Nucl. Instrum. Methods Phys. Res. B* **268**, 192–199 (2010).

41. Korschinek, G. *et al.* A new value for the half-life of Be-10 by heavy-ion elastic recoil detection and liquid scintillation counting. *Nucl. Instrum. Methods Phys. Res. B* **268**, 187–191 (2010).

42. Nishiizumi, K. Preparation of Al-26 AMS standards. *Nucl. Instrum. Methods Phys. Res. B* **223-224**, 388–392 (2004).

43. Xu, S., Dougans, A. B., Freeman, S., Schnabel, C. & Wilcken, K. M. Improved Be-10 and Al-26-AMS with a 5 MV spectrometer. *Nucl. Instrum. Methods Phys. Res. B* **268**, 736–738 (2010).

44. Xu, S., Freeman, S. P. H. T., Sanderson, D., Shanks, R. P. & Wilcken, K. M. Cl can interfere with Al3 + AMS but B need not matter to Be measurement. *Nucl. Instrum. Methods Phys. Res. B* **294**, 403–405 (2013).

45. Xu, S., Freeman, S. P. H. T., Rood, D. H. & Shanks, R. P. Al-26 interferences in accelerator mass spectrometry measurements. *Nucl. Instrum. Methods Phys. Res. B* **333**, 42–45 (2014).

46. Codilean, A. T. *et al.* Single-grain cosmogenic Ne-21 concentrations in fluvial sediments reveal spatially variable erosion rates. *Geology* **36**, 159–162 (2008).

47. Middleton, J. L., Ackert, Jr R. P. & Mukhopadhyay, S. Pothole and channel system formation in the McMurdo Dry Valleys of Antarctica: new insights from cosmogenic nuclides. *Earth Planet. Sci. Lett.* **355**, 341–350 (2012).

48. Balco, G. & Shuster, D. L. Production rate of cosmogenic Ne-21 in quartz estimated from Be-10, Al-26, and Ne-21 concentrations in slowly eroding Antarctic bedrock surfaces. *Earth Planet. Sci. Lett.* **281**, 48–58 (2009).

49. Marrero, S. *et al.* Cosmogenic nuclide systematics and the CRONUScalc program. *Quat. Geochronol* **31**, 160–187 (2016).

50. Borchers, B. *et al.* Geological calibration of spallation production rates in the CRONUS-Earth project. *Quat. Geochronol* **31**, 188–198 doi:10.1016/j.quageo.2015.01.009 (2016).

51. Lifton, N., Sato, T. & Dunai, T. J. Scaling in situ cosmogenic nuclide production rates using analytical approximations to atmospheric cosmic-ray fluxes. *Earth Planet. Sci. Lett.* **386**, 149–160 (2014).

52. Stone, J. O. Air pressure and cosmogenic isotope production. *J. Geophys. Res. Solid Earth* **105**, 23753–23759 (2000).

53. Suganuma, Y., Miura, H., Zondervan, A. & Okuno, J. East Antarctic deglaciation and the link to global cooling during the Quaternary: evidence from glacial geomorphology and Be-10 surface exposure dating of the Sor Rondane Mountains, Dronning Maud Land. *Quat. Sci. Rev.* **97**, 102–120 (2014).

54. Amidon, W. H., Rood, D. H. & Farley, K. A. Cosmogenic He-3 and Ne-21 production rates calibrated against Be-10 in minerals from the Coso volcanic field. *Earth Planet. Sci. Lett.* **280**, 194–204 (2009).

55. Vermeesch, P. *et al.* Cosmogenic He-3 and Ne-21 measured in quartz targets after one year of exposure in the Swiss Alps. *Earth Planet. Sci. Lett.* **284**, 417–425 (2009).

56. Niedermann, S. *et al.* Cosmic-ray produced Ne-21 in terrestrial quartz—the neon inventory of Sierra-Nevada quartz separates. *Earth Planet. Sci. Lett.* **125**, 341–355 (1994).

57. Fretwell, P. *et al.* Bedmap2: improved ice bed, surface and thickness datasets for Antarctica. *Cryosphere* **7**, 375–393 (2013).

58. Bindschadler, R. *et al.* Getting around Antarctica: new high-resolution mappings of the grounded and freely-floating boundaries of the Antarctic ice sheet created for the International Polar Year. *Cryosphere* **5**, 569–588 (2011).

59. De Keyser, M. *The International Antarctic Weather Forecasting Handbook* (eds. Turner, J. & Pendlebury, S.) (British Antarctic Survey, Cambridge, 2004). Available at http://www.antarctica.ac.uk/met/momu/International_Antarctic_Weather_Forecasting_Handbook/updatePatriot Hills.php (cited 2008).

Acknowledgements

The research was funded by the UK Natural Environment Research Council grant numbers NE/I025840/1, NE/I027576/1, NE/I024194/1 and NE/I025263/1. We thank the British Antarctic Survey for logistical support and Scott Webster, Malcolm Airey and Phil Stevens for field support. Reed Scherer checked till samples for marine diatoms. Claire Todd, John Stone and Mike Bentley provided helpful discussions. We thank Martin Siegert and the British Antarctic Survey for use of the RES data.

Author contributions

A.S.H., J.W. and D.E.S. conceived the project and carried out the fieldwork and analysis with S.A.D., S.M.M. K.W. and M.J.W. The cosmogenic analysis was by A.S.H., S.P.H.T.F., F.M.S. and S.M.M., E.J.S. contributed analysis of ice-sheet model results. All contributed to the writing of the paper.

Additional information

Competing financial interests: The authors declare no competing financial interests.

24

End-Cretaceous extinction in Antarctica linked to both Deccan volcanism and meteorite impact via climate change

Sierra V. Petersen[1], Andrea Dutton[2] & Kyger C. Lohmann[1]

The cause of the end-Cretaceous (KPg) mass extinction is still debated due to difficulty separating the influences of two closely timed potential causal events: eruption of the Deccan Traps volcanic province and impact of the Chicxulub meteorite. Here we combine published extinction patterns with a new clumped isotope temperature record from a hiatus-free, expanded KPg boundary section from Seymour Island, Antarctica. We document a $7.8 \pm 3.3\,°C$ warming synchronous with the onset of Deccan Traps volcanism and a second, smaller warming at the time of meteorite impact. Local warming may have been amplified due to simultaneous disappearance of continental or sea ice. Intra-shell variability indicates a possible reduction in seasonality after Deccan eruptions began, continuing through the meteorite event. Species extinction at Seymour Island occurred in two pulses that coincide with the two observed warming events, directly linking the end-Cretaceous extinction at this site to both volcanic and meteorite events via climate change.

[1] Department of Earth & Environmental Sciences, University of Michigan, 2534 C.C. Little Building, 1100 North University Avenue, Ann Arbor, Michigan 48109, USA. [2] Department of Geological Sciences, University of Florida, 241 Williamson Hall, PO Box 112120, Gainesville, Florida 32611, USA. Correspondence and requests for materials should be addressed to S.V.P. (email: sierravp@umich.edu).

The cause of the Cretaceous–Paleogene (KPg) mass extinction remains controversial due to difficulties in separating the influence of two closely timed potential causal events. Some suggest that the eruption of the massive Deccan Traps volcanic province in India caused species extinction[1] through trace metal toxicity[2] or negative effects of volatiles emitted during the eruption (for example, CO_2, SO_2)[3]. Others cite the impact of the massive Chicxulub meteorite as the cause[4,5]. Recent work has suggested that the impact event might have triggered and accelerated eruption of the Deccan Traps[6,7], further complicating the cause–effect relationship. Distinguishing between the effects of these two potential causal events can be difficult because many KPg boundary (KPB) sites have hiatuses, insufficient temporal resolution, and lack of species continuity across the boundary.

Seymour Island, Antarctica (64°17′ S, 56°45′ W) is particularly well-suited for studying the KPB interval due to its expanded section, continuous sedimentation, and abundant, exceptionally preserved macrofossils, including some species and genera that survive across the KPB[8,9]. Previous efforts to link temperature change and biotic turnover at Seymour Island relied on untested assumptions about the value and stability of the oxygen isotopic composition of seawater ($\delta^{18}O_w$) to calculate temperature[10]. These assumptions can now be avoided (and tested) using the carbonate clumped isotope paleothermometer, a new proxy that can measure temperature independent of $\delta^{18}O_w$, and can therefore directly calculate $\delta^{18}O_w$ for each sample[11].

Here we present a record of high latitude ocean temperature and $\delta^{18}O_w$ from Seymour Island that covers the last few million years of the Cretaceous and crosses the KPB (from ~69–65.5 Ma), measured on well-preserved bivalve shells. We find evidence of climate change at the onset of Deccan volcanism and again at the KPB that align with two previously identified pulses of extinction[10], indicating the complete end-Cretaceous extinction at this site is due to the combined effects of the volcanic and meteorite kill mechanisms.

Results

Stable and clumped isotope analysis of fossil bivalves.
Twenty-nine well-preserved shells, representing five species of bivalve (*Lahillia larseni*, *Cucullaea antarctica*, *Cucullaea ellioti*, *Eselaevitrigonia regina* and *Nordenskjoldia nordenskjoldi*) were selected from the Zinsmeister collection[9], and were analysed for $\delta^{18}O$, $\delta^{13}C$, and the clumped isotope composition, Δ_{47} (see Methods). These shells represent the best-preserved individuals from a larger sample set of 116 shells that were analysed for only $\delta^{18}O$ and $\delta^{13}C$ (see Methods). Shell preservation was determined by X-ray diffraction, cathodoluminescence and trace element screening (see Methods). $\delta^{18}O_w$ values were calculated for each clumped isotope sample using the Δ_{47}-derived temperature, measured $\delta^{18}O$ of the shell, and an equilibrium temperature relationship for aragonite[12]. Shells come from Units 9 and 10 of the López de Bertodano Formation, which represent an outer to inner shelf environment with a water depth of <200 m (ref. 8). Most shells were sampled in two locations, the umbo (hinge) and the ventral margin (see Methods). An age model was constructed using published magnetostratigraphic data[10] and was corroborated with published[13] and new $^{87}Sr/^{86}Sr$ measurements (see Methods).

Cretaceous–Paleogene temperature at Seymour Island.
The Δ_{47}-derived temperatures for each species and shell position are shown in Fig. 1a. Temperatures warm from ~5 °C to ~14 °C between 68.7 and 67.8 Ma. After ~1 Myr of sustained warm (~9–12 °C) temperatures, a gradual cooling back to ~4 °C occurs from 66.9 to 66.25 Ma. Approximately 150 kyr before the KPg boundary, a marked warming of ~7.8 ± 3.3 °C occurs. After the initial warming, temperatures decline until a second smaller warming pulse occurs at the KPB (1.1 ± 2.7 °C), followed by continued decline until pre-event levels are reached ~200–400 kyr after the initial spike. Although the second warming is minor when comparing the horizon means, individual samples reach temperatures near those seen at the peak of the earlier, larger warming spike (Fig. 1a).

The mean ocean temperature for the entire section is 7.7 ± 3.4 °C (1 s.d.), consistent with a terrestrial temperature estimate of 7 °C from fossil wood[14] and a soil temperature estimate of 10–13 °C from branched tetraethers[15]. Maastrichtian ocean temperatures from Seymour Island have been previously estimated at anywhere from 5 to 16 °C based on $\delta^{18}O$ of micro- and macro-fossils[10,16–19], but these estimates assume a fixed $\delta^{18}O_w$ value, which is not appropriate for this location (see below). The long-term temperature pattern is mirrored in all species and shell positions (Fig. 1a; Supplementary Fig. 1), and is robust across different methods of calculating horizon means (see Methods; Supplementary Fig. 2). The pattern is consistent with terrestrial climate reconstructions based on fungal palynomorphs and pollen grains, which document warm, humid conditions during Chron 30 N, bracketed by cooler conditions before and after[20]. Similar pre-KPB warming in Chron 29R, albeit of lower magnitude, has been seen before, with ~5 °C of warming observed in a terrestrial section in North America[21] and 2–3 °C of warming recorded in multiple open-marine records[22–24].

Patterns in position-specific isotopic and temperature data.
Position-specific measurements of the umbo and ventral margin can provide additional information about seasonal variability. Within a single shell, the difference between the temperature recorded in the umbo and ventral margin can be as great as 7–8 °C or as little as 0 °C. All specimens of *Lahillia* before the warming spike show consistent offsets between the two positions, with temperatures ~3–9 °C warmer, $\delta^{18}O_w$ values ~1–3‰ higher, and $\delta^{13}C$ values ~1–4‰ higher in the umbo than in the ventral margin (Fig. 2; Supplementary Fig. 3). In contrast, *Cucullaea* specimens have $\delta^{13}C$ values ~2–6‰ lower in the umbo than in the ventral margin (Fig. 2b). Position-specific differences in temperature and $\delta^{18}O_w$ for *Cucullaea* are less consistent, but are generally in the opposite direction of those seen in *Lahillia* (Fig. 2; Supplementary Fig. 4).

We suggest these position-specific differences in temperature, $\delta^{18}O_w$, and $\delta^{13}C$ are due to different seasonal aliasing early and late in the life cycle of the bivalve. Bivalves cease shell growth during reproduction[25], or when under thermal stress (either too hot[26] or cold[27], depending on the taxon), and shell growth slows and becomes more seasonally restricted later in life[28]. For example, Eocene-age *Cucullaea* shells from Seymour Island were shown to cease growth during the warmest months of each year, based on high-resolution $\delta^{18}O$ measurements[26]. The umbo, which represents early life, should therefore average a larger number of months than the ventral margin, which represents later life when shell growth slows and ceases during reproduction months. For *Lahillia*, the cooler temperature recorded in the ventral margin suggests that shell growth ceased in summer later in life. For *Cucullaea*, the opposite is true, with shell growth biased towards warmer months later in life. The magnitude of the temperature difference between the two positions (0–9 °C) represents a minimum estimate of the seasonal amplitude at this latitude. Position-specific differences can also be seen in $\delta^{18}O$ and $\delta^{13}C$ in shells not measured for clumped isotopes (Supplementary Figs 5–7).

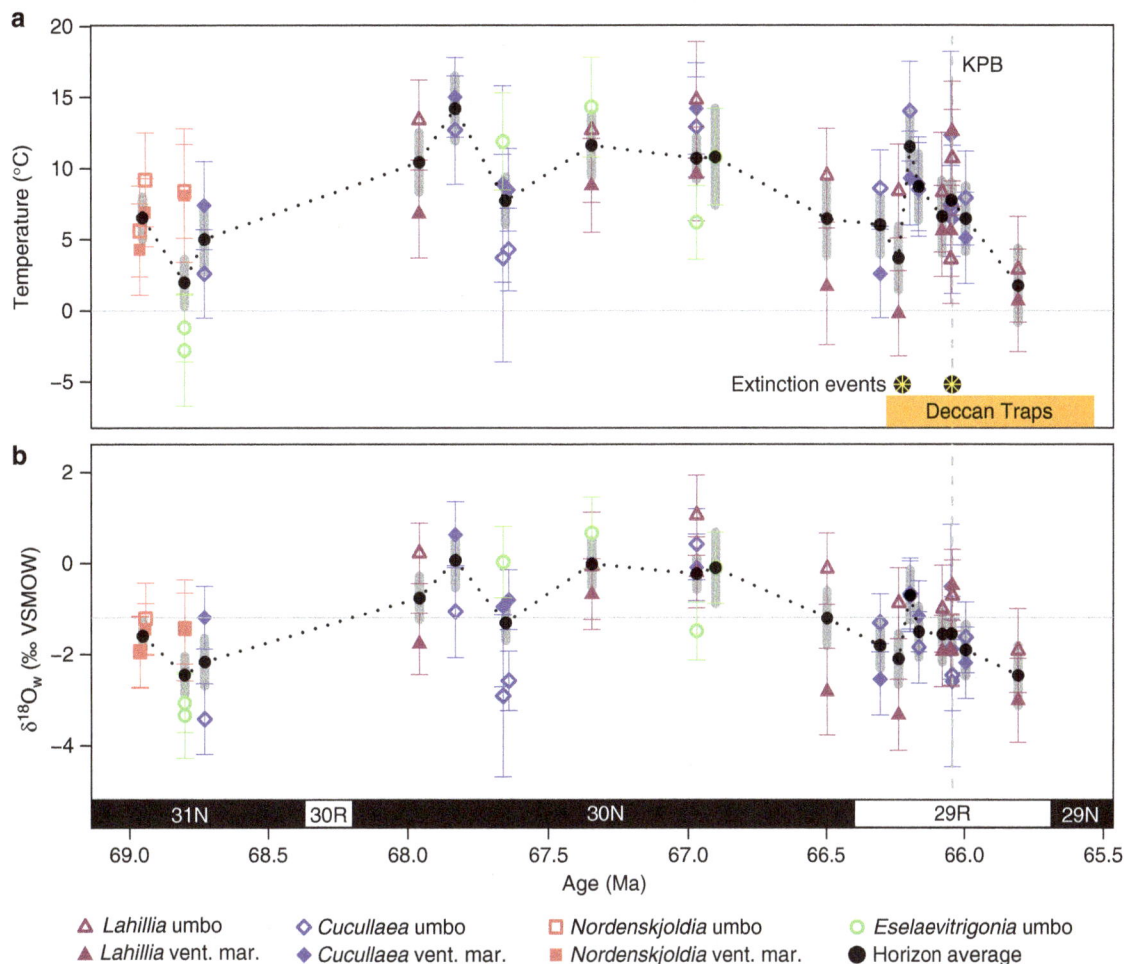

Figure 1 | Late Cretaceous temperature and $\delta^{18}O_w$ at Seymour Island. (a) Δ_{47}-derived temperature and (b) $\delta^{18}O_w$ versus age. The timing of Deccan Traps volcanism[7,33] (orange rectangle) and two extinction pulses[10] (black/orange stars) are shown for comparison. Coloured error bars on individual points represent 1 s.e. of the mean of many replicates or the long-term reproducibility of carbonate standards, whichever is larger (see Methods). Horizon averages in **a,b** were calculated as the inverse variance weighted mean, to account for variable reproducibility of samples, with thick grey error bars showing the inverse variance weighted error on the mean (see Methods). Black line connecting horizon means is dashed to represent extrapolation of temperature and $\delta^{18}O_w$ trends between measured horizons. Samples within ± 3.5 m stratigraphic position of another sample were combined into a single horizon for horizon means (see Methods). Grey horizontal line in **a** represents the freezing point (0 °C) and in **b** represents the ice free, latitude-adjusted seawater value of − 1.2‰ (ref. 30). Vertical grey dashed line denotes the KPg boundary (labelled KPB)[34]. Age model construction is described in Methods. Data for this figure can be found in Supplementary Data 5 and 6.

High variability in $\delta^{18}O_w$ within and between shells. Individual sample $\delta^{18}O_w$ values vary from − 3.4 to +1.0‰ through the section, and $\delta^{18}O_w$ is highly correlated to temperature (Fig. 1; Supplementary Fig. 8; Supplementary Discussion). The mean $\delta^{18}O_w$ value for the entire section is − 1.3 ± 0.8‰ (1 s.d.). This agrees with the value predicted for an ice-free world (− 1.0‰; ref. 29) or for an ice-free world adjusted for the modern latitudinal isotopic gradient (− 1.2‰; ref. 30), two $\delta^{18}O_w$ values typically assumed when converting carbonate $\delta^{18}O$ to temperature. However, the large range and temporal variability in $\delta^{18}O_w$, on the order of ± 1–2‰, suggests that the typical assumptions of constant $\delta^{18}O_w$ are incorrect for this location and introduce bias into temperatures calculated from $\delta^{18}O$. Figure 3 compares the clumped isotope temperature record to temperature calculated from carbonate $\delta^{18}O$ for shells from this study and from Tobin *et al.*[10], assuming a constant $\delta^{18}O_w$ value of − 1.0‰. Both $\delta^{18}O$-derived temperature records fail to capture the temperature structure seen in the clumped isotope record on both million-year and sub-million-year timescales. For example, both $\delta^{18}O$-derived temperature records do not identify the

coldest temperatures (< 5 °C) around 68.8, 66.4 and 65.7 Ma (Fig. 3). This is because the positively correlated variation in temperature and $\delta^{18}O_w$ have opposite effects on shell $\delta^{18}O$, masking both signals and resulting in minimally varying, intermediate $\delta^{18}O$ values. If a fixed value for $\delta^{18}O_w$ is assumed, the resulting $\delta^{18}O$-derived temperature record has only around half the variability of the 'true' clumped isotope temperature record, which allows for variations in $\delta^{18}O_w$ (7 °C versus 12.5 °C temperature range for shells from this study). Therefore, temperatures derived from $\delta^{18}O$ of micro- and macro-fossils from Seymour Island that assume a fixed $\delta^{18}O_w$ value[10,16–19] should only be taken on average, and any time series that do not account for changing $\delta^{18}O_w$ should be treated with caution and assumed to underestimate true temperature variability.

Discussion

We propose that the observed $\delta^{18}O_w$ variability reflects the influence of continental runoff. Precipitation over land is more depleted in ^{18}O than ocean water, sometimes by more than

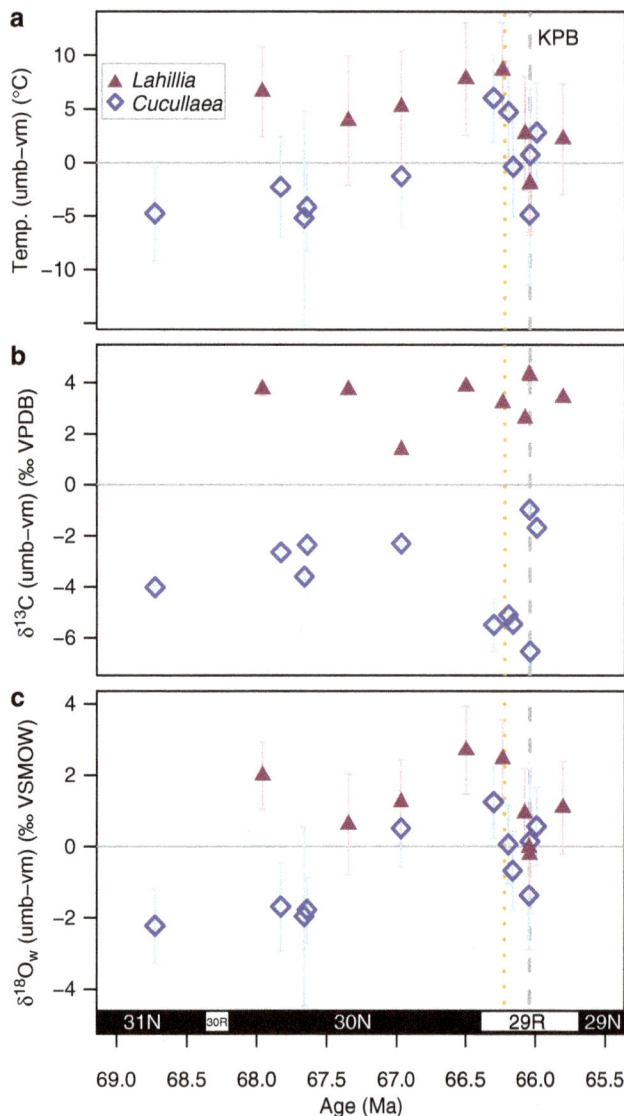

Figure 2 | Position-specific differences. Differences in (**a**) temperature, (**b**) $\delta^{13}C$ and (**c**) $\delta^{18}O_w$ for *Lahillia* and *Cucullaea* versus age. All differences calculated as the umbo minus the ventral margin value (umb-vm), with the horizontal grey line at 0 representing no difference between positions. Error bars are 1 s.e. for temperature and $\delta^{18}O_w$ and 1 s.d. for $\delta^{13}C$, propagated through the difference calculation. *Lahillia* consistently shows warmer temperatures, and higher $\delta^{13}C$ and $\delta^{18}O_w$ in the umbo relative to the ventral margin (+ difference). *Cucullaea* generally shows the opposite, with warmer temperatures, and higher $\delta^{13}C$ and $\delta^{18}O_w$ in the ventral margin relative to the umbo (- difference), although this pattern is less consistent. After the pre-KPB warming spike, the position-specific differences in temperature and $\delta^{18}O_w$ for both *Lahillia* and *Cucullaea* decrease to within error of zero and remain low until the end of the study interval. Differences in $\delta^{13}C$ remain fairly constant through the section. Vertical grey dashed line represents the KPg boundary (labelled KPB)[34] and orange dotted line represents the timing of the first extinction event[10]. Age model construction is described in Methods.

10–15‰, so any appreciable contribution of runoff would decrease $\delta^{18}O_w$ values in shelf waters. The correlation between $\delta^{18}O_w$ and temperature, with the coldest temperatures correlating with the lowest $\delta^{18}O_w$ values, suggests a climatological control on runoff delivery. Precipitation becomes more depleted in ^{18}O as air temperature decreases. In addition, colder temperatures would

promote the accumulation of snowpack in winter and the concentrated delivery of isotopically depleted meltwater during the spring melt, reducing $\delta^{18}O_w$ below expected open-marine values. Terrestrial carbon, depleted in $\delta^{13}C$ relative to marine carbon, could accompany the spring melt, explaining the co-occurrence of low $\delta^{18}O_w$ and low $\delta^{13}C$. This hypothesized seasonal meltwater delivery is consistent with the temperature differences observed between positions in *Lahillia* (*Cucullaea*), with lower (higher) $\delta^{18}O_w$ and $\delta^{13}C$ in the ventral margin, the position skewed more towards colder (warmer) months.

The coldest temperatures recorded in the Seymour Island section (at 68.8, 66.4 and 65.7 Ma) are near the freezing point, implying the possibility of sea ice formation near Seymour Island. Dinoflagellate cyst abundance has previously suggested that sea ice formed at this time[20]. With temperatures near freezing at sea level at the tip of the Antarctic Peninsula, the interior of Antarctica was likely cold enough to sustain year-round glaciers, particularly at high elevation. Global sea level reconstructions record two sea level falls in the late Maastrichtian (KMa4 and KMa5)[31]. Sequence boundaries occur at 66.8 and 68.8 Ma, with sea level minima closely following[31]. The close temporal alignment of sea level lowstands and coldest Antarctic temperatures suggests that the observed sea level fluctuations could be driven by continental ice accumulation on Antarctica.

The ∼8 °C pre-KPB warming observed in the Seymour Island record is larger than similar warming events seen at other marine sites[22-24]. If sea ice or continental ice were present during the coldest interval immediately preceding the warming spike (66.5–66.3 Ma), the disappearance of this ice during warming could accentuate local climate change at Seymour Island through the ice-albedo feedback, the same process that is currently amplifying anthropogenic climate change in the Arctic by roughly a factor of two[32]. In addition, if this warming was accompanied by melting of continental ice, the reduction in global seawater $\delta^{18}O_w$ would mute marine temperature change calculated from $\delta^{18}O$ assuming constant $\delta^{18}O_w$ (refs 22–24). One terrestrial record derived from plant assemblages, which are immune to changing $\delta^{18}O_w$ values, indicated a ∼5 °C warming at this time[21], larger than the 2–3 °C observed at open-marine sites[22-24]. $\delta^{18}O$-derived temperature records from Seymour Island[10] also record a pre-KPB warming of ∼5 °C, but the warming event begins earlier, is more gradual, and is of lower magnitude because this proxy method does not account for concurrent increases in $\delta^{18}O_w$ at this time (Fig. 3).

The majority of Deccan Traps volcanism occurred during Chron 29R, and the oldest formation (Jawhar) has been dated near its base to 66.288 ± 0.027 Ma (ref. 33) and 66.38 ± 0.05 Ma (ref. 34) using two different dating methods. The temporal alignment of the pre-KPB warming event at Seymour Island with the onset of Deccan Traps volcanism suggests a genetic link between volcanic release of CO_2 and the abrupt increase in temperature. The eruptions that occurred between the onset of Chron 29R volcanism and the KPg boundary emitted anywhere from 270 to 900 p.p.m. CO_2 (assuming 14 Tg CO_2 emitted per km^3 lava erupted[35], and all CO_2 remaining in the atmosphere, see Methods) onto a background atmospheric concentration of ∼360–380 p.p.m. (ref. 36). The timing of transient increases in marine carbonate dissolution driven by the emitted CO_2 suggest the main phase of degassing took place beginning at the onset of Chron 29R, lasting less than 200 kyr (ending before the KPB)[37]. Using the same volume-to-CO_2 conversion, the remaining (post-KPB) volcanism potentially emitted another 825–900 p.p.m. The magnitude of the second, weaker warming event at the KPB only appears as 1.1 °C in the horizon mean, but individual samples show a potentially larger warming (Fig. 1a). Though the volume of lava erupted

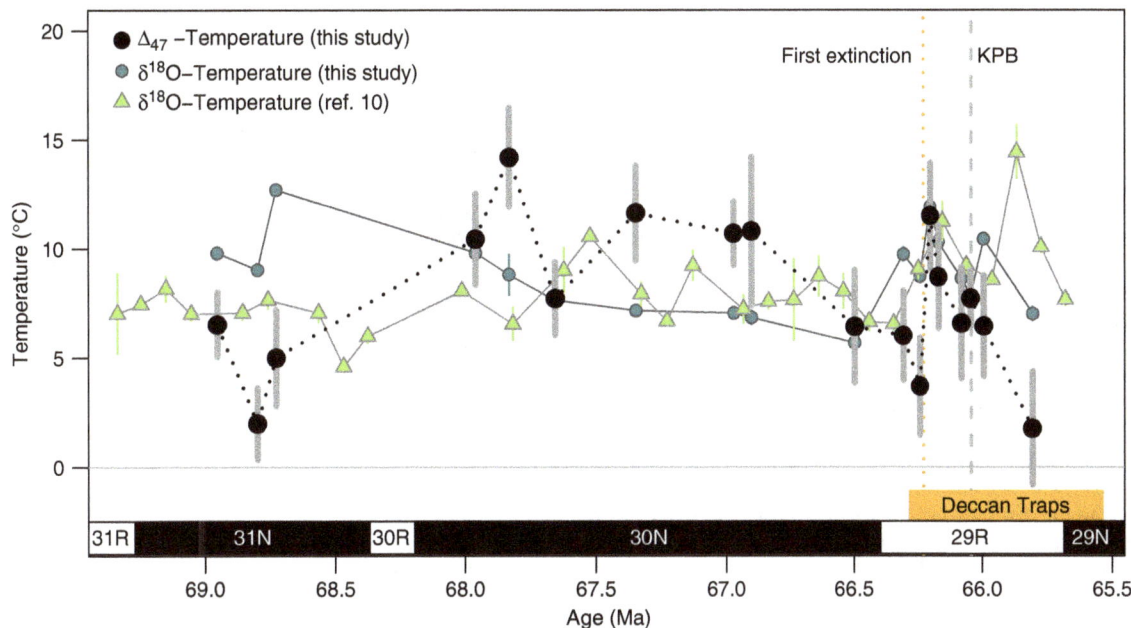

Figure 3 | Comparison of Δ_{47}-derived and $\delta^{18}O$-derived temperature records. Horizon mean Δ_{47}-derived temperature shown in black with grey error bars (same as Fig. 1a). Horizon mean was calculated as the inverse variance weighted mean, to account for variable reproducibility of samples, with error bars showing the inverse variance weighted error on the mean (see Methods). Line connecting horizon means is dashed to represent extrapolation of temperature trends between measured horizons. Samples within ± 3.5 m stratigraphic position of another sample were combined into a single horizon for horizon means (see Methods). $\delta^{18}O$-derived temperature calculated from horizon mean $\delta^{18}O$ (this study) shown in dark turquoise circles with error bars representing 1 s.d. on $\delta^{18}O$, propagated through the $\delta^{18}O$-T-$\delta^{18}O_w$ relationship[12]. Published $\delta^{18}O$-derived temperature record from Tobin et al.[10], with age model updated to match this study, shown in light green triangles with error bars representing 1 s.e. on the mean of all sample replicates within a given time bin. Both $\delta^{18}O$-derived temperature records assume a $\delta^{18}O_w$ value of − 1.0‰ (ice-free, globally constant)[29]. If $\delta^{18}O_w$ was instead assumed to be − 1.2‰ (ice-free, latitude-adjusted)[30], the shape of the $\delta^{18}O$-derived curves would be identical but shifted ∼1 °C colder. The two $\delta^{18}O$-derived temperature curves miss the million-year cool/warm/cool temperature structure, underestimate the magnitude of the pre-KPB warming event and infer an earlier timing for the onset of warming, and do not capture the full range of temperatures seen in the Δ_{47}-derived temperature record. Vertical grey dashed line represents the KPg boundary (labelled KPB)[34] and orange dotted line represents the timing of the first extinction event[10]. Age model construction is described in Methods.

(and therefore probably CO_2 emitted) was larger after the KPB, the climatological effects of these emissions were potentially reduced due to higher background CO_2 levels and an emissions duration roughly twice as long[7,32]. Additionally, the character of eruptions and the geochemical signature of the lava changed at the KPB[6], which could be correlated with changes in the CO_2 content and degassing rate. Feedbacks triggered by the meteorite impact may have also played a role in setting temperatures at and after the KPB.

Modelling studies using differing CO_2 estimates and injection durations predict 0.5–4 °C of global warming[37–40]. If the sea ice-albedo feedback had the same amplifying impact in Maastrichtian Antarctica as it is having today in the Arctic, the local temperature change could have been as much as double the global value[32]. Roughly 200–400 kyr after the initial large warming event, temperatures at Seymour Island had returned to pre-event levels. This rate of cooling is roughly consistent with the e-folding timescale of silicate weathering (∼ 240 kyr (ref. 41)), supporting the interpretation that this climate change is the result of changes in atmospheric CO_2, which would be scrubbed from the atmosphere through this long-term carbon cycle feedback (as well as by other short-term processes).

Species extinction at Seymour Island occurred in two statistically distinguishable pulses[10], one at the KPg boundary and another one 40 m below, calculated to be at 66.23 Ma in this age model (Fig. 1a). Early studies of Seymour Island also noted a drop in species diversity ∼30 m below the KTB[8]. This earlier extinction pulse coincides with the large warming event seen in our record (66.24–66.20 Ma) and with the onset of Chron 29R Deccan Traps

volcanism[7,33]. The two extinction intervals are of similar magnitude (14 and 10 species eliminated, respectively), but differ in their extinction patterns, suggesting different kill mechanisms. In the earlier (Deccan) extinction, 9 of 10 species were benthic, whereas in the second (Chicxulub, KPB) extinction, only 6 of 14 species were benthic[10]. Only one ammonite species became extinct in the first event, whereas all remaining (six) ammonite species became extinct at the KPg boundary[10,42].

The warm temperatures reached at the peak of warming alone could not have caused the first pulse of extinction. Many of the affected species were also present during Chron 30N (ref. 10), when temperatures were equivalently warm. Instead, it is possible that the rate of warming, as opposed to the maximum temperatures reached, led to extinction. The rate of warming during the Deccan temperature spike appears to be more rapid than at any other temperature change seen in the record (Fig. 1a), although this may be an artifact of the variable sampling resolution. Other indirect effects of volcanic emissions could also have played a role in the extinction, such as trace metal toxicity[2], ocean acidification from increased atmospheric CO_2, or short-term cooling and acid rain from emissions of SO_2 (ref. 3).

The lowest $\delta^{13}C$ values are observed in the interval between the pre-KPB and KPB events. Low $\delta^{13}C$ values have been measured at the KPB from and nearby Seymour Island before[16,43], which could be an indication of a reduction in surface ocean productivity, as has been previously suggested[44–47]. The difference in temperature between the umbo and ventral margin in *Lahillia*, a minimum estimate of seasonality, decreased from 5-8 °C to near 0 °C after the pre-KPB warming pulse and remained low until the end of the study interval. *Cucullaea* also shows little to no temperature difference between shell positions after the pre-KPB warming event. This could indicate a reduction in seasonality or a reduced growing season due to environmental stresses occurring after the onset of volcanism.

Roughly half of the total species extinction at Seymour Island (10 of 24 species) occurred before the Chicxulub impact event and can be temporally and climatologically linked to Deccan Traps volcanism, indicating the volcanic kill mechanism definitively contributed to the end-Cretaceous extinction. The remaining extinctions (14 of 24 species) took place at the KPg boundary, at the time of the Chicxulub impact, but also the beginning of the post-KPB phase of Deccan volcanism, which was potentially triggered and/or accelerated by the impact itself[6,7]. It is therefore impossible at this point to separate the influences of the meteorite and the additional volcanism in driving the second phase of extinction or in causing the second warming pulse. Nevertheless, the full magnitude of the end-Cretaceous extinction at this site can be attributed to the combined influence of Deccan Traps volcanism and the meteorite impact event, previously described as a press-pulse extinction mechanism[48,49]. Importantly, the pre-KPB warming, which itself correlated with significant extinction, may have increased ecosystem stress, making the ecosystem more vulnerable to collapse when the meteorite hit. This sequence of events may have combined into a 'one-two punch' that produced one of the largest mass extinctions in Earth history.

Methods

Environment and sample selection. Samples are from the W.J. Zinsmeister collection, formerly housed at Purdue University, and now located at the Paleontological Research Institute in Ithaca, NY, USA. Specimens were collected from Seymour Island from the López de Bertodano and Sobral Formations, covering the Late Cretaceous to early Paleocene. The stratigraphic framework used in this study is described by Zinsmeister et al.[50] and has a stratigraphic resolution of ± 3.5 m. Samples come from Units 9 and 10 of the López de Bertodano Formation, part of the upper beds (Units 7–10, Molluscan Units), as defined by Macellari[8]. The Molluscan Units contain diverse and abundant bivalve, gastropod, and ammonite macrofauna. Since the Cretaceous, these sediments are thought to have experienced only mild diagenetic conditions[51]. The depositional environment is interpreted to be inner to outer depths, a semi-enclosed or quiet environment, with water depths less than 200 m, on the basis of faunal assemblages, grain size and sorting analysis, diversity of macrofauna, and abundance of glauconite[8]. Interpretations of water depth variation through the López de Bertodano Fm. disagree, with some suggesting a shallowing-upward[52] and others a deepening-upward[8] trend through the section. Overall, the homogeneous nature of the Molluscan Units 7–9 suggests[8,42,52], at most, minor variation in water depth over this interval. Although there is some evidence of methane seeps on Seymour and the surrounding islands, it is restricted to lower in the stratigraphy (Units 1–6)[53]. There is no indication of methane seeps in the upper López de Bertodano Formation on Seymour Island[43].

Bivalve shells (116) were selected from the Zinsmeister collections at Purdue for use in the PhD thesis of author Andrea Dutton[54] with the aim of including as many stratigraphic horizons as possible for maximum temporal resolution (the 'full sample set'). Samples were selected from six species (*L. larseni, C. antarctica, C. ellioti, E. regina, D. drygalskiana* and *N. nordenskjoldi*) to provide internal consistency between horizons. All these species were thought to live in shallow waters and be infaunal suspension feeders, with the exception of *Nordenskjoldia*, which is thought to be an epibyssate suspension feeder[8].

A subset of 29 shells were selected from the full sample set for clumped isotope analysis, representing five of the six species in the original study (eliminating *D. drygalskiana*; the 'clumped isotope sample set'). These were again selected for maximum stratigraphic coverage and species overlap, but were filtered to only include the best-preserved samples, as indicated by low trace element concentrations (see below). A list of samples in the full sample set and those sub-selected for clumped isotope analysis can be found in Supplementary Data 1.

Sample preparation and extraction. Bulk samples of 50–100 µg were drilled from the umbo, mid-shell and ventral margin of all shells in the full sample set for stable isotope and trace element analysis. Samples were drilled to average over multiple growth bands to represent typical mean annual conditions. The umbo position was centred near the cusp of the umbo. The ventral margin ('back') was within 1 cm of the edge of the shell. The mid-shell position was half way between the umbo and ventral margin. In the case of broken, partial shells, it was not always possible to sample the ventral margin and/or mid-shell position. Broken and partial shells were not selected for the clumped isotope sample set.

The 29 shells selected for clumped isotope analysis were re-drilled in two positions, the umbo and ventral margin. Sampling locations were directly adjacent to earlier drill locations. Sample size requirements for clumped isotope analysis are much larger than for stable isotope analysis[11], so 20–30 mg of sample material was collected at each position and homogenized. During drilling for clumped isotope analysis, the drill was operated at the lowest possible rotation speed (1,000 r.p.m.) and pressure was only applied for a few seconds at a time to minimize frictional heating. Based on tests with other samples in our lab, at this drill speed, we do not expect to see any resetting effects. Most shells were drilled from the exterior at a position that was clear of external encrustation or fractures. In a few cases, shells were cut in half and were drilled on the cut interior surface. No correlation was seen between measured values and interior versus exterior drilling. Six shells of the genus *Eselaevitrigonia* were only sampled in the umbo position because invariant stable isotopic compositions between positions suggested that identical environmental conditions were being recorded at all points on the shell (see below), and some shells were broken such that the ventral margin was no longer present.

Trace element analysis. Elemental analysis of Sr, Mg, Ca, Fe and Mn was carried out on powdered splits of all bulk samples using either an inductively coupled plasma optical emission spectrophotometer (ICP-OES) or an inductively coupled plasma mass spectrometer (HRICP-MS) depending upon the amount of powdered carbonate available for analysis. Analysis was performed in the Keck Elemental Geochemistry Laboratory at the University of Michigan (Director: K.C. Lohmann). Smaller samples were run on HRICP-MS owing to the lower detection limits attainable on this instrument. By comparison with lab standards and of replicate analyses, precision of both types of analyses was maintained at better than 3%. Because these samples were too small to allow for efficient weight determination, calculation of cation concentrations within the carbonate was achieved by assuming that Sr, Mg, Ca, Fe, Mn are carbonate-bound cations and that these five elements account for 100% of the carbonate-bound cations. Trace element data can be found in Supplementary Data 1.

Assessment of shell preservation. Many lines of evidence (lack of smectite-illite transition, mineralogical and vitrinite-reflectance data) suggest a maximum burial depth of less than 1 km and a maximum burial temperatures of less than 80 °C for the López de Bertodano Formation[51,55]. Visual inspection shows shell preservation is excellent, with many shells still displaying pristine mother-of-pearl sheen, indicating the presence of primary aragonite. This was confirmed by conducting X-ray diffraction of powdered material from seven shells representing each of the six species in the full sample set, taken from across the full range of the stratigraphic section. These were shell numbers C1556E (*C. antarctica*), L776E (*L. larseni*), D1110I (*D. dryganskalia*), E1109E (*E. regina*), N1614C (*N. nordenskjoldia*), L1161A2 (*L. larseni*) and C1430 (*C. ellioti*). In all cases, the mineralogy was completely aragonitic and there was no indication of partial conversion to calcite.

Thick or thin sections were made for nine shells (sample numbers C757C2, L757B, L1480B, L1480E, C1467A, L1529A, C915A2, L1430A1, and L1609). Primary textures, including excellent preservation of growth band structures, were observed in shell carbonate from these sections (Supplementary Fig. 9). These sections were also viewed under cathodoluminescence, which can detect elevated amounts of Mn^{2+} that may become incorporated in diagenetic carbonate due to post-depositional water–rock interaction. Shells were non-luminescent, except in rare instances where fractures cut across the shell. This observation is particularly notable because it indicates that pore-fluids in the host sediment were reducing and enriched in Mn^{2+}, yet they did not affect bulk shell chemistry. Therefore, alteration is likely only a concern in areas where the shells are fractured. Fractured areas were avoided in sample extraction.

Shell preservation was also assessed through trace element composition. Acceptable levels of Fe and Mn for a well-preserved shell have been determined differently by different authors[17,56]. Conservative trace element thresholds of 500 p.p.m. for Fe and 200 p.p.m. for Mn were set, and samples with concentrations above these were deemed 'potentially altered' and were excluded from further analysis. Of 339 total bulk samples, 42 individual bulk samples (representing 32 unique bivalve shells) had elevated trace element concentrations, indicating overall excellent preservation of samples at Seymour Island.

Shells selected for clumped isotope analysis all had Fe and Mn concentrations below the defined thresholds in the umbo and ventral margin areas. In three cases (N1416B, C772A and C1109A), the trace element composition was only measured at one of the positions on the shell but both positions were sampled for clumped isotopes. In three other cases (C757C-umbo, C915A1-umbo and C915A1-back), the trace element measurements came from the opposite half of the bisected shell, in a mirror image position to where the shell was sampled for clumped isotope analysis. For two shells (L757 and L1430A1), the trace element samples were taken from the outside of the shell, whereas the clumped isotope samples were taken from the inside after the shell was cut in half.

Stable and clumped isotope analysis. Stable carbon and oxygen isotope data were generated for all bulk samples from the full sample set using an automated Kiel device coupled to a Finnigan MAT 251 isotope ratio mass spectrometer in the

University of Michigan Stable Isotope Laboratory (Director: K.C. Lohmann). Before reaction with phosphoric acid in the Kiel device, individual powdered carbonate samples were first roasted in vacuo at 200 °C for 1 h to remove volatile contaminants. Precision of the data was maintained at better than 0.1‰ by calibration to carbonate standards (both internal lab standards and NBS-19), measured daily. All carbonate stable isotope compositions are reported relative to Vienna Peedee Belemnite (VPDB). Stable isotope data for bulk samples can be found in Supplementary Data 1 and are shown in Supplementary Figs 5–7.

Samples in the 'clumped isotope sample set' were measured for clumped isotopic composition at the University of Michigan Stable Isotope Laboratory. Carbonate powders were prepared on an offline sample preparation device described by Defliese et al.[57]. Samples measured in 2015 were cleaned using the 'WarmPPQ' configuration, whereas samples measured in 2014 used the 'ColdPPQ' configuration, and were corrected accordingly for fractionations introduced by the Porapak trap[58].

Standard gases heated to 1,000 °C and equilibrated with water at 25 °C were measured alongside samples and were used to define the absolute reference frame[59]. An acid fractionation factor of 0.067‰ was used, chosen to reflect the 75 °C reaction temperature[57]. Two carbonate standard materials, Carrara Marble and Joulter's Cay Ooids, were measured alongside samples. Accepted Δ_{47} values for these two standards are 0.414‰ and 0.704‰, respectively[57,58]. Over the three measurement sessions presented here, the mean values for Carrara and Ooids were 0.423 ± 0.005‰ (1 s.e., $n = 15$) and 0.712 ± 0.007‰ (1 s.e., $n = 13$). Raw voltages are converted into delta values following Huntington et al.[60], including the same defined values for λ and for the isotopic composition of VPDB and VSMOW, and Δ_{47} values were converted to temperature using equation 6 of Defliese et al.[57].

The stable isotopic composition of the powders in the 'clumped isotope sample set' was measured simultaneously with the measurement of the clumped isotopic composition. Oxygen isotope data were corrected for fractionation during acid digestion using an acid fractionation factor of 1.00836 for aragonitic samples, determined empirically using the aragonitic Ooids carbonate standard, separately calibrated to NBS 18 and NBS-19 (ref. 58). Oxygen isotopic composition of water was calculated using the aragonite-water fractionation of Kim et al.[12], and is reported relative to VSMOW. Measured clumped isotopic and stable isotopic compositions of 'clumped isotope sample set' shells can be found in Supplementary Data 2–5. Carbonate and gas standard data can be found in Supplementary Data 2–4.

Calculation of sample and horizon averages. Average $\delta^{13}C$ and $\delta^{18}O$ values are calculated as the mean of many ($n = 2$–5) replicates, with error taken as 1 s.d. Average values for Δ_{47}, temperature, and $\delta^{18}O_w$ are taken as the mean of n replicates with the error taken as 1 s.e. on the mean, as is typical in the clumped isotope community (Supplementary Data 5). $\delta^{18}O_w$ is first calculated individually for each replicate from each $\delta^{18}O$ and Δ_{47}-derived temperature pair, then mean $\delta^{18}O_w$ is taken as the mean of many replicates, with 1 s.e. error. In some cases, the calculated 1 s.e. is smaller than the long-term performance of our carbonate standards. We determine 'external error' on Δ_{47} for a given sample to be the larger of (1) the calculated 1 s.e. of the mean of n replicates or (2) the standard deviation of all replicates of Carrara (0.019‰) divided by the square-root of n replicates (Supplementary Data 5). For two to five replicates, this gives a 1 s.e. of 0.013‰, 0.011‰, 0.009‰ and 0.008‰. The 'external error' on temperature is calculated as half of T(mean Δ_{47}) − extSE.) − T(mean Δ_{47} + extSE), where $T(x)$ is the Δ_{47} − Temperature calibration function and 'extSE' is the external s.e. replacing the calculated s.e. Based on typical 1 s.e. values on $\delta^{18}O_w$ for a sample with a Δ_{47}-error of 0.013‰, 0.011‰, 0.009‰ and 0.008‰, the 'external error' on $\delta^{18}O_w$ is assigned to be 0.92‰, 0.78‰, 0.64‰ and 0.57‰ for $n = 2$–5 replicates, respectively. External error is used in all figures and calculations for Supplementary Data 6–8.

Most (11 of 18) horizons are represented by two samples taken from different positions on a single shell. Where more than one shell is combined (6 of 18 horizons), all samples are treated equally, regardless of position or shell. Only one horizon is represented by a single position on a single shell (E. regina at 66.9 Ma). Due to consistently observed patterns within species, we interpret differences between shell positions to reflect environmental aliasing, not scatter around a mean value in a noisy data set. Therefore, the error calculated on a horizon average does not represent certainty on the average value of a noisy data set, but instead represents the spread in data from different shell positions, possibly an indicator of seasonality, combined with some degree of measurement noise. We do not know which species or shell position best represents mean annual temperature, so we felt combining all points equally was the fairest way to treat the data.

Horizon means (Supplementary Data 6–8) are calculated using equation (1) as the inverse variance weighted mean of all samples from a given horizon, combining samples within the uncertainty of the stratigraphy (± 3.5 m)[50].

$$\text{Inverse variance weighted mean} = \bar{y} = \frac{\sum \frac{y_i}{(\sigma_i)^2}}{\sum \frac{1}{(\sigma_i)^2}} \qquad (1)$$

Error on the horizon mean is calculated using equation (2) as error on an inverse variance weighted mean, to maintain consistency.

$$\text{Inverse variance weighted mean error} = \sigma^2 = \frac{1}{\sum \frac{1}{(\sigma_i)^2}} \qquad (2)$$

Inverse variance weighted mean was selected to account for the variable reproducibility of samples and not overweight the mean towards a poorly

reproducing sample (for example, Sample C1109A at 67.7 Ma). Supplementary Fig. 2 compares multiple methods for calculating horizon mean (for example, inverse variance weighted mean, traditional mean, average of shell averages). The use of traditional mean versus inverse variance weighted mean does not change the long-term pattern, interpretations, or conclusions. The largest difference is seen around 67.7 Ma where there are two poorly reproducing samples.

Taking the average of positions within each shell before combining shells is mathematically equivalent to averaging all points if using the inverse variance weighted mean. Averaging within-shell first only affects four horizons mean minimally at that, if using the traditional mean. We chose to combine adjacent samples based on stratigraphic position, but a composite record could also have been created by combining samples into equal-time bins. Equal-time bins combine samples from before and after the warming spike, resulting in a 'smoothed' temperature record. If the bin size is small enough (50 kyr), two warming pulses are still observed.

Age model construction. To construct an age model using the recently published magnetostratigraphic data from Seymour Island[10], the stratigraphic heights of chron reversals were converted to the Zinsmeister stratigraphic framework[50] based on the established position of the KPg boundary in each (1,059 m for this study, 865 for Tobin et al.[10]) and assuming equal unit thickness. The stratigraphic height of a point in the Zinsmeister framework is therefore calculated as the height in the Tobin framework plus the difference in the KPg boundary positions (1,059−865 = 194 m). This assumption is supported by the position of the Unit 8–Unit 9 boundary, which is at 652 m in the Tobin framework[10]. The age model conversion would predict that this Unit boundary should be found at 846 m in the Zinsmeister framework, and it is described as ~850 m by Zinsmeister[50]. Based on this close agreement, when comparing extinction events observed in the Tobin framework with climatic events from this study, any events occurring between the base of Unit 8 and the KPB should be in agreement to within ~4 m.

The ages of the relevant chron reversals were taken from the most recent Geological Time Scale (2012)[61], and the age the KPg boundary was taken as 66.043 ± 0.043 Ma (ref. 34). The position of chron reversals was taken as the average of the stratigraphic heights of the first sample listed above and below the reversal position[10]. This is different from the age model used by Tobin et al.[10], which used the 2004 edition of the Geologic Time Scale with the age of the KPg boundary at 65.5 Ma. To plot the $\delta^{18}O$-derived temperature record from Tobin et al.[10] (Fig. 2), ages for chron reversals were updated to the 2012 Geological Time Scale[61].

Sample ages were linearly interpolated between age model tie points, listed in Supplementary Table 1. There is inherent uncertainty in magnetostratigraphy-based age models due to this interpolation and uncertainties in the construction of the Geological Time Scale[61]. Although not explicitly accounted for, this uncertainty should be considered when comparing events in our age model with absolute U-Pb or $^{40}Ar/^{39}Ar$ ages for the Deccan Traps[7,33].

Strontium isotope analysis. Ten shells were analysed for strontium isotopes for use in corroborating the age model construction. Analyses were performed in the Biogeochemistry and Environmental Isotope Geochemistry Laboratory at the University of Michigan (Director: J.D. Blum). Powdered samples of bivalve aragonite were dissolved in nitric acid and strontium was subsequently separated using ion-exchange chromatography with Eichrom Sr-specific resin. The effluent was dried and loaded onto a tungsten filament for isotope ratio analysis using a Finnigan MAT 262 thermal ionization mass spectrometer. The $^{86}Sr/^{88}Sr$ ratio was normalized to 0.1194 and measured $^{87}Sr/^{86}Sr$ ratios were corrected for the difference between the long-term laboratory average of NIST 987 (0.710252 ± 0.000026 (2σ)) and the accepted value of 0.710248. Analytical uncertainty was calculated using the standard deviation of NIST 987 over a 12-month period (2σ = 0.000026).

One sample (L1480B) produced a strontium isotopic composition that is not possible to match to the LOWESS curve at any point between 65 and 69 Ma. L1480B also displays anomalously high Sr/Ca values (10–20 mmol mol^{-1}), suggesting that it may have been diagenetically altered. Another sample (L1480E) from the same locality produced a reasonable strontium isotope value, so L1480B was ignored. L1480B was also avoided for clumped isotope analysis. Strontium isotope data measured in this study can be found in Supplementary Table 2.

Corroboration of age model with Sr-isotope data. Strontium isotopic data was not used in the creation of the age model described above. These data can, therefore, be used to assess the validity of the age model constructed based on magnetostratigraphy and stratigraphic correlation. In addition to new strontium isotope measurements made on bivalve samples from this study (Supplementary Table 2), strontium isotope data from other Maastrichtian bivalves from the Zinsmeister collection have been previously published (Supplementary Table 3)[13]. Because they come from the same collection, their stratigraphic positions can be directly compared. Ages were calculated for all published samples using the age model described above based on their stratigraphic position in the Zinsmeister stratigraphic framework[50]. Supplementary Fig. 10 shows a comparison of measured

^{87}Sr/^{86}Sr values and calculated ages with the most recent seawater strontium isotope curve (LOWESS 5)[62]. There is good agreement between the bivalve data and the seawater curve, lending strong credence to the magnetostratigraphic age model and the stratigraphic conversion between the Zinsmeister framework and the Tobin framework.

Stable isotope trends in the full sample set. In the paper we discuss systematic differences in δ^{13}C, temperature, and δ^{18}O$_w$ between the umbo and back position seen in samples in the clumped isotope sample set (Figs 1 and 2; Supplementary Figs 3,4 and 11). Supplementary Fig. 11 also shows the relationship between position and shell δ^{18}O. Systematic position-specific behaviour in δ^{18}O is complicated due to the competing influences of δ^{18}O$_w$ and temperature, which vary together in a way to oppositely influence δ^{18}O (Supplementary Fig. 8). Despite this, *Cucullaea* still consistently shows higher δ^{18}O in the ventral margin position, and *Lahillia* generally shows higher δ^{18}O in the umbo position.

Although we cannot look at trends in temperature and δ^{18}O$_w$ in the full sample set, we can still look at variation in δ^{13}C and δ^{18}O. In the full sample set, *Lahillia* and *Cucullaea* maintain the position-specific relationships in δ^{13}C and δ^{18}O seen in the limited clumped isotope sample set (Supplementary Figs 5–7). *Cucullaea* has the highest δ^{13}C and δ^{18}O values in the ventral margin, whereas *Lahillia* has the highest δ^{13}C and δ^{18}O values in the umbo. The consistency in position-specific relationships in the larger data set suggests that the same position-specific behaviour in temperature and δ^{18}O$_w$ is likely present in all samples, even if not measured for clumped isotopes.

Nordenskjoldia and *Dozyia* show similar position-specific relationships in δ^{13}C and δ^{18}O to *Cucullaea* and *Lahillia*, respectively, but of lower total magnitude. *Eselaevitrigonia* shows a much smaller variation in δ^{18}O than *Cucullaea* or *Lahillia* and δ^{18}O does not correlate to δ^{13}C as in the other species. This lower variability in stable isotopes suggests consistent environmental and growth conditions between early and late life for this species. For this reason, *Eselaevitrigonia* was only sampled in the umbo position. This decision is supported by the small differences in temperature and δ^{18}O$_w$ seen in *Nordenskjolia*, another species showing small variations in stable isotopes (Fig. 1).

Calculations of CO$_2$ emissions from Deccan Traps volcanism. Richards *et al.*[6] tentatively place the KPB between the Bushe and Poladpur formations in the Deccan volcanic sequence due to evidence for a hiatus in volcanism and fracturing and faulting in the lower Bushe formation that did not extend into the higher Poladpur formation. This placement is consistent with existing dating of the Deccan Traps, which unfortunately has a sampling gap in the critical region preventing direct placement of the KPB[7,33]. Assuming this stratigraphic placement of the KPB, we determined an estimate of the minimum and maximum volume of lava erupted between the onset of Chron 29R Deccan volcanism and the KPB, which includes all formations between (and including) Jawhar and Bushe (Kalsubai and Lonavala subgroups).

Estimates of the volume of lava erupted in this interval range from 150×10^3 km^3 (ref. 6) to 500×10^3 km^3 (ref. 35). Using an emission rate of 14 Tg CO$_2$ km^{-3} lava[34], a modern atmospheric mass of 5.15×10^{21} g and an atmospheric density of 28.966 g mol^{-1} air, these lava volumes equate to 270–900 p.p.m. CO$_2$ emitted to the atmosphere. Compare this to end-Maastrichtian atmospheric CO$_2$ levels, which are thought to be ∼360–380 p.p.m. CO$_2$ (ref. 36). The exact timing of lava extrusion during this interval is unknown, but a lack of red boles in the older portion of the section implies very few hiatuses[7,63]. The transient timing of marine carbonate dissolution suggests the main phase of degassing occurred beginning shortly after the onset of Chron 29R and ending within 200 kyr (ref. 37).

We also estimate CO$_2$ emissions during the interval between the KPB and the end of Chron 29R volcanism, which includes the Poladpur to Mahabaleshwar formations (Wai subgroup) based on our hypothetical placement of the KPB. The post-KPB portion of Deccan volcanism emitted another 461×10^3 km^3 (ref. 6) to 500×10^3 km^3 (ref. 35) of lava, equivalent to 830–900 p.p.m. CO$_2$ added to the atmosphere. The Wai subgroup eruptions contain more red boles[7,63], suggesting hiatuses between eruptive events. Based on published dates[7,33], the duration of the post-KPB eruptions is roughly twice the length of the pre-KPB portion (∼500 kyr versus 250 kyr), although this does not account for potential hiatuses within the eruptive sequence. Although the total volume of lava erupted (and therefore CO$_2$ emitted, based on our assumptions) was larger post-KPB, the climatological impacts might have been smaller due to higher background CO$_2$ levels and a longer duration of eruption. Additionally, the emission rate of 14 Tg CO$_2$ km^{-3} lava[35] is an average value. Changes in the character of eruptions and the geochemical signature of the lava before and after the Bushe/Poladpur contact (the hypothesized KPB)[6] could coincide with changes in the lava's CO$_2$ content and therefore emission rate.

References

1. Courtillot, V. *et al.* Deccan flood basalts and the Cretaceous/Tertiary boundary. *Nature* **333**, 843–846 (1988).
2. Vogt, P. R. Evidence for global synchronism in mantle plume convection, and possible significance for geology. *Nature* **240**, 338–342 (1972).
3. Self, S., Schmidt, A. & Mather, T. A. Emplacement characteristics, time scales, and volcanic gas release rates of continental flood basalt eruptions on Earth. *Geol. Soc. Am. Bull.* **505**, 319–337 (2014).
4. Alvarez, L. W., Alvarez, W., Asaro, F. & Michel, H. V. Extraterrestrial cause for the Cretaceous-Tertiary extinction. *Science* **208**, 1095–1108 (1980).
5. Schulte, P. *et al.* The chicxulub asteroid impact and mass extinction at the Cretaceous-Paleogene boundary. *Science* **327**, 1214–1218 (2010).
6. Richards, M. A. *et al.* Triggering of the largest Deccan eruptions by the Chicxulub impact. *Geol. Soc. Am. Bull.* http://dx.doi.org/10.1130/B31167.1 (2015).
7. Renne, P. R. *et al.* State shift in Deccan volcanism at the Cretaceous-Paleogene boundary, possibly induced by impact. *Science* **350**, 76–78 (2015).
8. Macellari, C. E. Stratigraphy, sedimentology, and paleoecology of Upper Cretaceous/Paleocene shelf-deltaic sediments of Seymour Island. *Geol. Soc. Am. Memoirs* **169**, 25–54 (1988).
9. Zinsmeister, W. J. & Macellari, C. E. Bivalvia (Mollusca) from Seymour Island, Antarctic Peninsula. *Geol. Soc. Am. Memoirs* **169**, 253–284 (1988).
10. Tobin, T. S. *et al.* Extinction patterns, δ^{18}O trends, and magnetostratigraphy from a southern high-latitude Cretaceous-Paleogene section: Links with Deccan volcanism. *Palaeogeogr. Palaeoclimatol. Palaeoecol.* **350**, 180–188 (2012).
11. Eiler, J. M. Paleoclimate reconstruction using carbonate clumped isotope thermometry. *Quat. Sci. Rev.* **30**, 3575–3588 (2011).
12. Kim, S. T., O'Neil, J. R., Hillaire-Marcel, C. & Mucci, A. Oxygen isotope fractionation between synthetic aragonite and water: influence of temperature and Mg^{2+} concentration. *Geochim. Cosmochim. Acta* **71**, 4704–4715 (2007).
13. McArthur, J. M., Thirwall, M. F., Engkilde, M., Zinsmeister, W. J. & Howarth, R. J. Strontium isotope profiles across K/T boundary sequences in Denmark and Antarctica. *Earth Planet. Sci. Lett.* **160**, 179–192 (1998).
14. Francis, J. E. & Poole, I. Cretaceous and early Tertiary climates of Antarctica: evidence from fossil wood. *Palaeogeogr. Palaeoclimatol. Palaeoecol.* **182**, 47–64 (2002).
15. Kemp, D. B. *et al.* A cool temperature climate on the Antarctic Peninsula through the latest Cretaceous to early Paleogene. *Geology* **42**, 583–586 (2014).
16. Barrera, E., Huber, B. T., Savin, S. M. & Webb, P.-N. Antarctic marine temperatures: late Campanian through Early Paleocene. *Paleoceanography* **2**, 21–47 (1987).
17. Ditchfield, P. W., Marshall, J. D. & Pirrie, D. High latitude palaeotemperature variation: new data from the Thithonian to Eocene of James Ross Island, Antarctica. *Palaeogeogr. Palaeoclimatol. Palaeoecol.* **107**, 79–101 (1994).
18. Elorza, J., Gomez Alday, J. J. & Olivera, E. B. Environmental stress and diagenetic modifications in inoceramids and belemnites from the Upper Cretaceous James Ross basin, Antarctica. *Facies* **44**, 227–242 (2001).
19. Dutton, A., Huber, B. T., Lohmann, K. C. & Zinsmeister, W. J. High-resolution stable isotope profiles of a Dimitobelid belemnite: Implications for paleodepth habitat and Late Maastrichtian climate seasonality. *Palaois* **22**, 642–650 (2007).
20. Bowman, V. C., Francis, J. E. & Riding, J. B. Late Cretaceous winter sea ice in Antarctica? *Geology* **41**, 1227–1230 (2013).
21. Wilf, P., Johnson, K. R. & Huber, B. T. Correlated terrestrial and marine evidence for global climate change before mass extinction at the Cretaceous-Paleogene boundary. *Proc. Natl Acad. Sci. USA* **100**, 599–604 (2003).
22. Li, L. & Keller, G. Maastrichtian climate, productivity and faunal turnovers in planktic foraminifera in South Atlantic DSDP sites 525A and 21. *Geology* **26**, 995–998 (1998).
23. MacLeod, K. G., Huber, B. T. & Isaza-Londoño, C. North Atlantic warming during global cooling at the end of the Cretaceous. *Geology* **33**, 437–440 (2005).
24. Westerhold, T., Rohl, U., Donner, B., McCarren, H. K. & Zachos, J. C. A complete high-resolution Paleocene benthic stable isotope record for the central Pacific (ODP Site 1209). *Paleoceanography* **26**, PA2216 (2011).
25. Harrington, R. J. Aspects of growth deceleration in bivalves: Clues to understanding the seasonal δ^{18}O and δ^{13}C record—A comment on Krantz et al. (1987). *Palaeogeogr. Palaeoclimatol. Palaeoecol.* **70**, 399–403 (1989).
26. Buick, D. P. & Ivany, L. C. 100 years in the dark: extreme longevity of Eocene bivalves from Antarctica. *Geology* **32**, 921–924 (2004).
27. Shone, B. R., Lega, J., Flessa, K. W., Goodwin, D. H. & Dettman, D. L. Reconstructing daily temperatures from growth rates of the intertidal bivalve mollusk Chione cortezi (northern Gulf of California, Mexico). *Palaeogeogr. Palaeoclimatol. Palaeoecol.* **184**, 131–146 (2002).
28. Ivany, L. C., Wilkinson, B. H. & Jones, D. S. Using stable isotopic data to resolve rate and duration of growth throughout ontogeny: an example from the surf clam, Spisula solidissima. *Palaios* **18**, 126–137 (2003).
29. Shackleton, N. J. & Kennett, J. P. Paleotemperature history of the Cenozoic and the initiation of Antarctic glaciation: oxygen and carbon isotope analyses in DSDP Sites 277, 279, and 281. *Initial Reports of the Deep Sea Drilling Project* **29**, 743–755 (1975).

30. Zachos, J. C., Stott, L. D. & Lohmann, K. C. Evolution of Early Cenozoic marine temperatures. *Paleoceanography* **9**, 353–387 (1994).

31. Haq, B. U. Cretaceous eustasy revisited. *Glob. Planet. Change* **113**, 44–58 (2014).

32. Screen, J. A. & Simmonds, I. The central role of diminishing sea ice in recent Arctic temperature amplification. *Nature* **464**, 1334–1337 (2010).

33. Schoene, B. *et al.* U-Pb geochronology of the Deccan Traps and relation to the end-Cretaceous mass extinction. *Science* **347**, 182–184 (2015).

34. Renne, P. R. *et al.* Time scales of critical events around the Cretaceous-Paleogene boundary. *Science* **339**, 684–687 (2013).

35. Self, S., Widdowson, M., Thordarson, T. & Jay, A. E. Volatile fluxes during flood basalt eruptions and potential effects on the global environment: a Deccan perspective. *Earth Planet. Sci. Lett.* **248**, 518–532 (2006).

36. Beerling, D. J., Lomax, B. H., Royer, D. L., Upchurch, G. R. & Kump, L. R. An atmospheric pCO$_2$ reconstruction across the Cretaceous-Tertiary boundary from leaf megafossils. *Proc. Natl Acad. Sci. USA* **99**, 7836–7840 (2002).

37. Henehan, M. J., Hull, P. M., Penman, D. E., Rae, J. W. B. & Schmidt, D. N. Biogeochemical signficiance of pelagic ecosystem function: an end-Cretaceous case study. *Phil. Trans. R. Soc. B* **371**, 1–9 (2016).

38. Caldeira, K. G. & Rampino, M. R. Deccan volcanism, greenhouse warming, and the Cretaceous/Tertiary boundary. *Geol. Soc. Am. Spec. Pap.* **247**, 117–124 (1990).

39. Caldeira, K. G. & Rampino, M. R. Carbon dioxide emissions from Deccan volcanism and a K/T boundary greenhouse effect. *Geophys. Res. Lett.* **17**, 1299–1302 (1990).

40. Dessert, C. *et al.* Erosion of Deccan Traps determined by river geochemistry: impact on the global climate and the Sr-87/Sr-86 ratio of seawater. *Earth Planet. Sci. Lett.* **188**, 459–474 (2001).

41. Colbourn, G. & Ridgwell, A. The time scale of the silicate weathering negative feedback on atmospheric CO$_2$. *Glob. Biogeochem. Cycles* **29**, 1–14 (2015).

42. Witts, J. D. *et al.* Evolution and extinction of Maastrichtian (Late Cretaceous) cephalopods from the López de Bertodano Formation, Seymour Island, Antarctica. *Palaeogeogr. Palaeoclimatol. Palaeoecol.* **418**, 193–212 (2015).

43. Tobin, T. S. & Ward, P. D. Carbon isotope (δ^{13}C) differences between Late Cretaceous ammonites and benthic mollusks from Antarctica. *Palaeogeogr. Palaeoclimatol. Palaeoecol.* **428**, 50–57 (2015).

44. Zachos, J. C., Arthur, M. A. & Dean, W. E. Geochemical evidence for suppression of pelagic marine productivity at the Cretaceous/Tertiary boundary. *Nature* **337**, 61–64 (1989).

45. D'Hondt, S., Donaghay, P., Zachos, J. C., Luttenberg, D. & Lindinger, M. Organic carbon fluxes and ecological recovery from the Cretaceous-Tertiary mass extinction. *Science* **282**, 276–279 (1998).

46. D'Hondt, S. Consequences of the Cretaceous/Paleogene mass extinction for marine ecosystems. *Annu. Rev. Ecol. Evol. Syst.* **36**, 295–317 (2005).

47. Birch, H. S., Coxall, H. K., Pearson, P. N., Kroon, D. & Schmidt, D. N. Partial collapse of the marine carbon pump after the Cretaceous-Paleogene boundary. *Geology* **44**, 287–290 (2016).

48. Arens, N. C. & West, I. D. Press-pulse: a general theory of mass extinction? *Paleobiology* **34**, 456–471 (2008).

49. Arens, N. C., Thompson, A. & Jahren, A. H. A preliminary test of the press-pulse extinction hypothesis: Palynological indicators of vegetation change preceding the Cretaceous-Paleogene boundary, McCone County, Montana, USA. *Geol. Soc. Am. Spec. Pap.* **503**, 209–227 (2014).

50. Zinsmeister, W. J. Late Maastrichtian short-term biotic events on Seymour Island, Antarctic Peninsula. *J. Geol.* **109**, 213–229 (2001).

51. Pirrie, D., Ditchfield, P. W. & Marshall, J. D. Burial diagenesis and pore-fluid evolution in a Mesozoic back-arc basin: the Marambio Group, Vega Island, Antarctica. *J. Sediment. Res.* **A64**, 541–552 (1994).

52. Crame, J. A., Francis, J. E., Cantrill, D. J. & Pirrie, D. Maastrichtian stratigraphy of Antarctica. *Cretaceous Res.* **25**, 411–423 (2004).

53. Little, C. T. S. *et al.* Late Cretaceous (Maastrichtian) shallow water hydrocarbon seeps from Snow Hill and Seymour Islands, James Ross Basin, Antarctica. *Palaeogeogr. Palaeoclimatol. Palaeoecol.* **418**, 213–228 (2015).

54. Dutton, A. *Extracting Paleoenvironmental Records from Molluscan Carbonate.* (PhD thesis, University of Michigan, 2002).

55. Palamarczuk, S. *et al.* Las Formaciones López de Bertodano y Sobral en la Isla Vicecomodoro Marambio, Antártida: IX Congreso Geológico Argentino. *Actas* **1**, 399–419 (1984).

56. Anderson, T. F., Popp, B. N., Williams, A. C., Ho, L.-Z. & Hudson, J. D. The stable isotopic records of fossils from the Peterborough Member, Oxford Clay Formation (Jurassic), UK: palaeoenvironmental implications. *J. Geol. Soc. London* **151**, 125–138 (1994).

57. Defliese, W. F., Hren, M. T. & Lohmann, K. C. Compositional and temperature effects of phosphoric acid fractionation on Δ$_{47}$ analysis and implications for discrepant calibrations. *Chem. Geol.* **396**, 51–60 (2015).

58. Petersen, S. V., Winkelstern, I. Z., Lohmann, K. C. & Meyer, K. W. The effects of Porapak trap temperature on δ^{18}O, δ^{13}C, and Δ$_{47}$ values in preparing samples for clumped isotope analysis. *Rapid. Commun. Mass Spectrom.* **30**, 199–208 (2016).

59. Dennis, K. J., Affek, H. P., Passey, B. H., Schrag, D. P. & Eiler, J. M. Defining an absolute reference frame for 'clumped' isotope studies of CO$_2$. *Geochim. Cosmochim. Acta* **75**, 7117–7131 (2011).

60. Huntington, K. W. *et al.* Methods and limitations of 'clumped' CO$_2$ isotope (Δ$_{47}$) analysis by gass-source isotope ratio mass spectrometry. *J. Mass Spectrom* **44**, 1318–1329 (2009).

61. Ogg, J. G. in *The Geologic Time Scale.* (eds Gradstein, F. M., Ogg, J. G., Schmitz, M. D. & Ogg, G. M.) vol. 1, chap. 5 (Elsevier, Oxford, 2012).

62. McArthur, J. M., Howarth, R. J. & Shields, G. A. in *The Geologic Time Scale.* (eds Gradstein, F. M., Ogg, J. G., Schmitz, M. D. & Ogg, G. M.) vol. 1, chap. 7 (Elsevier, Oxford, 2012).

63. Chenet, A.-L. *et al.* Determination of rapid Deccan eruptions across the Cretaceous-Tertiary boundary using paleomagnetic secular variation: 2. Constraints from analysis of eight new sections and synthesis for a 2500-m-thick composite section. *J. Geophys. Res.* **144**, B06103 (2009).

Acknowledgements

We thank W.J. Zinsmeister for contributing his fossil samples and expertise, as well as L. Wingate and H. Ochoa for laboratory support. This work was supported by funding from NSF-OCE-PRF #1420902, NSF-EAR #1123733, NSF-OPP #9318212 and NSF-PLR #9980538.

Author contributions

S.V.P. made clumped isotope measurements, interpreted data and wrote manuscript. A.D. made stable isotope, trace element, and strontium isotope measurements and helped write manuscript. K.C.L. oversaw all measurements and helped write manuscript.

Additional information

25

Aviation effects on already-existing cirrus clouds

Matthias Tesche[1,†], Peggy Achtert[1,†], Paul Glantz[1] & Kevin J. Noone[1]

Determining the effects of the formation of contrails within natural cirrus clouds has proven to be challenging. Quantifying any such effects is necessary if we are to properly account for the influence of aviation on climate. Here we quantify the effect of aircraft on the optical thickness of already-existing cirrus clouds by matching actual aircraft flight tracks to satellite lidar measurements. We show that there is a systematic, statistically significant increase in normalized cirrus cloud optical thickness inside mid-latitude flight tracks compared with adjacent areas immediately outside the tracks.

---

[1] Department of Environmental Science and Analytical Chemistry (ACES), Stockholm University, SE-10691 Stockholm, Sweden. † Present addresses: School of Physics, Astronomy and Mathematics, University of Hertfordshire, Hatfield AL10 9AB, UK (M.T.); Institute for Climate and Atmospheric Science, School of Earth and Environment, University of Leeds, Leeds LS2 9JT, UK (P.A.). Correspondence and requests for materials should be addressed to K.J.N. (email: kevin.noone@aces.su.se).

A ir traffic is known to have an immediate and noticeable effect on clouds in the upper troposphere. New clouds that form due to aircraft effluent are called contrails[1,2], and may develop into more persistent and widespread contrail cirrus. Boucher[3] was the first to realize that aviation might have a strong influence on the occurrence rate of cirrus clouds. Previous studies of contrail optical properties are either based on passive remote sensing in which contrails are identified as linear features in scenes of brightness temperature differences[4–6] or modelling studies in which contrails are formed when favourable meteorological conditions are reached[7]. The life cycle of contrails and aviation-induced cirrus, their radiative forcing and feedback on natural clouds have been studied by treating them as an independent cloud class in a climate model[8]. The study by Iwabuchi et al.[9] is the only one so far that has used height-resolved observations from space-borne lidar measurements to investigate the physical and optical properties of contrails. In their approach, the authors used passive MODIS (moderate resolution imaging spectroradiometer) observations to identify contrails for a subsequent detailed analysis of CALIOP (cloud-aerosol lidar with orthogonal polarization) observations.

In general, aviation-induced clouds (that is, contrails and contrail cirrus) have been found to be optically thin[10,11], and their climatic effects have been estimated to be minor[4,12–15] even when considering their entire life cycle[8,16]. The effect of contrails embedded in natural cirrus is a mechanism that currently has neither been studied nor assessed for its radiative effect on climate[8,15–17].

While optically thick cirrus clouds have a net cooling effect on surface temperature, optically thin cirrus clouds, like greenhouse gases, can have a warming effect[15,18]. Aircraft emissions and contrails at cirrus altitudes have the potential to either cause optically thin cirrus clouds to form (that would have a warming effect on surface temperatures) or increase the optical thickness of existing clouds (or induce new optically thick clouds), thus, causing a net cooling effect. Enhanced observations of the effects of aircraft on cirrus cloud properties are needed to help bound and quantify these possible effects.

The aim of this study is to test the hypothesis that contrails formed within natural cirrus clouds have no measurable immediate effect on cirrus optical depth inside and outside flight tracks in the upper troposphere. We combine data of aircraft flight tracks with spaceborne lidar observations to investigate the effect of aviation on the optical thickness of already-existing cirrus clouds. We detect a statistically significant 22% increase in normalized cirrus optical thickness in mid-latitude air traffic flight tracks compared with adjacent areas outside the flight tracks.

Results

Data sources. We have used commercially available flight track data from FlightAware.com for aircraft serving the major connections between the west coast of the United States and Hawaii in the years 2010 and 2011. Actual flight data (measurements from the aircraft) are received from Air Navigation Service Providers and Automatic Dependent Surveillance—Broadcast receivers. In intervals when no data is received from the aircraft itself, positions are interpolated between the last two reported positions. We consider commercial airline connections between Seattle (KSEA), San Francisco (KSFO), Los Angeles (KLAX) and Honolulu (PHNL).

We derive information about cirrus optical thickness (COT), cloud base and top height, cloud geometrical depth (vertical extent) and mean extinction coefficient from observations with the CALIOP instrument aboard the cloud-aerosol lidar and infrared pathfinder satellite observations (CALIPSO) satellite[19]. Details on the selection of CALIPSO data used in this study are given in the Methods section.

Approach. Figure 1 illustrates our approach. Typical flight tracks for connections between Seattle (KSEA), San Francisco (KSFO), Los Angeles (KLAX) and Honolulu (PHNL) are shown as thick

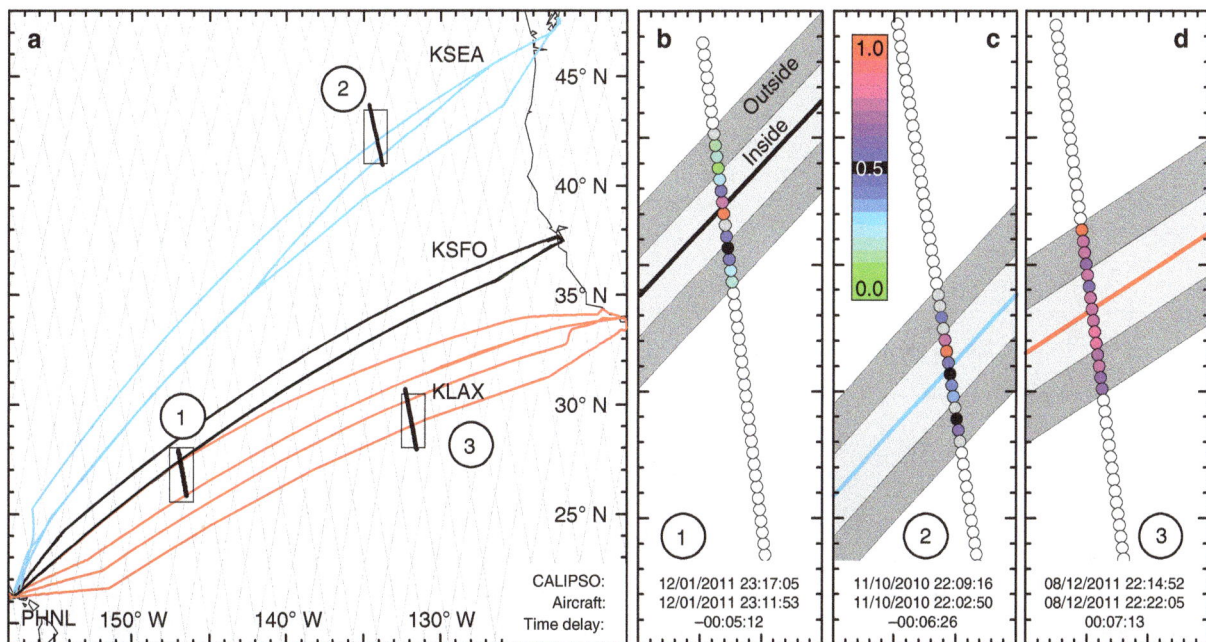

Figure 1 | Overview of the analysis approach. (**a**) Typical aircraft flight tracks (coloured lines) and CALIPSO satellite trajectories (grey lines for 16-day cycle, black lines for example cases). (**b–d**) Close-up of three example overpasses indicated in **a** with values of normalized cirrus optical thickness (coloured dots) and illustration of the inner and outer track (light and dark grey shading, respectively). White and grey dots in **b–d** refer to data that have not been considered in the analysis and do not fulfil the quality assurance criteria, respectively. Times of CALIPSO and aircraft overpasses are given at the bottom of **b–d**. Negative and positive time delay values indicate that the aircraft arrived at the scene before and after, respectively, the satellite overpass.

coloured lines. CALIPSO orbits are indicated as thin grey lines in the figure. The inset shows normalized COT (nCOT; see below) at 532 nm across each of the flight corridors between Los Angeles, San Francisco, Seattle and Honolulu. In cases 1 and 2 aircraft had passed the area < 30 min before the CALIPSO overpass. In case 3, CALIOP observed the location of the flight track before the passage of the aircraft. For these cases cirrus clouds were present at the flight level of the aircraft. For the cases 1 and 2 where the aircraft arrived before the satellite overpass, CALIOP nCOT was clearly larger for the inner part of the flight track compared with clouds present on either side—creating a 'plane track' signature caused by an embedded contrail or another effect on the cloud caused by the aircraft.

Categories for data analysis. We accumulated data for these three air traffic corridors during 2010 and 2011 in which the absolute difference between aircraft and CALIPSO arrival times was 30 min or less. We classified the data into four categories illustrated in Fig. 2: (I) inside flight track, ahead of aircraft; (II) outside flight track, ahead of aircraft; (III) inside flight track, behind aircraft; and (IV) outside flight track, behind aircraft.

If our hypothesis that aircraft have no observable effect on cirrus cloud properties is true, then there should be no statistically

significant differences in COT or in nCOT between the four categories shown in Fig. 2. If the hypothesis is false, and aircraft emissions do have an impact on cloud properties, clouds in category III should have different nCOT compared with the other categories.

Normalization of cirrus cloud optical thickness. As a result of natural variability in the COT of unperturbed cirrus clouds, our COT data set is skewed towards large values. The histograms in Fig. 3 show the distribution of 720 5-km CALIOP data points with a maximum cloud geometrical depth of 2.5 km. The distribution of absolute COT in Fig. 3a has a skewness of 1.57. We normalize COT with respect to the maximum CALIOP value for the overpass. Figure 3b shows the frequency distribution for the same data after normalization. The mean and median values are now essentially equal, and the distribution has a skewness value of 0.18. The distributions of COT and nCOT for cases with maximum cirrus geometrical depths different from 2.5 km have very similar shape and skewness (not shown). The rightmost column in Fig. 3b bears further explanation. nCOT values in these data vary from a minimum of 0.03 to a maximum of 1. The column limits in Fig. 3b go from 0 to 0.0999 for the first column, 0.1 to 0.1999 for the second and 0.9 to 0.9999 for the next-to-last column. Since the true COT for each of the overpasses we accumulated will have at least one maximum value, each pass for nCOT will have at least one value of unity. The rightmost column shows these unity values, which are too large to be included in the interval from 0.9 to 0.9999. The rightmost column should not be interpreted as values larger than one, but rather values of identically one.

The normalization ensures that the data are more normally distributed (a requirement for the statistical tests we use), and the analysis is not biased by a few large values. If our hypothesis that aircraft have no measurable effect on cirrus cloud optical properties is true, there should be no statistically significant differences in mean COT and mean nCOT between these four categories. Otherwise, the mean values for category III should be different than for the other categories, which should not exhibit any differences in mean values.

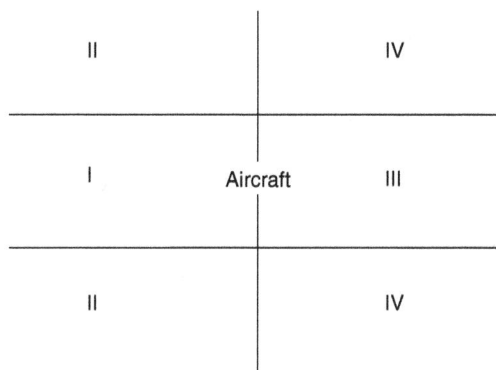

Figure 2 | Categories for data analysis. Category I: inside the flight track, ahead of the aircraft. Category II: outside the flight track, ahead of the aircraft. Category III: inside the flight track, behind the aircraft. Category IV: outside the flight track, behind the aircraft.

Effects of advection and other aircraft. We use ERA-Interim wind speed v and direction in combination with speed and heading of the aircraft at the location and height of the crossing between flight track and CALIPSO overpass to account for the effect of advection. The displacement D perpendicular to the

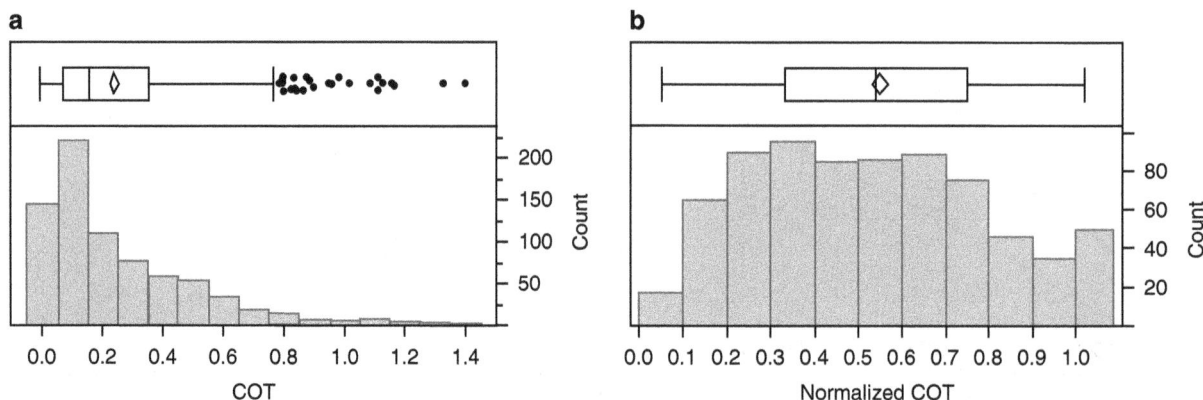

Figure 3 | Frequency distribution of cirrus optical thickness. Distribution of absolute COT (**a**) and nCOT (**b**) for cases where the maximum cloud layer depth is 2.5 km. The data come from all four categories. The box-and-whisker plots show the 25 and 75% quartiles as the box, with the median value being the vertical line inside the box and outliers shown as dots. The diamond inside the box indicates the mean value (vertical vertices) and ± 95% confidence interval (horizontal vertices) for the data. The number of observations is shown on the right vertical axis. Note that the highest bin in **b** only contains values of 1.0.

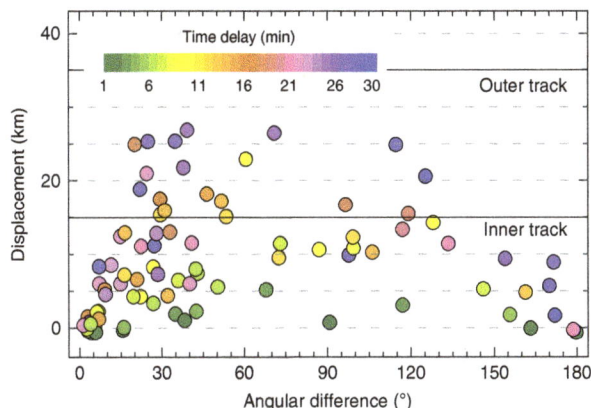

Figure 4 | Effect of advection. The effect of advection presented as displacement perpendicular to the flight track versus the angular difference between aircraft heading and wind direction. The colour coding refers to the absolute time difference between aircraft and CALIPSO.

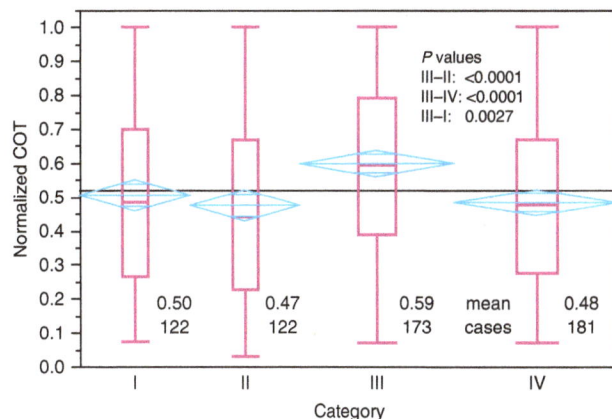

Figure 5 | Cirrus cloud optical thickness per category for a maximum cirrus geometrical depth of 2.5 km. The magenta box-and-whisker plots show the quantiles for the data in each category from a one-way analysis of variance using the JMP software package. Mean diamonds (cyan) indicate the 95% confidence intervals for the mean values of each of the categories. If the upper and lower horizontal lines overlap, there is no statistically significant difference in means. Numbers in the lower part of the figure give the mean value and number of observations (that is, CALIPSO L2 5-km points) in each category. The horizontal grey line represents the overall mean value.

flight track is derived as $D = \Delta t \, v \sin \Delta dir$, where Δdir is the angular difference between aircraft heading and wind direction and Δt is the absolute value of the time delay between aircraft and CALIPSO overpass (Fig. 4). We then omit cases that are associated with a displacement larger than 30 km perpendicular to the flight track. We cannot simply calculate how far the emissions from the aircraft would be moved in the time interval between the satellite overpass and the aircraft passage given a constant wind and 'move' our observations there. We only have observations along the line of the CALIPSO orbit, as illustrated in panels b–d in Fig. 1 and in Supplementary Fig. 3. Let us choose the CALIPSO/flight track crossing point as the origin for our coordinate system. A hypothetical wind vector is shown as a thick purple arrow in Supplementary Fig. 3. Given this wind vector, material emitted by the aircraft at the crossing point (1) would be advected to point (2) in the time interval between the passage of the aircraft and the time that CALIPSO observes this point. Point (2) is not on the CALIPSO track. To perform this sort of advection calculation properly for this system, we would need to first calculate the wind vector component along the CALIPSO track to find point (3), then use the wind vector to calculate the origin of the air that CALIPSO would have observed at this new point (which is point (4) in the illustration). Given the CALIPSO/ flight track geometry, the lag time between the satellite and air-craft passing point (1) and the wind vector, this new point of origin may or may not be on the flight track aft of the aircraft. In any case, this iterative calculation would be necessary for each crossing.

In addition, winds in the atmosphere are not constant. Turbulence (even for a constant wind speed) can redistribute emissions around the centreline calculated in the manner described above, and uncertainties in the ERA-Interim winds also need to be taken into account. To use this approach we would at a minimum need to perform Gaussian plume dispersion calculations for a line source, and augment these with estimates of the effects of uncertainties in the ERA-Interim wind fields. This plume advection approach is illustrated as a diffuse horizontal line in Supplementary Fig. 3. The result of these calculations would be a flight corridor with a new location, centred around a particular line, but with a probability distribution of location.

Figure 4 shows how far advection would have moved any emissions from the aircraft perpendicular to the flight track in the time interval between the passage of the aircraft and the satellite overpass. In the majority of cases, advection does not move material outside what we define as the inner flight track, and

Table 1 | Statistical significance between all categories pairs with $P \leq 0.05$, subdivided with respect to maximum cirrus geometrical depth.

Maximum cirrus geometrical depth (km)	Significant differences between categories
2.0	III − IV, III − II, III − I
2.5	III − IV, III − II, III − I
3.0	III − IV, III − II
4.0	III − IV
5.0	III − IV
6.0	III − IV

The tests were performed for all category pairs.

therefore a more sophisticated calculation is not needed in this approach.

To account for the effect of other aircraft on the same flight track we omit cases in which the delay between any previous aircraft and our flight of interest was < 30 min. Supplementary Fig. 4 illustrates how advection may influence the properties of the air in the categories we use for analysis. Further details on the effect of advection on our findings are provided in the Supplementary Discussion, as well as in Supplementary Tables 1 and 2.

Findings for clouds with different geometrical thickness. The results of this analysis are shown in Fig. 5. For brevity we present the results for cases in which the maximum cirrus depth was 2.5 km. The mean nCOT for category III (0.59) is significantly higher than for the other three categories (III − II: $P < 0.0001$; III − IV: $P < 0.0001$; III − I: $P = 0.0027$). In terms of true COT, the category 3 mean value was 0.30, while the means of the other categories were as follows: I, 0.27; II, 0.26; IV, 0.26. Thus, the mean COT for category III was 14% higher than the other categories, though statistically significant only at the 93% confidence level due to the skewness of the data. Differences between the other categories were not statistically significant. We

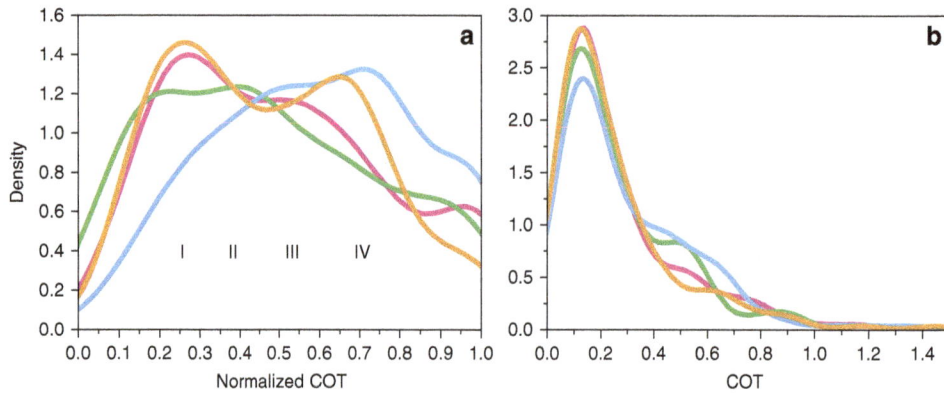

Figure 6 | Distribution of cirrus optical thickness per category. Frequency of observation of nCOT (**a**) and absolute COT (**b**) for all cases, with maximum cirrus geometrical depth of 2.5 km (that is, the data presented in Fig. 2). Colours refer to the different categories. Category I (magenta): inside the flight track, ahead of the aircraft. Category II (green): outside the flight track, ahead of the aircraft. Category III (blue): inside the flight track, behind the aircraft. Category IV (orange): outside the flight track, behind the aircraft.

examined cases for different maximum cloud layer depths (Supplementary Fig. 1). Table 1 reveals that the difference in nCOT for clouds inside and outside the flight track aft of the aircraft persists for all cases of maximum cirrus geometrical depth. The difference between category III and the other categories ahead of the aircraft is significant only for geometrically thin clouds.

Figure 6 gives the frequency of observation of both COT and nCOT for the case of 2.5 km maximum cirrus geometrical depth. nCOT in category III shows fewer low values and a larger proportion of high values than the other categories. This difference in the frequency distributions for the different categories decreases for increasing maximum cirrus geometrical depth (not shown). The distributions become essentially indistinguishable from each other by a maximum cirrus geometrical depth of 5.0 km.

We have falsified our hypothesis, and observed a detectable 22% increase in nCOT in flight tracks for cases where the aircraft was 30 min or less ahead of the satellite overpass. We examined our data for differences in COT between day- and night-time CALIPSO overpasses. We did not observe any significant day–night differences in nCOT, and therefore have included both kinds of observations in our analysis (Supplementary Fig. 2).

Discussion

Air traffic corridors are far more prevalent in the northern hemisphere than in the southern hemisphere, so we anticipate any climatic effects these embedded contrails may have will be more pronounced there. Even though cloudiness may already be changed by earlier aircraft, we can isolate the effect of a single aircraft on cloud properties. Since the effect of aircraft on cloud properties may well last longer than 30 min the overall effect of air traffic on cloud properties may be larger than the values estimated here. Estimating the climatic effects of embedded contrails is beyond the scope of this paper; however, given the broad coverage of air traffic corridors in the northern hemisphere, embedded contrails as identified in this study are potentially an important and not yet considered contributor to the non-CO_2 effects of aviation on climate[17].

Further work will be needed to quantify the effect identified in this study. Initially, detailed radiative transfer modelling is needed to assess the impact of an increase in COT on the Earth's radiative budget. From the modelling perspective, future studies will need to estimate the magnitude of the observed effect on a global scale and assess its contribution to the overall non-CO_2

effects of aviation on climate. The increase in cirrus optical depth may result from the emitted soot in the first few seconds within the plume. Soot particles are not efficient ice nuclei. They rather form droplets when water saturation is reached in the plume and freeze subsequently[20]. Hence, the effect on the microphysics of the cirrus is an open question, and will require detailed microphysical modelling to address.

Methods

Overall approach. Our empirical approach is based on the ship tracks methodology[21]. Ship tracks occur in marine stratocumulus clouds in locations where cloud albedo increases due to particulate emissions from ships[22,23]. Here we apply a similar observational approach—we investigate the optical and geometrical properties of cirrus clouds within known flight tracks (that is, a region affected by aircraft exhaust and contrails) to the ones immediately adjacent on either side of the flight track (that is, the unperturbed region outside of the main flight routes). To apply this approach we first need to identify regions in which we are likely to encounter cirrus clouds that are influenced by air traffic. Favourable regions for this study should fulfil the following criteria: the aircraft should be flying below the tropopause and at cirrus level, tropical convection should be of minor influence on cloud formation (that is, the region should lie outside the ITCZ band), and the area should be a sufficiently large so that CALIPSO will have multiple transects in a reasonable amount of time. These criteria rule out the intercontinental polar routes, since in these regions the aircraft are often flying above the tropopause level. The most promising routes are in the northern hemisphere mid-latitudes. Data on aircraft location are used to identify flight tracks.

Spaceborne lidar observations. CALIOP is an elastic-backscatter lidar that emits linearly polarized laser light at 532 and 1,064 nm and features three measurement channels. The CALIPSO lidar has been operational since June 2006. An overview of the instrument and retrieval algorithms can be found in Winker et al.[19]. CALIPSO is a polar orbiting satellite with a return cycle of 16 days.

The spatial and optical properties of cirrus clouds as derived from CALIOP observations have been carefully validated with airborne lidar measurements[24–27]. CALIOP observations became an invaluable tool for regional and global studies on the occurrence and properties of cirrus[28–30] and sub-visual cirrus[31,32]. CALIPSO's location in the A-Train satellite constellation furthermore enables the addition of further co-located measurements to the lidar data for advanced cirrus studies[29,32].

CALIOP observations in the area 15–55° N and 115–155° W for the years 2010 and 2011 have been considered in this study. We use the level 2 version 3.01 (January 2010–October 2011) and 3.02 (November 2011 to December 2011) 5-km Cloud Layer products. Data are provided with a horizontal resolution of 5 km and refer to layers starting with the uppermost detected feature.

CALIOP data quality assurance. We only considered CALIOP observations for which cirrus (feature type = cloud, sub-type = cirrus[33], uppermost two layers are considered) occurs at the altitude of a coinciding flight track. For quality assurance we also require these features to be detected at 5-km averaging intervals (Horizontal_Averaging = 5) and to show an Extinction_QC_Flag_532 of zero[34]. Note that data points with an Extinction_QC_Flag_532 other than zero are virtually absent in our data set. A detailed description of the CALIPSO products can be found in the CALIPSO's Data User Guide[35].

CALIOP data selection. We only consider the seven 5-km CALIOP data points to the north and south of the flight track, respectively. In addition, we require that at least half of these points (that is, 7 out of 14) show valid COT that fulfils our quality assurance criteria. This ensures that we observe a signal corresponding to homogeneous cirrus rather than natural variability of cirrus optical properties. We consider the three closest CALIPSO observations on either side of the aircraft track to form the inner part of the flight track (Fig. 1). Depending on the angle between the flight track and CALIPSO ground path, the inner part of the flight track spreads over up to 15 km to the north and south of the aircraft track. This distance agrees with the typical contrail width of about 5 km[9] and allows compensating for minor advection effects caused by the 30-min (maximum) time interval between the aircraft and CALIPSO overpasses. Points (4)–(7) in Fig. 1b–d represent the unperturbed conditions outside of the flight track. This approach provides us with an overall number of data points that is balanced with respect to the inner and outer parts of the flight track.

Difference between CALIOP night-time and day-time observations. Owing to the influence of daylight noise CALIOP feature detection sensitivity is higher at night than during day time. We therefore examined the data for differences in day- and night-time measurements. Examples of the mean nCOT for cases in which the maximum cirrus geometrical depth was 2.5 and 5.0 km are shown in Supplementary Fig. 2. We did not observe any significant day–night differences in normalized COT, and therefore have included both kinds of observations in our analysis.

References

1. Appleman, H. The formation of exhaust contrails by jet aircraft. *Bull. Am. Meteorol. Soc.* **34**, 14–20 (1953).
2. Stubenrauch, C. J. & Schumann, U. Impact of air traffic on cirrus coverage. *Geophys. Res. Lett.* **32**, L14813 (2005).
3. Boucher, O. Air traffic may increase cirrus cloudiness. *Nature* **397**, 30–31 (1999).
4. Minnis, P., Schumann, U., Doelling, D. R., Gierens, K. M. & Fahey, D. W. Global distribution of contrail radiative forcing. *Geophys. Res. Lett.* **26**, 1853–1856 (1999).
5. Minnis, P., Palikonda, R., Walter, P. J., Ayers, J. K. & Mannstein, H. Contrail properties over the eastern North Pacific from AVHRR data. *Meteorol. Z.* **14**, 515–523 (2005).
6. Duda, D. P., Minnis, P., Khlopenkov, K., Chee, T. L. & Boeke, R. Estimation of 2006 Northern Hemisphere contrail coverage using MODIS data. *Geophys. Res. Lett.* **40**, 612–617 (2013).
7. Schumann, U. A contrail cirrus prediction model. *Geosci. Model Dev.* **5**, 543–580 (2012).
8. Burkhardt, U. & Kärcher, B. Global radiative forcing from contrail cirrus. *Nat. Clim. Change* **1**, 54–58 (2011).
9. Iwabuchi, H., Yang, P., Liou, K. N. & Minnis, P. Physical and optical properties of persistent contrails: climatology and interpretation. *J. Geophys. Res.* **117**, D06215 (2012).
10. Kübbeler, M. *et al.* Thin and subvisible cirrus and contrails in a subsaturated environment. *Atmos. Chem. Phys.* **11**, 5853–5865 (2011).
11. Voigt, C. *et al.* Extinction and optical depth of contrails. *Geophys. Res. Lett.* **38**, L11806 (2011).
12. Sassen, K., Griffin, M. K. & Dodd, G. C. Optical scattering and microphysical properties of subvisual cirrus clouds, and climatic implications. *J. Appl. Meteorol.* **28**, 91–98 (1989).
13. Sassen, K. Contrail-cirrus and their potential for regional climate change. *Bull. Am. Meteorol. Soc.* **78**, 1885–1903 (1997).
14. Schumann, U. Formation, properties and climatic effects of contrails. *C. R. Phys.* **6**, 549–565 (2005).
15. IPCC. in *Climate Change 2013: The Physical Science Basis—Contribution of Working Group I to the Fifth Assessment Report of the Intergovernmental Panel on Climate Change* 1535 (eds Stocker, T. F. *et al.*) (IPCC, 2013).
16. Burkhardt, U. & Kärcher, B. Process-based simulation of contrail cirrus in a global climate model. *J. Geophys. Res.* **114**, D16201 (2009).
17. Lee, D. *et al.* Aviation and global climate change in the 21st century. *Atmos. Environ.* **43**, 3520–3537 (2009).
18. Hartmann, D. L., Ockert-Bell, M. E. & Michelsen, M. L. The effect of cloud type on Earth's energy balance: global analysis. *J. Clim.* **5**, 1281–1304 (1992).
19. Winker, D. M. *et al.* Overview of the CALIPSO mission and CALIOP data processing algorithms. *J. Atmos. Oceanic Technol.* **26**, 2310–2323 (2009).
20. Kärcher, B. *et al.* The microphysical pathway to contrail formation. *J. Geophys. Res.* **120**, 7893–7927 (2015).
21. Durkee, P. A., Noone, K. J. & Bluth, R. T. The monterey area ship track experiment. *J. Atmos. Sci.* **57**, 2523–2541 (2000).
22. Noone, K. J. *et al.* A case study of ships forming and not forming tracks in moderately polluted clouds. *J. Atmos. Sci.* **57**, 2729–2747 (2000).
23. Noone, K. J. *et al.* A case study of ship track formation in a polluted marine boundary layer. *J. Atmos. Sci.* **57**, 2748–2764 (2000).
24. McGill, M. J. *et al.* Airborne validation of spatial properties measured by the CALIPSO lidar. *J. Geophys. Res.* **112**, D20201 (2007).
25. Mioche, G. D. *et al.* Validation of the CALIPSO-CALIOP extinction coefficients from *in situ* observations in midlatitude cirrus clouds during the CIRCLE-2 experiment. *J. Geophys. Res.* **115**, D00H25 (2010).
26. Yorks, J. E. *et al.* Airborne validation of cirrus cloud properties derived from CALIPSO lidar measurements: spatial properties. *J. Geophys. Res.* **116**, D19207 (2011).
27. Hlavka, D. L. *et al.* Airborne validation of cirrus cloud properties derived from CALIPSO lidar measurements: optical properties. *J. Geophys. Res.* **117**, D09207 (2012).
28. Nazaryan, H., McCormick, M. P. & Menzel, W. P. Global characterization of cirrus clouds using CALIPSO data. *J. Geophys. Res.* **113**, D16211 (2008).
29. Sassen, K., Wang, Z. & Liu, D. Global distribution of cirrus clouds from CloudSat/Cloud-Aerosol lidar and infrared pathfinder satellite observations (CALIPSO) measurements. *J. Geophys. Res.* **113**, D00A12 (2008).
30. Meenu, S., Rajeev, K. & Parameswaran, K. Regional and vertical distribution of semitransparent cirrus clouds over the tropical Indian region derived from CALIPSO data. *J. Atmos. Sol. Terr. Phys.* **73**, 1967–1979 (2011).
31. Sun *et al.* A study of subvisual clouds and their radiation effect with a synergy of CERES, MODIS, CALIPSO, and AIRS data. *J. Geophys. Res.* **116**, D22207 (2011).
32. Martins, E., Noel, V. & Chepfer, H. Properties of cirrus and subvisible cirrus from nighttime Cloud-Aerosol Lidar with Orthogonal Polarization (CALIOP), related to atmospheric dynamics and water vapor. *J. Geophys. Res.* **116**, D02208 (2011).
33. Vaughan, M. A. *et al.* Fully automated detection of cloud and aerosol layers in the CALIPSO lidar measurements. *J. Atmos. Oceanic Technol.* **26**, 2034–2050 (2009).
34. Young, S. A. & Vaughan, M. A. The retrieval of profiles of particulate extinction from Cloud-Aerosol Lidar Infrared Pathfinder Satellite Observations (CALIPSO) data: Algorithm description. *J. Atmos. Oceanic Technol.* **26**, 1105–1119 (2009).
35. CALIPSO's User Guide. Available at http://www-calipso.larc.nasa.gov/ resources/calipso_users_guide/index.php (2015), accessed 1 June 2016.

Acknowledgements

CALIPSO data used in this study were obtained from the NASA Langley Research Center Atmospheric Science Data Center (http://eosweb.larc.nasa.gov). The ERA-Interim data were obtained from the European Centre for Medium-Range Weather Forecasts (ECMWF) data server. This study has been funded by the Swedish Research Council grant D0612701. We thank Ulrike Burkhardt and three anonymous referees for their comments on the paper.

Author contributions

K.J.N. and M.T. developed the idea that led to the paper; P.A. and M.T. processed and interpreted the CALIOP data; K.J.N. performed the statistical analysis; P.G. provided meteorological data to assess the effect of advection and as input for future studies to estimate climatic effects; M.T. and K.J.N. wrote the original manuscript. All authors contributed to the final editing of the manuscript and the design of the figures.

Additional information

Competing financial interests: The authors declare no competing financial interests.

PERMISSIONS

LIST OF CONTRIBUTORS

Daniel C. Anderson, Timothy P. Canty and Russell R. Dickerson
Department of Atmospheric and Oceanic Science, University of Maryland, College Park, Maryland 20742, USA

Julie M. Nicely
Department of Chemistry and Biochemistry, University of Maryland, College Park, Maryland 20742, USA

Ross J. Salawitch
Department of Atmospheric and Oceanic Science, University of Maryland, College Park, Maryland 20742, USA
Department of Chemistry and Biochemistry, University of Maryland, College Park, Maryland 20742, USA
Earth System Science Interdisciplinary Center, University of Maryland, College Park, Maryland 20742, USA

Thomas F. Hanisco and Anne M. Thompson
NASA Goddard Space Flight Center, Greenbelt, Maryland 20771, USA

Glenn M. Wolfe
NASA Goddard Space Flight Center, Greenbelt, Maryland 20771, USA
Joint Center for Earth Systems Technology, University of Maryland Baltimore County, Baltimore, Maryland 21250, USA

Eric C. Apel, Kirk Ullmann, Teresa L. Campos, Douglas E. Kinnison, Andrew J. Weinheimer, Rebecca S. Hornbrook and Samuel R. Hall
Atmospheric Chemistry Observation and Modeling Laboratory, National Center for Atmospheric Research, Boulder, Colorado 80305, USA

Elliot Atlas and Daniel D. Riemer
Department of Atmospheric Sciences, Rosenstiel School of Marine and Atmospheric Science, University of Miami, Miami, Florida 33149, USA

Thomas Bannan, Carl Percival and Michael Le Breton
Centre for Atmospheric Science, School of Earth, Atmospheric, and Environmental Science, The University of Manchester, Manchester M13 9PL, UK

Stephane Bauguitte
Facility for Airborne Atmospheric Measurements, Cranfield MK43 0JR, UK

Nicola J. Blake
Deparment of Chemistry, University of California, Irvine, California 92697, USA

James F. Bresch
Mesoscale and Microscale Meteorology Laboratory, National Center for Atmospheric Research, Boulder, Colorado 80305, USA

Lucy J. Carpenter
Wolfson Atmospheric Chemistry Laboratories, Department of Chemistry, University of York, York YO10 5DD, UK

Mathew Evans
Wolfson Atmospheric Chemistry Laboratories, Department of Chemistry, University of York, York YO10 5DD, UK
National Centre for Atmospheric Science, Department of Chemistry, University of York, York YO10 5DD, UK

Barbara J.B. Stunder and Mark D. Cohen
NOAA Air Resources Laboratory, College Park, Maryland 20740, USA

Adam Vaughan and James D. Lee
National Centre for Atmospheric Science, Department of Chemistry, University of York, York YO10 5DD, UK

Alfonso Saiz-Lopez
Department of Atmospheric Chemistry and Climate, Institute of Physical Chemistry Rocasolano, CSIC, Madrid 28006, Spain

Rafael P. Fernandez
Department of Atmospheric Chemistry and Climate, Institute of Physical Chemistry Rocasolano, CSIC, Madrid 28006, Spain
Department of Natural Science, National Research Council (CONICET), FCEN-UNCuyo, Mendoza 5501, Argentina

Brian H. Kahn
Jet Propulsion Laboratory, California Institute of Technology, Pasadena, California 91109, USA

Neil R.P. Harris
Department of Chemistry, Cambridge University, Cambridge CB2 1EW, UK

Jean-Francois Lamarque
Atmospheric Chemistry Observation and Modeling Laboratory, National Center for Atmospheric Research, Boulder, Colorado 80305, USA
Climate and Global Dynamics Laboratory, National Center for Atmospheric Research, Boulder, Colorado 80305, USA

Leonhard Pfister
Earth Sciences Division, NASA Ames Research Center, Moffett Field, California 94035, USA

R. Bradley Pierce
NOAA/NESDIS Center for Satellite Applications and Research, Madison, Wisconsin 53706, USA

Bror F. Jönsson
Department of Geosciences, Princeton University, Princeton, New Jersey 08544, USA
College of Earth, Ocean and Atmospheric Sciences, Oregon State
University, Corvallis, Oregon 97331-5503, USA
The Stockholm Resilience Centre, Stockholm University, 118 14 Stockholm, Sweden

Chia-Ying Lee
International Research Institute of Climate and Society, Columbia University, Palisades, New York 10964, USA

Michael K. Tippett
Department of Applied Physics and Applied Mathematics, Columbia University, New York 10027, USA
Center of Excellence for Climate Change Research, Department of Meteorology, King Abdulaziz University, Jeddah 21589, Saudi Arabia

Suzana J. Camargo
Division of Ocean and Climate Physics, Lamont-Doherty Earth Observatory, Columbia University, Palisades, New York 10964, USA

Adam H. Sobel
Department of Applied Physics and Applied Mathematics, Columbia University, New York 10027, USA
Division of Ocean and Climate Physics, Lamont-Doherty Earth Observatory, Columbia University, Palisades, New York 10964, USA

Manfred Milinski, Dirk Semmann and Ralf Sommerfeld
Department of Evolutionary Ecology, Max-Planck-Institute for Evolutionary Biology, August-Thienemann-Strasse 2, 24306 Plön, Germany

Christian Hilbe
Department of Organismic and Evolutionary Biology, Department of Mathematics, Program for Evolutionary Dynamics, Harvard University, One Brattle Square, Cambridge,
Massachusetts 02138, USA
Institute of Science and Technology Austria, Am Campus 1, Klosterneuburg 3400, Austria

Jochem Marotzke
Max Planck Institute for Meteorology, Department "The Ocean in the Earth System", 20146 Hamburg, Germany

Ayako Yamamoto
Department of Atmospheric and Oceanic Sciences, McGill University, 805 Sherbrooke StreetWest, Montreal, Quebec, Canada H3A 2K6

Jaime B. Palter
Department of Atmospheric and Oceanic Sciences, McGill University, 805 Sherbrooke StreetWest, Montreal, Quebec, Canada H3A 2K6
Graduate School of Oceanography, University of Rhode Island, Narragansett Bay Campus, Narragansett, Rhode Island 02882, USA

Alexey Portnov, Sunil Vadakkepuliyambatta, Jürgen Mienert and Alun Hubbard
CAGE—Centre for Arctic Gas Hydrate, Environment and Climate, Department of Geology, UiT The Arctic University of Norway, 9037 Tromsø, Norway

Nicholas P. Foukal and M. Susan Lozier
Nicholas School of the Environment, Duke University, Durham, North Carolina 27708, USA

Andrew M.W. Newton, Mads Huuse and Simon H. Brocklehurst
School of Earth, Atmospheric and Environmental Sciences, University of Manchester, Manchester M13 9PL, UK

T.A. Ronge, R. Tiedemann, F. Lamy and P. Köhler
Alfred-Wegener-Institut Helmholtz-Zentrum für Polar-und Meeresforschung, Department for Marine Geology, PO Box 120161, Bremerhaven 27515, Germany

B.V. Alloway
School of Geography, Environment and Earth Sciences, Victoria University of Wellington, PO Box 600, 6012 Wellington, New Zealand

R. De Pol-Holz
GAIA Antárctica Universidad de Magellanes, Department of Paleclimatology, Oceanography, Punta Arenas 01855, Chile

K. Pahnke
Max Planck Research Group—Marine Isotope Geochemistry, Institute for Chemistry and Biology of the Marine Environment, Department of Marine Isotope Geochemistry, Carl von Ossietzky University, PO Box 2503, Oldenburg 26111, Germany

J. Southon
School of Physical Science, Department of Earth Science, University of California, Irvine, California 92697-4675, USA

L. Wacker
Laboratory of Ion Beam Physics (HPK), Eidgenössische Technische Hochschule, Schafmattstrasse 20, Zürich 8093, Switzerland

Hauke Schmidt and Jochem Marotzke
Max Planck Institute for Meteorology, Bundesstrasse 53, Hamburg 20146, Germany

Max Popp
Max Planck Institute for Meteorology, Bundesstrasse 53, Hamburg 20146, Germany
Program in Atmospheric and Oceanic Sciences, Princeton University, 300 Forrestal Road, Sayre Hall, Princeton, New Jersey 08544, USA
NOAA's Geophysical Fluid Dynamics Laboratory, Princeton, New Jersey, USA

María T. Hernández-Sánchez, Saúl González-Lemos, Lorena Abrevaya, Ana Mendez-Vicente and Heather M. Stoll
Geology Department, Oviedo University, Arias de Velasco s/n, 33005 Oviedo, Asturias, Spain

Clara T. Bolton
Geology Department, Oviedo University, Arias de Velasco s/n, 33005 Oviedo, Asturias, Spain
Aix-Marseille University, CNRS, IRD, CEREGE UM34, 13545 Aix en Provence, France

Miguel-Ángel Fuertes and José-Abel Flores
Grupo de Geociencias Oceánicas, Geology Department, University of Salamanca, Salamanca 37008, Spain

Ian Probert
CNRS, Sorbonne Universités-Université Pierre et Marie Curie (UPMC) Paris 06, FR2424, Roscoff Culture Collection, Station Biologique de Roscoff, Place Georges Teissier, 29680 Roscoff, France

Liviu Giosan
Department of Geology and Geophysics, Woods Hole Oceanographic Institution, 266 Woods Hole Road, MS# 22, Woods Hole, Massachusetts 02543-1050, USA

Joel Johnson
University of New Hampshire, Department of Earth Sciences, 56 College Road, James Hall, Durham, New Hampshire 03824-3589, USA

Julia Gottschalk and Luke C. Skinner
Godwin Laboratory for Palaeoclimate Research, Earth Sciences Department, University of Cambridge, Downing Street, Cambridge CB2 3EQ, UK

Jörg Lippold, Hendrik Vogel and Samuel L. Jaccard
Institute of Geological Sciences and Oeschger Center for Climate Change Research, University of Bern, Baltzerstr. 1-3, Bern 3012, Switzerland

Norbert Frank
Institute of Environmental Physics, University of Heidelberg, Im Neuenheimer Feld 229, Heidelberg 69120, Germany

Claire Waelbroeck
Laboratoire des Sciences du Climat et de l'Environnement, LSCE/IPSL, CNRS-CEA-UVSQ, Université de Paris-Saclay, Domaine du CNRS, bât. 12, Gif-sur-Yvette 91198, France

Michael K. Tippett
Department of Applied Physics and Applied Mathematics, Columbia University, New York, New York 10027, USA
Center of Excellence for Climate Change Research, Department of Meteorology, King Abdulaziz University, Jeddah 21589, Saudi Arabia

Joel E. Cohen
Laboratory of Populations, Rockefeller University, New York, New York 10065, USA
The Earth Institute, Columbia University, New York, New York 10027, USA

Philippe Ciais and Frédéric Chevallier
Laboratoire des Sciences du Climat et de l'Environnement, LSCE/IPSL, CEA-CNRS-UVSQ, Université Paris-Saclay, F-91191 Gif-sur-Yvette, France

Ana Bastos
Laboratoire des Sciences du Climat et de l'Environnement, LSCE/IPSL, CEA-CNRS-UVSQ, Université Paris-Saclay, F-91191 Gif-sur-Yvette, France
Instituto Dom Luiz, IDL, Faculdade de Ciências, Universidade de Lisboa, Lisboa 1749-016, Portugal

Célia M. Gouveia and Ricardo M. Trigo
Instituto Dom Luiz, IDL, Faculdade de Ciências, Universidade de Lisboa, Lisboa 1749-016, Portugal

Ivan A. Janssens
Department of Biology, University of Antwerp, Universiteitsplein 1, 2610 Wilrijk, Belgium

Josep Peñuelas
CREAF, Cerdanyola del Vallès, Catalonia, 08193 Barcelona, Spain
CSIC, Global Ecology Unit CREAF-CSIC-UAB, Cerdanyola del Vallès, Catalonia, 08193 Barcelona, Spain

Christian Rödenbeck
Max Planck Institute for Biogeochemistry, Jena 07701, Germany

Shilong Piao
Department of Ecology, College of Urban and Environmental Sciences, Peking University 5 Yiheyuan Road, Haidian District, Beijing 100871, China

Pierre Friedlingstein
College of Engineering, Mathematics and Physical Sciences, University of Exeter, Exeter EX4 4QF, UK

Steven W. Running
Numerical Terradynamic Simulation Group, University of Montana, Missoula, Montana 59812, USA

Zhi-Gang Shao
Guangdong Provincial Key Laboratory of Quantum Engineering and Quantum Materials, SPTE, South China Normal University, Guangzhou 510006, China

Peter D. Ditlevsen
Centre for Ice and Climate, Niels Bohr Institute, University of Copenhagen, Juliane Maries Vej 30, Copenhagen 2100, Denmark

Qian Yang and Timothy H. Dixon
School of Geosciences, University of South Florida, 4202 E Fowler Avenue, Tampa, Florida 33620, USA

Paul G. Myers
Department of Earth and Atmospheric Sciences, University of Alberta, 1-26 ESB, Edmonton, Alta, Canada T6G 2E3

Jennifer Bonin and Don Chambers
College of Marine Science, University of South Florida, St. Petersburg, Florida 33701, USA.

M.R. van den Broeke
Institute for Marine and Atmospheric Research Utrecht, Utrecht University, P.O. Box 80.005, 3508 TA, Utrecht, The Netherlands

Hanno Sandvik
Centre for Biodiversity Dynamics, Department of Biology, Norwegian University of Science and Technology, 7491 Trondheim, Norway

Robert T. Barrett
Department of Natural Sciences, Tromsø University Museum, PO Box 6050 Langnes, 9037 Tromsø, Norway

Kjell Einar Erikstad
Centre for Biodiversity Dynamics, Department of Biology, Norwegian University of Science and Technology, 7491 Trondheim, Norway
Norwegian Institute for Nature Research, FRAM— High North Research Centre for Climate and the Environment, 9296 Tromsø, Norway

Tone K. Reiertsen and Geir Helge Systad
Norwegian Institute for Nature Research, FRAM— High North Research Centre for Climate and the Environment, 9296 Tromsø, Norway

Nigel G. Yoccoz
Norwegian Institute for Nature Research, FRAM— High North Research Centre for Climate and the Environment, 9296 Tromsø, Norway
Department of Arctic and Marine Biology, University of Tromsø, PO Box 6050 Langnes, 9037 Tromsø, Norway

Mari S. Myksvoll, Jofrid Skarjhamar, Mette Skern-Mauritzen and Frode Vikebø
Institute of Marine Research and Hjort Centre for Marine Ecosystem Dynamics, PO Box 1870 Nordnes, 5817 Bergen, Norway

Tycho Anker-Nilssen and Svein-Håkon Lorentsen
Norwegian Institute for Nature Research, PO Box 5685 Sluppen, 7485 Trondheim, Norway

Zunli Lu, Xiaoli Zhou, Kristina M. Gutchess and Wanyi Lu
Department of Earth Sciences, Syracuse University, Syracuse, New York 13244, USA

Babette A.A. Hoogakker, Luke Jones and Rosalind E.M. Rickaby
Department of Earth Sciences, University of Oxford, Oxford OX1 3AN, UK

Claus-Dieter Hillenbrand
British Antarctic Survey, Cambridge CB3 0ET, UK

Ellen Thomas
Department of Geology and Geophysics, Yale University, New Haven, Connecticut, USA

Rob Raiswell, Rob J. Newton and Fiona Gill
Cohen Laboratories, School of Earth and Environment, University of Leeds, Leeds LS2 9JT, UK

Stefanie Lutz and Liane G. Benning
Cohen Laboratories, School of Earth and Environment, University of Leeds, Leeds LS2 9JT, UK
GFZ German Research Centre for Geosciences, Telegrafenberg, Potsdam 14473, Germany

Alexandre M. Anesio
Bristol Glaciology Centre, School of Geographical Sciences, University of Bristol, Bristol BS8 1SS, UK

Arwyn Edwards
Institute of Biological, Environmental and Rural Sciences (IBERS), Aberystwyth University, Aberystwyth SY23 3FL, UK
Interdisciplinary Centre for Environmental Microbiology, Aberystwyth University, Aberystwyth SY23 3FL, UK

P. Porada and C. Beer
Department of Environmental Science and Analytical Chemistry (ACES), Stockholm University, 10691 Stockholm, Sweden
Bolin Centre for Climate Research, Stockholm University, 10691 Stockholm, Sweden

T.M. Lenton
Earth System Science Group, College of Life and Environmental Sciences, University of Exeter, Laver Building (Level 7), North Park Road, Exeter EX4 4QE, UK

A. Pohl
Laboratoire des Sciences du Climat et de l'Environnement, LSCE/IPSL, CEA-CNRS-UVSQ, Université Paris-Saclay, F-91191 Gif-sur-Yvette, France

Y. Donnadieu
Laboratoire des Sciences du Climat et de l'Environnement, LSCE/IPSL, CEA-CNRS-UVSQ, Université Paris-Saclay, F-91191 Gif-sur-Yvette, France
Aix-Marseille Université, CNRS, IRD, CEREGEUM34, 13545 Aix en Provence, France

U. Pöschl and B. Weber
Max Planck Institute for Chemistry, PO Box 3060, 55020 Mainz, Germany

L. Mander
Department of Environment, Earth and Ecosystems, The Open University, Milton Keynes, Buckinghamshire MK7 6AA, UK

A. Kleidon
Max Planck Institute for Biogeochemistry, PO Box 10 01 64, 07701 Jena, Germany

Samuel L. Jaccard
Institute of Geological Sciences and Oeschger Centre for Climate Change Research, University of Bern, 3012 Bern, Switzerland

Olivier Cartapanis
Institute of Geological Sciences and Oeschger Centre for Climate Change Research, University of Bern, 3012 Bern, Switzerland
Department of Earth and Planetary Sciences, McGill University, Montreal, Canada H3A 2A7

Daniele Bianchi
Department of Earth and Planetary Sciences, McGill University, Montreal, Canada H3A 2A7
School of Oceanography, University of Washington, Seattle, Washington 98105, USA.
Department of Atmospheric and Oceanic Sciences, University of California Los Angeles, Los Angeles, California 90095-1565, USA

Eric D. Galbraith
Department of Earth and Planetary Sciences, McGill University, Montreal, Canada H3A 2A7
Institució Catalana de Recerca i Estudis Avancats (ICREA), 08010 Barcelona, Spain
Institut de Ciència i Tecnologia Ambientals and Department of Mathematics, Universitat Autonoma de Barcelona, 08193 Barcelona, Spain

Richard Davy and Igor Esau
Nansen Environmental and Remote Sensing Center and Bjerknes Centre for Climate Research, Thormøhlensgt. 47, 5006 Bergen, Norway

Armin Köhl and Detlef Stammer
Institute of Oceanography, Center for Earth System Research and Sustainability (CEN), University of Hamburg (UHH), Hamburg 20146, Germany

Chuanyu Liu
Institute of Oceanography, Center for Earth System Research and Sustainability (CEN), University of Hamburg (UHH), Hamburg 20146, Germany
Key Lab of Ocean Circulation and Waves (KLOCAW), Institute of Oceanology, Chinese Academy of Sciences (IOCAS), Nanhai Road 7, Qingdao 266071, China
Function Laboratory for Ocean and Climate Dynamics, Qingdao National Laboratory for Marine Science and Technology (QNLM), Qingdao 266237, China

Fan Wang
Key Lab of Ocean Circulation and Waves (KLOCAW), Institute of Oceanology, Chinese Academy of Sciences (IOCAS), Nanhai Road 7, Qingdao 266071, China
Function Laboratory for Ocean and Climate Dynamics, Qingdao National Laboratory for Marine Science and Technology (QNLM), Qingdao 266237, China

Zhiyu Liu
State Key Laboratory of Marine Environmental Science (MEL) and Department of Physical Oceanography, College of Ocean and Earth Sciences, Xiamen University, Xiamen 361102, China

David E. Sugden, Andrew S. Hein and Shasta M. Marrero
School of GeoSciences, University of Edinburgh, Drummond Street, Edinburgh EH8 9XP, UK

John Woodward,   Kate Winter and Matthew J. Westoby
Department of Geography, Northumbria University, Ellison Place, Newcastle upon Tyne NE1 8ST, UK

Stuart A. Dunning
Department of Geography, Northumbria University, Ellison Place, Newcastle upon Tyne NE1 8ST, UK
Department of Geography, School of Geography, Politics and Sociology, Newcastle University, Newcastle upon Tyne NE1 7RU, UK

Eric J. Steig
School of GeoSciences, University of Edinburgh, Drummond Street, Edinburgh EH8 9XP, UK
Quaternary Research Center and Department of Earth and Space Sciences, University of Washington, Seattle, Washington 98195, USA

Stewart P.H.T. Freeman and Finlay M. Stuart
Scottish Universities Environmental Research Centre, Rankine Avenue, East Kilbride G75 0QF, UK

Sierra V. Petersen and Kyger C. Lohmann
Department of Earth & Environmental Sciences, University of Michigan, 2534 C.C. Little Building, 1100 North University Avenue, Ann Arbor, Michigan 48109, USA

Andrea Dutton
Department of Geological Sciences, University of Florida, 241 Williamson Hall, PO Box 112120, Gainesville, Florida 32611, USA

Matthias Tesche
Department of Environmental Science and Analytical Chemistry (ACES), Stockholm University, SE-10691 Stockholm, Sweden
School of Physics, Astronomy and Mathematics, University of Hertfordshire, Hatfield AL10 9AB, UK(M.T.)

Peggy Achtert
Department of Environmental Science and Analytical Chemistry (ACES), Stockholm University, SE-10691 Stockholm, Sweden
Institute for Climate and Atmospheric Science, School of Earth and Environment, University of Leeds, Leeds LS2 9JT, UK (P.A.)

Paul Glantz and Kevin J. Noone
Department of Environmental Science and Analytical Chemistry (ACES), Stockholm University, SE-10691 Stockholm, Sweden

Index

www.ingramcontent.com/pod-product-compliance
Lightning Source LLC
Chambersburg PA
CBHW080254230326
41458CB00097B/4460